T0180900

Lecture Notes in Computer Science

Lecture Notes in Bioinformatics 13919

The series Lecture Notes in Bioinformatics (LNBI) was established in 2003 as a topical subseries of LNCS devoted to bioinformatics and computational biology.

The series publishes state-of-the-art research results at a high level. As with the LNCS mother series, the mission of the series is to serve the international R & D community by providing an invaluable service, mainly focused on the publication of conference and workshop proceedings and postproceedings.

Ignacio Rojas · Olga Valenzuela ·
Fernando Rojas Ruiz · Luis Javier Herrera ·
Francisco Ortuño
Editors

Bioinformatics and Biomedical Engineering

10th International Work-Conference, IWBBIO 2023
Meloneras, Gran Canaria, Spain, July 12–14, 2023
Proceedings, Part I

Editors
Ignacio Rojas ⓘ
University of Granada
Granada, Spain

Fernando Rojas Ruiz ⓘ
University of Granada
Granada, Spain

Francisco Ortuño ⓘ
University of Granada
Granada, Spain

Olga Valenzuela ⓘ
University of Granada
Granada, Spain

Luis Javier Herrera ⓘ
University of Granada
Granada, Spain

ISSN 0302-9743   ISSN 1611-3349 (electronic)
Lecture Notes in Bioinformatics
ISBN 978-3-031-34952-2   ISBN 978-3-031-34953-9 (eBook)
https://doi.org/10.1007/978-3-031-34953-9

LNCS Sublibrary: SL8 – Bioinformatics

This Springer imprint is published by the registered company Springer Nature Switzerland AG
The registered company address is: Gewerbestrasse 11, 6330 Cham, Switzerland

Preface

We are proud to present the final set of accepted full papers for the 10th "International Work-Conference on Bioinformatics and Biomedical Engineering" (IWBBIO 2023), held in Gran Canaria, Spain, during July 12–14, 2023.

IWBBIO 2023 provided a discussion forum for scientists, engineers, educators, and students about the latest ideas and realizations in the foundations, theory, models, and applications for interdisciplinary and multidisciplinary research encompassing disciplines of computer science, mathematics, statistics, biology, bioinformatics, and biomedicine.

The aim of IWBBIO 2023 was to create a friendly environment that could lead to the establishment or strengthening of scientific collaborations and exchanges among attendees, and therefore IWBBIO 2023 solicited high-quality original research papers (including significant work in progress) on any aspect of bioinformatics, biomedicine, and biomedical engineering.

New computational techniques and methods in machine learning; data mining; text analysis; pattern recognition; data integration; genomics and evolution; next-generation sequencing data; protein and RNA structure; protein function and proteomics; medical informatics and translational bioinformatics; computational systems biology; modelling and simulation; and their application in the life science domain, biomedicine, and biomedical engineering were especially encouraged. The list of topics in the call for papers also evolved, resulting in the following list for the present edition:

1. **Computational proteomics**. Analysis of protein-protein interactions. Protein structure modelling. Analysis of protein functionality. Quantitative proteomics and PTMs. Clinical proteomics. Protein annotation. Data mining in proteomics.
2. **Next-generation sequencing and sequence analysis**. De novo sequencing, resequencing, and assembly. Expression estimation. Alternative splicing discovery. Pathway analysis. Chip-seq and RNA-Seq analysis. Metagenomics. SNPs prediction.
3. **High performance in bioinformatics**. Parallelization for biomedical analysis. Biomedical and biological databases. Data mining and biological text processing. Large-scale biomedical data integration. Biological and medical ontologies. Novel architecture and technologies (GPU, P2P, Grid, etc.) for bioinformatics.
4. **Biomedicine**. Biomedical computing. Personalized medicine. Nanomedicine. Medical education. Collaborative medicine. Biomedical signal analysis. Biomedicine in industry and society. Electrotherapy and radiotherapy.
5. **Biomedical engineering**. E-Computer-assisted surgery. Therapeutic engineering. Interactive 3D modelling. Clinical engineering. Telemedicine. Biosensors and data acquisition. Intelligent instrumentation. Patient monitoring. Biomedical robotics. Bio-nanotechnology. Genetic engineering.
6. **Computational systems for modelling biological processes**. Inference of biological networks. Machine learning in bioinformatics. Classification for biomedical

data. Microarray data analysis. Simulation and visualization of biological systems. Molecular evolution and phylogenetic modelling.

7. **Healthcare and diseases**. Computational support for clinical decisions. Image visualization and signal analysis. Disease control and diagnosis. Genome-phenome analysis. Biomarker identification. Drug design. Computational immunology.

8. **E-Health**. E-Health technology and devices. E-Health information processing. Telemedicine/E-Health application and services. Medical image processing. Video techniques for medical images. Integration of classical medicine and E-Health.

9. **COVID-19**. A special session was organized in which different aspects, fields of application, and technologies that have been applied against COVID-19 were analyzed.

After a careful peer review and evaluation process (each submission was reviewed by at least 2, and on the average 3.2, Program Committee members or additional reviewers), 79 papers were accepted, according to the recommendations of reviewers and the authors' preferences, to be included in the LNBI proceedings.

During IWBBIO 2023 several Special Sessions were carried out. Special Sessions are a very useful tool in order to complement the regular program with new and emerging topics of particular interest for the participating community. Special Sessions that emphasized multidisciplinary and transversal aspects as well as cutting-edge topics were especially encouraged and welcomed, and in this edition of IWBBIO 2023 the following were received:

– **SS1. High-Throughput Genomics: Bioinformatic Tools and Medical Applications.**
Genomics is concerned with the sequencing and analysis of an organism's genome. It is involved in the understanding of how every single gene can affect the entire genome. This goal is mainly afforded using the current, cost-effective, high-throughput sequencing technologies. These technologies produce a huge amount of data that usually require high-performance computing solutions and opens new ways for the study of genomics, but also transcriptomics, gene expression, and systems biology, among others. The continuous improvements and broader applications of sequencing technologies is producing a continuous new demand for improved high-throughput bioinformatics tools.

In this context, the generation, integration, and interpretation of genetic and genomic data is driving a new era of healthcare and patient management. Medical genomics (or genomic medicine) is an emerging discipline that involves the use of genomic information about a patient as part of clinical care with diagnostic or therapeutic purposes to improve the health outcomes. Moreover, it can be considered a subset of precision medicine that has an impact in the fields of oncology, pharmacology, rare and undiagnosed diseases, and infectious diseases. The aim of this Special Session was to bring together researchers in medicine, genomics, and bioinformatics to translate medical genomics research into new diagnostic, therapeutic, and preventive medical approaches. Therefore, we invited authors to submit original research, new tools or pipelines, as well as update and review articles on relevant topics, such as (but not limited to):

- Tools for data pre-processing (quality control and filtering)
- Tools for sequence mapping
- Tools for the comparison of two read libraries without an external reference
- Tools for genomic variants (such as variant calling or variant annotation)
- Tools for functional annotation: identification of domains, orthologues, genetic markers, and controlled vocabulary (GO, KEGG, InterPro,etc.)
- Tools for gene expression studies and tools for Chip-Seq data
- Integrative workflows and pipelines

Organizers: **M. Gonzalo Claros**, *University of Málaga, Spain.*
Javier Pérez Florido, *Fundación Progreso y Salud, Spain.*
Francisco M. Ortuño, *University of Granada, Spain.*

– **SS2. Feature Selection, Extraction, and Data Mining in Bioinformatics: Approaches, Methods, and Adaptations.**
Various applications of bioinformatics, system biology, and biophysics measurement data mining require proper, accurate, and precise preprocessing or data transformation before the analysis itself. Here, the most important issues are covered by the feature selection and extraction techniques to translate the raw data into the inputs for the machine learning and multi-variate statistic algorithms. This is a complex task; it requires reducing the problem dimensionality, removal of redundant of irrelevant data, without affecting significantly the present information. The methods and approaches are often conditioned by the physical properties of the measurement process, mathematically congruent description and parameterization, as well as biological aspects of specific tasks. With the current increase of artificial intelligence methods adoption into bioinformatics problem solutions, it is necessary to understand the conditionality of such algorithms, to choose and use the correct approach and avoid misinterpretations, artefacts, and aliasing affects. The adoption often uses already existing knowledge from different fields, and direct application might underestimate the required conditions and corrupt the analysis results. This special session saw discussion on the multidisciplinary overlaps, development, implementation, and adoption of feature and selection methods for datasets of biological origin in order to set up a pipeline from the measurement design through signal processing to knowledge obtaining. The topic covered theoretical questions, practical examples, and results verifications.

Organizer: **Jan Urban**, *University of South Bohemia, Czech Republic*

– **SS3. Sensor-Based Ambient Assisted Living Systems and Medical Applications.**
Many advancements in medical technology are possible, including developing intel ligent systems for the treatment, diagnosis, and prevention of many healthcare issues. The field of surgery is now experiencing an increase in intelligent systems. Medical rules should be in place during the creation of these technologies to ensure market acceptance. Hospital systems can be connected to mobile devices or other specialized equipment, boosting patient monitoring.

Technology-based tools may monitor, treat, and reduce several health-related issues. The system and concepts discussed here can use sensors found in mobile devices and other sensors found in intelligent environments, and sensors utilized with other equipment. The advancements in this area right now will be incredibly beneficial for treating various ailments.

The main points of this topic are the presentation of cutting-edge, active projects, conceptual definitions of devices, systems, services, and sensor-based advanced healthcare efforts.

The special session covered, but was not limited to:

- Assistive technology and adaptive sensing systems
- Diagnosis and treatment with mobile sensing systems
- Healthcare self-management systems
- M-Heath, eHealth, and telemedicine systems
- Body-wearer/implemented sensing devices
- Sensing vital medical metrics
- Sensing for persons with limited capabilities
- Motion and path-tracking medical systems
- Artificial intelligence with sensing data
- Patient empowerment with technological equipment
- Virtual and augmented reality in medical systems
- Mobile systems usability and accessibility
- Medical regulations in mobile systems
- Medical regulations and privacy

Organizer: **Ivan Miguel Pires**, *Universidade da Beira Interior, Portugal.*
Norberto Jorge Gonçalves, *Universidade de Trás-os-Montes e Alto Douro, Portugal.*
Paulo Jorge Coelho, *Polytechnic Institute of Leiria, Portugal.*

- **SS4. Analysis of Molecular Dynamics Data in Proteomics.**
Molecular dynamics (MD) simulations have become a key method for exploring the dynamic behavior of macromolecules and studying their structure-to-function relationships. In proteomics, they are crucial for extending the understanding of several processes related to protein function, e.g., protein conformational diversity, binding pocket analysis, protein folding, ligand binding, and its influence on signaling, to name a few. Nevertheless, the investigation of the large amounts of information generated by MD simulations is a far from trivial challenge.

This special session addressed new research on computational techniques and machine learning (ML) algorithms that can provide efficient solutions for the diverse problems that entail the analysis of MD simulation data in their different areas of application.

Topics of interest included, but were not limited to:

- Sampling techniques in MD simulations
- Potential energy surfaces
- Detection of rare events
- Transition pathways analytics
- Visualization techniques for MD
- Feature representations for molecular structures

- Deep learning architectures for MD simulations
- Generative models for MD

Organizer: **Caroline König**, *Universitat Politècnica de Catalunya, Spain.*
Alfredo Vellido, *Universitat Politècnica de Catalunya.*

- **SS5. Image Visualization and Signal Analysis.**
Signal processing focuses on analysing, modifying and synthesizing signals such as sound, images, and biological measurements. Signal processing techniques can be used to improve transmission, storage efficiency, and subjective quality and also to emphasize or detect components of interest in a measured signal and are a very relevant topic in medicine.

Any signal transduced from a biological or medical source could be called a biosignal. The signal source could be at the molecular level, cell level, or a systemic or organ level. A wide variety of such signals are commonly encountered in the clinic, research laboratory, and sometimes even at home. Examples include the electrocardiogram (ECG), or electrical activity from the heart; speech signals; the electroencephalogram (EEG), or electrical activity from the brain; evoked potentials (EPs, i.e., auditory, visual, somatosensory, etc.), or electrical responses of the brain to specific peripheral stimulation; the electroneurogram, or field potentials from local regions in the brain; action potential signals from individual neurons or heart cells; the electromyogram (EMG), or electrical activity from the muscle; the electroretinogram from the eye; and so on.

From the other side, medical imaging is the technique and process of creating visual representations of the interior of a body for clinical analysis and medical intervention, as well as visual representation of the function of some organs or tissues (physiology). Medical imaging seeks to reveal internal structures hidden by the skin and bones, as well as to diagnose and treat disease. Medical imaging also establishes a database of normal anatomy and physiology to make it possible to identify abnormalities. Although imaging of removed organs and tissues can be performed for medical reasons, such procedures are usually considered part of pathology instead of medical imaging.

Organizer: **L. Wang**, *University of California, San Diego, USA.*

- **SS6. Computational Approaches for Drug Design and Personalized Medicine.**
With continuous advancements of biomedical instruments and the associated ability to collect diverse types of valuable biological data, numerous recent research studies have focused on how to best extract useful information from the big biomedical data currently available. While drug design has been one of the most essential areas of biomedical research, the drug design process for the most part has not fully benefited from the recent explosive growth of biological data and bioinformatics algorithms. With the incredible overhead associated with the traditional drug design process in terms of time and cost, new alternative methods, possibly based on computational approaches, are very much needed to propose innovative ways to propose effective

drugs and new treatment options. Employing advanced computational tools for drug design and precision treatments has been the focus of many research studies in recent years. For example, drug repurposing has gained significant attention from biomedical researchers and pharmaceutical companies as an exciting new alternative for drug discovery that benefits from computational approaches. This new development also promises to transform healthcare to focus more on individualized treatments, precision medicine, and lower risks of harmful side effects. Other alternative drug design approaches that are based on analytical tools include the use of medicinal natural plants and herbs as well as using genetic data for developing multi-target drugs.

Organizer: **Hesham H. Ali**, *University of Nebraska at Omaha.*

It is important to note that for the sake of consistency and readability of the book, the presented papers are classified under 15 chapters. The organization of the papers is in two volumes arranged basically following the topics list included in the call for papers. The first volume (LNBI 13919), entitled "Bioinformatics and Biomedical Engineering. Part I", is divided into nine main parts and includes the contributions on:

1. Analysis Of Molecular Dynamics Data In Proteomics
2. Bioinformatics
3. Biomarker Identification
4. Biomedical Computing
5. Biomedical Engineering
6. Biomedical Signal Analysis
7. Computational Support for Clinical Decisions
8. COVID-19 Advances in Bioinformatics and Biomedicine

The second volume (LNBI 13920), entitled "Bioinformatics and Biomedical Engineering. Part II", is divided into seven main parts and includes the contributions on:

1. Feature Selection, Extraction, and Data Mining in Bioinformatics
2. Genome-Phenome Analysis
3. Healthcare and Diseases
4. High-Throughput Genomics: Bioinformatic Tools and Medical Applications
5. Image Visualization And Signal Analysis
6. Machine Learning in Bioinformatics and Biomedicine
7. Medical Image Processing
8. Next-Generation Sequencing and Sequence Analysis
9. Sensor-Based Ambient Assisted Living Systems And Medical Applications

This 10th edition of IWBBIO was organized by the University of Granada. We wish to thank our main sponsor as well as the following institutions: Department of Computer Engineering, Automation and Robotics, CITIC-UGR from the University of Granada, International Society for Computational Biology (ISCB) for their support and grants. We also wish to thank the editors in charge of various international journals for their interest in editing special issues from a selection of the best papers of IWBBIO 2023. At IWBBIO 2023, there were two awards (best contribution award and best contribution from student participant) sponsored by the editorial office of Genes, an MDPI journal.

We would also like to express our gratitude to the members of the different committees for their support, collaboration, and good work. We especially thank the Program Committee, the reviewers, and Special Session organizers. We also want to express our gratitude to the EasyChair platform. Finally, we wish to thank the staff of Springer, for their continuous support and cooperation.

April 2023

Ignacio Rojas
Olga Valenzuela
Fernando Rojas
Luis Javier Herrera
Francisco Ortuño

Organization

Conference Chairs

Ignacio Rojas   University of Granada, Spain
Olga Valenzuela   University of Granada, Spain
Fernando Rojas   University of Granada, Spain
Luis Javier Herrera   University of Granada, Spain
Francisco Ortuño   University of Granada, Spain

Steering Committee

Miguel A. Andrade   University of Mainz, Germany
Hesham H. Ali   University of Nebraska at Omaha, USA
Oresti Baños   University of Granada, Spain
Alfredo Benso   Politecnico di Torino, Italy
Larbi Boubchir   University of Paris 8, France
Giorgio Buttazzo   Superior School Sant'Anna, Italy
Gabriel Caffarena   University CEU San Pablo, Spain
Mario Cannataro   University Magna Graecia of Catanzaro, Italy
Jose María Carazo   Spanish National Center for Biotechnology (CNB), Spain
Jose M. Cecilia   Universidad Católica San Antonio de Murcia (UCAM), Spain
M. Gonzalo Claros   University of Malaga, Spain
Joaquin Dopazo   Fundacion Progreso y Salud, Spain
Werner Dubitzky   University of Ulster, UK
Afshin Fassihi   Universidad Católica San Antonio de Murcia (UCAM), Spain
Jean-Fred Fontaine   University of Mainz, Germany
Humberto Gonzalez   University of the Basque Country (UPV/EHU), Spain
Concettina Guerra   Georgia Tech, USA
Roderic Guigo   Pompeu Fabra University, Spain
Andy Jenkinson   Karolinska Institute, Sweden
Craig E. Kapfer   Reutlingen University, Germany
Narsis Aftab Kiani   European Bioinformatics Institute (EBI), UK
Natividad Martinez   Reutlingen University, Germany
Marco Masseroli   Politecnico di Milano, Italy

Federico Moran	Complutense University of Madrid, Spain
Cristian R. Munteanu	University of A Coruña, Spain
Jorge A. Naranjo	New York University Abu Dhabi, UAE
Michael Ng	Hong Kong Baptist University, China
Jose L. Oliver	University of Granada, Spain
Juan Antonio Ortega	University of Seville, Spain
Fernando Rojas	University of Granada, Spain
Alejandro Pazos	University of A Coruña, Spain
Javier Perez Florido	Fundación Progreso y Salud, Spain
Violeta I. Pérez Nueno	Inria Nancy Grand Est, LORIA, France
Horacio Pérez-Sánchez	Universidad Católica San Antonio de Murcia (UCAM), Spain
Alberto Policriti	Università di Udine, Italy
Omer F. Rana	Cardiff University, UK
M. Francesca Romano	Superior School Sant'Anna, Italy
Yvan Saeys	Ghent University, Belgium
Vicky Schneider	The Genome Analysis Centre (TGAC), UK
Ralf Seepold	HTWG Konstanz, Germany
Mohammad Soruri	University of Birjand, Iran
Yoshiyuki Suzuki	Tokyo Metropolitan Institute of Medical Science, Japan
Shusaku Tsumoto	Shimane University, Japan
Renato Umeton	Dana-Farber Cancer Institute and Massachusetts Institute of Technology, USA
Jan Urban	University of South Bohemia, Czech Republic
Alfredo Vellido	Polytechnic University of Catalunya, Spain
Wolfgang Wurst	GSF National Research Center of Environment and Health, Germany

Program Committee and Additional Reviewers

Heba Afify	MTI University, Egypt
Fares Al-Shargie	American University of Sharjah, United Arab Emirates
Jesus Alcala-Fdez	University of Granada, Spain
Hesham Ali	University of Nebraska at Omaha, USA
Georgios Anagnostopoulos	Florida Institute of Technology, USA
Cecilio Angulo	Universitat Politècnica de Catalunya, Spain
Masanori Arita	National Institute of Genetics, Japan
Gajendra Kumar Azad	Patna University, India
Hazem Bahig	Ain Sham University, Egypt
Oresti Banos	University of Granada, Spain

Ugo Bastolla   Centro de Biologia Molecular "Severo Ochoa",
                                     Spain
Payam Behzadi   Islamic Azad University, Iran
Sid Ahmed Benabderrahmane   University of Edinburgh, UK
Alfredo Benso   Politecnico di Torino, Italy
Anna Bernasconi   Politecnico di Milano, Italy
Mahua Bhattacharya   Indian Institute of Information Technology and
                                     Management, Gwalior, India
Paola Bonizzoni   Università di Milano-Bicocca, Italy
Larbi Boubchir   University of Paris 8, France
Gabriel Caffarena   University CEU San Pablo, Spain
Mario Cannataro   University Magna Graecia of Catanzaro, Italy
Jose Maria Carazo   National Center for Biotechnology, CNB-CSIC,
                                     Spain
Francisco Carrillo Pérez   Universidad de Granada, Spain
Claudia Cava   IBFM-CNR, Italy
Francisco Cavas-Martínez   Technical University of Cartagena, Spain
Ting-Fung Chan   Chinese University of Hong Kong, China
Kun-Mao Chao   National Taiwan University, Taiwan
Chuming Chen   University of Delaware, USA
Javier Cifuentes Faura   University of Murcia, Spain
M. Gonzalo Claros   Universidad de Málaga, Spain
Darrell Conklin   University of the Basque Country, Spain
Alexandre G. De Brevern   Université Denis Diderot Paris 7 and INSERM,
                                     France
Javier De Las Rivas   CiC-IBMCC, CSIC/USAL/IBSAL, Spain
Paolo Di Giamberardino   Sapienza University of Rome, Italy
Maria Natalia Dias Soeiro   University of Porto, Portugal
  Cordeiro
Marko Djordjevic   University of Belgrade, Serbia
Joaquin Dopazo   Fundación Progreso y Salud, Spain
Mohammed Elmogy   Mansoura University, Egypt
Gionata Fragomeni   Magna Graecia University, Italy
Pugalenthi Ganesan   Bharathidasan University, India
Hassan Ghazal   Mohammed I University, Morocco
Razvan Ghinea   University of Granada, Spain
Luis Gonzalez-Abril   University of Seville, Spain
Humberto Gonzalez-Diaz   UPV/EHU, IKERBASQUE, Spain
Morihiro Hayashida   National Institute of Technology, Matsue College,
                                     Japan
Luis Herrera   University of Granada, Spain
Ralf Hofestaedt   Bielefeld University, Germany

Jingshan Huang	University of South Alabama, USA
Xingpeng Jiang	Central China Normal University, China
Hamed Khodadadi	Khomeinishahr Branch, Islamic Azad University, Iran
Narsis Kiani	Karolinska Institute, Sweden
Dongchul Kim	University of Texas Rio Grande Valley, USA
Tomas Koutny	University of West Bohemia, Czech Republic
Konstantin Krutovsky	Georg-August-University of Göttingen, Germany
José L. Lavín	Neiker Tecnalia, Spain
Chen Li	Monash University, Australia
Hua Li	Bio-Rad, USA
Li Liao	University of Delaware, USA
Zhi-Ping Liu	Shandong University, China
Francisco Martínez-Álvarez	Universidad Pablo de Olavide, Spain
Marco Masseroli	Politecnico di Milano, Italy
Roderick Melnik	Wilfrid Laurier University, Canada
Enrique Muro	Johannes Gutenberg University, Germany
Kenta Nakai	Institute of Medical Science, University of Tokyo, Japan
Isabel Nepomuceno	University of Seville, Spain
Dang Ngoc Hoang Thanh	University of Economics Ho Chi Minh City, Vietnam
Anja Nohe	University of Delaware, USA
José Luis Oliveira	University of Aveiro, Portugal
Yuriy Orlov	Institute of Cytology and Genetics, Russia
Juan Antonio Ortega	University of Seville, Spain
Andres Ortiz	University of Malaga, Spain
Francisco Manuel Ortuño	Fundación Progreso y Salud, Spain
Motonori Ota	Nagoya University, Japan
Mehmet Akif Ozdemir	Izmir Katip Celebi University, Turkey
Joel P. Arrais	University of Coimbra, Portugal
Paolo Paradisi	ISTI-CNR, Italy
Taesung Park	Seoul National University, South Korea
Alejandro Pazos	University of A Coruña, Spain
Antonio Pinti	I3MTO Orléans, France
Yuri Pirola	Univ. degli Studi di Milano-Bicocca, Italy
Joanna Polanska	Silesian University of Technology, Poland
Alberto Policriti	University of Udine, Italy
Hector Pomares	University of Granada, Spain
María M Pérez	University of Granada, Spain
Julietta V. Rau	Istituto di Struttura della Materia, Italy
Khalid Raza	Jamia Millia Islamia, India

Jairo Rocha	University of the Balearic Islands, Spain
Ignacio Rojas-Valenzuela	University of Granada, Spain
Fernando Rojas	University of Granada, Spain
Ignacio Rojas	University of Granada, Spain
Luca Roncati	University of Modena and Reggio Emilia, Italy
Gregorio Rubio	Universitat Politècnica de València, Spain
Irena Rusu	University of Nantes, France
Michael Sadovsky	Institute of Computational Modelling of SB RAS, Russia
Emmanuel Sapin	University of Colorado Boulder, USA
Beata Sarecka-Hujar	Medical University of Silesia in Katowice, Poland
Jean-Marc Schwartz	University of Manchester, UK
Jose A. Seoane	Vall d'Hebron Institute of Oncology, Spain
Xuequn Shang	Northwestern Polytechnical University, China
Surinder Singh	Panjab University, Chandigarh, India
Sónia Sobral	Universidade Portucalense, Portugal
Jiangning Song	Monash University, Australia
Joe Song	New Mexico State University, USA
Natarajan Sriraam	M.S. Ramaiah Institute of Technology, India
Jiangtao Sun	Beihang University, China
Wing-Kin Sung	National University of Singapore, Singapore
Prashanth Suravajhala	Amrita University Kerala, India
Martin Swain	Aberystwyth University, UK
Sing-Hoi Sze	Texas A&M University, USA
Alessandro Tonacci	IFC-CNR, Italy
Carolina Torres	University of Granada, Spain
Marcos Roberto Tovani Palone	University of São Paulo, Brazil
Shusaku Tsumoto	Shimane University, Japan
Renato Umeton	Massachusetts Institute of Technology, USA
Jan Urban	University of South Bohemia, Czech Republic
Olga Valenzuela	University of Granada, Spain
Alfredo Vellido	Universitat Politècnica de Catalunya, Spain
Jianxin Wang	Central South University, China
Jiayin Wang	Xi'an Jiaotong University, China
Lusheng Wang	City University of Hong Kong, China
Ka-Chun Wong	City University of Hong Kong, China
Fang Xiang Wu	University of Saskatchewan, Canada
Phil Yang	George Mason University, USA
Zhongming Zhao	University of Texas Health Science Center at Houston, USA
Shanfeng Zhu	Fudan University, China

Contents – Part I

Biomedical Signal Analysis

Computational Proteomics

COVID-19 Advances in Bioinformatics and Biomedicine

Contents – Part II

High-Throughput Genomics: Bioinformatic Tools and Medical Applications

Image Visualization and Signal Analysis

Machine Learning in Bioinformatics and Biomedicine

Analysis of Molecular Dynamics Data in Proteomics

Recognition of Conformational States of a G Protein-Coupled Receptor from Molecular Dynamic Simulations Using Sampling Techniques

Mario Alberto Gutiérrez-Mondragón[1]([✉]),
Caroline König[1], and Alfredo Vellido[1,2]

[1] Computer Science Department, and Intelligent Data Science and Artificial Intelligence (IDEAI-UPC) Research Center, Universitat Politècnica de Catalunya, Barcelona, Spain
{mario.alberto.gutierrez,caroline.leonore.konig}@upc.edu,
avellido@cs.upc.edu
[2] Centro de Investigación Biomédica en Red (CIBER), Madrid, Spain

Abstract. Protein structures are complex and dynamic entities relevant to many biological processes. G-protein-coupled receptors in particular are a functionally relevant family of cell membrane proteins of interest as targets in pharmacology. Nevertheless, the limited knowledge about their inherent dynamics hampers the understanding of the underlying functional mechanisms that could benefit rational drug design. The use of molecular dynamics simulations and their analysis using Machine Learning methods may assist the discovery of diverse molecular processes that would be otherwise beyond our reach. The current study builds on previous work aimed at uncovering relevant motifs (groups of residues) in the activation pathway of the $\beta 2$-*adrenergic* ($\beta_2 AR$) receptor from molecular dynamics simulations, which was addressed as a multi-class classification problem using Deep Learning methods to discriminate active, intermediate, and inactive conformations. For this problem, the interpretability of the results is particularly relevant. Unfortunately, the vast amount of intermediate transformations, in contrast to the number of re-orderings establishing active and inactive conditions, handicaps the identification of relevant residues related to a conformational state as it generates a class-imbalance problem. The current study aims to investigate existing Deep Learning techniques for addressing such problem that negatively influences the results of the predictions, aiming to unveil a trustworthy interpretation of the information revealed by the models about the receptor functional mechanics.

Keywords: Deep Learning · G Protein-Coupled Receptors · Molecular Dynamics · Imbalanced data · Class Imbalanced Methods · Imbalanced Metrics · Explainability · Layer-Wise Relevance Propagation

I. Rojas et al. (Eds.): IWBBIO 2023, LNBI 13919, pp. 3–16, 2023.
https://doi.org/10.1007/978-3-031-34953-9_1

1 Introduction

Molecular Dynamics (MD) simulations have become a powerful tool for unveiling the inner flexibility and polymorphic character of numerous protein structures, including G protein-coupled receptors (GPCRs). The latter are relevant to pharmacology as druggable targets, owing to their influence on diverse physiological processes in the human body [18]. From a structural perspective, GPCRs are composed of seven alpha-helical elements embedded in the cell membrane connected by three extracellular and three intracellular loops [22]. Like other biomolecules, these receptors are highly complex in their inner dynamics, i.e., over time, they experience multiple transformations observed at an atomic level as subtle structural changes that influence the functional response [17].

MD-based experimentation leads to structural blueprints that help deciphering the multifaceted functionality of GPCR receptors [23]. Understanding the complex mechanism of signal transmission to the cell constitutes a significant research endeavor in both academy and the pharmaceutical industry [4]. GPCRs bind a range of signaling partners (from hormones to neurotransmitters) that distinctively transform the receptor structure, modulating their activation-deactivation mechanism. The investigation of the ligand-binding process has provided evidence that GPCR receptors constitute more than simple two-state structures [12]. For example, MD simulations of the inactive-active state of the β_2AR receptor allowed observing how a small molecule (so-called agonist) modulates the transition between conformational states (clusters of active, inactive, and intermediate conformations) and further induces nonexistent conformational states when it is absent [5]. On a detailed level, these structural rearrangements can be subtle but critical in understanding the functional properties of GPCRs as a step towards the development of novel treatments for an array of human pathologies.

This study focuses on advancing previous research in which we found relevant motifs (residues) related to the receptor activation pathway of the inactive state of the β_2AR receptor when bound to the full agonist BI-167107, from Google Exacycle MD simulations [15]. The Layer-Wise Relevance Propagation (LRP) algorithm was used to identify critical residues for the prediction of active, intermediate, and inactive conformational states of a Deep Learning (DL) model [8], therefore increasing the interpretability of the model. Overall, the proposed model found relevant motifs for active and inactive states, but the classification of intermediate states was broadly inaccurate, leading to a negative impact on the global model performance. The cause for this was, most likely, the highly imbalanced class distribution of the data, hampered by the high amount of intermediate conformational states (roughly 96% of the data). Further investigation of the class imbalance problem and the methods available for addressing it is indispensable for successfully applying the proposed methodology. In previous work [7] and [8], we considered random undersampling (RUS) for handling such imbalanced distribution. Nevertheless, discarding data removes a considerable amount of information that could be significant for model learning. Therefore, this study aims to systematically investigate more effectual resampling methods for addressing the class imbalance problem. In particular, we investigate over-sampling, undersampling, and synthetic sampling methods, as well as algorithm-based strategies.

To conclude the study, we discuss the influence of the class-imbalance correcting methods on the interpretability study.

2 Materials and Methods

For our analysis, we consider $\beta_2 AR$ simulations starting from the inactive state and bound to the full agonist BI-167107, obtained from Google's Exacycle cloud computing platform [15]. The dataset includes 10,000 short simulations (of about 6ns average length) that we used for training a one-dimensional Convolutional Neural Network (1D-CNN) for identifying motifs discerning conformational states.

For conformational state denotation, we computed the distances between alpha-carbon atoms in arginine 131 (R131) and leucine 272 (L272) residues from Helix 3 (H3) and Helix 6 (H6). A value above or equal to 14Å indicates active conformations; if lower than or equal to 8.5Å, it refers to inactive conditions; otherwise, the structure refers to intermediate rearrangements. Broadly, the computation of distances in each simulation step of the MD trajectories yields the data distribution of Table 1. Intermediate transformations comprise 96.01% of the dataset, resulting in a highly imbalanced distribution that could lead to biased predictions.

Table 1. Dataset distribution per conformational state.

Conformational State	# Samples
active	1610
intermediate	259,557
inactive	9181
Total:	270,348

In previous work, we resorted to a random undersampling strategy to limit the impact of intermediate conformations. As mentioned in the introduction, here we try to overcome this limitation by employing several alternative sampling strategies to obtain a more balanced data distribution and improve the conformational space analysis.

There is a wide variety of methods available for solving the data-imbalance problem [14]. However, we have discarded many of them due to their over-complexity and high computational cost. Our experiments concern standard and straightforward techniques such as random undersampling (RUS) and random oversampling (ROS), as well as more sophisticated methods, namely, undersampling-based approaches such as the NearMiss Algorithm; synthetic data-based techniques such as the Synthetic-Based Oversampling Technique (SMOTE) and the Adaptative Synthetic (ADASYN) method; hybrid methods like SMOTEENN (SMOTE combining with Edited Nearest Neighbor undersampling method); and algorithm-based approaches like the Weighted Loss method.

2.1   Class Imbalance Correction Methods

Class imbalance is an inherent characteristic of our domain application. As already mentioned, a receptor exhibits numerous transformations related to active, intermediate, and inactive conformational conditions. Out of this, intermediate conformations are extremely prevalent. Learning from such skewed distribution for producing meaningful knowledge is a challenge for the DL-based model used for studying the transition pathways.

Simple RUS and ROS strategies were first considered. The RUS method discards random samples from the majority class, while ROS randomly duplicates instances in minority groups. Despite their potential effectiveness, these methods offer no guarantee of obtaining an optimal solution [9]. Moreover, by randomly deleting examples, there is no guarantee that the most valuable information of the majority class will be retained. Likewise, randomly duplicating data points from minority groups could easily lead to overfitting.

More advanced methods were thus considered. The Near-Miss algorithm refers to a collection of undersampling methods that use the K-nearest neighbors (KNN) classifier for removing examples based on the distances from majority to minority class examples [20]. In this study, we have dropped majority class examples with minimum distance to each minority class example, i.e., we bypass intermediate samples that could represent minor structural differences by contrast with active and inactive states. On the negative side, a high amount of intermediate reorderings that could be valuable for fully understanding the activation mechanism is still lost.

Oversampling methods that generate synthetic samples would be valuable for preserving relevant information in the transition pathway. SMOTE is a suitable method for generating multiple slightly different points from minority classes [2], reducing bias from the replacement strategy of the ROS method. SMOTE produces synthetic observations from the minority groups using the KNN algorithm. The general idea is to draw new points by interpolating random samples between the minority groups and their k nearest neighbors (typically 5) till the data is proportionally balanced. A general downside of this method is that the conformational space representation could be ambiguous or noisy as majority classes are not considered, causing a possible overlap among samples. Thus, variants based on SMOTE could be considered to address its limitations. In particular, we focus on the Adaptive Synthetic Sampling (ADASYN) technique, which generates samples based on density [10]. This means more synthetic data are created in regions of the feature space where the density of minority examples is low, thus creating a bit more scattered feature space than the one generated through SMOTE.

Some research suggests that a hybrid undersampling-oversampling strategy could be beneficial [1]. For that, combinations of SMOTE and undersampling techniques have been proposed [6]. Here, we focus on the SMOTEENN strategy. SMOTE creates synthetic samples of minority classes to balance the distribution. Then, the Edited Nearest Neighbor (ENN) *cleans* the feature space, reducing overfitting and improving learning. Although SMOTEENN has been shown to improve on other techniques, its computational cost and long balancing process restricts its application and generalization.

Algorithm-based methods can be more practical in problems such as the one addressed in this paper, where a large amount of data in the active-inactive instances must be over-sampled for equal distribution, and losing a considerable amount of the information from intermediate conformations cannot be accepted. Thus, we have explored a cost-sensitive learning strategy in which penalties are assigned to each class [19]. Misclassifications from minority classes assume a higher cost during the model training. In our experiments, the weighting scheme follows $class\_weight = 1 - \frac{n_c}{n_t}$, where n_c refers to the number of instances in the class and n_t to the number of instances in the dataset. Using class weights forces the model to keep track of underrepresented classes without altering the data distribution, avoiding any bias or information lost, providing correct performance in a non-complex manner, and saving time and computational resources.

Other hybrid-based approaches consider merging data-level and algorithm-level methods for addressing class imbalanced problems [16]. Nevertheless, we have not considered these methods here mainly because of the high computational resources required.

2.2 Experimental Setup

As mentioned, a multi-classification problem involving active, intermediate, and inactive conditions in the protein structure was addressed in [8] for identifying relevant motifs in the transition pathway. In particular, the proposed methodology follows two main directions; the first relies on training a 1D-CNN model on the MD simulations of the inactive $\beta_2 AR$ structure with the full agonist BI-167107; the second is related to uncovering relevant motifs by analyzing the model predictions using the Layer-Wise Relevance Propagation (LRP) algorithm.

Figure 1 illustrates the suggested methodology, including the computation of the center of mass in one of the simulation steps related to a conformational condition. This sample is re-dimensioned to 846 × 1 (282 residues × 3 coordinates in the center of mass), scaled in the range (0, 1) and inputted to the 1D-CNN. As for the CNN architecture definition, our experimentation followed an empirical strategy by increasingly adding layers and convolution filters until a higher number of layers revealed no significant improvement on the classification and its interpretability.

The model prediction is investigated for retrieving relevant motifs in that particular simulation step. The contribution of the residues results from averaging individual computed values from the LRP algorithm in each position of the center of mass (X, Y, Z). The whole process is repeated for each step simulation in the 10,000 MD trajectories. The final contribution map (which reveals significant residues) is derived from averaging the contribution of the residues related to a conformational condition in the dataset.

Further details on the experimental setup and the guidelines for designing and optimizing the neural network architecture can be consulted in [8]. Here, we focus on improving the model predictions by handling the dataset imbalanced distribution using sampling techniques.

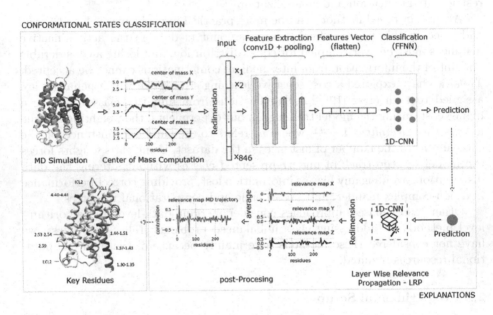

Fig. 1. Overall proposed workflow for relevant motif recognition in the activation pathway of the $\beta_2 AR$ when bound the full agonist BI-167107.

2.3   Evaluation Metrics for Imbalanced Data Problems

As already mentioned, a correct diagnosis of the proposed DL-based model performance under resampling methods is fundamental for effectively judging its generalization. In most ML-related problems, the so-called Accuracy and Error Rate are the evaluation metrics of choice. Nonetheless, in problems with imbalanced datasets, high accuracies can be produced as a consequence of the predictions being dominated by the majority classes [11]. Alternatively, we make use of Precision and Recall (also known as sensitivity) as the basis for more suitable metrics for estimating the model results in imbalanced learning problems. They can be computed as $precision = \frac{TP}{TP+FP}$ and $recall = \frac{TP}{TP+FN}$ and derived from the confusion matrix (a table-type representation of predicted and actual values) in terms of combinations of True Positives, True Negatives, and errors Type I (False positive) and Type II (False Negative) [14]. To ease the discussion of the relevance of the metrics in our experiments, we assembled our multi-classification problem as a series of binary problems following a One-*vs*-Rest (OVR) approach.

From these, we obtain the F-measure, which is preferable for assessing the robustness of the model as it balances the effects of Recall and Precision through the use of the β coefficient in the equation:

$$F - measure = \frac{(1 + \beta^2) \times Recall \times Precision}{\beta^2 \times Recall + Precision} \tag{1}$$

In this work, we use $\beta = 1$ (the so-called F1-score) to give equal significance to Precision and Recall. It means we aim for a predictor with a high rate of positive instances correctly predicted but also being precise, i.e., we intend to avoid confusion among conformational states, even if the structural transformations are subtle. Although the F1-measure is acknowledged as a reliable metric for evaluating ML-based models in both binary and multi-classification problems, it could still be insufficient when working with severe skewed data distribution [3]. Therefore, we also contemplate the use of a more robust metric, the Matthews Correlation Coefficient (MCC). It incorporates an analysis of the predictions from both negative and positive samples to compute the classifier effectiveness, and it is calculated as:

$$MCC = \frac{TP \times TN - FP \times FN}{\sqrt{(TP + FP)(TP + FN)(TN + FP)(TN + FN)}} \tag{2}$$

It confers the same importance to the four terms in the confusion matrix, producing more informative scores in the range $(-1, 1)$. A classifier performing random predictions will score near 0, the center of the range.

3   Results and Discussion

As previously described, this study aims at unraveling three conformational states (active, intermediate, and inactive) in the $\beta_2 AR$ receptor from MD simulations. The results comparing the performance of the 1D-CNN model for the different sampling methods can be found in Table 2.

Table 2. Summary of classification results.

Method	%Precision	%Recall	%F1-score	%Accuracy	MCC
Unbalanced	78.9332	45.6969	39.1644	45.6969	0.1854
RUS	76.4164	76.7272	76.5396	76.7272	0.6511
NearMiss	73.0825	74.0000	73.0931	74.0000	0.6138
ROS	83.7330	82.2424	82.3530	82.2424	0.7386
SMOTE	87.6269	83.1993	83.6683	83.1993	0.7569
ADASYN	90.1738	88.8699	89.0832	88.8699	0.8284
SMOTEENN	91.5593	**90.6688**	**90.8289**	**90.6688**	**0.8543**
Weighted Loss	**95.9836**	80.7428	86.4711	80.7428	0.7923

The significance of the use of sampling techniques is glaring, particularly in scenarios like the one under study, where the data skewed distribution is severe. If not used, the model effectiveness is the worst, as demonstrated by the MCC score computed when data is imbalanced. The confusion matrices displayed in Fig. 2 stress this result. Evidently, the model predictions are entirely biased and dominated by the majority class (the intermediate state) when the imbalance is not controlled. By contrast, the models using any of the sampling methods yield substantial improvements.

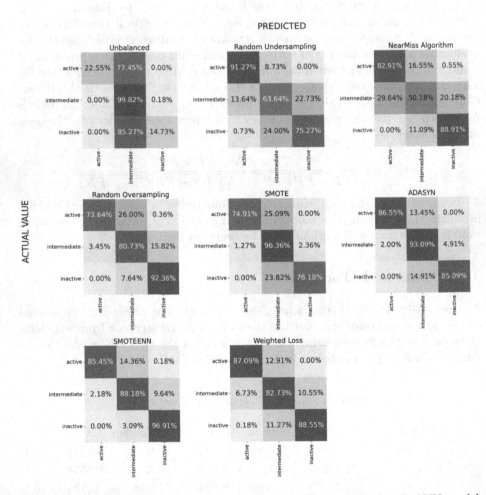

Fig. 2. Multi-label confusion matrices mapping predictions from the 1D-CNN model to actual conformational states (true values). The rows represent the predictions, and the columns the actual values. Each quadrant includes the data percentage.

Individual performance per conformational state in terms of Precision, Recall, and F1-score is visually summarised in Fig. 3. Note the degree of improvement when considering a basic method such as RUS. In the same vein, a more complex undersampling technique (NearMiss) does not provide a superior improvement in comparison. Nonetheless, both undersampling methods produce a proper trade-off between Precision and Recall. Likewise, we must stress that the high proportion of intermediate states discarded limits its accurate identification.

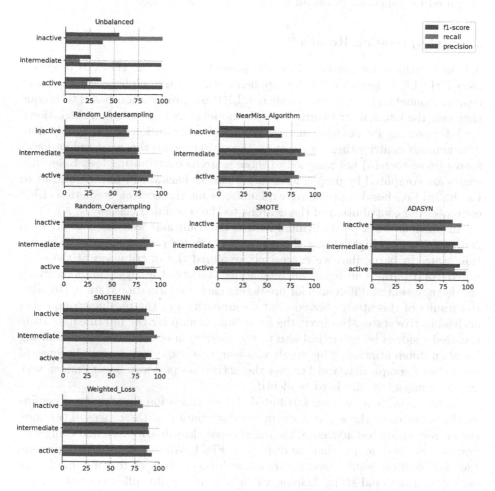

Fig. 3. Evaluation performance per conformational state.

If we consider the ROS results, the improvement in the identification of intermediate conditions is evident, and so is the significant improvement in the discrimination of inactive-active states. This pattern is repeated by SMOTE, in which the discrimination of the intermediate state is very accurately. The ADASYN method, on the other hand, seems to yield more even results in terms of Recall and Precision for the three conformational states. Overall, by sampling a vast amount of the intermediate instances we have increased the feature space, causing an improvement in conformational states identification. However, the risk of overfitting by overlapping has also increased.

When combining undersampling and oversampling methods using the SMO-TEENN strategy, the performance in almost all evaluation metrics is much better than that of other sampling methods. The less computationally demanding Weighted Loss method yields an intermediate performance.

3.1   Interpretation Results

Aiming to achieve interpretability of the predictions from the 1D-CNN proposed model, the LRP algorithm [21] was applied to the dataset while considering only conformational states correctly predicted. LRP is a propagation-based technique that uses the internal structure of the DL model (topology, weights, activations, etc.) for scoring its neurons in terms of their contribution to the prediction. The neurons contributing the most to the output score the most, while lower scores (close to zero) are assigned to those neurons contributing less. Relevance scores are computed by propagating target classes backward from the output to the input, i.e., besides providing explanations for the correct prediction, LRP estimates the contribution of the neurons to the remaining classes [13].

In previous work [8], individual neuron contributions were used to create a *contribution map* highlighting the relevant motifs for predicting any conformation state. In particular, we extensively explored the distribution of the computed contributions utilizing an intelligible statistical analysis for identifying subtle yet essential differences for predicting each conformational state. Broadly, the results of this study showed clear dissimilarities in the residues comprising active-inactive states. However, the contribution map for the intermediate state revealed residues being critical also to the prediction of inactive states, limiting its plain differentiation. This result was somehow expected, mainly because of the dataset complexity, and because the activation pathway occurs under very subtle changes that are hard to identify.

In the same vein, we also attributed the identification complexity of conformational states to the severe skew in the distribution of the classes. Therefore, the current study has addressed the imbalanced class distribution and computed new contribution maps, that we display in Fig. 4. We expected that improving the classification results would enrich the identification of relevant motifs for each conformational state, helping us to obtain a plain differentiation on the transformations in structure. However, in most experiments, we found no critical differences in the identified residues related to a conformational state. This outcome could mean either that the 1D-CNN model has exhausted its capabilities for searching relevant information in the MD trajectories under study, or that the existing relevance differences were already obtained even with strongly imbalanced classes and that they reveal a very gradual evolution from inactive to intermediate states, in contrast with a more brisk evolution from intermediate to active states. Despite this, the improvement of the classification results achieved is still crucial, as it guarantees that the model is capable of captur-

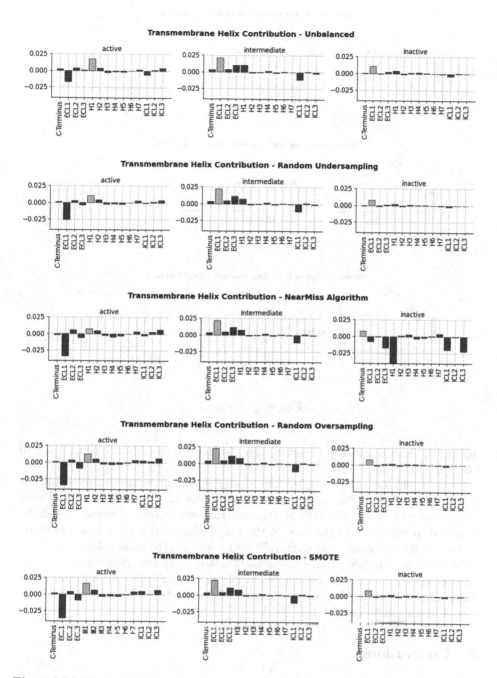

Fig. 4. LRP Computed Contributions per transmembrane region using different sampling techniques. Color red highlight the most representative region. (Color figure online)

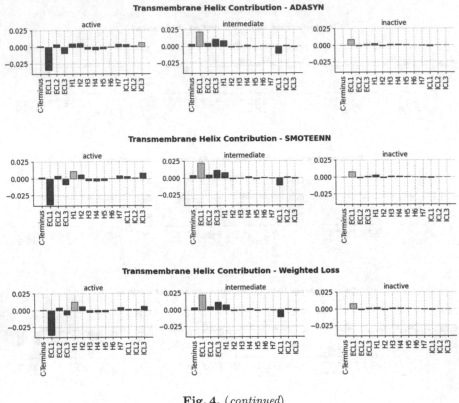

Fig. 4. (*continued*)

ing transformations in the structure despite the rearrangements being hard to discern.

This could be confirmed by the contribution maps per transmembrane domain displayed in Fig. 4. At first, it seems like all methods computed very similar relevances for the GPCR regions. However, subtle differences can be observed, particularly in the case of the maps computed for the inactive state, where the H1 region and the Extracellular Loop-1 are the most relevant regions. This outcome requires further investigation and expertise domain to uncover its usefulness, i.e., deeper analysis of the β2-adrenergic structure information is required to determine the trustworthiness of the predictions from the 1D-CNN model under any imbalanced method.

4   Conclusions

In this brief paper, we have shown that the discrimination of conformational states of a GPCR molecule from MD simulation data can be substantially improved by tackling the class-imbalance problem using methods that do not necessarily require undersampling.

Furthermore, we have also provided evidence that the impact of different GPCR regions on that state discrimination is very heterogeneous as quantified by the LRP algorithm, thus providing a layer of interpretability to the results. In future research, we aim to go into further detail and analyze such impact at the residue level.

In general terms, it might be concluded that the more suitable technique for addressing the class imbalance problem in our study domain is Weighted Loss because, while it shows one of the best classification performances, it does not alter the distribution of the data -that is, the structural information of the receptor is retained- and the corresponding LRP-based interpretation is almost identical to that of SMOTEENN.

References

1. Batista, G.E., Prati, R.C., Monard, M.C.: A study of the behavior of several methods for balancing machine learning training data. ACM SIGKDD Explor. Newsl. **6**(1), 20–29 (2004)
2. Chawla, N.V., Bowyer, K.W., Hall, L.O., Kegelmeyer, W.P.: Smote: synthetic minority over-sampling technique. J. Artif. Intell. Res. **16**, 321–357 (2002)
3. Chicco, D., Jurman, G.: The advantages of the Matthews correlation coefficient (MCC) over F1 score and accuracy in binary classification evaluation. BMC Genomics **21**, 1–13 (2020)
4. Congreve, M., de Graaf, C., Swain, N.A., Tate, C.G.: Impact of GPCR structures on drug discovery. Cell **181**(1), 81–91 (2020)
5. Durrant, J.D., McCammon, J.A.: Molecular dynamics simulations and drug discovery. BMC Biol. **9**(1), 1–9 (2011)
6. Fernández, A., Garcia, S., Herrera, F., Chawla, N.V.: Smote for learning from imbalanced data: progress and challenges, marking the 15-year anniversary. J. Artif. Intell. Res. **61**, 863–905 (2018)
7. Gutiérrez-Mondragón, M.A., König, C., Vellido, A.: A deep learning-based method for uncovering GPCR ligand-induced conformational states using interpretability techniques. In: Rojas, I., Valenzuela, O., Rojas, F., Herrera, L.J., Ortuño, F. (eds.) IWBBIO 2022. LNCS, vol. 13347, pp. 275–287. Springer, Cham (2022). https://doi.org/10.1007/978-3-031-07802-6_23
8. Gutiérrez-Mondragón, M.A., König, C., Vellido, A.: Layer-wise relevance analysis for motif recognition in the activation pathway of the β2-adrenergic GPCR receptor. Int. J. Mol. Sci. **24**(2), 1155 (2023)
9. Hasanin, T., Khoshgoftaar, T.: The effects of random undersampling with simulated class imbalance for big data. In: 2018 IEEE International Conference on Information Reuse and Integration (IRI), pp. 70–79. IEEE (2018)
10. He, H., Bai, Y., Garcia, E.A., Li, S.: Adasyn: adaptive synthetic sampling approach for imbalanced learning. In: 2008 IEEE International Joint Conference on Neural Networks (IEEE World Congress on Computational Intelligence), pp. 1322–1328. IEEE (2008)
11. He, H., Garcia, E.A.: Learning from imbalanced data. IEEE Trans. Knowl. Data Eng. **21**(9), 1263–1284 (2009)
12. Hollingsworth, S.A., Dror, R.O.: Molecular dynamics simulation for all. Neuron **99**(6), 1129–1143 (2018)

13. Holzinger, A., Saranti, A., Molnar, C., Biecek, P., Samek, W.: Explainable AI methods-a brief overview. In: Holzinger, A., Goebel, R., Fong, R., Moon, T., Müller, K.R., Samek, W. (eds.) xxAI 2020. LNCS, vol. 13200, pp. 13–38. Springer, Cham (2022). https://doi.org/10.1007/978-3-031-04083-2_2
14. Johnson, J.M., Khoshgoftaar, T.M.: Survey on deep learning with class imbalance. J. Big Data **6**(1), 1–54 (2019). https://doi.org/10.1186/s40537-019-0192-5
15. Kohlhoff, K.J., et al.: Cloud-based simulations on google exacycle reveal ligand modulation of GPCR activation pathways. Nat. Chem. **6**(1), 15–21 (2014)
16. Krawczyk, B.: Learning from imbalanced data: open challenges and future directions. Prog. Artif. Intell. **5**(4), 221–232 (2016). https://doi.org/10.1007/s13748-016-0094-0
17. Latorraca, N.R., Venkatakrishnan, A., Dror, R.O.: GPCR dynamics: structures in motion. Chem. Rev. **117**(1), 139–155 (2017)
18. Lefkowitz, R.J.: Historical review: a brief history and personal retrospective of seven-transmembrane receptors. Trends Pharmacol. Sci. **25**(8), 413–422 (2004)
19. Ling, C.X., Sheng, V.S.: Cost-sensitive learning and the class imbalance problem. Encyclopedia Mach. Learn. **2011**, 231–235 (2008)
20. Mani, I., Zhang, I.: KNN approach to unbalanced data distributions: a case study involving information extraction. In: Proceedings of Workshop on Learning from Imbalanced Datasets, vol. 126, pp. 1–7. ICML (2003)
21. Montavon, G., Binder, A., Lapuschkin, S., Samek, W., Müller, K.R.: Layer-wise relevance propagation: an overview. In: Explainable AI: Interpreting, Explaining and Visualizing Deep Learning, pp. 193–209 (2019)
22. Rosenbaum, D.M., Rasmussen, S.G., Kobilka, B.K.: The structure and function of g-protein-coupled receptors. Nature **459**(7245), 356–363 (2009)
23. Torrens-Fontanals, M., Stepniewski, T.M., Aranda-García, D., Morales-Pastor, A., Medel-Lacruz, B., Selent, J.: How do molecular dynamics data complement static structural data of GPCRs. Int. J. Mol. Sci. **21**(16), 5933 (2020)

Identification of InhA-Inhibitors Interaction Fingerprints that Affect Residence Time

Magdalena Ługowska[(✉)] and Marcin Pacholczyk

Department of Systems Biology and Engineering, Silesian University of Technology,
Akademicka 16, Gliwice, Poland
magdalena.lugowska@polsl.pl

Abstract. Drug development is a complex process that remains subject to risks and uncertainties. In its early days, much emphasis was placed on the equilibrium binding affinity of a drug to a particular target, which is described by the equilibrium dissociation constant (K_d). However, there are a large number of drugs that exhibit non-equilibrium binding properties. For this reason, optimization of other kinetic parameters such as dissociation constants (k_{off}) and association constants (k_{on}) is becoming increasingly important to improve accuracy in measuring *in vivo* efficacy. To achieve this, the concept of residence time between drug and target (τ) was developed to account for the continuous elimination of the drug, the absence of equilibrium conditions, and the conformational dynamics of the target molecules. Residence time has been shown to be a better estimate of drug lifetime potency than equilibrium binding affinity and is recognized as a key parameter in drug development. However, because residence time is only one measure of drug potency, it provides only a limited picture of binding kinetics and affinity.

A machine-learning algorithm was proposed to identify molecular features affecting protein-ligand binding kinetics for a set of similar compounds. Molecular dynamics simulations of τRAMD results were used as model input. The study confirmed that τRAMD provides information about the characteristics of the dissociation pathway since the obtained dissociation trajectories can be used to identify the interactions that occur and the conformational changes of the system at subsequent time points. The proposed algorithm made it possible to obtain information on protein-ligand contacts that are specific to their residence times.

Keywords: drug residence time · interaction fingerprints · binding kinetics

1 Introduction

For a drug to be effective, it must not only have the right concentration, but it must also have the ability to bind to the target molecule. One of the most commonly used selection criteria in drug design is the equilibrium binding affinity

I. Rojas et al. (Eds.): IWBBIO 2023, LNBI 13919, pp. 17–31, 2023.
https://doi.org/10.1007/978-3-031-34953-9_2

of a small chemical molecule (the ligand) to its molecular target (the receptor). This affinity is a measure of the persistence of the binding and the strength of the effect of the ligand on the receptor. Ligand potency is an important determinant of ligand activity. This parameter defines the potential of a ligand to efficiently activate a receptor for the production of a strong response *in vivo*. Thus, affinity is a measure to quantify the efficacy of a drug and to determine the benefit of the interactions that occur between the drug molecule and its target. Therefore, drug design protocols are mainly based on molecules with high binding affinity. However, this approach does not always translate into higher drug efficacy under *in vivo* conditions because many drugs have non-equilibrium binding properties. Besides pharmacokinetic properties, binding and dissociation rates of drugs can be measured to predict their biological activity profile under *in vivo* conditions [1, 2]. The concept of drug residence time in a molecular target has been introduced, which takes into account the conformational dynamics of target molecules that affect drug binding and dissociation. One important observation is that this time is sometimes better correlated with *in vivo* drug effectiveness than binding affinity [3]. Therefore, ligand residence time at the target site (τ) has become a reliable determinant of drug efficacy that is considered along with reaction kinetic parameters in drug discovery programs. However, it is important to keep in mind that the use of residence time as the sole measure of drug efficacy can be a limited picture of reaction kinetics [4].

1.1   Ligand-Receptor Binding Kinetics

A substrate-ligand (L), which is a small chemical molecule that is acted upon by an enzyme-receptor (R), begins the series of successive biochemical reactions that occur in the cell. When the ligand (molecule that is the initial stimulus) meets a specific protein receptor, cellular pathways are initiated outside the cell. The molecules come in tightly matched pairs. The receptor recognizes only one (or a few) specific ligands, and the ligand binds only one (or a few) target receptors. When the ligand binds to the receptor, the shape or activity of the receptor changes, allowing it to transmit a signal or to directly induce a change in the cell.

Understanding and fully describing receptor-ligand (RL) binding kinetics, as well as the molecular determinants of this fit, is an important part of drug design.

Enzymatic reactions can be characterized by relevant kinetic (reaction rate constants) and thermodynamic (reaction equilibrium constants) parameters. Kinetic parameters include:

- k_{on}, the association rate constant ($M^{-1}s^{-1}$), indicating the rate of formation of the receptor-ligand complex
- k_{off}, the dissociation rate constant (s^{-1}), describing the rate of ligand release from the receptor-binding site, and
- τ, the residence time (s), a measure of the time the ligand spends in the receptor-binding site (the lifetime of the receptor-ligand complex) defined as the inverse of the dissociation rate constant.

The process of binding a ligand to a receptor is accompanied by a change in free energy described by the Gibbs potential as a function of enthalpic and entropic factors. The enthalpic factor may be related to the formation and/or breaking of hydrogen bonds, ionic interactions, or hydrophobic interactions, among others. The entropic factor refers to the changes in the number of degrees of freedom after the formation of a complex.

$$\Delta G \equiv G(R + L) - G(RL) = -k_b T ln(K_d) \Rightarrow K_d = e^{\frac{\Delta G}{k_b T}} \tag{1}$$

where ΔG - change in free energy of Gibbs bond, k_b - Boltzmann's constant, T - temperature. K_d, the equilibrium constant of the dissociation process, is a thermodynamic parameter that quantifies the strength of the interaction in the RL system andcan be expressed in terms of kinetic parameters through the following relationship:

$$K_d = \frac{k_{off}}{k_{on}} \tag{2}$$

In the simplest case, the reaction of a ligand with a receptor is a one-step process:

$$R + L \underset{k_{-1}}{\overset{k_1}{\rightleftarrows}} RL \tag{3}$$

where $k_1 = k_{on}$, and $k_{-1} = k_{off}$ (Fig. 1).

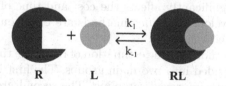

Fig. 1. Schematic representation of the ligand-receptor one-step binding reaction.

However, this model is not always sufficient to describe the interactions of drugs with their targets, which often involve multistep binding and dissociation. This led to the development of a two-step model of binding kinetics that accounts for conformational changes leading to increased complementarity between molecules.

$$R + L \underset{k_{-1}}{\overset{k_1}{\rightleftarrows}} RL^* \underset{k_{-2}}{\overset{k_2}{\rightleftarrows}} RL \tag{4}$$

where k_1, k_{-1}, k_2, k_{-2} kinetic constants, R - receptor, L ligand, RL^* - transient receptor-ligand complex and RL - final protein-ligand complex (Fig. 2).

In this model, a free drug encounters its target in a conformational state defined by a binding pocket that is suboptimal for the structure of the drug molecule. The initial phase of binding is an association process that forms an encounter complex (RL^*) defined by the association rate constant (k_1), the dissociation rate constant (k_{-1}), and the equilibrium dissociation constant (K_d) [13,14]. Initial binding is followed by another step in which the system must overcome the energy barriers created by conformational changes of the receptor and the ligand to form a new stable state (RL) in which the binding pocket adopts a conformation more suitable for the drug molecule.

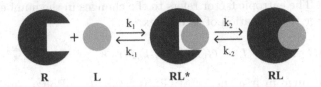

Fig. 2. Schematic representation of the ligand-receptor two-step binding reaction.

1.2   Computer Methods to Determine Binding Kinetic Parameters

With the increasing interest in residence time and the importance of drug binding kinetics at the target binding site, *in silico* methods are becoming increasingly important. This is especially true when commonly used experimental methods are often time-consuming and costly. In addition, the use of computational methods that predict residence time and characterize reaction kinetics can support personalized medicine. Patient-specific simulations can speed up a physician's decision to select the optimal drug from several potential candidates. What's more, such calculations can be performed on compounds that have not yet been synthesized, which significantly affects the cost and time of research. It should be noted that the developed *in silico* methods are based on experimental data, which can confirm their reliability.

Computational methods for the estimation of residence time and other kinetic parameters can be divided into two main groups. The first is a set of molecular dynamics methods with improved sampling. The second are methods based on machine learning, often also using molecular dynamics simulations.

Machine learning techniques have been used for molecular structural analysis, prediction of dynamic behavior, investigation of molecular dynamics trajectories, and molecular dynamics sampling [5]. Rather than predicting the dissociation of a ligand or its residence time in a target, these methods are most commonly used to predict binding affinity [for a detailed review, see elsewhere: [6]].

Most ML-based residence time prediction methods have been applied to HIV-1 and HSP90 protease inhibitors. For the first time, the three-dimensional Vol-Surf lattice and partial least squares (PLS) modeling were used to develop predictive models of binding kinetics for HIV-1 and HSP90 inhibitors dataset [7].

Energetic and conformational features from molecular modeling were combined with a multitarget machine learning (MTML) approach to classify the HSP90 inhibitor system, with classes representing ligand binding kinetics.

A quantitative structure kinetic relationships (QSKR) for the dissociation rate constant (k_{off}) were determined using the COMBINE method, which was used as a tool to determine binding affinity [8]. In addition, the energy of intermolecular interactions obtained from molecular dynamics simulations can be imprinted into the molecular structure. It was shown that the dissociation rate is a function of the first half of the total dissociation process. A regression model was then developed based on a systematic analysis of protein-ligand binding interactions in dissociation trajectories. It cannot be excluded that the predictive power of these methods is overestimated because the test set did not contain structurally distinct ligands (they were closely structurally related to the training set).

The QSKR model for predicting the dissociation rate constant (k_{off}) of a ligand based on the structure of the receptor-ligand complex was applied to a larger and more diverse data set (406 ligands) [9]. The authors found that the model showed good predictive accuracy on external test sets that included multiple targets as well as a single target.

Random forest approach focused on structural descriptors from molecular dynamics simulations was used to demonstrate their importance in predicting kinetic rate constants for different receptor-ligand complexes [10].

Similarly, to identify new receptor-ligand contacts that place the molecular system in transient states, a method for transient state analysis (MLTSA) was developed [11]. These novel interactions have been shown to contain key features that determine the kinetics of binding. However, the method has been applied to a small data set: cyclin-dependent kinase 2 (CDK2) and its two inhibitors.

The PCA-ML method for clustering dcTMD trajectories into clusters reflecting binding pathways and the RMSD-based clustering method for grouping pathways by their mean Euclidean distance are some of the newer methods for analyzing binding mechanisms in receptor-ligand system simulations. However, they are limited by the need for human intervention in the initial selection of pathways during model training or in deciding the boundaries of neighborhood networks [12].

1.3   Research Objective

Molecular simulations are an important tool for describing the dissociation pathway, predicting kinetic parameters, including residence time, and determining structural features. In order to observe the occurrence of rare events during the simulation and to reduce the computational complexity, simplifications such as enhanced sampling are often used in these approaches. Short-lived events, such as the rearrangement of atoms in a molecule during the induced fitting step, are inaccurately described by these simplifications. This can be seen as a limitation of the simplifications. On the other hand, classical molecular dynamics allows us to understand these fast, important events. However, it cannot be applied to longer time scales, such as the residence time of a drug in a target, which can range from a few seconds to hours. In addition, the accuracy of the simulations is not stable, although they are often used as input to machine learning-based algorithms. The reason for these studies is the need for more efficient and accurate methods to analyse receptor-ligand binding kinetics.

The aim of this work is to propose drug residence time solution to analyse structural interactions in the binding process of the InhA protein and its inhibitors.

2   Mycobacterium Enoyl Acyl Carrier Protein Reductase (InhA)

The global fight against the spread of multidrug-resistant tuberculosis (MDR-TB) has created an urgent need for new chemotherapeutic agents. The enoyl acyl carrier protein (ACP) reductase (InhA) is clinically one of the few targets for TB drug discovery. It is involved in fatty acid synthesis, mainly mycolic acid biosynthesis in *M. tuberculosis* [15] (Fig. 3).

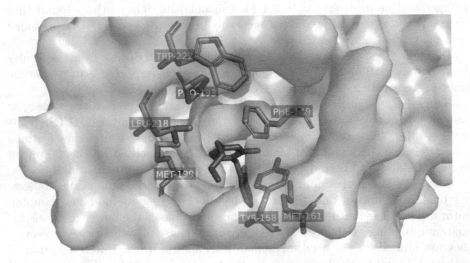

Fig. 3. Crystallographic structure of InhA enzyme (example: PDB complex ID: 4OYR) visualized with PyMOL. The structure of the enzyme is shown in a "surface view" oriented towards the ligand binding pocket. Amino acids shown in the "sticks" representation that make up the active site of the enzyme: Phe149, Tyr158, Met161, Met199, Pro193, Leu218, and Trp222. In the binding pocket is the ligand (1US).

The enzymes are structurally unique, with deeper drug binding pockets in the active site. This makes the enzyme a unique target for antitubercular drug development. InhA is also the primary molecular target of the most potent anti-tuberculosis drug, isoniazid. Since this drug has no human orthologue, resistance problems can be avoided. This example shows that one of the most effective ways to combat tuberculosis is to inhibit Mycobacterium enoyl acyl carrier protein reductase (InhA).

The study used 11 complexes of InhA inhibitors for which both experimental kinetic measurements and crystallographic structures were available (see Table 1).

Table 1. Summarized data for 11 InhA-inhibitor complexes. Experimental residence time, τ_{exp} (min), was obtained from PDBrt.

Complex PDB ID	Ligand ID	τ_{exp} (min)
2X22	TCU	24
2X23	TCU	24
4OIM	JUS	50
4OXY	1TN	27
4OYR	1US	90
5COQ	TCU	30
5MTP	53K	94
5MTQ	XT3	119
5MTR	XT0	106
5UGT	XTW	220
5UGU	XTV	194

3 Methods

The dissociation trajectories obtained from τRAMD simulations were used as input data for the analysis of interactions occurring between molecules.

τRAMD is an enhanced sampling method for molecular dynamics simulations. It was developed to calculate the relative residence time of pharmacological compounds in their molecular targets and to study the dissociation pathway of ligands from receptor binding sites [17].

τRAMD simulations, in which a small randomly oriented force is applied to the center of mass of the ligand to accelerate its removal from the receptor active site, are performed on receptor-ligand systems immersed in a solvent. After a given time, the ligand movement is checked. A random change in force direction occurs if the change in position was less than a predefined threshold distance. When the ligand leaves the receptor binding site, the simulation ends. This condition is defined by specifying the ligand distance from the binding site. Simulation time is dependent on the residence time of the ligand in the target. Ligands with a longer residence time will require more time to leave the target (simulated time) or more force to leave the target in a given simulated time. No prior knowledge of the dissociation path or extensive parameterization is required for this method. The only parameter that is carefully defined is the magnitude of the force, which must not interfere with the calculated relative residence time, which must be small enough to avoid simple pull of the ligand out the binding site and large enough to facilitate the dissociation process. Due to the above mentioned characteristics, τRAMD is an efficient and relatively simple tool to estimate the relative residence time of drugs for molecular studies.

The MD simulation procedure with an additional force to accelerate the ligand output (τRAMD) was performed according to the published protocol [17].

For each system, the molecular dynamics simulation steps were repeated 5 times using NAMD software, which was treated as start files for τRAMD. From each starter file, a set of 10 dissociation trajectories was generated, resulting in a total of 50 dissociation simulations for each system.

From the τRAMD dissociation trajectory, receptor-ligand interactions were extracted as follows: (i) the position of the ligand center of mass and the coordinates of the atoms that make up the entire system were obtained from each frame (timepoint) of the trajectory and stored in separate .pdb files using a tcl script written for the VMD tool and executed from a Python script, (ii) the obtained coordinates of the position of the atoms in space were used as input for the identification of ligand-receptor interactions using the RDKit and ProLif libraries of Python (interaction classes: Hydrophobic, π-stacking, π-cation and cation-π, anionic and cationic, and H-bond donor and acceptor); (iii) each interaction was marked as "1" if an interaction was observed or "0", i.e., (v) on the basis of the frequency of occurrence of the interactions, a threshold was set which allowed the separation of the bound state from the transient and fully released states - for the purpose of the subsequent evaluation, those states of the system in which an interaction was detected in at least 20% of a single dissociation pathway have been removed.

4   Results

4.1   Ligand Similarity

The Tanimoto similarity [16] was used to calculate binary fingerprints for a set of ligands. Measure ranges between 0 and 1. $S_{A,B} = 1$ means that A and B are similar, not equal. It is assumed that a value of 0.85 ($S_{A,B} >= 0.85$) is the threshold for determining the high similarity of two chemical structures that they will have similar biological activity.

The ligand similarity analysis was performed using the RDKit library in Python. The following steps were included in the analysis protocol: (i) create a list of SMILES strings from which 2D molecular structures of the ligands were produced; (ii) drop ligands with the same SMILES string; (iii) generate molecular fingerprints for each molecule using the RDKit generator; (iv) compute the Tanimoto similarity measure for each ligand pair; (v) create similarity matrix that presents data only once without repetition; (vi) find a suitable similarity score threshold above which a given percentage of the compound pairs are considered similar, (vii) cluster the obtained results using the Linking Hierarchy Cluster (LHC).

Analysis of Table 2 revealed that two randomly selected compounds have a similarity score of 0.34 on average. If two ligands have a Tanimoto measure score of 0.35, which is near the average, the random compounds have a 42% chance of having a similarity score above 0.35. Therefore, it can be concluded that the two compounds are not similar to each other. The 95% percentile value is 0.79. This means that this value can be considered as a similarity score threshold, above which a certain percentage of compound pairs are determined to be similar. This is approximately 18% of the compound pairs.

Based on the similarity of the fingerprints, the presented approach was able to identify several compounds clusters. The similarity within the groups is confirmed by the identified ligand groups (Fig. 4).

Table 2. Summary of ligand similarity analysis using Tanimoto coefficient.

InhA-inhibitors	
Number of fingerprints	9
Total compound pairs	81
Mean $S_{A,B}$	0.34
$S_{A,B} >= 0.85$	13 (16.05%)
$S_{A,B} >= 0.75$	14 (17.28%)
$S_{A,B} >= 0.65$	17 (21%)
$S_{A,B} >= 0.55$	21 (25.92%)
$S_{A,B} >= 0.45$	28 (34.57%)
$S_{A,B} >= 0.35$	34 (41.98%)
$S_{A,B} >= 0.2$	45 (55.56%)
95% of compound pairs	78
score at 95% percentile	0.79

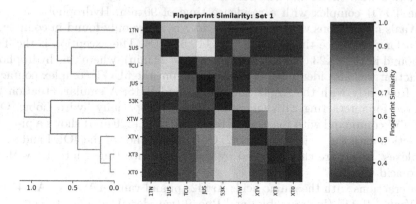

Fig. 4. Heatmaps of molecular similarity by Tanimoto similarity index for inhibitors of InhA.

4.2 Drug-Like Ligand Properties that Affect Residence Time

Feature Generation. 983 775 simulation snapshots (a snapshot is the next observed conformational change in the system) and 832 identified contacts were initially extracted from τRAMD simulations. The number of features was reduced in the next step by: (i) removal of the bound state on the assumption that the interaction describing this state occurs in at least 20% of the snapshots of a single dissociation trajectory; (ii) removal of noise, which is defined as a very rare event that does not affect the dissociation rate (the set threshold is the occurrence of contact in less than 5% of all trajectories of a given complex). Dissociation transients were identified along with relevant events using this approach. The data was reduced to 24 features and 35986 snapshots for further analysis. Table 3 lists all identified interaction fingerprints.

Table 3. Frequencies of identified ligand-amino acid interactions for the tested InhA protein inhibitors.

A preliminary identification of amino acids likely to affect drug residence time in a molecular target can be made by analyzing Table 3. Only in the 5MTQ complex, characterized by one of the longer residence times (119 min), compound XT3 showed hydrophobic interactions with the amino acids Ala33, Cys242, Gln31, Ile9, Leu245, Leu4, Lys7, Phe96, and Trp248, and van der Waals interactions with the amino acids Cys242, Leu4, Ser246, and Trp248. A hydrophobic interaction with the amino acid Arg41 was only observed with the ligand 1US in the 4OYR complex with a residence time of 90 min. Hydrophobic and van der Waals interactions with the amino acid Arg42 are only found in compounds with a long residence time (5MTR, 5UGT, 5UGU). The exception is the TCU compound in the 2X23 complex (residence time 24 min), where this hydrophobic interaction was also identified. XTW ligand from the 5UGT complex contacted most frequently with the Arg42 amino acid and Ile15. A similar situation was observed for interacting with amino acid Asp41, especially hydrophobic. Only the XTW compound with the longest residence time (220 min) shows a pi-cation interaction with the amino acid Arg42. Only compounds of the 4OIM and 5COQ complexes, which are characterized by short residence times, interact with the amino acid Gln99.

Interactions with the amino acids Arg42 (pi and van der Waals), Asp41 (van der Waals), Ile115 (hydrophobic) and Phe40 (van der Waals) are thought to be particularly characteristic of a compound with long residence times (the longest in the dataset studied) in complex with the InhA enzyme.

Identification of Key Interaction Fingerprints. Analysis of the table of interactions of the studied complexes using PCA allowed isolation of the main factors for all InhA inhibitor systems studied. Figure 5 shows how the primary variables correlate with the principal components.

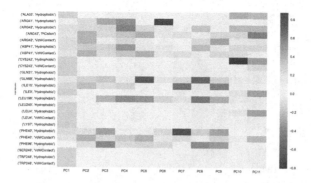

Fig. 5. Features correlation with the principal components.

The position of the samples in the coordinate system defined by the principal components after PCA analysis and k-means data clustering is shown in Fig. 6. The elbow method was used to determine the number of clusters. This method performs k-means clustering on the data set for a range of k values (in the thesis, a range of 2 to 8). It then calculates the average score for all clusters for each value of k. The default calculation is the distortion score, which is the sum of the squared lengths of each point from their associated midpoint. The point of the elbow at which the rate of decrease changes is then detected.

Figure 6a shows groups of receptor-ligand contacts and Fig. 6b - ligands. The expected result was separate clusters consisting of ligands with similar residence times, or interaction fingerprints specific to the given residence time lengths. The optimal number of clusters for grouping the identified interaction fingerprints is 5 and for ligand grouping - 4.

Table 4 provides a detailed list of the ligands that make up a given cluster, along with their residence time information. Group 0 includes compounds with a residence time of less than 100 min. Ligands with residence times greater than 100 min are grouped as 1, 2, and 3. A separate cluster was identified for the 5UGT complex containing the ligand with the longest residence time in the group.

Table 5 provides a detailed summary of the contacts that form a given cluster, along with information about the receptor-ligand system in which they were observed. The hydrophobic interaction with the amino acid Phe96 occurs in most of the complexes and therefore forms a separate group 2. This interaction does not affect the residence time of the ligand. The interactions of groups 0 and 3 occur only in systems with a ligand with a long residence time, and these are interactions that can be considered to be characteristic of compounds with a long residence time. Group 1 consists of interactions that, with the exception of two complexes with a relatively short residence time ligand: 4OIM and 2X23, have mostly been identified in complexes with long residence time compounds. Group 4 is characterized by contact identified only for compounds with short residence times, while additionally there is an interaction identified for both a compound with short and those with long residence times.

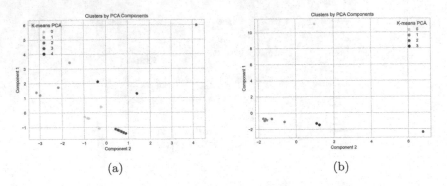

(a)   (b)

Fig. 6. Sample projection onto the space defined by the first two principal factors.

Table 4. Quantitative summary of the size of each ligand group.

Cluster ID	Cluster size	receptor-ligand PDB IDs with residence times that belong to cluster
0	7	2X22 (24 min), 2X23 (24 min), 4OIM (50 min), 4OXY (27 min), 4OYR (90 min), 5COQ (30 min), 5MTP (94 min)
1	1	5MTQ (119 min)
2	1	5UGT (220 min)
3	2	5UGU (194 min), 5MTR (106 min)

A projection of the weights onto the space defined by the first two principal components is made in order to examine which factors are responsible for the differentiation of the samples.

Each variable (interaction) is represented in the form of a vector, which direction and length determine how much the variable influences each principal component. Thus, it may be concluded that the largest contribution to the formation of the first component (the values of the coefficients are the highest) is the van der Waals interaction with the amino acid Trp248. Hydrophobic contacts with Leu196, Arg41, Gln99 and Phe96 also contribute to the first principal component. The last three together with hydrophobic interactions with Phe40, Arg42, Asp41, Ile15 and van der Waals interactions with Arg42 and Phe40 form the second principal component.

From the weight projections, it can be concluded that hydrophobic interactions with Gln99 and Arg41 are highly correlated. The hydrophobic interactions with the amino acids Phe40, Arg42, Asp41 and Ile15 and the van der Waals interactions with Arg42 and Phe40 are also positively correlated with each other. This interaction set is negatively correlated with hydrophobic interactions with Gln99, Arg41 and Phe96. The hydrophobic interaction between the ligand and Leu196 is negatively correlated with the van der Waals interaction with amino

Table 5. Quantitative summary of the size of each interaction fingerprint group.

Cluster ID	Cluster size	Interaction fingerprints that belong to cluster	receptor-ligand PDB IDs with residence times in which interaction occured
0	5	ARG41 Hydrophobic ARG42 PiCation ASP41 VdWContact PHE40 VdWContact LEU196 Hydrophobic	4OYR (90 min) 5UGT (220 min) 5UGT (220 min) 5UGT (220 min) 5UGU (194 min)
1	4	ILE15 Hydrophobic ARG42 Hydrophobic Arg42 VdWContact ASP41 Hydrophobic	4OIM (50 min), 5MTR (106 min), 5UGT (220 min), 5UGU (194 min) 2X23 (24 min), 5MTR (106 min), 5UGT (220 min) 5MTR (106 min), 5UGT (220 min), 5UGU (194 min) 2X23 (24 min), 5MTR (106 min), 5UGT (220 min)
2	1	Phe96 Hydrophobic	2X22 (24 min), 4OIM (50 min), 5COQ (30 min), 5MTQ (119 min), 5MTR (106 min), 5UGU (194 min)
3	12	SER246 VdWContact LYS7 Hydrophobic LEU4 VdWContact Leu4 Hydrophobic ALA33 Hydrophobic ILE9 Hydrophobic TRP248 Hydrophobic GLN31 Hydrophobic CYS242 VdWContact Cys242 Hydrophobic LEU245 Hydrophobic TRP248 VdWContact	5MTQ (119 min)
4	2	GLN99 Hydrophobic PHE40 Hydrophobic	4OIM (50 min), 5COQ (30 min) 5COQ (30 min), 5MTR (106 min), 5UGT (220 min), 5UGU (194 min)

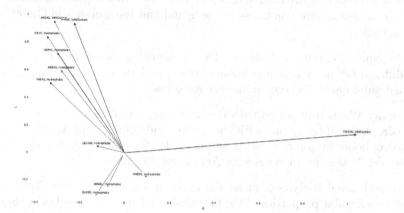

Fig. 7. Projection of the weights on the space of the first two principal components.

acid Trp248 which shows no correlation with other interactions. The interaction with Trp248 was only observed with the ligand in the 5MTQ complex (119 min residence time) and with Leu196 with the ligand in the 5UGU complex (194 min) (for the occurrence of each interaction see Table 5) (Fig. 7).

5   Discussion and Conclusions

To identify the molecular properties of the InhA protein and its inhibitors during the dissociation process characteristic of a given residence time, an approach that takes into account the transition states of the system and the interactions that take place between the protein molecule and the ligand molecule has been proposed. The approach requires no prior knowledge of the binding mechanism and is based on PCA analysis and k-means clustering. The identified groups form interaction fingerprints characteristic of ligands with a specific residence time, as well as groups that distinguish ligands based on the length of residence time in a molecular target.

A set of features was generated from the dissociation trajectories of the studied complexes. It was assumed that interactions occurring in at least 20% of a single trajectory describe the bound state, and that interactions occurring in less than 5% of all trajectories do not affect the residence time. Features meeting the above assumptions were removed from the feature set in the subsequent analysis.

Analyzing the frequency of occurrence of a given interaction in the complexes allowed identifying key amino acids that are likely to have a significant impact on differentiating the tested compounds by residence time:

- the hydrophobic interaction with the amino acid Leu196, as well as the van der Waals with Phe40, Asp41, Arg42 and the π-cation with Arg42 promote longer residence times, as they were identified only in complexes with the ligands with the longest residence times in the studied data set (106, 194 and 220 min),
- for compounds with relatively short residence times (30 and 50 min), the hydrophobic interaction between the ligand and the amino acid Gln99 is characteristic.

Principal Component Analysis (PCA) identified the factors responsible for the differentiation of ligand residence times, and thus to verify the previously defined amino acids. These factors are as follows:

- van der Waals interaction with the amino acid Trp248,
- hydrophobic interactions with the amino acids Gln99 and Arg41,
- hydrophobic interactions with amino acids Phe40, Arg42, Asp41, Ile15 and van der Waals interactions with Arg42 and Phe40.

An additional analysis of molecular descriptors, which allow a deeper insight into the molecular properties of the protein-ligand system, would probably be an interesting extension of the method. This will be an area for further development of the method.

References

1. Copeland, R.A.: The drug-target residence time model: a 10-year retrospective. Nat. Rev. Drug Discovery **15**, 87–95 (2016). https://doi.org/10.1038/nrd.2015.18
2. Vauquelin, G.: Effects of target binding kinetics on in vivo drug efficacy: koff, kon and rebinding. Br. J. Pharmacol. 2319–2334 (2016)
3. Copeland, R.A., Pompliano, D.L., Meek, T.D.: Drug-target residence time and its implications for lead optimization. Nat. Rev. Drug Discovery **5**, 730–739 (2006)
4. Folmer, R.H.: Drug target residence time: a misleading concept. Drug Discovery Today **23**, 12–16 (2018). https://doi.org/10.1016/j.drudis.2017.07.016
5. Wang, Y., Ribeiro, J.M.L., Tiwary, P.: Machine learning approaches for an analyzing and enhancing molecular dynamics simulations. Curr. Opin. Struct. Biol. **61**, 139–145 (2020)
6. Dhakal, A., McKay, C., Tanner, J.J., Cheng, J.: Artificial intelligence in the prediction of protein-ligand interactions: recent advances and future direction (2022)
7. Qu, S., Huang, S., Pan, X., Yang, L., Mei, H.: Constructing interconsistent, reasonable, and predictive models for both the kinetic and thermodynamic properties of HIV-1 protease inhibitors. J. Chem. Inf. Model. **56**, 2061–2068 (2016)
8. Ganotra, G.K., Wade, R.C.: Prediction of drug-target binding kinetics by comparative binding energy analysis. ACS Med. Chem. Lett. **9**, 1134–1139 (2018)
9. Su, M., Liu, H., Lin, H., Wang, R.: Machine-learning model for predicting the rate constant of protein-ligand dissociation. Wuli Huaxue Xuebao/Acta Physico-Chimica Sinica 3 (2020)
10. Amangeldiuly, N., Karlov, D., Fedorov, M.V.: Baseline model for predicting protein-ligand unbinding kinetics through machine learning. J. Chem. Inf. Model. **60**, 5946–5956 (2020)
11. Badaoui, M., et al.: Combined free-energy calculation and machine learning methods for understanding ligand unbinding kinetics. J. Chem. Theory Comput. **18**, 2543–2555 (2022)
12. Bray, S., Tanzel, V., Wolf, S.: Ligand unbinding pathway and mechanism analysis assisted by machine learning and graph methods (2022)
13. Gabdoulline, R.R., Wade, R.C.: On the protein-protein diffusional encounter complex (1999)
14. Gabdoulline, R.R., Wade, R.C.: Biomolecular diffusional association Gabdoulline and Wade 205 (2022)
15. Prasad, M.S., Bhole, R.P., Khedekar, P.B., Chikhale, R.V.: Mycobacterium enoyl acyl carrier protein reductase (InhA): a key target for antitubercular drug discovery. Bioorg. Chem. **115**, 105242 (2021)
16. Bajusz, D., Racz, A., Heberger, K.: Why is tanimoto index an appropriate choice for fingerprint-based similarity calculations? J. Cheminformatics **7**, 1–13 (2015)
17. Kokh, D.B., et al.: Estimation of drug-target residence times by τ-random acceleration molecular dynamics simulations. J. Chem. Theory Comput. **14**, 3859–3869 (2018)

Bioinformatics

Biohformatics

Validation of Height-for-Age and BMI-for-Age Z-scores Assessment Using Android-Based Mobile Apps

Valerii Erkudov[1]([⊠]) [iD], Sergey Lytaev[1] [iD], Kenjabek Rozumbetov[2] [iD], Andrey Pugovkin[3] [iD], Azat Matchanov[2] [iD], and Sergey Rogozin[1] [iD]

[1] St. Petersburg State Pediatric Medical University, St. Petersburg 194100, Russia
verkudov@gmail.com, physiology@gpmu.org
[2] Berdakh Karakalpak State University, Nukus, Uzbekistan 230100
[3] Saint Petersburg Electrotechnical University «LETI», St. Petersburg 197022, Russia

Abstract. The WHO's standards for analysis and presentation of anthropometric data is widely recognized as the best system. This study was aimed to assess the agreement of calculation of height-for-age and BMI-for-age Z-scores according to WHO growth charts using of two randomly selected mobile apps in two cohorts of European and Asian subjects. In 1,347 adolescents aged 13 to 17 years, boys and girls, living in St. Petersburg, Northwest of the Russia (European cohort, 663 subjects) and Nukus, Uzbekistan (Asian cohort 684 subjects) measured body weight and stature. Each child's height-for-age Z-score and BMI-for-age Z-scores were calculated based on WHO Child Growth Standards PC version WHO AntroPlus software and mobile applications. It was performed a Blend-Altman analysis to assess the consistency of Z-scores calculated using WHO AntroPlus software and two mobile apps. The results of the evaluation of the consistency of height-for-age and BMI-for-age Z-scores obtained using WHO AntroPlus software and mobile applications showed that Android-based mobile applications systematically overestimate Z-scores. However, this error is much smaller than any clinically significant deviations of the physical development evaluation parameters both in the European and Asian cohort volunteers. Thus, it can be claimed that it is possible to use Android-based mobile applications for growth monitoring and BMI and WHO AntroPlus software for children from 5 to 19. This is especially relevant to the need for «field» monitoring of children's growth and development performed by general practitioners visiting patients at home.

Keywords: WHO growth charts · BMI · Stature · Z-scores · AntroPlus software · Mobile apps · Blend-Altman analysis

1 Introduction

The growth monitoring as is part of preventive child health programs, as both stunted stature [36] and obesity [3] are regarded as relatively early signs of poor health [3, 24]. The need for continuous monitoring of growth and body mass index (BMI) generates

I. Rojas et al. (Eds.): IWBBIO 2023, LNBI 13919, pp. 35–47, 2023.
https://doi.org/10.1007/978-3-031-34953-9_3

a discussion about the standards for their evaluation [9, 29]. In 2006, the World Health Organization (WHO) released universal child growth standards, intended to describe the optimal growth of children [10]. Height-for-age and BMI-for-age Z-scores were calculated using a reference group of children from almost all over the world living in favorable conditions [10]. By 2011, most countries, including Uzbekistan, had adopted the WHO standards. Russia was among the countries that did not officially adopt these standards [10], but leading pediatricians still recommend it [17]. Although some countries insist on applying using standards, created on the basis of the national sample populations [5, 29], Z-score WHO's standards is widely recognized as the best system for analysis and presentation of anthropometric data and it is the most appropriate descriptor of malnutrition [12, 22] and overweight [2, 23].

On the basis of the WHO international standards there were developed program applications for PC: WHO Antro and WHO AntroPlus for younger and older children respectively [19].These applications make work much easier, but only if the physician has access to an individual workstation on a technically serviceable personal computer, which is not the case in all clinics. In addition, pediatricians are often faced with the need to assess growth and BMI in «field» studies. For example, this may be a home visit to a healthy child or a phone consultation to assess physiological weight loss 2–3 days after birth or to monitor breastfeeding. This issue can be solved by developing mobile apps using WHO grows charts [13]. According to approximate estimates hundreds of thousands health apps available worldwide [34, 37]. However, it is reported that, baby care apps not always fail to provide adequate information on normal growth and development later in childhood and adolescence [13]. Following these considerations we aimed to assess the agreement of calculation of height-for-age and BMI-for-age Z-scores according to WHO growth charts using of two randomly selected mobile apps and the «golden standard» method – PC version WHO AntroPlus software. As a result of the evaluation, the question of the reliability of Z-scores calculated using mobile and PC version in two cohorts of European (St. Petersburg, Russia) and Asian (Nukus, Uzbekistan) subjects will be solved. We hypothesized that the consistency of the data would differ because of the ethnic diversity of physical development of subjects living in Europe and Asia [7, 28, 38].

To the best of our knowledge, this is the first study that scientifically substantiates the consistency of a computer-based and mobile version of a Z-index estimator application. Other authors have focused on comparing different assessment systems: national standards or Z scores that were computed according to the Centers for Disease Control and Prevention [5, 32, 39].

Thus, present research was aimed to assess the agreement of calculation of height-for-age and BMI-for-age Z-scores according to WHO growth charts using of two randomly selected mobile apps in two cohorts of European and Asian subjects.

2 Methods

2.1 Subjects

The study according to the guidelines of the Declaration of Helsinki was performed and by the local ethics committee of the St. Petersburg State Pediatric Medical University (protocol no. 15/1, 10 Jan 2017) was approved. During the 5-year period from 2017 to 2022, 1,347 anonymous profiles of adolescents aged 13 to 17 years living in St. Petersburg, Northwest of the Russia (European cohort) and Nukus, Uzbekistan (Asian cohort) were obtained. The sample included 663 (314 boys, 349 girls) and 684 (312 boys, 372 girls) volunteers in the European and Asian cohorts respectively. Data were collected at visits to pediatric clinics for preventive examinations and included date of birth, date of visit, anthropometric measures, and brief information about health status and lifestyle.

2.2 Anthropometric Measurements

The study was performed by two or more qualified anthropometrists in the spring-summer period, at a comfortable temperature in the morning hours in an isolated room with sufficient light. The volunteer was dressed in underwear. Body weight was measured on floor electronic medical scales SECA 869, Germany, stature – using a folding mobile stadiometer SECA 217, Germany.

2.3 Z-score Calculation

Each child's height-for-age Z-score and BMI-for-age Z-scores were calculated based on WHO Child Growth Standards PC version WHO AntroPlus software [19]. Android-based mobile applications for calculating Z-scores were found in the Google Play store using the keywords «WHO Growth Charts», «WHO Growth Standards», «WHO Anthro», «WHO Z-scores», meeting the following criteria:

1. App description informing that the manufacturer uses WHO Child Growth Standards for children from 5 to 19 years old as the Z-scores calculation methodology;
2. Availability of a fast visit mode for the anthropometric calculator when you do not need to create a report card for each child;
3. Available for download for over 5 years, manufacturer is constantly updating their product;
4. The app has been downloaded over 10,000 times by users;
5. User comments and reviews with usage reports;
6. The app rating is over 4.5.

At least two applications met all or nearly all of these conditions. We chose Anthro-Calc v. 2.24, by Daniel L. Metzger, release date 12.11.2003, last updated 12.29.2013, >10,000 downloads and rating 4.8 based on 109 user reviews [4], and Growth Charts CDC WHO Percentile v. 3.1.21229.01, by Osama Orabi, release date 01.12.2015, last updated 17.08.2021, >100 thousand downloads, rating 4.8 based on 905 user reviews [18]. Each child's height and body weight data were entered into specific forms of selected mobiles to calculate eight-for-age Z-score and BMI-for-age Z-scores.

2.4 Statistical Analysis

In order to find out whether the anthropometric data of Asian cohort and European cohort volunteers are indeed different, Z-scores were compared statistical in these groups using the Mann-Whitney U test. Comparison of deviations of body weight and stature from the WHO Child Growth Standards in subjects from both groups was carried out by assessing the homogeneity of the distribution of «average» (from -1 to $+1$), «above average» (from $+1$ to $+2$), «below average» (from -2 to -1), «high» (from $+2$ to $+3$), «low» (from -3 to -2), «extremely high» (more than $+3$) «extremely low» (less than -3) Z scores. Fisher's exact test for contingency table 37×7 with the calculation of the proportion of children with one or another deviation in body weight and height was used. Estimations were carried out using statistical software, Past version 2.17, Oslo, 2012 and algorithm of statistical data processing StatXact-8 with Cytel Studio software shell version 8.0.0., USA, 2007. All continuous data are presented as means (M) and 95% confidence intervals (CI). Categorical data were presented as proportions with 95% CI.

After checking data against the normal distribution using Shapiro-Wilk test, we performed a Blend-Altman analysis to assess the consistency of Z-scores calculated using WHO AntroPlus software («golden standard») and two mobile apps. The Blend-Altman analysis evaluates bias, -1.96 and $+1.96$ limits of agreement (LOA) and their 95% confidence intervals.

Checking and analysis of differences in data obtained with both approaches was provided according to the principles proposed by us [26, 30, 33] and other authors [1, 11, 14]:

- The Blend-Altman plot system does not say if the agreement is sufficient or suitable to use a one method instead of the other. It simply quantifies the bias and a range of agreement, within which 95% of the differences between one measurement and the other are included.
- Only analytical, biological or clinical goals could define whether the bias and agreement interval is too wide or sufficiently narrow.
- 95% CI of the bias illustrates the magnitude of the systematic difference. If the bias not equal to zero, there is a significant systematic difference, and the second method constantly under- or over- estimates compared to the first one
- The bias value is significant, if the zero line of equality is not within the its confidence interval since the random variable estimated by this interval will never take a zero value
- 95% CI of bias and of the agreement limits simply describe a possible error in the estimate, due to a sampling error. The greater the number of samples used for the evaluation of the difference between the methods, the narrower will be the CIs, both for the mean difference and for the agreement limits.
- 95% of dots on the Bland-Altman plot must fall within the -1.96 and + 1.96 LOA

Bland-Altman analysis was performed using Statistica data analysis software version 10, StatSoft inc, USA, 2011.

3 Outcomes

Adolescents from the European cohort showed statistically significantly lower values of the Z-score for height-for-age and BMI-for-age by comparison with volunteers from the Asian cohort (Table 1). The distribution of height and BMI deviations was heterogeneous and statistically significantly different in the examinees from both groups (Table 2). The frequency of low values of Height-for-age and BMI-for-age Z-scores is higher in the Asian cohort and, therefore, depends on its ethnic characteristics and place of residence.

Table 1. Comparison of height-for-age and BMI-for-age Z-scores in adolescents from the two study groups (M; 95% CI)

Z-score	European cohort	Asian cohort	p-value
Height-for-age	0.63 (0,57; 0,69)	$-0,17$ (-0.23; -0.11)	1.797×10^{-58}
BMI-for-age	0.06 (-0.007; -0.13)	-0.26 (-0.34; -0.18)	3.062×10^{-9}

Table 2. The proportions of height and BMI deviations and their 95% CI in adolescents from the two study groups

Group	Extremely low	Low	Below average	Average	Above average	High	Extremely high
Proportions of height deviations*							
European cohort	0 (0; 0.008)	0 (0; 0.008)	0.03 (0.019; 0.06)	0.62 (0.57; 0.67)	0.31 (0.26; 0.36)	0.04 (0.02; 0.06)	0 (0; 0.008)
Asian cohort	0.0015 (1.05×10^{-5}; 0,01)	0.02 (0.009; 0.04)	0.15 (0.11; 0.18)	0.73 (0.69; 0.79)	0.09 (0.07; 0.13)	0.003 (0.0002; 0.014)	0 (0; 0.008)
Proportions of BMI deviations**							
European cohort	0 (0; 0.008)	0.009 (0.02; 0.06)	0.10 (0.07; 0.13)	0.75 (0.71; 0.80)	0.12 (0.09; 0.16)	0.02 (0.008; 0.04)	0 (0; 0.008)
Asian cohort	0.009 (0.002; 0.02)	0.038 (0.02; 0.06)	0.20 (0.16; 0.24)	0.63 (0.58; 0.68)	0.10 (0.07; 0.13)	0.03 (0.01; 0.05)	0.003 (0.0002; 0.01)

Note: * $p = 9.078 \times 10^{-10}$; ** $p = 1.62 \times 10^{-9}$

The Shapiro–Wilk test showed that height-for-age and BMI-for-age Z-scores in adolescents from European cohort (p = 0.994) and Asian cohort (p = 0.9943) have a normal distribution (Fig. 1).

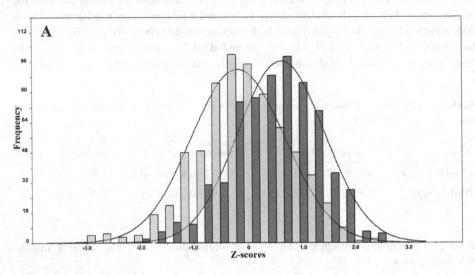

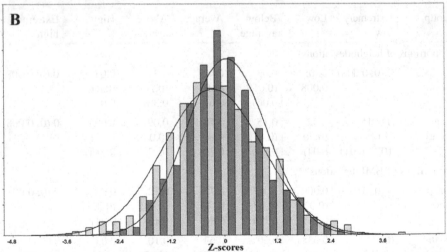

Fig. 1. Histogram of the distribution of height-for-age (A) and BMI-for-age Z-scores (B) in volunteers from the European cohort (light gray bars) and Asian cohort (dark gray bars). A, B – values along the y-axis are similar.

Bland-Altman analysis represents that the use of both mobile app algorithms in both groups systematically overestimates Z-scores, as indicated by the negative bias values (Table 3, 4). However, the magnitude of the obtained bias of both European and Asian cohorts is less than the reported clinical deviations of Z-scores by a 100–1000 times (Table 3, 4). The Z-score variation ± 1 SD has been shown to be clinically significant for discrete estimates of stature and BMI deviations [10, 17]. Some studies have reported that a change in height-for-age Z-score of 0.3–0.5 [35] may be associated and BMI-for-age Z-score of 0.2–0.4 with impaired growth [20]. Values of zero are not included in the 95% CI for bias in all comparisons except the BMI-for-age consistency assessment using AnthroCalc in the European cohort and using Growth Charts CDC WHO Percentile in the Asian cohort (Table 3, 4). Given the extremely low values of bias, this result cannot be taken into account as questioning the consistency of the Z-scores methods. The narrow limits of -1.96 and $+1.96$ LOA and their 95%CI are noteworthy: the difference $(-1.96) - (+1.96)$ LOA in all comparison cases does not exceed 0.05, 95%CI - 0.03. All this indicates insignificant sampling error [14]. Evaluation of the Bland-Altman plot shows that in all plots, 98.5% to 99.5% of the points are within -1.96 and $+1.96$ LOA (Fig. 2, 3).

Table 3. Bland-Altman analysis comparing height-for-age and BMI-for-age Z-scores calculated using the WHO standards (AntroPlus) and mobile app AnthroCalc in European and Asian cohorts (bias, -1.96; $+1.96$ LOA and its 95%CI)

Z-scores	Height-for-age Z-scores			BMI-for-age Z-scores		
Parameter	Bias	-1.96 LOA	$+1.96$ LOA	Bias	-1.96 LOA	$+1.96$ LOA
European cohort	-0.001 $(-0.01;$ $0.01)$	-0.33 $(-0.36;$ $-0.31)$	0.33 $(0.31;$ $0.35)$	-0.02 $(-0.03;$ $-0.002)$	-0.36 $(-0.38;$ $-0.34)$	0.33 $(0.31;$ $0.35)$
Asian cohort	-0.0006 $(-0.002;$ $0.0003)$	-0.026 $(-0.030;$ $-0.023)$	0.023 $(0.022;$ $0.024)$	-0.005 $(-0.13;$ $0.003)$	-0.20 $(-0.22;$ $0.19)$	0.19 $(0.18;$ $0.21)$

Table 4. Bland-Altman analysis comparing height-for-age and BMI-for-age Z-scores calculated using the WHO standards (AntroPlus) and mobile app Growth Chart CDC WHO in European and Asian cohorts (bias, −1.96; + 1.96 LOA and its 95%CI)

Z-scores	Height-for-age Z-scores			BMI-for-age Z-scores		
Parameter	Bias	−1.96 LOA	+1.96 LOA	Bias	−1.96 LOA	+1.96 LOA
European cohort	-0.004 (−0.008; 0.0003)	−0.11 (−0.123; −0.10)	0.10 (0.09; 0.11)	−0.02 (−0.04; 0.004)	−0.61 (−0.65; −0.57)	0.57 (0.53; 0.61)
Asian cohort	−0.004 (−0.01; 0.003)	−0.17 (−0.19; −0.16)	0.17 (0.16; 0.18)	−0.02 (−0.03; −0.01)	−0.35 (−0.37; −0.32)	0.30 (0.28; 0.32)

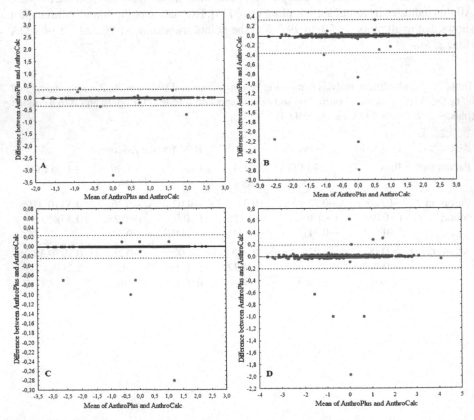

Fig. 2. Bland-Altman plot comparing height-for-age (A,C) and BMI-for-age Z-scores (B, D) calculated using the WHO standards (AntroPlus) and mobile app AnthroCalc in European (A,B) and Asian (C, D) cohorts.

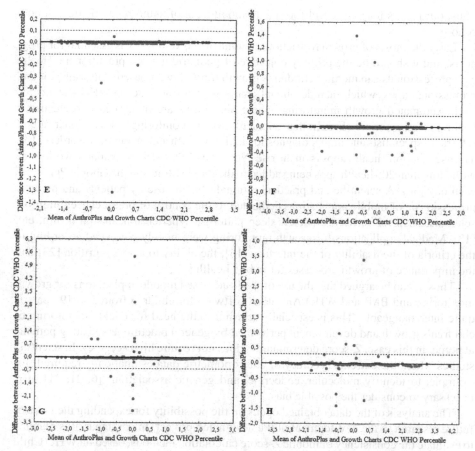

Fig. 3. Bland-Altman plot comparing height-for-age (E, G) and BMI-for-age Z-scores (F, H) calculated using the WHO standards (AntroPlus) and mobile app Growth Charts CDC WHO Percentile in European (E, F) and Asian (G, H) cohorts.

4   Discussion

All of the above facts indicate that mobile anthropometric data assessment applications using the WHO Child Growth Standards can replace the benchmark method of calculating both height-for-age and BMI-for-age Z-scores using the WHO Child Growth Standards PC version WHO AntroPlus software in subjects of different ethnicity and place of residence. Apparently, the methods used to calculate Z-scores are robust to the ethnic diversity of anthropometric data. It should be noted that this method is the methodology of choice for describing the ethnic diversity of physical development in subjects of Asian and European populations. According to a WHO expert consultation Asians are overweight and obese at lower BMIs than Europeans because they are more obese in adults [15] and adolescents [16] when overweight and less physically active.

Thus, our results may be useful for «field» monitoring of anthropometric data in both cohorts.

Early diagnosis of growth restrictions, malnutritions or obesity is essential for prognosis, and it should be the primary objective of pediatric general practitioners. Health care professionals can monitor children's growth through web-based platforms [25] and professional apps, which include official charts such as the Z-score WHO's standards [34]. Automated growth monitoring, where algorithms are integrated into electronic health records, is more efficient than standard growth monitoring, with a higher rate of referral to specialists and higher diagnostic yield of growth and metabolic disorders [13]. The use of mobile health apps is on the rise. Now over 300,000 were available worldwide with more than 200 health apps being added to the Apple Store and the Google Play store each day [36]. A scientific and practical rationale for the use by patients and medical professionals of mobile applications dedicated to the control of breast cancer, multiple sclerosis, physical activity, diabetes, sleep patterns, hypertension, mental health, etc. [13]. Medical applications for growth monitoring were mainly compared according to the criteria of the usability of the interface [13], the ability to assess nutrition [23] and the importance of growth processes for public health [27].

Thus, it can be argued that the use of Android-based mobile applications for growth monitoring and BMI and WHO AntroPlus software for children from 5 to 19 years is quite interchangeable. This is especially relevant to the need for «field» monitoring of children's growth and development performed by general practitioners visiting patients at home. In this case, Z-score determination does not require high accuracy. In scientific studies where the extreme accuracy of anthropometric data estimation is required, for example, to identify molecular, endocrine, and genetic associations [6, 21, 31], it is necessary to consider the possible bias.

The analysis of the data obtained opens up the possibility for expanding the tasks of future research in the development of the designated project. In particular, it is planned to evaluate the consistency of mobile Z-score calculation methods, based on WHO Child Growth Standards in children 0–5 years old. These standards use a different algorithm to calculate Z-scores and describe the growth of children living in favorable circumstances who during infancy were fed in ways that are thought to be conducive to optimal long-term health [10].

5   Conclusion

1. The results of the evaluation of the consistency of height-for-age and BMI-for-age Z-scores obtained using WHO AntroPlus software («golden standard») and mobile applications also using WHO international standards showed that in none of the eight comparison cases the bias was equal to zero.
2. Android-based mobile applications systematically overestimate Z-scores. It suggests that a very likely insignificant error in the Z-scores calculation algorithm can occur during the creation of program code supported by the operating system of the mobile device.
3. Discribed errors are much smaller than any clinically significant deviations of the physical development evaluation parameters.

4. There was no found the difference in the consistency of the mobile and the reference methods of Z-scores calculation in in the European and Asian cohort volunteers.

6 Limitations

Due to the objective of this study was to assess the consistency of the Z-score calculation methods and not to compare physical development parameters in different groups; we did not introduce very strict criteria for the inclusion of volunteers in the sample. At the beginning of the study, the database included 1430 subjects, and then we excluded 83 volunteers. These were data from study participants with acute or chronic diseases, orthopedic, endocrine pathology, edema, and who were athletes or lived in rural areas. Validation of Z-score methods in children with medical conditions, athletes, or residing in different territories should constitute a separate research task.

Conflict of Interests. This project has exclusively scientific goals and objectives. The authors deny receiving funding from software developers.

References

1. Abu-Arafeh, A., Jordan, H., Drummond, G.: Reporting of method comparison studies: a review of advice, an assessment of current practice, and specific suggestions for future reports. Br. J. Anaesth **117**(5), 569–575 (2016)
2. Al-Hazzaa, H., et al.: Prevalence of overweight and obesity among saudi children: A comparison of two widely used international standards and the national growth references. Front. Endocrinol. (Lausanne) **13**, 954755 (2022)
3. Andolfi, C., Fisichella, P.: Epidemiology of obesity and associated comorbidities. J. Laparoendosc. Adv. Surg. Tech A **28**(8), 919–924 (2018)
4. AnthroCalc – Apps on Google Play. https://play.google.com/store/apps/details?id=appinv entor.ai_dlmetzger58.AnthroCalc. Accessed 30 Dec 2022
5. Asif, M., et al.: Establishing height-for-age z-score growth reference curves and stunting prevalence in children and adolescents in pakistan. Int. J. Environ. Res. Public Health **19**(19), 12630 (2022)
6. Aslam, M., Mazhar, I., Ali, H., Ismail, T., Matłosz, P., Wyszyńska, J.: Establishing height-for-age Z-score growth reference curves and stunting prevalence in children and adolescents in Pakistan. Int. J. Environ. Res. Public Health **19**(19), 12630 (2022)
7. Braz, V., Lopes, M.H.B.M.: Evaluation of mobile applications related to nutrition. Public Health Nutr. **22**(7), 1209–1214 (2019)
8. Chang, H., Yang, S., Wang, S., Su, P.: Associations among IGF-1, IGF2, IGF-1R, IGF-2R, IGFBP-3, insulin genetic polymorphisms and central precocious puberty in girls. BMC Endocr. Disord. **18**(1), 66 (2018)
9. Cho, J., Choi, J., Suh, J., Ryu, S., Cho, S.: Comparison of anthropometric data between Asian and caucasian patients with obstructive sleep apnea: a meta-analysis. Clin. Exp. Otorhinolaryngol. **9**(1), 1–7 (2016)
10. de Onis, M., Onyango, A., Borghi, E., Siyam, A., Blössner, M., Lutter, C.: WHO multicentre growth reference study group: worldwide implementation of the WHO child growth standards. Public Health Nutr. **15**(9), 1603–10 (2012)
11. Doğan, N.: Bland-Altman analysis: a paradigm to understand correlation and agreement. Turk. J. Emerg. Med. **18**(4), 139–141 (2018)

12. Duan, Y., et al.: Association between dairy intake and linear growth in Chinese pre-school children. Nutrients 12(9), 2576 (2020)
13. Fernandez-Luque, L., Labarta, J., Palmer, E., Koledova, E.: Content analysis of apps for growth monitoring and growth hormone treatment: systematic search in the android app store. JMIR Mhealth Uhealth 8(2), e16208 (2020)
14. Giavarina, D.: Understanding Bland Altman analysis. Biochem. Med. (Zagreb) 25(2), 141–151 (2015)
15. Gong, S., Wang, K., Li, Y., Zhou, Z., Alamian, A.: Ethnic group differences in obesity in Asian Americans in California, 2013–2014. BMC Public Health 21(1), 1589 (2021)
16. Grgic, O., et al.: Skeletal maturation in relation to ethnic background in children of school age: the Generation R Study. Bone 132, 115180 (2020)
17. Gritsinskaya, V., Novikova, V., Khavkin, A.: Features of linear growth of pupils with different levels of physical development. Clin. Pract. Pediat. 17(1), 79–83 (2022)
18. Growth Chart CDC WHO Percentile – Apps on Google Play. https://play.google.com/store/apps/details?id=com.LetsStart.GrowthChart, Accessed 30 Dec 2022
19. Growth reference data for 5–19 years. https://www.who.int/tools/growth-reference-data-for-5to19-years. Accessed 30 Dec 2022
20. Hermanussen, M., Assmann, C., Tutkuviene, J.: Statistical agreement and cost-benefit: comparison of methods for constructing growth reference charts. Ann. Hum. Biol. 37(1), 57–69 (2010)
21. Hwang, I., et al.: Gene polymorphisms in leptin and its receptor and the response to growth hormone treatment in patients with idiopathic growth hormone deficiency. Endocr. J. 68(8), 889–895 (2021)
22. Kamruzzaman, M., et al.: The anthropometric assessment of body composition and nutritional status in children aged 2–15 years: a cross-sectional study from three districts in Bangladesh. PLoS ONE 16(9), e0257055 (2021)
23. Khadilkar, V., Shah, N.: Evaluation of children and adolescents with obesity. Indian J. Pediatr. 88(12), 1214–1221 (2021)
24. Lamb, M., et al.: Anthropometric proxies for child neurodevelopment in low-resource settings: length- or height-for-age, head circumference or both? J. Dev. Orig. Health Dis. 18, 1–9 (2022)
25. Loftus, J., et al.: Individualised growth response optimisation (iGRO) tool: an accessible and easy-to-use growth prediction system to enable treatment optimisation for children treated with growth hormone. J. Pediatr. Endocrinol. Metab. 30(10), 1019–1026 (2017)
26. Lytaev, S.: Modern neurophysiological research of the human brain in clinic and psychophysiology. In: Rojas, I., Castillo-Secilla, D., Herrera, L.J., Pomares, H. (eds.) BIOMESIP 2021. LNCS, vol. 12940, pp. 231–241. Springer, Cham (2021). https://doi.org/10.1007/978-3-030-88163-4_21
27. Martín-Martín, J., et al.: Evaluation of android and apple store depression applications based on mobile application rating scale. Int. J. Environ. Res. Public Health 18(23), 12505 (2021)
28. Mittal, M., Gupta, P., Kalra, S., Bantwal, G., Garg, M.: Short stature: understanding the stature of ethnicity in height determination. Indian J. Endocrinol. Metab. 25(5), 381–388 (2021)
29. Natale, V., Rajagopalan, A.: Worldwide variation in human growth and the World Health Organization growth standards: a systematic review. BMJ Open 4(1), e003735 (2014)
30. Orel, V., et al.: Ways of economical production in medical institution risk management. Lect. Notes Comput. Sci. 13320, 237–248 (2022)
31. Osman, W., Tay, G., Alsafar, H.: Multiple genetic variations confer risks for obesity and type 2 diabetes mellitus in Arab descendants from UAE. Int. J. Obes. (Lond.) 42(7), 1345–1353 (2018)
32. Poulimeneas, D., et al.: Comparison of international growth standards for assessing nutritional status in cystic fibrosis: the GreeCF study. J. Pediatr. Gastroenterol. Nutr. 71(1), e35–e39 (2020)

33. Pugovkin, A.P., Erkudov, V.O., Lytaev, S.A.: Methodological approaches to the comparison of left ventricular stroke volume values measured by ultrasonic technique or estimated via transfer functions. In: Rojas, I., Castillo-Secilla, D., Herrera, L.J., Pomares, H. (eds.) BIOME-SIP 2021. LNCS, vol. 12940, pp. 112–120. Springer, Cham (2021). https://doi.org/10.1007/978-3-030-88163-4_11

34. Sankilampi, U., Saari, A., Laine, T., Miettinen, P.J., Dunkel, L.: Use of electronic health records for automated screening of growth disorders in primary care. JAMA **310**(10), 1071–1072 (2013)

35. Wallenborn, J., Levine, G., Carreira Dos Santos, A., Grisi, S., Brentani, A., Fink, G.: Breastfeeding, physical growth, and cognitive development. Pediatrics **147**(5), e2020008029 (2021)

36. Webb, P., Stordalen, G., Singh, S., Wijesinha-Bettoni, R., Shetty, P., Lartey, A.: Hunger and malnutrition in the 21st century. BMJ **361**, k2238 (2018)

37. Xu, W., Liu, Y.: mHealthApps: a repository and database of mobile health apps. JMIR Mhealth Uhealth **3**(1), e28 (2015)

38. Yerkudov, V., et al.: Anthropometric characteristics of young adults in areas with different ecological risks in the Aral Sea Region Uzbekistan. Hum. Ecol. **10**, 45–54 (2020)

39. Zong, X., Li, H.: Construction of a new growth references for China based on urban Chinese children: comparison with the WHO growth standards. PLoS ONE **8**(3), e59569 (2013)

An Algorithm for Pairwise DNA Sequences Alignment

Veska Gancheva(✉) ⓘ and Hristo Stoev

Faculty of Computer Systems and Control, Technical University of Sofia, 8 Kliment Ohridski, 1000 Sofia, Bulgaria
vgan@tu-sofia.bg

Abstract. A challenge in data analysis in bioinformatics is to offer integrated and modern access to the progressively increasing volume of data, as well as efficient algorithms for their processing. Considering the vast databases of biological data available, it is extremely important to develop efficient methods for processing biological data. A new algorithm for arranging DNA sequences based on the suggested CAT method is proposed, consisting of an algorithm for calculating a CAT profile against the selected reference sequences and an algorithm for comparing two sequences, based on the calculated CAT profiles. Implementation steps, inputs and outputs are defined. A software implementation of the proposed method for arranging biological sequences CAT has been designed and developed. Experiments have been carried out using different data sets to align DNA sequences based on CAT method. An analysis of the experimental results have been done in terms of collisions, speed and effectiveness of the proposed solutions.

Keywords: Bioinformatics · Biological Data Sequences · DNA Sequences Alignment

1 Introduction

Bioinformatics is one of the most rapidly developing and promising sciences in recent years, which makes it possible to carry out scientific experiments using computer models and simulations based on efficient methods, algorithms and means of storage, management, analysis and interpretation of a huge amount of biological data. A major challenge in biological data analysis is to offer integrated and modern access to the progressively increasing volume of data in multiple formats, as well as efficient algorithms and tools for their search and processing.

A major issue in biological data processing is the search for homologous sequences in a database. Algorithms such as Needleman-Wunsch [1] and Smith-Waterman [2], which accurately determine the degree of similarity between two sequences, take a long time to apply to all entries in large datasets. For faster searches in large databases, scientists apply heuristic methods and algorithms that significantly speed up the search time, but reduce the quality of the results obtained. FASTA is a software package for DNA and protein sequence alignment that introduces heuristic methods for sequence alignment -

I. Rojas et al. (Eds.): IWBBIO 2023, LNBI 13919, pp. 48–61, 2023.
https://doi.org/10.1007/978-3-031-34953-9_4

querying the entire database. BLAST is one of the most widely used sequence search tools [3, 4]. The heuristic algorithm it uses is much faster than other approaches, such as computing an optimal alignment. The BLAST algorithm is more time efficient than FASTA, searching only the more significant sequences, but with comparable sensitivity. Even the parallel execution of the above algorithms is limited by the hardware systems [5–13]. The metaheuristic method for multiple sequence ordering adopts the idea of generating a favorite sequence, after which all other sequences from the database are compared with the favorite sequence [14]. In this way, the favorite sequence becomes a benchmark for the rest of the sequences in the database. Some problems arise when using this approach, such as entering new data into the database or deleting some of the existing records.

Since the favorite sequence is generated based on the existing records: (1) Changing the data causes the favorite sequence to be recalculated. (2) Each of the sequences in the database must be compared again with the newly calculated favorite sequence to obtain a new result, which consumes computational time and resources. (3) There is a different favorite sequence for each database, and this can lead to problems in merging different databases, especially in big data, where there is a collection of many different database structures and access methods.

To improve the idea of the existing heuristic algorithms, the work is related to propose improvements in the following topics:

1. Constant favorite sequence – i.e. not depend on the data in the database and remain the same when changed;
2. Avoiding comparisons or reducing the number of comparisons with the favorite sequence during search in the database (for each sequence a complex comparison algorithm is applied against the favorite sequence)
3. Unification of sequence favorites for all databases.

The purpose of the research is to propose a new algorithm for pairwise DNA sequence alignment based on a new efficient and unified method for DNA sequences alignment utilizing trilateration method. The goal is to suggest solution for three main issues in biological sequence alignment: (1) creating constant favorite sequence, (2) reducing the number of comparisons with the favorite sequence, and (3) unifying favorite sequence by defining benchmark sequences.

2   Related Work

At the heart of the idea of a favorite sequence is to find a starting point - a benchmark against which the rest of the data in the database can be analyzed. Or, looking at it mathematically, sequence favorite could be represented as a function of N unknowns (speaking of DNA the unknowns are the 4 bases: adenine, thymine, guanine, and cytosine), then represent the remaining database entries again as functions of the same variables. In such a case, the similarity comparison would represent the distance of the individual sequence to the favorite sequence. In other words, determine the location of a point described by the sequence function relative to another point defined by the favorite sequence function. When comparing to a sequence favorite, there is a set of points (the database entries) and since there is no coordinate system, a point is generated somewhere around the center of the cloud of points that is used as a reference (sequence favorite). But if some kind of coordinate system is introduced, or three or more reference points are found, then it would be possible, by means of elementary analytical geometry, or in particular trilateration, to determine the positions of the points relative to each other, which will reflect the match rate between the records in the database. Also, to eliminate the need of sequence favorite calculation.

A new method for aligning DNA sequences, called CAT, based on the trilateration method, has been proposed [15]. Three constant benchmarks have been established for the application of trilateration, which creates a constant favorite sequence - i.e. it does not depend on the data in the database and remains the same when it changes.

Three benchmark sequences are defined: C-benchmark, A-benchmark and T-benchmark (CAT method):

ACGTACGTACGTACGTACGTACGTACGTACGTACGTAC.......   – A-Benchmark
TGCATGCATGCATGCATGCATGCATGCATGCATGCATG......   – T-Benchmark
CGATCGATCGATCGATCGATCGATCGATCGATCGATCG...... – C-Benchmark

Once the three constant benchmarks for applying trilateration are established, issue (1) is eliminated. Constant sequence favorite – ie. Independent of the records in the database and to remain the same when the data set is changed.

Since the constant benchmark sequences are determined (i.e. they do not depend on either the data or their count), this allows comparisons to be made at the very beginning - when the sequences are uploaded into the database and this is metadata information accompanying each sequence. This way, sequences won't need to be compared during lookup (which is the slowest operation), but instead only the metadata information generated during the upload will be compared. By establishing the benchmark sequences, issue (3) unification of favorite sequences for all databases is also eliminated. There are now unified sequences that are standardized for all databases using the described alignment algorithm. When two sequences have the same profiles, this means that they have regions with the same alignment, and one hundred percent complete matching of one sequence on the other can be expected.

For the evaluation of two sequences, it is necessary to calculate the distance of segment S1S2 in Fig. 1.

$$S_1 S_2 = \sqrt{|AD_1 - AD_2|^2 + |h_1 - h_2|^2} \tag{1}$$

For now, $\triangle AS1T$ is considered, then analogous calculations and reasoning are performed for AS2T. What is known about $\triangle AS1T$ is the sides AT = |1|, AS1 = distance from S1 to A-benchmark (it is known), S1T = distance from S1 to T-benchmark. Use the cosine theorem to find $\angle TAC$, then side AD1:

$$S_1 T^2 = AS_1{}^2 + AT^2 - 2.AS_1.AT.cos(\alpha_1) \tag{2}$$

$$cos(\alpha_1) = \frac{AS_1{}^2 + AT^2 - S_1 T^2}{2.AS_1.AT} \tag{3}$$

$$AD_1 = AS_1.cos(\alpha_1) \tag{4}$$

$$h_1 = \sqrt{AS_1^2 - (AS_1.cos(\alpha_1))^2} \tag{5}$$

The calculations for triangle AS2T are analogous. After substituting the values found, a value for S1S2 is obtained.

$$S_1S_2 = \sqrt{(S_1D_1 - S_1D_2)^2 + (AD_1 - AD_2)^2} \tag{6}$$

The smaller value obtained for the intercept, the greater the probability of a complete match, expressed as a percentage. This allows the database to be quickly searched for sequences with a certain percentage of similarity, which can later be aligned and compared with more accurate algorithms such as Needleman-Wunsch or Smith-Waterman.

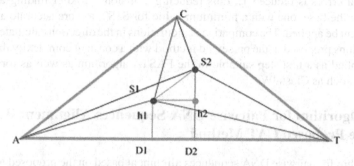

Fig. 1. Calculation of the distance between two profiles.

It is possible to occur collisions in such proposed DNA sequence alignment method based on trilateration, i.e.:

1. More than one sequence of the same length to get the same values for AD and h:

   - Due to the nature of benchmark sequences and the fact that the real sequence projects at most a quarter of the bases onto the benchmark, i.e. with benchmark ACGT and projection of G at the second position, there are no matches and no value is accumulated for match rate.

2. During S1S2 calculation, the same values are obtained:

   - Because of statistical errors accumulated when calculating AD and h.
   - Because of rounding in calculations due to the range of data types, this cannot be avoided even with the use of more precise types.

To minimize collisions, the precision in calculations for AD and h should be increased by adding the dependency on neighboring bases. Similar to local alignment, when the current base of the benchmark sequence does not match the current base of the real sequence, additional points can be added or subtracted, depending on whether a neighboring base match is. After Needleman-Wunsch alignment, the places where the bases do not match appear as gaps "_" positions and are given different points accordingly. A similar principle can be applied in the proposed method. If a base and index match is given value of 1. If there is a mismatch, a neighboring base from the benchmark sequence is checked and given a value of 0.05 or −0.05 depending on whether it is left or right adjacent (whether it is possible to appear empty position before or after the base) and − 0.003 if it is across a base (neighbor of the adjacent).

Example of benchmark ACGT base at position 2 G

ACGT

XGXX

G at position 2 corresponds to C from the benchmark sequence and instead of 0 a match value of 0.05 can be given because it is adjacent to the right and in Needleman-Wunsch ordering it has the following alignment:

ACG_T

X_GXX

In this way, the precision of AD and h calculation is increased, the accumulation of statistical errors is reduced 1., thus reducing collisions 2. After finding a suitable sequence in the base, one with a minimum value for S1S2, a more accurate alignment algorithm can be applied. The comparison calculations in the direction calculate the CAT of two sections proposed in the presented method with a constant complexity that makes it to be applied as a first step suitable in the FASTA algorithm as well as for multiple alignments such as ClustalW.

3   An Algorithm for Pairwise DNA Sequences Alignment Based on the Proposed CAT Method

The algorithm for pairwise DNA sequences alignment based on the proposed alignment method can be divided into two stages. The first stage is the calculation of a CAT profile against the selected benchmark sequences. For each of the benchmarks, the profile of the input sequence is calculated to create a complete CAT profile of the input sequence (Fig. 2 and Fig. 3).

Algorithm:	Calculation of CAT profile for DNA sequence
Input:	DNA sequence as string (AGGTGCCGGTÖ Ö .)
Output:	CAT profile: {C:{D,H},A:{D,H},T:{D,H}}
Processing Steps:	
Step1:	**For each benchmark:** Count exact matches of the input DNA string to calculate the distance to that benchmark.
Step2:	**For each benchmark:** Calculate Cos(sequence benchmark distance, benchmark to benchmark distance) Calculate H(calculated benchmark cos) Calculate D(calculated benchmark cos)

Fig. 2. Algorithm for calculation of CAT profile against the selected benchmark sequences.

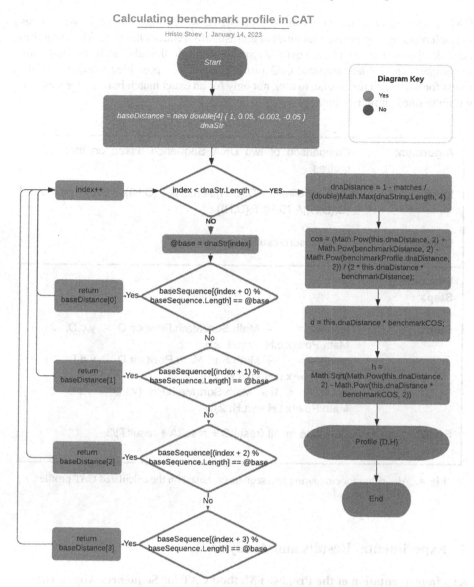

Fig. 3. Block diagram of an algorithm for calculating a sequence profile against a benchmark profile.

This operation is done once when entering the sequence in the database and the result is stored as accompanying information about the sequence. The algorithm has linear complexity O(n).

The second stage is a comparison against the calculated CAT profiles (Fig. 4). This operation is performed repeatedly when evaluating the similarity of two or more sequences from the database. The DNA sequence alignment algorithm based on the

CAT method has a constant complexity of O(24), which makes it very fast and easy to implement by computing machines and drastically reduces the time when searching large databases. For even better database speed, multiple threads can be run in parallel to compare against the computed CAT profiles. Also, it is possible to define similarity limits for the search results, i.e. to look not only for an exact match but also for similarity within defined similarity limits.

Algorithm:	Comparison of two DNA Sequence based on their CAT profiles
Input:	CAT profile1: {C:{D,H},A:{D,H},T:{D,H}}, CAT profile2: {C:{D,H},A:{D,H},T:{D,H}}
Output:	Comparison result in %
Processing Steps:	
Step1:	resultC = 1 - Math.Sqrt(Math.Pow(x.c.D - y.c.D, 2) + Math.Pow(x.c.H - y.c.H, 2)); resultA = 1 - Math.Sqrt(Math.Pow(x.a.D - y.a.D, 2) + Math.Pow(x.a.H - y.a.H, 2)); resultT = 1 - Math.Sqrt(Math.Pow(x.t.D - y.t.D, 2) + Math.Pow(x.t.H - y.t.H, 2));
Step2:	Calculate result (resultC + resultA + resultT)/3

Fig. 4. Algorithm for comparing two sequences, based on the calculated CAT profiles.

4   Experimental Results and Analysis

4.1   Implementation of the Proposed Method CAT for Sequences Alignment

The purpose of the experiments is to evaluate experimentally the effectiveness of the designed algorithm based on the proposed CAT method for DNA sequence alignment. For this purpose, a program implementation implemented in the C# programming language have been developed (Fig. 5).

Benchmark - base class for representing the benchmark sequence abstraction.

BenchmarkRepo - contains the predefined benchmark sequences.

BenchmarkProfile - an abstraction for plotting a DNA sequence against a benchmark sequence. Here the base parameters for the CAT comparison method, Cos, D, H, are calculated.

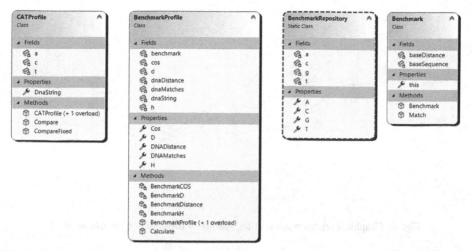

Fig. 5. Class diagram of the implementation of the algorithm based on the CAT method for sequences alignment.

CatProfile - an abstraction for representing a DNA sequence, with pre-calculated parameters for each benchmark sequence from the CAT method.

Example code realization of CAT Method, under license protection of GNU General Public License v3.0, can be found in GitHub at: https://github.com/HristoS/CATSequen ceAnalysis.

4.2 Collision Analysis

In order to investigate the reliability of the CAT method, it is necessary to investigate the possibilities of collisions in a full combination, i.e. all permutations for a given sequence length. This leads to the question of how unique the CTA profiles are and how much the accumulation of statistical errors affects the reliability of the method. For this purpose, all possible permutations of sequences with lengths 10, 11, 12, 13 and 15 were generated (Table 1 column DNA length). For each set, 100 sequences were taken at random and compared to the entire set. It was counted how many of them after comparison with CAT give the result 100% (1) (Table 1 column average cat matches) and how many of them actually have identical CAT profiles (Table 1 column average actual matches). The percentage of collisions is calculated as the number of all permutations (Table 1 column Total permutations) for a sequence of a certain length is divided by the average result of the number of comparisons with CAT giving 100% (1) and identical CAT profiles, the results are respectively in columns collision % 0 – 100 and actual collision % 0 – 100 from Table 1. Graphical representation of the results of the collision investigation is shown in Fig. 6.

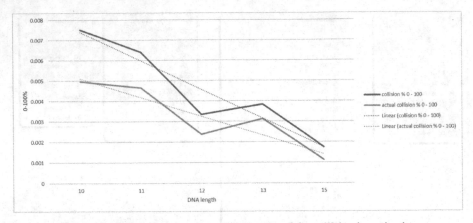

Fig. 6. Graphical representation of the results of the collision investigation.

Table 1. Collision Comparison Results

DNA length	average cat matches	average actual matches	Total permutations	collision % 0 - 100	actual collision % 0 - 100
10	78.73	52.05	1048576	0.007508278	0.004963875
11	268.7	195.35	4194304	0.006406307	0.004657507
12	562.27	400.15	16777216	0.00335139	0.00238508
13	2575.8	2099.11	67108864	0.003838241	0.003127918
15	18747.4	12052.89	1073741824	0.001745988	0.001122513

From the graph it is clear that the average values of the collisions obtained are below 0.005 percent, i.e. less than 5 hundredths of a percent and the trend is decreasing.

But what fraction of the resulting collisions lead to a good result after Needleman-Wunsch ordering? A Needleman-Wunsch ranking was done for each of the 100 sequences against its collisions from the previous experiment, and the results were averaged. On the graph in Fig. 7. The percentage ratio of averaged post-order Needleman-Wunsch scores for sequences of different lengths is depicted. For example, for a sequence of length 11, which after alignment by the Needleman-Wunsch algorithm, has 50% matches of the perfect alignment, making up 40.85% of the collisions obtained in CAT processing. 12.33% of the collisions produced by CAT processing after alignment with the Needleman-Wunsch algorithm have 60% matches of the ideal alignment.

It can be seen from the graph that about more than half of the collisions actually yield a good alignment after applying a more precise algorithm. Or in other words, when searching for an exact match of a sequence of a given length, out of all possible permutations using the CAT method, one can very quickly narrow down the possibilities to about 0.005% and the sequence searched is within the range of sequences found by the CAT method, and half of the sequences detected by the CAT method give a good result after applying a more precise method, around and above 50%.

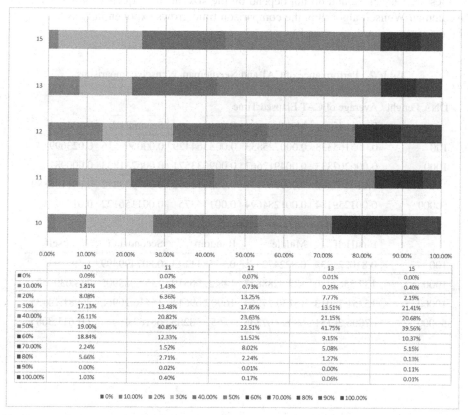

	10	11	12	13	15
■ 0%	0.09%	0.07%	0.07%	0.01%	0.00%
■ 10.00%	1.81%	1.43%	0.73%	0.25%	0.40%
■ 20%	8.08%	6.36%	13.25%	7.77%	2.19%
▦ 30%	17.13%	13.48%	17.85%	13.51%	21.41%
■ 40.00%	26.11%	20.82%	23.63%	21.15%	20.68%
■ 50%	19.00%	40.85%	22.51%	41.75%	39.56%
■ 60%	18.84%	12.33%	11.52%	9.15%	10.37%
■ 70.00%	2.24%	1.52%	8.02%	5.08%	5.15%
■ 80%	5.66%	2.71%	2.24%	1.27%	0.13%
■ 90%	0.00%	0.02%	0.01%	0.00%	0.11%
■ 100.00%	1.03%	0.40%	0.17%	0.06%	0.01%

■0% ■10.00% ■20% ▦30% ■40.00% ■50% ■60% ■70.00% ■80% ■90% ■100.00%

Fig. 7. The percentage of post-aligned averaged Needleman-Wunsch scores for sequences of different lengths that collided according to CAT profile comparisons.

4.3 Performance Analysis

To examine the speed of CAT profile comparisons, the following experiment was performed. 100 sequences of length 100, 1000, 10000 and 50000 were generated. For these sequences, the CAT profile was pre-calculated. The CAT profiles and the Needle-man-Wunsch algorithm of each sequence was compared with itself, and the execution time of each comparison was noted. The average results are shown in Table 2.

It is clear from Table 2 that the times for the comparisons made with the CAT profiles are very close and do not depend on the size of the sequence. While with the Needleman-Wunsch algorithm, the comparison time grows exponentially as the length of the sequence increases (Fig. 8).

Table 2. Performance of CAT and Needleman-Wunsch comparisons.

DNA Lenght	Average of CAT Elapsed Time				
	FirstHalf	Middle	Random	SecondHalf	WithItself
100	0.005165333	0.000725833	0.008364429	0.000569616	0.023699
1000	0.006207333	0.004917667	0.009033571	0.006231833	0.000981
10000	0.010463345	0.009767736	0.008311662	0.009993412	0.001188889
50000	0.001259184	0.001284694	0.00134375	0.001156122	0.0174
	Average of Needleman Wunsch Elapsed Time				
	FirstHalf	Middle	Random	SecondHalf	WithItself
100	0.810224833	0.8224725	0.875811143	0.807499499	1.59419
1000	137.5561462	137.9288827	148.2296519	136.8088905	238.639932
10000	15351.73013	15130.27244	16907.69611	15416.20968	26981.14435
50000	67806.26151	68099.07336	78652.58283	68162.06354	116611.3085

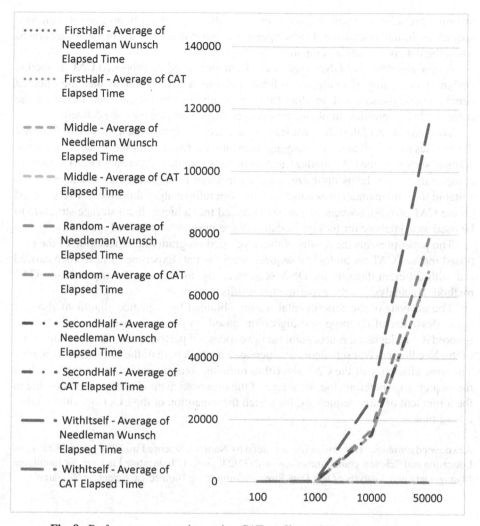

Fig. 8. Performance comparison using CAT profiles and Needleman-Wunsch.

5 Conclusion

The proposed new method for DNA sequences alignment, called CAT, based on the trilateration method, is experimentally verified. Three constant benchmarks have been established for the application of trilateration, which creates a constant favorite sequence - ie. Independent of the records in the database and remains the same when the set is changed.

Since the benchmark sequences established are constant (i.e. they do not depend either on the data or on their number), this allows the comparisons to be made at the very beginning – when the sequences is uploaded into database and this to be metadata information, accompanying each sequence. In this way, there is no need to compare sequences during lookup (the slowest operation), but instead only the metadata generated when the data is entered is compared.

A new algorithm for DNA sequences alignment based on proposed CAT method is designed, consisting of an algorithm for calculating a CAT profile against the selected benchmark sequences and an algorithm for comparing two sequences, based on the calculated CAT profiles. Implementation steps, inputs and outputs are defined.

The generation of the CAT profiles is done once during the data upload, which allows the profiles to be used as accompanying metadata information for the sequences. Search comparisons with the CAT method are minimized and have a constant O(24) algorithm complexity, which helps optimize searches in large biological datasets and makes it suitable for implementation as a first step in more refined algorithms like FASTA. Based on the CAT profiles, sequences can be organized into a hierarchical storage structure to be used as a database for biological data storage in search-optimized systems.

The paper presents the results of the developed program implementation of the proposed method CAT for biological sequences alignment. Experiments have been carried out with different datasets for DNA sequence alignment using the triplet-based CAT method. An analysis of the experimental results was made.

The analysis of the experimental results obtained by sequence alignment shows a small deviation of the proposed algorithm based on the CAT method, which can be ignored if this deviation is acceptable at the expense of performance. The execution time of the Needleman-Wunsch algorithm increases as the length of the sequences increases. The time efficiency of the CAT algorithm remains constant regardless of the length of the sequences. Therefore, the advantage of the proposed method is the fast processing in the alignment of large sequences, for which the execution of the exact algorithms takes a long time.

Acknowledgements. This research was funded by National Science Fund, Bulgarian Ministry of Education and Science, grant number KP-06-N37/24, project "Innovative Platform for Intelligent Management and Analysis of Big Data Streams Supporting Biomedical Scientific Research".

References

1. Needleman, S., Wunsch, C.: A General method applicable to the search for similarities in the amino acid sequence of two proteins. J. Mol. Biol. **48**, 443–453 (1970)
2. Smith, T., Waterman, M.: Identification of common molecular subsequences. J. Molecul. Biol. **147**, 195–197 (1981)
3. Altschul, S., et al.: Basic local alignment search tool. J. Molecul. Biol. **215**(3) 1990
4. Altschul, S., et al.: Gapped BLAST and PSIBLAST: a new generation of protein database search programs. Nucleic Acids Res. **25**, 3389–3402 (1997)
5. Hoksza, D., Svozil, D.: Multiple 3D RNA structure superposition using neighbor joining. IEEE/ACM Trans. Comput. Biol. Bioinf. **12**(3), 520–530 (2015)
6. Ebedes, J., Datta, A.: Multiple sequence alignment in parallel on a workstation cluster. Bioinformatics **20**(7), 1193–1195 (2004)
7. Mikhailov, D., et al.: Performance optimization of ClustalW: parallel ClustalW, HT Clustal, and MULTICLUSTAL, White paper, 2001, Silicon Graphics, Mountain View, CA

8. Thompson, J., Higgins, D., Gibson, T.: ClustalW: improving the sensitivity of progressive multiple sequence alignment through sequence weighting, position-specific gap penalties and weight matrix choice. Nucleic Acids Res. **22**(22), 4673–4680 (1994)
9. Cheetham, J., et al.: Parallel ClustalW for PC clusters. In: Proceedings of International Conference on Computational Science and its Applications, Montreal, Canada (2003)
10. Zhang, F., Qiao, X.Z., Liu, Z.Y.: A parallel Smith-Waterman algorithm based on divide and conquer. In: Proceedings of the Fifth International Conference on Algorithms and Architectures for Parallel Processing ICA3PP 2002 (2002)
11. Farrar, M.: Striped Smith-Waterman speeds database searches six times over other SIMD implementations. Bioinformatics **23**(2), 156–161 (2007)
12. Sharma, C., Agrawal, P., Gupta, P.: Multiple sequence alignments with parallel computing. In: Proceedings of IJCA International Conference Advantages in Computer Engineering Application ICACEA, no. 5, pp. 16–21, Mar. 2014
13. Sathe, S.R., Shrimankar, D.D.: Parallelizing and analyzing the behavior of sequence alignment algorithm on a cluster of workstations for large datasets. Int. J. Comput. Appl. **74**(21), 1–13 (2013)
14. Borovska, P., Gancheva, V., Landzhev, N.: Massively parallel algorithm for multiple biological sequences alignment. In: Proceeding of 36th IEEE International Conference on Telecommunications and Signal Processing (TSP), pp. 638–642 (2013). 10.1109/ TSP.2013.6614014
15. Gancheva, V., Stoev, H.: DNA sequence alignment method based on trilateration. In: Rojas, I., Valenzuela, O., Rojas, F., Ortuño, F. (eds.) IWBBIO 2019. LNCS, vol. 11466, pp. 271–283. Springer, Cham (2019). https://doi.org/10.1007/978-3-030-17935-9_25

Multiallelic Maximal Perfect Haplotype Blocks with Wildcards via PBWT

Paola Bonizzoni$^{(\boxtimes)}$ (iD), Gianluca Della Vedova (iD), Yuri Pirola (iD),
Raffaella Rizzi (iD), and Mattia Sgrò (iD)

DISCo, University of Milano - Bicocca, viale Sarca 336, Milano, Italy
`paola.bonizzoni@unimib.it`

Abstract. Computing maximal perfect blocks of a given panel of haplotypes is a crucial task for efficiently solving problems such as polyploid haplotype reconstruction and finding identical-by-descent segments shared among individuals of a population. Unfortunately, the presence of missing data in the haplotype panel limits the usefulness of the notion of perfect blocks.

We propose a novel algorithm for computing maximal blocks in a panel with missing data (represented as *wildcards*). The algorithm is based on the Positional Burrows-Wheeler Transform (PBWT) and has been implemented in the tool Wild-pBWT, available at https://github.com/AlgoLab/Wild-pBWT/. Experimental comparison showed that Wild-pBWT is 10–15 times faster than another state-of-the-art approach, while using a negligible amount of memory.

Keywords: Positional Burrows-Wheeler Transform · Haplotype blocks · Approximate pattern matching

1 Introduction

The positional Burrows-Wheeler Transform (PBWT) has been introduced by Durbin [5] to reduce the memory needed to index a large collection of haplotypes and to quickly perform pattern matching queries over those haplotypes, becoming widely used when working on large haplotype datasets. With improved haplotype phasing in large cohorts, genomes of large populations are resolved at the haplotype level, meaning that their genotypes is determined at a set of single nucleotide polymorphisms (SNP) sites. As a consequence, haplotypes of whole-genome sequence data collected from hundreds of thousands of individuals for projects such as the UK Biobank [7] and TOPMed projects [13] are becoming available for several tasks.

A trivial representation of those haplotypes is as a large $M \times N$ binary bidimensional matrix X, where each row is an individual (or a sample) and each column is a SNP locus. Typically, the value 0 indicates no change from the reference (the so-called major allele) and any other value represents a distinct change from the reference. The Positional Burrows-Wheeler Transform (PBWT)

I. Rojas et al. (Eds.): IWBBIO 2023, LNBI 13919, pp. 62–76, 2023.
https://doi.org/10.1007/978-3-031-34953-9_5

is a data structure that represents the matrix X and it is the most widely used data structure that couples space and time efficiency with the ability to quickly query large haplotype data, such as those stored in biobanks and used in computational pangenomics [2]. Most notably, the PBWT allows to query a haplotype panel, extracting all pattern occurrences in linear time – classical methods require quadratic time [3]. Therefore, PWBT-based algorithms have been widely incorporated in state-of-the-art statistical phasing and imputation tools [12]. Computing *maximal perfect haplotype block* (MHB) of a given panel is a key step for solving more complex problems, such as polyploid haplotype reconstruction [9] and finding identical-by-descent (IBD) [11] segments shared among individuals of a population. A maximal haplotype block is defined as a set of rows and an interval of columns from a start to an end column, such that the substring delimited by the start and the end column is the same in all specified rows. By definition, MHBs are identity-by-state segments, possibly shared across multiple chromosomes. Detection of genomic regions under positive selection is another problem that is attacked by using the notion of maximal perfect haplotype blocks [8]. An algorithm for computing MHBs has been proposed in Cunha [4], but the first efficient algorithm has been proposed by [1] using the positional PBWT: the PBWT allows to get a linear time algorithm for computing MHBs, more precisely a computation time $\mathcal{O}(NM + b)$, where b is the total number of blocks produced – we recall that the input is a $M \times N$ matrix.

The fact that the definition of maximal perfect blocks forbids mismatches renders MHBs vulnerable to sequencing errors and *missing data*, thus limiting its applications. For example, in positive selection, the presence of missing data may break up blocks leading to the underestimation of selection strength. For this reason, the notion of MHBs with wildcards has been proposed in [15]: this new definition extends the panel with a new symbol $*$, called *wildcard*, which can be replaced with any other symbol – wildcards correspond to missing data. A block consists therefore of a set of rows and an interval of columns such that the corresponding substrings might differ only for wildcards. If we consider only characters that are not wildcards, we obtain the *consensus* string of the block – note that the consensus string of a block is unique if and only if all columns of the block contain at least a character that is not a wildcard. The algorithm proposed in [15] is based on a tree data structure (a *trie*) and can be applied only to biallelic matrices, that is, where the only possible symbols are 0, 1, and $*$.

In this paper, we overcome the limitation to biallelic matrices of [15] and we propose a novel algorithm for computing MHBs with wildcards that is based on the PBWT. The algorithm is implemented in the tool Wild-pBWT available at https://github.com/AlgoLab/Wild-pBWT/. Our experimental analysis shows that Wild-pBWT can compute haplotype block more efficiently than an approach essentially based on backtracking. Indeed, the results show that Wild-pBWT is 10–15 times faster than WildHap [15] on all instances, never requiring more than 5 min even on matrices with 1 billion entries with 5% wildcards.

Also in terms of memory usage the comparison is even more impressive, since WildHap uses at least 75 times more RAM than Wild-pBWT. Indeed, Wild-pBWT never required more than 25 MB of RAM. We conclude our experimental analysis on some large multiallelic matrices (between 125 and 500 million entries), where Wild-pBWT never takes more than 7 min and 55 MB of memory.

2   Finding *Maximal Blocks* with *Wildcards*

A *haplotype* is an N-long string over the ordered alphabet $\{0, 1, \ldots, t-1\}$, where t is the number of alleles allowed for each SNP. A *panel* X of haplotypes is a set of M haplotypes $\{x_0, x_1, \ldots, x_{M-1}\}$ each of length N. The positions on each haplotype are indexed from 0 to $N-1$. In the following, we will refer to $[k_1, k_2)$ and $x[k_1 : k_2)$ as the interval of positions from k_1 to $k_2 - 1$ and the substring of haplotype x starting at k_1 and ending at $k_2 - 1$, respectively. Two haplotypes x_i and x_j have a *match* in $[k_1, k_2)$ iff $x_i[k_1 : k_2) = x_j[k_1 : k_2)$. In other words, a match involves two substrings in the same positions of the two haplotypes. Furthermore, the match $x_i[k_1 : k_2) = x_j[k_1 : k_2)$ is *left-maximal* (*right-maximal*, respectively) if it is either $k_1 = 0$ or $x_i[k_1 - 1] \neq x_j[k_1 - 1]$ (either $k_2 = N$ or $x_i[k_2] \neq x_j[k_2]$, resp.). In other words, it cannot be extended on the left (right, respectively).

A *block* [4] in a panel X is a triple (X_B, k_1, k_2) such that X_B is a subset of the rows of X, $[k_1, k_2)$ is an interval of the columns of X, and for each pair of haplotypes of X_B there is a match in $[k_1, k_2)$ (*i.e.*, the substring starting at k_1 and ending at $k_2 - 1$ is the same for all haplotypes of X_B). The block is *maximal* if three conditions hold: (1) $k_2 = N$ or $(X_B, k_1, k_2 + 1)$ is not a block, (2) $k_1 = 0$ or $(X_B, k_1 - 1, k_2)$ is not a block, (3) there is no any other haplotype $x \notin X_B$ such that $(X_B \cup x, k_1, k_2)$ is still a block.

A haplotype with *missing data* is a string over alphabet $\Sigma_* = \{0, 1, \ldots, t - 1, *\}$, where each missing SNP is represented by the *wildcard* *. The notions of *match* and *maximal block* can be extended to a panel X with missing data [15], by extending the notion of equality between two symbols such that each symbol $\sigma \in \Sigma$ is equal to the wildcard *. In the following, $\approx$ will be used to denote the *extended equality*, such that $\sigma_1 \approx \sigma_2$ iff $\sigma_1 = \sigma_2$ or at least one of σ_1 and σ_2 is equal to *. Consequently, a *match with wildcards* of two haplotypes x_i and x_j, are two substrings $s_1 = x_i[k_1 : k_2)$ and $s_2 = x_j[k_1 : k_2)$, such that $s_1[i] \approx s_2[i]$ for each i in the interval $[0, k_2 - k_1)$. With a slight abuse of language we can say that $x_i[k_1 : k_2) \approx x_j[k_1 : k_2)$.

Now, we give the definition of *maximal blocks* with *wildcards*. A Maximal Haplotype Block with *wildcards* (MHBw) is a triple (X_B, k_1, k_2) that satisfies the following conditions:

1. there exists a string $s \in \Sigma^{k_2 - k_1}$, called *extended consensus*, such that $s \approx x[k_1, k_2), \forall x \in X_B$;
2. $k_1 = 0$ or there exists two rows $x, y \in X_B$ such that $x[k_1 - 1] \neq y[k_1 - 1]$ (**extended** left-maximality);

$$
\begin{aligned}
x_0 &= 1\ 1\ 2\ 2\ \boxed{*\ *}\ 0\ \boxed{1\ *\ 1} \\
x_1 &= *\ 0\ 0\ 2\ |0\ *|2\ 1\ 1\ 2 \\
x_2 &= \boxed{0\ 0\ 2\ 1}\ |*\ 0|\ 1\ \boxed{1\ 1\ 1} \\
x_3 &= \boxed{0\ *\ 2\ 1\ 2}\ *\ 2\ 2\ 1\ *
\end{aligned}
$$

Fig. 1. Example of three maximal blocks in a panel of four haplotypes with *wildcards*. Each block is depicted with a colored box. (Color figure online)

3. $k_2 = N$ or there exists two rows $x, y \in X_B$ such that $x[k_2] \not\approx y[k_2]$ (right-maximality);
4. there does not exists a row $h \in X \setminus X_B$ such that $s \approx h[k_1, k_2)$ (row-maximality).

Figure 1 depicts an example of three *maximal blocks* for four haplotypes. Observe that a *wildcard* included in different blocks may be resolved as different symbols of the alphabet. For example, the *wildcard* (of haplotype x_2) included in the yellow and green blocks of Fig. 1 can be resolved as symbols 2 (yellow block, where the consensus is 002120) and 0 (green block, where the consensus is 00). Moreover, we point out that this definition is slightly different from that given in [15]. Indeed, we allow a given SNP position inside a block to contain a * for each haplotype involved, that is, columns of all symbols * are allowed inside a block according to our definition. As a consequence, we may have blocks where each one of the symbols is a *wildcard*.

3 The Algorithm

3.1 Positional BWT Extended to *Wildcards*

The Positional Burrows-Wheeler Transform (PBWT) is a data structure introduced in [5] for representing a panel $X = \{x_0, x_1, \ldots, x_{M-1}\}$ of M haplotypes with N biallelic sites and has been extended to the multiallelic case [10]. The PBWT has been designed in order to find matches among the haplotypes of X, or of an external haplotype with respect to the panel X. We recall that the PBWT consists of a pair of arrays a_k and d_k (called *prefix* and *divergence arrays*, respectively). More in detail, let k be an index between 0 and N, also referred as *column*. Given the haplotype $x \in X$, its *prefix at position* k is the k-long prefix $x[0 : k)$, we denote $\mathsf{pref}(x, k)$. The *reversed prefix at position* k will be the reverse of $\mathsf{pref}(x, k)$ (that is, the string $x[k - 1] \ldots x[0]$) and will be denoted by $\mathsf{revpref}(x, k)$. With a slight abuse of notation, $x[k_1 : k_2)$, with $k_1 \geq k_2$, is assumed to be the empty string. Hence, $\mathsf{pref}(x, 0) = \mathsf{revpref}(x, 0)$ is the empty string. At this point, let x_i, x_j be two haplotypes of X. Then, x_i is co-lexicographically smaller than x_j at column k (*k-smaller*), iff $\mathsf{revpref}(x_i, k)$ is lexicographically smaller than $\mathsf{revpref}(x_j, k)$ or $i < j$ when $\mathsf{revpref}(x_i, k) = \mathsf{revpref}(x_j, k)$. The *prefix array* a_k is a permutation of indices $0, 1, \ldots, M - 1$, such that $a_k[i] = j$

iff x_j is the i-th haplotype in the co-lexicographic ordering of the prefixes of the haplotypes up to position $k-1$ (referred in the following as k-smaller haplotype). Observe that, at each column k, an ordering (called k-order) of the haplotypes is produced. The *divergence array* d_k is such that $d_k[i]$ is the starting position of the left-maximal match, ending at position $k-1$, between the i-th and $(i-1)$-th haplotypes in the k-order. Observe that $a_0 = \langle 0, 1, \ldots, M-1 \rangle$, since the ordering at $k = 0$ is $\langle x_0, x_1, \ldots, x_{M-1} \rangle$ (revpref$(x_i, 0)$ is the empty prefix, $0 \le i \le M-1$). The haplotype $x_{a_k[i]}$ is denoted by y_i^k and y^k denotes the array of symbols at position k of the k-ordered haplotypes, such that $y^k[i] = x_{a_k[i]}[k]$. Note that $d_k[i] = k$ means that no match ending at position $k-1$ exists between haplotypes y_i^k and y_{i-1}^k.

At this point, we introduce the extension of the PBWT for representing a panel X of haplotypes with *wildcards*. Recall that a haplotype with *wildcards* is a string over the ordered alphabet $\Sigma_* = \{0, 1, \ldots, t-1, *\}$, where the number of positions g such that $x[g] = *$ (missing SNP information) is denoted as W_x. Given an integer $F \le W_x$, let us consider a tuple $\tau = (\sigma_1, \ldots, \sigma_F) \in \Sigma^F$. We call F-*expansion* of x (or simply expansion) with τ, the string obtained by replacing each wildcard at position g_i with the symbol σ_i. In other words, the first F wildcards of x are replaced by a symbol of Σ. Such expansion might wildcards only after the position g_F. Given an F-expansion h_F and an F_1-expansion h_{F_1} of the same haplotype x, such that $F_1 > F$, then h_{F_1} is an *extension* of h_F if $h_F[: g_F + 1] = h_{F_1}[: g_F + 1]$. For example, string $h = \texttt{01212*2*1}$ is a 2-expansion of haplotype $\texttt{0*2*2*1}$, where the first $F = 2$ *wildcards* have been replaced with symbol 1 and $\texttt{0121202*1}$ is an extension of h.

Let us denote by $H(x, F)$ the set of the expansions obtained by replacing the first F *wildcards* with all possible tuples of Σ^F. The main idea of our paper is that we can extend most of the original definitions and algorithms on the PBWT to the wildcards, only considering expansions instead of haplotypes. Given a position $k < N$, we call k-expansion-set of haplotype x the set $H(x, F)$ such that g_F is the largest position of a *wildcard* smaller than or equal to k. Note that the expansions of the N-expansion-set ($k = N$) do not contain any *wildcard* since every position g_i ($1 \le i \le W_x$) has been replaced with a symbol of Σ. Given a column k, let X_k be the panel obtained by the union of the k-expansions-sets of the haplotypes in X. Note that, given a haplotype x, the number of expansions of the set $H(x, F)$ ($F \le W_x$) is $|\Sigma|^F$. Hence, $|X_k|$ increases exponentially with k. A main contribution of our paper is a *collapsing* procedure that limits such exponential growth.

The arrays a_k and d_k (for a given column k) of the extended (with wildcards) PBWT, are defined as the prefix and divergence arrays constructed for the panel X_k, only on the set of k-expansions. More precisely, $a_k[i] = j$ iff the i-th expansion (in the order at position k) has been obtained from the input haplotype x_j. As in [5], the i-th k-smaller expansion will be denoted as y_i^k and y^k will be used in order to refer to the array of symbols at position k of the k-ordered expansions. Observe that symbol $y^k[i]$ (*i.e.*, the symbol at position k of the i-th expansion) is equal to $x_{a_k[i]}[k]$. arrays a_{k+1} and d_{k+1} we collapse each

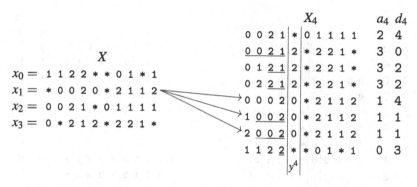

X

$x_0 = 1\ 1\ 2\ 2\ *\ *\ 0\ 1\ *\ 1$

$x_1 = *\ 0\ 0\ 2\ 0\ *\ 2\ 1\ 1\ 2$

$x_2 = 0\ 0\ 2\ 1\ *\ 0\ 1\ 1\ 1\ 1$

$x_3 = 0\ *\ 2\ 1\ 2\ *\ 2\ 2\ 1\ *$

Fig. 2. Example of *prefix* and *divergence arrays* at column 4 for a panel X (on the left) of four triallelic haplotypes with *wildcards* ($\Sigma = \{0, 1, 2\}$). Arrays a_4 and d_4 are reported before *collapsing*. Red characters refer to symbols replacing a *wildcard*; only *wildcards* before column 4 have been filled with an alphabet symbol. The arrows point to the three expansions obtained from haplotype x_1 by filling its first *wildcard* (that is before column 4) with the symbols in $\{0, 1, 2\}$; observe that a_4 contains the same haplotype index 1 in the three positions related to such expansions. The underlined substrings are the left-maximal matches ending at position 3 between an expansion and the previous one in the 4-order. Their starting positions are given by the *divergence array* d_4. Two vertical bars delimit the permutation y^4 of the symbols at position 4 of the ordered expansions. (Color figure online)

interval of consecutive expansions of the same input haplotype, keeping at most two expansions.

Figure 2 represents an example of *prefix* and *divergence array* for a panel with wildcards (Fig. 3).

3.2 Building the Extended PBWT

In this paper, we extend the algorithm in [10], that builds the multiallelic PBWT, allowing *wildcards* in the input panel. We also show how to reduce the space used to store the extended PBWT, while maintaining its ability to compute the maximal blocks. We recall that the algorithm in [10] computes the array a_{k+1} from a_k (for increasing k) with a single scan of y^k, listing the characters at position k of the haplotypes according to their k-order. This procedure is essentially a pass of radix sort. Similarly, our procedure (see Algorithm 1) scans y^k, listing the characters at position k of the expansions of X_k (that is, computed considering the column smaller than k) and further *expands* each expansion that has a wildcard at position k, by creating l copies and replacing the wildcard with each of the characters of the alphabet $\{0, 1, \ldots, t-1\}$. Note that an input haplotype x can contain more than one wildcard, therefore several expansions can originate (from x) throughout the iterations from $k = 0$ to $k = N$. For this reason, we distinguish between *input haplotypes* – which are in the original panel X – and *expansions* – which are the result of replacing some wildcards with symbols in Σ.

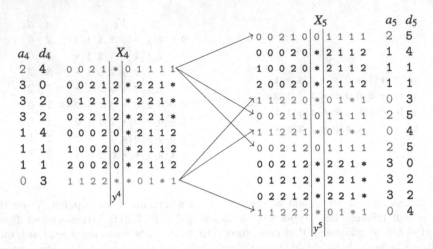

Fig. 3. Computing array a_5 from a_4. All the expansions related to a_4, whose character at position 4 is 0, are placed in the 5-order (array a_5) before the expansions whose character at position 4 is 1, which in turn are placed before the expansions whose character at position 4 is 2. The reciprocal order (array a_5), among the expansions having the same symbol at position 4, is maintained with respect to array a_4. Note that $t = 3$ new expansions are created for the first (green) expansion, obtained from haplotype x_2, and the last (blue) expansion, obtained from haplotype x_0, in the 4-order. (Color figure online)

We summarize the radix sort pass over all k-ordered expansions to compute the $(k + 1)$-order. First we initialize t empty lists $A(\sigma)$ for each $\sigma \in \Sigma$. Then, we scan the expansions by going through the array a_k. When considering the i-th expansion, that is one corresponding to the haplotype $x_{a_k[i]}$, let σ be its symbol at position k (that is, $\sigma = x_{a_k[i]}[k]$). If $\sigma \neq *$, then we append $a_k[i]$ to the list $A[\sigma]$ – in this case no expansion is performed – otherwise we append $a_k[i]$ to all list $A[\cdot]$ – in this case the expansion is performed. Finally, the array a_{k+1} is obtained by concatenating the lists $A[\sigma]$, in increasing order of σ. The divergence array d_{k+1} is obtained similarly, noticing that $d_{k+1}[i] = k + 1$ if i is the first element of a list $A[\sigma]$, otherwise its value is the same as in d_k.

The time complexity is similar to the non-wildcard case, depending also on the number w of wildcards that are present at column k. Prefix array are computed as shown in Algorithm 1 and the pass at column k requires $\mathcal{O}(|a_k| + (t-1)|w|_k)$ time. We recall that the size of the array a_k is not fixed. The following proposition allows us to delete some expansions from the $(k + 1)$-order without losing maximal blocks.

Proposition 1. *Let i, $i+1$ be two consecutive indices of the $(k+1)$-order such that $a_{k+1}[i] = a_{k+1}[i + 1]$, that is, the corresponding expansions y_i^{k+1} and y_{i+1}^{k+1} originate from the same input haplotype. Given $k_1 > k+1$, let j be an index such that $y_j^{k_1}$ is an extension of y_i^{k+1}. Then, $y_{j+1}^{k_1}$ is an extension of y_{i+1}^{k+1}. Moreover, if*

Algorithm 1. Building a_{k+1} and d_{k+1} from a_k and d_k.

```
1: build the empty lists A[c], D[c], p[c] for each c ∈ Σ
2: for c ← 0 to t − 1 do
3:     p[c] ← k + 1
4: for i ← 0 to |a_k| − 1 do
5:     h ← a_k[i]                            ▷ x_h is the current haplotype
6:     σ ← x_h[k]
7:     for c ← 0 to t − 1 do
8:         if d_k[i] > p[c] then
9:             p[c] ← d_k[i]
10:    if σ ≠ * then
11:        append a_k[i] to A[σ] and p[σ] to D[σ]
12:        p[σ] ← 0
13:    else
14:        for c ← 0 to t − 1 do
15:            append a_k[i] to A[c] and p[c] to D[c]
16:            p[c] ← 0
17: a_{k+1} ← concatenate(A[0], A[1], ..., A[t − 1])
18: d_{k+1} ← concatenate(D[0], D[1], ..., D[t − 1])
```

$j+1$ is an index such that the $y_{j+1}^{k_1}$ is extension of y_{i+1}^{k+1}, then $y_j^{k_1}$ is an extension of y_i^{k+1}.

In other words, Proposition 1 implies that when we have an interval of consecutive expansions originating from the same input haplotype in the $(k+1)$-order, they cannot be separated at later stages of the algorithm (*i.e.*, in a k_1-order such that $k_1 > k + 1$). Since a maximal block can contain at most one occurrence of each input haplotype, we do not need to keep the entire interval. Indeed, we collapse such interval and retain at most the first and the last expansion of such interval, thus reducing the size of the stored array, while maintaining the set of maximal blocks of the input panel X.

Now, we describe how we collapse an interval $[b, e]$ of consecutive (in the $(k+1)$-order) expansions of the same input haplotype, where b and e are positions of a_{k+1} (the pseudocode is Algorithm 2). Let $d_{\max}$ be $\max_{b \leq i \leq e}\{d_k[i]\}$, that is the maximum divergence in the interval $[b, e]$ and let max be the position between b and e that achieves such maximum value. Note that in a multi-allelic panel, more than one positions could achieve such maximum value; we arbitrarily choose one of them as max. We distinguish the following two cases: (1) $d_{k+1}[b] < d_{\max}$ and (2) $d_{k+1}[b] = d_{\max}$.

In case (1) all the haplotypes in the interval $[b + 1, e - 1]$ are discarded, thus retaining only the two expansions at position b and e. The divergence $d_{k+1}[e]$ is updated to $d_{\max}$. In this way, the blocks including the two haplotypes with indices $a_{k+1}[e]$ and $a_{k+1}[e+1]$ as well as the blocks including the two haplotypes with indices $a_{k+1}[b − 1]$ and $a_{k+1}[b]$ are maintained. In case (2), we remove all the expansions in the interval $[b + 1, e]$ and only the expansion at position b

---

Algorithm 2. Collapse procedure for a_{k+1}, d_{k+1}

1: $A, D \leftarrow$ empty lists
2: $b \leftarrow 0$
3: $e \leftarrow 1$
4: **while** $b < |a_{k+1}|$ **do**
5:     **while** $a_{k+1}[e] = a_{k+1}[e-1]$ **do**
6:         $e \leftarrow e + 1$
7:     $e \leftarrow e - 1$
8:     Append $a_{k+1}[b]$ to A and $d_{k+1}[b]$ to D
9:     **if** $e - b \geq 1$ **then**   ▷ collapse the interval $[b, e]$
10:        $max \leftarrow \arg\max_{b \leq i \leq e}\{d_{k+1}[i]\}$
11:        $d_{max} \leftarrow d_{k+1}[max]$
12:        **if** $d_{k+1}[b] < d_{max}$ **then**
13:            Append $a_{k+1}[b]$ to A and $d_{k+1}[max]$ to D
14:        $b \leftarrow e + 1$
15:        $e \leftarrow b + 1$
16: $a_{k+1} \leftarrow A$
17: $d_{k+1} \leftarrow D$

---

is retained. This guarantees that the blocks including the two haplotypes with indices $a_{k+1}[b-1]$ and $a_{k+1}[b]$ as well as the blocks including the haplotypes with indices $a_{k+1}[e]$ and $a_{k+1}[e+1]$ are retained.

3.3 Computing the Maximal Blocks

Given k, the maximal blocks ending at position k are computed by scanning arrays a_k and d_k (see Algorithm 3). For each position i between 0 and $|a_k| - 1$, we set $k_1 = d_k[i]$ as the starting position of a putative block and perform an up and down search around position i, in order to add, to a set B (initially empty), all the haplotype indices contained in $a_k[b, e]$, such that b and e are the smallest position before i and the biggest position after i (respectively), such that $d_k[j] \leq k_1$, for each $j \in [b, e]$.

The triple (X_B, k_1, k), where X_B is equal to $X|B$ (i.e., the subset of the input haplotypes whose index is in B), satisfies conditions (2) and (4) for a MHBw (i.e., left-maximality and row-maximality). The right maximality for (X_B, k_1, k) is verified if y^k (column of symbols at position k in the k-order) contains at least two distinct symbols of the alphabet Σ (wilcards excluded) in the interval $[b, e]$. In such case, the triple (X_B, k_1, k) is reported as a maximal block. Since our algorithm deals with haplotype expansions, we use a support array h_k in order to avoid the computation of duplicated blocks, i.e., blocks starting at the same k_1 position and referred to the same set of haplotypes. We point out that t bitvectors $bv[\sigma]$ (a vector for each symbol $\sigma \in \{0, 1, \ldots, t-1\}$) of length $|a_k|$ are used in order to speed up the right maximality check (for simplicity, such bitvectors are not shown in Algorithm 3). Precisely, $bv[\sigma][i]$ is set to 1 if $y^k[i]$ is equal to σ or is a *wildcard*, otherwise $bv[\sigma][i]$ is set to 0. The *rank* procedure [14]

Algorithm 3. Report all Maximal Blocks for a column k

1: Build the array h_k of $|a_k|$ values set to -1
2: **for** $i \leftarrow 0$ to $|a_k| - 1$ **do**
3:     $B \leftarrow \emptyset$ ▷ empty set
4:     $\sigma \leftarrow y^k[i]$
5:     add $a_k[i]$ to B
6:     $b \leftarrow i$; $e \leftarrow i + 1$
7:     **while** $b > 0$ and $d_k[b] \leq d_k[i]$ **do**
8:         $b \leftarrow b - 1$
9:         add $a_k[b]$ to B
10:         **if** $\sigma = *$ **then**
11:             $\sigma \leftarrow y^k[b]$
12:     **if** $d_k[i] \neq h_k[b]$ **then**
13:         **while** $e < |a_k|$ and $d_k[e] \leq d_k[i]$ **do**
14:             add $a_k[e]$ to B
15:             **if** $\sigma = *$ **then**
16:                 $\sigma \leftarrow y^k[e]$
17:             $e \leftarrow e + 1$
18:         $e \leftarrow e - 1$
19:         **if** $\sigma \neq *$ **then**
20:             **if** $|B| \geq 2$ **then**
21:                 **if exists** $\sigma_1 \in y^k[b, e], \sigma_1 \neq \sigma$ **then**   ▷ right-maximality
22:                     $h_k[b] \leftarrow d_k[i]$
23:                     $X_B \leftarrow X|B$
24:                     Report $(X_B, d_k[i], k)$ as maximal block

---

is exploited to check whether y^k contains in $[b, e]$ at least two distinct symbols of Σ. Indeed, $\text{rank}_{bv[\sigma]}(e + 1) - \text{rank}_{bv[\sigma]}(b)$ is equal to $(e - b + 1)$ if, for all the positions $i \in [b, e]$, $y^k[i]$ is equal to σ or $*$. To check whether y^k contains at least two distinct symbol of Σ in the interval $[b, e]$, it is sufficient to verify that condition $\text{rank}_{bv[\sigma]}(e + 1) - \text{rank}_{bv[\sigma]}(b) < e - b + 1$ holds.

4 Experimental Results

The method described in this paper has been implemented in C++ and it is available at https://github.com/AlgoLab/Wild-pBWT/ under the GNU GPL v3 License. The tool, called Wild-pBWT, takes as input a haplotype panel and returns the maximal blocks of the input haplotype panel. The program uses the implementation of bitvectors and of the associated rank and select operations provided by sdsl-lite [6].

The experimental evaluation of Wild-pBWT is divided in two parts. In the first part, we experimentally compared Wild-pBWT with WildHap [15] which, to the best of our knowledge, is the only other known method to find blocks in haplotype panels with wildcards. We recall that WildHap is based on a trie data structure and it is restricted to a binary alphabet, thus, in this part we focused only on biallelic haplotype panels. In the second part, instead, we considered multiallelic panels. Since to the best of our knowledge, no other method is able to compute blocks in multiallelic haplotype panels with wildcards, in the second part we focused on how Wild-pBWT performs on t-ary alphabets (with $t > 2$).

All the experiments were performed on a 64 bit Linux server equipped with four 8-core Intel® Xeon 2.30 GHz processors and 256 GB of RAM. The comparison was performed using a single thread.

Comparison with WildHap. The comparison with WildHap has been performed on 6 biallelic randomly-generated haplotype panels with different number of haplotypes (5k and 10k) and different number of SNPs (25k, 50k, 100k). We then randomly inserted wildcards, simulating therefore missing data, to each of these panels with increasing probability (0%, 0.5%, 1%, 2%, 5%), obtaining a total of 30 biallelic panels (6 w/o wildcards, 24 with wildcards).

We evaluated both implementations in terms of blocks found, running time, and memory usage. Since WildHap does not output the blocks but only their total number, we decided to compare only the number of blocks found. For this reason, Wild-pBWT has been run with -c y flag in order to only count the blocks (and not reporting them, since reporting blocks is an I/O intensive tasks and that part could affect the running times).

Table 1 reports the results obtained in the comparison. In terms of number of computed blocks, Wild-pBWT and WildHap compute the same number of blocks on panels without wildcards (wildcard probability equal to 0%), supporting the correctness of the implementation. To the contrary, Wild-pBWT computes more blocks than WildHap in panels with wildcards (wildcard probability grater than 0). However, this is expected since Wild-pBWT distinguishes blocks by their representative sequence (see Sect. 2 for details). Manual inspection supported the correctness of both implementations also on these instances.

In terms of running times, the results show that Wild-pBWT is 10–15 times faster than WildHap, despite the fact that Wild-pBWT computes more blocks than WildHap. We remark that Wild-pBWT required less than 5 min (296 s) on the largest panel with 5% of missing data, while WildHap required almost 70 min. The improvement in the running time is likely due to the fact that Wild-pBWT does not need to backtrack.

Also in terms of peak memory usage, Wild-pBWT appears to be significantly more efficient than WildHap. Indeed, Wild-pBWT never required more than 22 MB of RAM, while WildHap required up to 2090 MB. The results empirically confirm that Wild-pBWT memory usage is proportional to the number of haplotypes and independent to the number of columns. In particular, observe that Wild-pBWT requires approximately 11–12 MB on the first three matrices, that have the same number of haplotypes (5k) and an increasing number of columns (from 25k to 100k). The same observation is true for the other three matrices, that have 10k haplotypes and an increasing number of columns (from 25k to 100k): on these panels Wild-pBWT requires approx. 20 MB, which is almost the double of the amount of memory required on the first three panels.

Table 1. Comparison between Wild-pBWT and WildHap on six simulated biallelic panels of different size and percentage of random wildcards in terms of number of computed blocks (in millions), running times (in seconds), and memory usage (in Megabytes).

Matrix	Wild. prob.	Total blocks		CPU time (sec)		Memory usage (MB)	
		Wild-pBWT	WildHap	Wild-pBWT	WildHap	Wild-pBWT	WildHap
Synthetic 1 $5k \times 25k$ 125 MB	0%	94M	94M	12	179	10	740
	0.5%	100M	99M	18	210	12	778
	1%	106M	104M	20	223	11	784
	2%	120M	116M	22	264	11	783
	5%	177M	159M	34	345	12	786
Synthetic 2 $5k \times 50k$ 250 MB	0%	188M	188M	24	336	10	899
	0.5%	200M	198M	37	432	11	962
	1%	212M	209M	39	496	12	963
	2%	239M	232M	44	508	12	967
	5%	350M	319M	73	694	12	975
Synthetic 3 $5k \times 100k$ 500 MB	0%	377M	377M	48	674	10	1156
	0.5%	400M	397M	74	893	11	1314
	1%	426M	419M	79	989	11	1321
	2%	482M	465M	89	995	12	1342
	5%	712M	638M	142	1386	12	1357
Synthetic 4 $10k \times 25k$ 250 MB	0%	188M	188M	24	486	18	898
	0.5%	201M	199M	37	592	20	957
	1%	215M	211M	40	670	20	965
	2%	246M	237M	46	733	20	968
	5%	375M	335M	75	1062	22	972
Synthetic 5 $10k \times 50k$ 500 MB	0%	377M	377M	49	998	18	1090
	0.5%	403M	399M	75	1328	20	1261
	1%	430M	423M	80	1323	21	1266
	2%	492M	475M	92	1491	21	1289
	5%	750M	671M	145	2155	22	1291
Synthetic 6 $10k \times 100k$ 1000 MB	0%	754M	754M	97	1990	18	1690
	0.5%	806M	799M	151	2419	20	2020
	1%	862M	847M	169	2569	20	2028
	2%	986M	950M	185	3098	21	2051
	5%	1500M	1342M	296	4151	22	2090

Multiallelic Panels. In this part we assessed how Wild-pBWT scales on multiallelic random haplotype panels. As noted above, we are not able to compare these results with those obtained by other approaches, since no other tool is able to compute maximal haplotype blocks with wildcards on multiallelic panels. This initial evaluation was performed on 6 multiallelic matrices with 5k haplotypes and varying number of SNPs (25k, 50k, and 100k), number of alleles (3 and 8), and increasing wildcard probability rate (0%, 0.5%, 1%, 2%, 5%). Overall there are 24 different instances (6 w/o wildcards, 18 with wildcards).

Table 2 reports the number of maximal haplotype blocks computed by Wild-pBWT on each instance, along with the time and memory used to perform the computation. While on triallelic panels (matrices Multi 1, Multi 2, Multi 3) we essentially observe the same trends already discussed in the comparison with WildHap, on 8-allelic panels we observe that the number of blocks significantly grows when the wildcard probability is equal to 5%. We believe that this fact

Table 2. Evaluation of Wild-pBWT on six simulated multiallelic panels (first column) with varying percentage of random wildcards in terms of number of computed blocks, running time (in seconds), and memory usage (in MBytes).

Matrix	Wild. prob.	Total blocks	CPU time (sec)	Memory usage (MB)
Multi 1 $5k \times 25k$ 125 MB $|\Sigma| = 3$	0%	77M	12	10
	0.5%	83M	19	12
	1%	90M	21	11
	2%	105M	25	12
	5%	173M	42	13
Multi 2 $5k \times 50k$ 250 MB $|\Sigma| = 3$	0%	154M	25	10
	0.5%	166M	39	11
	1%	180M	43	11
	2%	211M	50	12
	5%	229M	57	12
Multi 3 $5k \times 100k$ 500 MB $|\Sigma| = 3$	0%	309M	50	10
	0.5%	333M	80	12
	1%	361M	85	12
	2%	423M	100	12
	5%	691M	169	13
Multi 4 $5k \times 25k$ 125 MB $|\Sigma| = 8$	0%	52M	15	10
	0.5%	60M	23	12
	1%	71M	27	13
	2%	96M	39	14
	5%	232M	104	20
Multi 5 $5k \times 50k$ 250 MB $|\Sigma| = 8$	0%	104M	29	10
	0.5%	121M	46	12
	1%	142M	59	13
	2%	192M	78	13
	5%	465M	188	28
Multi 6 $5k \times 100k$ 500 MB $|\Sigma| = 8$	0%	208M	58	10
	0.5%	243M	99	12
	1%	284M	109	13
	2%	384M	146	15
	5%	931M	393	54

can be easily explained since wildcards are expanded to each of the different 8 alleles, hence we can form, even by chance, more blocks. The larger number of blocks and expansions implies also an increase in the running time and memory usage.

We point out that, even for the largest panel (8-allelic panel with 5k haplotypes and 100k SNPs and wildcard probability equal to 5%), Wild-pBWT required approximately 5 min and a negligible amount of RAM (54 MB) to complete the computation. Finally, we remark that this evaluation should be intended only as an initial assessment of Wild-pBWT on multiallelic panels, leaving a more in-depth analysis of real multiallelic data as a future work.

5   Conclusions

In this paper we presented a novel method that finds all maximal perfect haplotype blocks of a multiallelic haplotype panel where some elements might be missing (represented as wildcards). The algorithm works by efficiently building the multiallelic Positional Burrows-Wheeler Transform of all the possible expansions of the missing data that are then collapsed when expansions of the same input haplotype fall in the same maximal block. On simulated data, we showed that this algorithm is more efficient than WildHap [15], a state-of-the-art approach for solving the problem, both in terms of running time (up to 15 times faster) and in terms of peak memory usage (our methods required up to 22 MB, while WildHap required up to 2 090 MB). Although the results obtained are promising, further testing on real haplotype panels has to be done.

Acknowledgements. This project has received funding from the European Union's Horizon 2020 research and innovation programme under the Marie Skłodowska-Curie grant agreement No. 872539.

References

1. Alanko, J., et al.: Finding all maximal perfect haplotype blocks in linear time. Algorithms Mol. Biol. **15** (2020). https://doi.org/10.1186/s13015-020-0163-6
2. Baaijens, J.A., et al.: Computational graph pangenomics: a tutorial on data structures and their applications. Nat. Comput. **21**(1), 81–108 (2022). https://doi.org/10.1007/s11047-022-09882-6
3. Bonizzoni, P., et al.: Compressed data structures for population-scale positional burrows–wheeler transforms. bioRxiv (2022). https://doi.org/10.1101/2022.09.16.508250
4. Cunha, L., Diekmann, Y., Kowada, L., Stoye, J.: Identifying maximal perfect haplotype blocks. In: Alves, R. (ed.) BSB 2018. LNCS, vol. 11228, pp. 26–37. Springer, Cham (2018). https://doi.org/10.1007/978-3-030-01722-4_3
5. Durbin, R.: Efficient haplotype matching and storage using the positional Burrows-Wheeler transform (PBWT). Bioinformatics **30** (2014). https://doi.org/10.1093/bioinformatics/btu014
6. Gog, S., Beller, T., Moffat, A., Petri, M.: From theory to practice: plug and play with succinct data structures. In: Gudmundsson, J., Katajainen, J. (eds.) SEA 2014. LNCS, vol. 8504, pp. 326–337. Springer, Cham (2014). https://doi.org/10.1007/978-3-319-07959-2_28
7. Halldorsson, B.V., et al.: The sequences of 150,119 genomes in the UK Biobank. Nature **607** (2022). https://doi.org/10.1038/s41586-022-04965-x
8. Kirsch-Gerweck, B., et al.: Haploblocks: efficient detection of positive selection in large population genomic datasets. Mol. Biol. Evol. **40** (2023). https://doi.org/10.1093/molbev/msad027
9. Moeinzadeh, M.H., et al.: Ranbow: a fast and accurate method for polyploid haplotype reconstruction. PLOS Comput. Biol. **16** (2020). https://doi.org/10.1371/journal.pcbi.1007843
10. Naseri, A., Zhi, D., Zhang, S.: Multi-allelic positional Burrows-Wheeler transform. BMC Bioinform. **20** (2019). https://doi.org/10.1186/s12859-019-2821-6

11. Naseri, A., et al.: RaPID: ultra-fast, powerful, and accurate detection of segments identical by descent (IBD) in biobank-scale cohorts. Genome Biol. **20** (2019). https://doi.org/10.1186/s13059-019-1754-8
12. Rubinacci, S., Delaneau, O., Marchini, J.: Genotype imputation using the positional Burrows-Wheeler transform. PLOS Genetics **16** (2020). https://doi.org/10.1371/journal.pgen.1009049
13. Taliun, D., et al.: Sequencing of 53,831 diverse genomes from the NHLBI TOPMed Program. Nature **590** (2021). https://doi.org/10.1038/s41586-021-03205-y
14. Vigna, S.: Broadword implementation of rank/select queries. In: McGeoch, C.C. (ed.) WEA 2008. LNCS, vol. 5038, pp. 154–168. Springer, Heidelberg (2008). https://doi.org/10.1007/978-3-540-68552-4_12
15. Williams, L., Mumey, B.: Maximal perfect haplotype blocks with wildcards. iScience **23** (2020). https://doi.org/10.1016/j.isci.2020.101149

GPU Cloud Architectures
for Bioinformatic Applications

Antonio Maciá-Lillo[(✉)][iD], Tamai Ramírez[iD], Higinio Mora[iD],
Antonio Jimeno-Morenilla[iD], and José-Luis Sánchez-Romero[iD]

University of Alicante, Alicante, Spain
a.macia@ua.es

Abstract. The world of computing is constantly evolving. The trends
that are shaping today's applications are Cloud computing and GPU
computing. These technologies allow bringing high performance computations to low power devices, when using a computing outsourcing
architecture. Following the trend, bioinformatic applications are looking
to take advantage of these paradigms, but there are challenges that have
to be solved. Data that these applications work with is usually sensible and has to be protected. Also, GPU usage in Cloud architectures
currently presents inefficiencies. This paper makes a review of the characteristics of Cloud computing outsourcing architectures, including the
security aspects, and GPU usage for these applications. The proposed
architecture includes GPU devices and tries to make efficient use of them.
The experiments show that it has the opportunity to increase parallelism
and reduce context switching costs when running different applications
concurrently on the GPU.

Keywords: GPU · Cloud · Computing Outsourcing

1 Introduction

The Cloud computing paradigm is changing how computations are done. Its ability to bring computations on demand with high scalable applications is framing
every field in computer science. The bioinformatics field is no exception. There
is interest in computing outsourcing architecture for bioinformatic applications.
With its ability to increase the computational capabilities of sensor and mobile
devices, its application would greatly benefit this field.

However, security is the main challenge in these applications [28]. Bioinformatic applications work with sensible data that has to be protected. To that end,
several techniques can be used to transfer data in Cloud bioinformatic applications, like classic or homomorphic encryption [28], secure multiparty computations [27], pseudo random functions [16] or software guard extensions [2].

Another aspect that has been important in bioinformatics is the use of GPU
devices to speedup computations. The GPU has been widely used in the field
of bioinformatics, specially in molecular level simulations [21], but the main

© The Author(s), under exclusive license to Springer Nature Switzerland AG 2023
I. Rojas et al. (Eds.): IWBBIO 2023, LNBI 13919, pp. 77–89, 2023.
https://doi.org/10.1007/978-3-031-34953-9_6

push is with the adoption of AI algorithms [12,15], as AI algorithms have great parallelism that can be exploited by GPU devices [10]. However, the use of GPU devices in Cloud architectures presents several inefficiencies. Sharing the GPU between applications and GPU virtualization can leave resources unused by its time multiplexing nature [17]. Also, the context switching overhead has to be taken into account.

This study makes a review of the characteristics of Cloud computing outsourcing architectures for bioinformatic applications, and gives proposals to solve the main challenges associated with them. The proposed architecture includes GPU devices and tries to make efficient use of them. This paper is organized as follows. After this introduction, a review of Cloud computing outsourcing architectures for bioinformatics, including the security aspects, and GPU usage for bioinformatic applications is performed. Then experiments are made to show how GPU concurrency works and how to use the GPU efficiently in these architectures. At last, the conclusions of this study are presented.

2   GPU Usage in Bioinformatics

The Graphics Processing Unit is a device originally intended for the specific processing of graphics. However, due to its inherent parallel nature, it has been used extensively in High Performance Computing (HPC) to accelerate algorithms in what is known as General Purpose GPU (GPGPU) computing.

The GPU has been widely used as an HPC device in bioinformatics, in the fields of sequence alignment, molecular dynamics, molecular docking, prediction and searching of molecular structures, simulation of spatio-temporal dynamics, deterministic simulation, stochastic simulation, spatial simulation and applications in systems biology [21]. The parallel nature of GPUs makes it ideal to develop applications like species-based systems, compartmentalized systems, individual-based systems, lattice-based methods and agent-based models [6].

Bioinformatic applications usually have to process big amounts of data, and rely on Big Data techniques. This processing has been improved with the introduction of the newest AI algorithms such as Machine Learning (ML) and deep neural networks (DNN), also known as Deep Learning (DL). Machine Learning in bioinformatics has been used to biomedical image recognition based on supervised methods, in this case, the data set preparation and its bifurcation are very important. However, Machine Learning in bioinformatics is not limited to image segmentation. Some of the application where Machine Learning is used are (1) genome annotation, (2) protein structure prediction, (3) gene expression analysis, (4) complex interaction modeling in biological systems, (5) and drug discovery [12]. As mentioned above, Deep Learning is especially formidable to handle Big Data and has demonstrated its power in promoting the bioinformatics field, concretely by employing DNN and convolutional neural networks (CNN) to manage important bioinformatics problems as identifying enzymes, gene expression regression, RNA-protein binding sites prediction or DNA sequence function prediction [15].

AI algorithms are also prone to be accelerated using the GPUs. Modern GPUs are even manufactured with specific characteristics to be used as an AI computing device [10]. Using GPUs in bioinformatics has several advantages. First, GPUs can significantly outperform CPUs in certain types of tasks, which is especially important for large-scale analyses. Second, GPUs can assist researchers in analyzing larger and more complex data sets than CPUs alone. Finally, GPUs are becoming more widely available, thanks to cloud-based GPU services and GPU-accelerated workstations. For example, the work in [9], shows an open-source design of the framework Galaxy[1] to offload the computational cost with GPU-support called GYAN. The key potential of this new version of Galaxy is that the authors added intelligent GPU-aware computation mapping and orchestration support to Galaxy, allowing researchers to execute the tools in both CPU (or) GPU based on the tool requirements. The results show that the framework improve the performance of the previous version of Galaxy.

3 Processing Outsourcing

Cloud computing is the offering of computation resources on demand with a pay per use model. The computations are offloaded to the Cloud server where the resources have a pay per use model, so Cloud applications are generally scalable. The advantages this paradigm offers are on demand resource offering, which allows lower costs and energy savings, and improved efficiency of resource management. However, it also presents some disadvantages, such as privacy and security, service continuity and service migration [25]. The Edge paradigm is an extension to the Cloud, where the computations are brought closer to where the data is gathered. The computations are performed in devices located in local network connectivity range, or even in the same sensor devices [19]. The advantages are cheaper computations and faster response times due to lower communications latency [8]. However, it is harder to manage Edge-Cloud applications as Edge devices have to be configured as well as the balancing of work from the Edge to the Cloud.

The externalization process to offload part of the computing load to the cloud allows increasing the capabilities of devices and other computing platforms at the edge. In this area, mobile cloud computing paradigm was initially designed to extend the battery life of IoT things and mobile devices [30]. However, this trend has evolved as a way to give enhanced performance and access to high performance computing resources [3,20].

The increased complexity and numerous offload possibilities to edge and cloud infrastructures introduces new variables that must be taken into account [18]. In this line, a distributed Cloud computing model incorporates the physical location of the service provider platforms as part of its definition [18]

---

[1] Galaxy is an open-source web-based framework for performing computational analyses in fields such as bioinformatics, genomics, proteomics, and others. It has an easy-to-use interface for running complex workflows and supports a variety of tools for genome assembly, annotation, and visualization, among other things.

providing additional capabilities, server proximity and increased data protection [26].

The field of bioinformatics has shown a keen interest in utilizing processing outsourcing architectures to meet the demand for intensive computations. In particular, computational biology and bioinformatics applications often involve large loads with considerable variability in size and duration. Therefore, efficient scheduling approaches are necessary to achieve maximal load balancing and overall system efficiency. These scheduling strategies can help optimize the use of high-performance computing resources, such as GPUs, to efficiently process complex data sets in a timely manner [29]. However, security is the main challenge, as the data that is worked on is usually sensitive, and can even be subject to government regulations. Working with Deep Learning algorithms in a cloud computing outsourcing architecture also has to protect the privacy of the training model [14]. A traditional approach to this issue is the use of encryption to protect the data. Several works make use of homomorphic encryption. This type of encryption allows performing mathematical operations on encrypted data. Therefore, it is possible to perform the computations on the server without decrypting the data. Suo et al. [28] present a privacy preserving computing outsourcing framework to solve the double digest problem that uses a homomorphic encryption technique that allows the cloud server side of the algorithm to work with encrypted data so that decryption is not needed on the server. Li et al. [14] proposes a privacy-preserving Deep Learning algorithm on homomorphic encrypted data for Cloud Computing architectures. Pramkaew et al. [24] also use homomorphic encryption to perform the SVD decomposition on large matrices in the Cloud server of the outsourcing architecture. However, lightweight homomorphic encryption is used, which consists in using a transform matrix as sparse as possible by using a random monomial matrix which has only one element in each row and column. Yan et al. [33], provide a comprehensive review of secure outsourcing based on homomorphic encryption. They mainly consider two factors in the schemes, security issues and efficiency performance. They also provide four other standard secure techniques: (1) secure multi-party computation, (2) pseudorandom functions, (3) software guard extensions, and (4) perturbation approaches, however the last one is not commonly used in outsourcing computing.

Secure multi-party computation (SMPC) is the collaboration of multiple parties to perform a computation while keeping their inputs private. This technique allows parties to perform computations without revealing their input data to one another, thereby improving the overall security of the computation. Smajlović et al. [27] introduces the framework Sequre for developing performant multi-party computation applications. This framework is based on Python programming language and translate the code to a secure MPc program. This framework preserve privacy for biomedical analysis. The authors evaluate the performance of this framework in three main applications: medical genetics, pharmacogenomics and metagenomics, the results showed that Sequre has a practical utility, performance and usability. Wu et al. [32], introduced an efficient server-aided multi-party computing (MPC) protocol that ensures security even when a semi-honest server and a majority of malicious client parties are present. When compared to previous works, the authors claimed considerable gains in protocol efficiency. In particular, in a two-party environment, the suggested protocol produced a fourfold speedup, while in a four-party situation, the protocol's efficiency increased by a factor of 83. The authors determined that their protocol is the most efficient server-assisted MPC protocol, with the same security assurances as all previously published studies in this field. Zhong et al. [34] evaluate the advantages of Blockchain in Secure Multi-Party Computation, highlighting that Blockchain is an ideal model due to its decentralization characteristics. In fact, the authors explain that using Blockchain, problems of fairness and scalability of SMPC can be aimed. However, Blockchain has challenges that have not been afforded such as balance between efficiency and privacy, so Blockchain needs further research in this field.

Pseudo-random functions are mathematical functions that produce seemingly random results from a fixed input. They are frequently used in cryptographic applications to prevent unauthorized access or tampering with sensitive data. Recent works, have been using this functions to secure the outsourced decision in Machine Learning algorithms, concretely, private evaluation of decision trees to transform a classification problem to a secure search problem and uses key encryption to protect data and classification privacy [16]. The aim is for a client with a feature vector to query an outsourced decision tree stored by a server, the result is that the client obtains the Machine Learning classification of their feature vector without the decision tree owner understanding what their input was [4].

Software Guard Extensions (SGX) [2] is an Intel-developed hardware-based security solution that allows apps to run in a secure enclave, which is a protected section of memory that is separated from the rest of the system. This technique can help protect sensitive data and calculations against unwanted access. Jie et al. [11], construct a strategy by using an untrusted high-performance GPU to transfer the parallelizable computation task. Furthermore, the strategy is supported by the use of MPC to protect the privacy of data holders and the correctness of the computation. In this case, the authors use this security technique to ensure the privacy-preserving outsourcing on Graph Neural Networks to fight against malicious servers. Chen et al. [5], proposed an outsourcing framework to ensure security against malicious attacks. As the authors present, the framework is the first SGX-based secure genetic testing framework implementations to enable efficient outsourced storage and processing. However, the framework has several limitations, such a memory limitation. When genomic datasets approach 4 GB, it is impossible to avoid costly data sealing and unsealing operations. Due to this, the authors suggest that further research is needed.

4   GPU Virtualisation in Cloud Architectures

GPU usage in Cloud architectures presents several inefficiencies. These come from the way service is offered. Cloud Service providers have two methods to offer GPU devices. They can offer exclusive use of the device, or they can use GPU virtualization. Full use of the device will leave unused the resources that are not used by the client at any given time. GPU virtualization is made via time multiplexing. It helps to reduce inefficiencies, but the resources that a kernel does not use are left unused.

In a previous work [17], a Cloud architecture is presented to share the GPU between several Cloud applications. It allows sharing the resources of the GPU between the applications at the same tame, which leads to higher GPU usage. The architecture uses Nvidia MPS technology to run trusted applications concurrently in the GPU, sharing its resources. The applications are run inside docker containers. The connection of each container to the GPU is done by sharing the UNIX socket that communicates with the MPS server between the host and the container. The security characteristics of the architecture are:

- Execution resource provisioning: Quality of Service (QoS) can be established by limiting the resources used by each application.
- Memory protection: The memory space of the applications running on the GPU is private and isolated from external processes.
- Error containment: A fatal error on an application running on the GPU is propagated to all applications that are running on the GPU at the same time. This behavior is why this architecture can only be used with trusted applications, where these errors have to be taken into account.

5   Architecture Characteristics

From the survey on Cloud usage for bioinformatic applications, a common computing outsourcing architecture is conceived, that takes into account the special needs of these types of applications. As bioinformatic applications benefit from the use of the GPU, the architecture model for the computing outsourcing architecture includes GPU devices. The model architecture for Cloud computing outsourcing is presented in Fig. 1. It consists of three layers. The first one consists of the devices that capture the data. The second layer consists of the relatively small computing devices that allow to perform some degree of computations and offload the rest to the Cloud. These devices are located near to where the data is gathered. Shall data be sensitive, there is no problem here because the transfer from the gathering devices to the Edge is done using a local network connection, so the transfer is secure. The third layer consists of the Cloud server. The data transfer between the Edge and the Cloud servers has to be secured using encryption, as it travels on the Internet. The computations needed to encrypt the data have to be made on the Edge devices. This is one of the benefits of having the Edge layer. Otherwise, encryption computations would have to be done on the low power gathering devices. Other security techniques seen such as Secure multi-party computation, pseudo random functions or software guard extensions can be implemented on the Cloud server, Edge, or on both collaboratively.

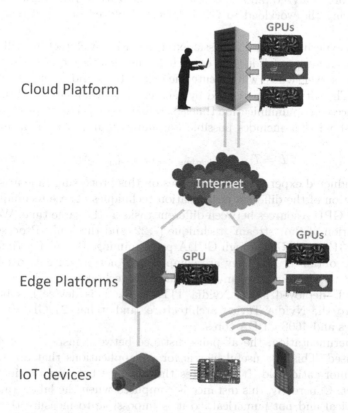

Fig. 1. Distributed Architecture with Edge and Cloud high performance computing

Bioinformatic applications would benefit from using GPUs, so a Cloud architecture built for these applications may want to use these devices. To avoid the inefficiencies of GPU devices in Cloud architectures, the architecture proposed in [17] can be used, but with concrete characteristics for these types of applications. The security characteristics of these applications allow having QoS mechanisms to leverage computing power between the different applications. Memory protection ensures application data is secured from external processes, which is vital for bioinformatic applications that work with sensitive data. However, the applications must be trusted as error containment characteristics might bring unexpected exceptions.

6   Experiments

In this paper, the main software parallelization techniques for dedicated processing devices are analyzed, and an effective parallelization framework is proposed
by outsourcing the workload to GPU devices deployed on the Cloud or Edge
servers.

For this experimentation, a classic externalization architecture will be used,
without the presence of Edge servers. Thus, the client's request will be transferred to the server, which will execute the calculations and return the results to
the client. The total time perceived by each client is formulated in Eq. 1, which
takes into account communication times (sending and receiving), processing time
and a threshold that includes possible fluctuations, and other non-predictable
delays.

$$T = T_{\text{Communication}} + T_{\text{Processing}} + \delta_T \tag{1}$$

The conducted experimentation focuses on this processing time arising from
the application of the different optimization techniques. These techniques allow
sharing the GPU resources between different tasks at the same time. We explore
the application of the "stream" technique [7, 22] and the Multi-Process Service
(MPS) [13, 31] on GPUs through CUDA programming. Processing time of each
technique is obtained, and they are compared to normal execution on a GPU
device, that is, without using any parallelization technique.

The GPU deployed is the Nvidia TITAN [1]. This device has been built
according to the Nvidia Turing architecture, and it has 24 GB VRAM, 576
Tensor cores and 4608 CUDA cores.

For experimentation, the all-pairs distance between instances of a dataset
has been used. This is a useful metric for AI applications that are then used
in the bioinformatic field. It calculates the distance between the instances of
the data set. Concretely, this distance is computed when the labels (attributes)
are categorical and not numerical, so it is impossible to measure the distance
between instances. The main idea of this distance is to evaluate if two categorical
labels are equal and in that case set the value of the distance to 0, instead if the
labels are different the distance between the instances is set to 1. It is important
to know that this distance method is very extreme, because it only has two
possible states (0 or 1).

The GPU implementation of the all-pairs distance algorithm by Payne. et
al. [23] has been used for the experiments. The implementation uses a block
for each pair of instances, so the amount of blocks used is $instances^2$. Inside a
block, every thread computes the distance of four attributes, using only internal
registers, and stores the result on a shared memory vector, with one element
per thread. Then, after a barrier to wait for all threads to finish, one of the
threads adds up the shared memory vector and writes the distance for that pair
of instances to the results vector.

In this first experiment, the all-pairs distance algorithm was run both in the
normal way and using the "streams" technique. Several instance sizes (batch
size) have been used to compute the algorithm, with and without the streams

technique (Fig. 2). The maximum number of available streams of a GPU depends on the internal architecture. For our Nvidia TITAN device, there are 128 streams for parallel programming. The Speedup shown in the experiment is computed by measuring only the kernel execution time. Therefore, the experiment shows the level of parallelism achieved by the different kernels depending on the amount of blocks spawned by each one.

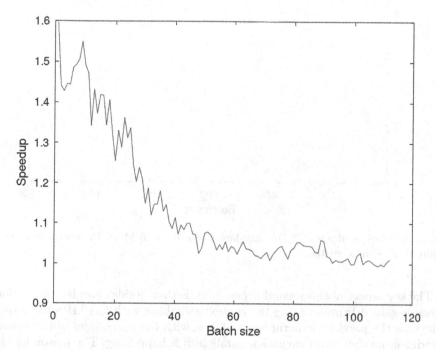

Fig. 2. Speedup results for CUDA standard execution and Stream execution of all-pairs algorithm

As shown in Fig. 2, the use of streams as an optimization mechanism improves the performance. The performance improvement goes down as the batch size increases. This happens as when the kernels use more and more blocks, the amount of parallelism is reduced because the resources of the GPU run out. For the all-pairs GPU algorithm, this is specially crucial. The amount of blocks grows at a square rate, so after a batch size of 40 no significant amount of parallelism occurs.

In the second experiment, the all-pairs function execution using Multi-Process Service technique to obtain Speedup is performed. Like the previous experiment, the function has been calculated on random attributes with different batch sizes. In contrast with the last experiment, in this one, each kernel is launched in a different operating system process. Whole process times are being measured. The fixed parameters of the kernels are 102400 attributes and 128 threads per block. Results are depicted in Fig. 3.

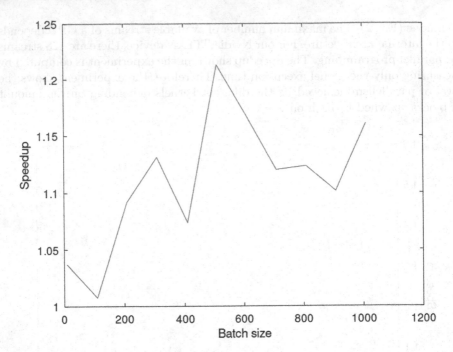

Fig. 3. Speedup results for CUDA standard execution and Multi-Process Service execution of all-pairs algorithm

The key aspect of this second experiment, is that besides parallelism, performance is gained from avoiding the context switching with the MPS technique. In fact, as the previous experiment confirms, with the batch sizes of the second experiment no significant execution parallelism is happening. The reason for the changes in the parameters of this experiment with respect to the first one is that as whole process time is being measured, the results obtained from smaller batch sizes and attributes were too small compared to the time of setting up the process.

7   Conclusions

The trend from bioinformatic applications will converge in Cloud computing outsourcing architectures with GPU devices. In this paper, the characteristic of this architecture have been studied. Security has been revealed as the biggest challenge, as bioinformatic applications usually work with sensitive data that has to be protected and has to comply with regulations. The proposed architecture is able to adapt to this security characteristics, and maintaining acceptable performance when using the necessary security measures.

To improve the efficiency of the GPU devices, the MPS server architecture can be used. The experiments show that it has the opportunity to increase parallelism and reduce context switching costs when running different concurrent applications on the GPU. However, in future work, it is needed to study GPU virtualization and improve on the efficiency of it on Cloud computing outsourcing architectures.

Acknowledgments. This work was supported by the Spanish Research Agency (AEI) under project HPC4Industry PID2020-120213RB-I00.

References

1. NVIDIA TITAN RTX is Here – nvidia.com. https://www.nvidia.com/en-us/deep-learning-ai/products/titan-rtx.html/. Accessed 24 Apr 2023
2. Anati, I., Gueron, S., Johnson, S., Scarlata, V.: Innovative technology for CPU based attestation and sealing. In: Proceedings of the 2nd International Workshop on Hardware and Architectural Support for Security and Privacy, vol. 13. ACM, New York (2013)
3. Atta-ur-Rahman, Dash, S., Ahmad, M., Iqbal, T.: Mobile cloud computing: a green perspective. In: Udgata, S.K., Sethi, S., Srirama, S.N. (eds.) Intelligent Systems, pp. 523–533. Springer, Cham (2021). https://doi.org/10.1007/978-981-33-6081-5_46
4. Blass, E.O., Kerschbaum, F., Mayberry, T.: Iterative oblivious pseudo-random functions and applications. In: Proceedings of the 2022 ACM on Asia Conference on Computer and Communications Security, pp. 28–41 (2022). https://doi.org/10.1145/3488932.3517403
5. Chen, F., et al.: Presage: privacy-preserving genetic testing via software guard extension. BMC Med. Genomics **10**(2), 77–85 (2017)
6. Dematté, L., Prandi, D.: GPU computing for systems biology. Brief. Bioinform. 323–333 (2010). https://doi.org/10.1093/bib/bbq006
7. Du, G., Jia, L., Wei, L.: A new algorithm of handwritten numeral recognition based on GPU multi-stream concurrent and parallel model. In: 2020 IEEE 2nd International Conference on Civil Aviation Safety and Information Technology (ICCASIT), pp. 232–236 (2020). https://doi.org/10.1109/ICCASIT50869.2020.9368829
8. Elouali, A., Mora Mora, H., Mora-Gimeno, F.J.: Data transmission reduction formalization for cloud offloading-based IoT systems. J. Cloud Comput. **12**(1), 48 (2023). https://doi.org/10.1186/s13677-023-00424-8
9. Gudukbay, G., et al.: GYAN: accelerating bioinformatics tools in galaxy with GPU-aware computation mapping. In: 2021 IEEE International Parallel and Distributed Processing Symposium Workshops (IPDPSW), pp. 194–203 (2021)
10. Hung, C.L., Tang, C.Y.: Bioinformatics tools with deep learning based on GPU. In: 2017 IEEE International Conference on Bioinformatics and Biomedicine (BIBM), pp. 1906–1908 (2017). https://doi.org/10.1109/BIBM.2017.8217950
11. Jie, Y., et al.: Multi-party secure computation with intel SGX for graph neural networks. In: ICC 2022 - IEEE International Conference on Communications, pp. 528–533 (2022). https://doi.org/10.1109/ICC45855.2022.9839282
12. Kumar, I., Singh, S.P.: Machine learning in bioinformatics. In: Singh, D.B., Pathak, R.K. (eds.) Bioinformatics, pp. 443–456. Academic Press (2022). https://doi.org/10.1016/B978-0-323-89775-4.00020-1
13. Li, B., Patel, T., Samsi, S., Gadepally, V., Tiwari, D.: MISO: exploiting multi-instance GPU capability on multi-tenant systems for machine learning. In: Proceedings of the 13th Symposium on Cloud Computing, pp. 173–189 (2022). https://doi.org/10.1145/3542929.3563510

14. Li, P., et al.: Multi-key privacy-preserving deep learning in cloud computing. Future Gener. Comput. Syst. 76–85 (2017). https://doi.org/10.1016/j.future.2017.02.006

15. Li, Y., Huang, C., Ding, L., Li, Z., Pan, Y., Gao, X.: Deep learning in bioinformatics: introduction, application, and perspective in the big data era. Methods 4–21 (2019). https://doi.org/10.1016/j.ymeth.2019.04.008

16. Liang, J., Qin, Z., Xiao, S., Ou, L., Lin, X.: Efficient and secure decision tree classification for cloud-assisted online diagnosis services. IEEE Trans. Dependable Secure Comput. 18(4), 1632–1644 (2021). https://doi.org/10.1109/TDSC.2019.2922958

17. Maciá-Lillo, A., Ribes, V.S., Mora, H., Jimeno-Morenilla, A.: Efficient GPU cloud architectures for outsourcing high-performance processing to the cloud (2022). https://www.researchsquare.com/article/rs-2120350

18. Mora, H., Mora Gimeno, F.J., Signes-Pont, M.T., Volckaert, B.: Multilayer architecture model for mobile cloud computing paradigm. Complexity e3951495 (2019). https://doi.org/10.1155/2019/3951495

19. Mora, H., Peral, J., Ferrández, A., Gil, D., Szymanski, J.: Distributed architectures for intensive urban computing: a case study on smart lighting for sustainable cities. IEEE Access 7, 58449–58465 (2019). https://doi.org/10.1109/ACCESS.2019.2914613

20. Mora Mora, H., Gil, D., Colom López, J.F., Signes Pont, M.T.: Flexible framework for real-time embedded systems based on mobile cloud computing paradigm. Mob. Inf. Syst. 2015, e652462 (2015). https://doi.org/10.1155/2015/652462

21. Nobile, M.S., Cazzaniga, P., Tangherloni, A., Besozzi, D.: Graphics processing units in bioinformatics, computational biology and systems biology. Brief. Bioinform. 18(5), 870–885 (2017). https://doi.org/10.1093/bib/bbw058

22. Novotný, J., Adámek, K., Armour, W.: Implementing CUDA Streams into AstroAccelerate - A Case Study (2021). https://doi.org/10.48550/arXiv.2101.00941

23. Payne, J.L., Sinnott-Armstrong, N.A., Moore, J.H.: Exploiting graphics processing units for computational biology and bioinformatics. Interdiscip. Sci. Comput. Life Sci. 2(3), 213–220 (2010). https://doi.org/10.1007/s12539-010-0002-4

24. Pramkaew, C., Ngamsuriyaroj, S.: Lightweight scheme of secure outsourcing SVD of a large matrix on cloud. J. Inf. Secur. Appl. 92–102 (2018). https://doi.org/10.1016/j.jisa.2018.06.003

25. Qian, L., Luo, Z., Du, Y., Guo, L.: Cloud computing: an overview. In: Jaatun, M.G., Zhao, G., Rong, C. (eds.) CloudCom 2009. LNCS, vol. 5931, pp. 626–631. Springer, Heidelberg (2009). https://doi.org/10.1007/978-3-642-10665-1_63

26. Qiu, T., Chi, J., Zhou, X., Ning, Z., Atiquzzaman, M., Wu, D.O.: Edge computing in industrial internet of things: architecture, advances and challenges. IEEE Commun. Surv. Tutor. 2462–2488 (2020). https://doi.org/10.1109/COMST.2020.3009103

27. Smajlović, H., Shajii, A., Berger, B., Cho, H., Numanagić, I.: Sequre: a high-performance framework for secure multiparty computation enables biomedical data sharing. Genome Biol. 24(1), 5 (2023). https://doi.org/10.1186/s13059-022-02841-5

28. Suo, J., Gu, L., Yan, X., Yang, S., Hu, X., Wang, L.: PP-DDP: a privacy-preserving outsourcing framework for solving the double digest problem. BMC Bioinform. 34 (2023). https://doi.org/10.1186/s12859-023-05157-8

29. Thavappiragasam, M., Kale, V., Hernandez, O., Sedova, A.: Addressing load imbalance in bioinformatics and biomedical applications: efficient scheduling across multiple GPUs. In: 2021 IEEE International Conference on Bioinformatics and Biomedicine (BIBM), pp. 1992–1999 (2021). https://doi.org/10.1109/BIBM52615.2021.9669317

30. Waheed, A., et al.: A comprehensive review of computing paradigms, enabling computation offloading and task execution in vehicular networks. IEEE Access 3580–3600 (2022). https://doi.org/10.1109/ACCESS.2021.3138219

31. Wu, H., Liu, W., Gong, Y., Jin, J.: Safe process quitting for GPU multi-process service (MPS). In: 2020 IEEE 40th International Conference on Distributed Computing Systems (ICDCS), pp. 1169–1170 (2020). https://doi.org/10.1109/ICDCS47774.2020.00125

32. Wu, Y., et al.: Generic server-aided secure multi-party computation in cloud computing. Comput. Stand. Interfaces **79**, 103552 (2022). https://doi.org/10.1016/j.csi.2021.103552

33. Yang, Y., et al.: A comprehensive survey on secure outsourced computation and its applications. IEEE Access **7**, 159426–159465 (2019). https://doi.org/10.1109/ACCESS.2019.2949782

34. Zhong, H., Sang, Y., Zhang, Y., Xi, Z.: Secure multi-party computation on blockchain: an overview. In: Shen, H., Sang, Y. (eds.) PAAP 2019. CCIS, vol. 1163, pp. 452–460. Springer, Singapore (2020). https://doi.org/10.1007/978-981-15-2767-8_40

Biomarker Identification

Novel Gene Signature for Bladder Cancer Stage Identification

Iñaki Hulsman[1]([⊠]), Luis Javier Herrera[1],
Daniel Castillo[1,2], and Francisco Ortuño[1]

[1] Department of Computer Engineering, Automatics and Robotics, University of
Granada. C.I.T.I.C., Periodista Rafael Gómez Montero, 2, 18014 Granada, Spain
hulsman@correo.ugr.es
[2] Fujitsu Technology Solutions S.A., CoE Data Intelligence, Cam. Cerro de los
Gamos, 1, Pozuelo de Alarcón 28224, Madrid, Spain

Abstract. This article presents a study that aimed to identify the stages
of bladder cancer based on gene expression data. The dataset used in the
study was obtained from the GDC repository and included 406 cases of
bladder cancer and 431 files from the TCGA-BLCA project. The study
categorized the cases into three classes based on disease stages: Stage 2,
Stage 3, and Stage 4. The methodology employed R programming language and the KnowSeq library for the study development. The authors
identified genes that showed significant differences in expression among
the classes and created a matrix of differentially expressed genes (DEG).
Machine learning models, including feature selection algorithms and classification models such as KNN and SVM, were constructed to predict the
bladder cancer stages. The results revealed that the mRMR feature selection algorithm performed the best, and the 8 most relevant genes were
used to build the classification models.

Keywords: Bladder Cancer · KnowSeq · RNA-Seq · cancer stage

1 Introduction

Bladder cancer (BLCA) is a prevalent type of cancer worldwide, with a significant
amount of research conducted to improve its identification and treatment. It is
responsible for approximately 200,000 deaths per year, and in 2020, an estimated
573,278 individuals were diagnosed with bladder cancer globally. Among men,
it is the fourth most frequent tumor, following lung, prostate, and colon cancer.
Early detection and accurate staging of bladder cancer are critical for effective
treatment and improved patient prognosis.

Despite its high incidence, bladder cancer does not have a high mortality rate
(ninth place among tumors with the highest mortality), as it is often a curable
oncological process. It is the tenth leading cause of death, ranking eighth among
men and fourteenth among women, with an average presentation age between

I. Rojas et al. (Eds.): IWBBIO 2023, LNBI 13919, pp. 93–102, 2023.
https://doi.org/10.1007/978-3-031-34953-9_7

60 and 70 years. It is more common in industrialized countries, particularly in individuals exposed to dyes, paints, and other related industries [6].

Bladder cancer originates when the healthy cells of the bladder lining undergo uncontrollable changes and proliferate, forming a mass known as a tumor. This process ranges from non-muscle-invasive and low-risk tumors to muscle-invasive tumors, which generally have a lower survival rate. Bladder cancer is classified into different stages depending on the extent of the tumor within the organ. Stage Ta refers to a tumor confined within the urothelium. In T1, the cancer has grown through the inner lining of the bladder and has reached into the lamina propria. In T2, the cancer has spread to the thick wall of the bladder muscle, also known as invasive or muscle-invasive cancer. At stage T3, the cancer has spread through the muscle wall into the fatty layer of tissue surrounding the bladder, or in men, to the prostate, and in women, to the uterus and vagina, or to regional lymph nodes. Finally, at stage T4, the tumor has spread within the pelvic or abdominal wall or has spread to lymph nodes outside the pelvis.

The 5-year survival rate for all individuals diagnosed with bladder cancer is 77%. However, survival rates depend on several factors, including the type and stage of bladder cancer. The 5-year survival rate for individuals diagnosed with bladder cancer that has not spread beyond the inner layer of the bladder wall is 96%, with about half of individuals diagnosed at this stage. If the tumor is invasive but has not yet spread outside the bladder, the 5-year survival rate is 70%, with approximately 33% of individuals diagnosed at this stage. If the cancer has spread to surrounding tissues or nearby lymph nodes or organs, the 5-year survival rate drops to 38%. If the cancer has spread to distant parts of the body, the 5-year survival rate is only 6%, with about 4% of individuals diagnosed at this stage [6].

The staging of bladder cancer based solely on histopathological evaluation can be subjective and may not always result in accurate tumor classification. Consequently, various studies have been conducted to identify molecular and biological biomarkers that could provide a more precise characterization of different stages of bladder cancer. Some of the most notable research in this area is outlined below:

In 2006, ISG15 was identified as a novel gene expression component associated with bladder cancer [1]. Subsequently, in 2013, 36 genes were discovered as potential molecular markers for predicting the transition from Ta-T1 to T1-T2 stages [3]. In 2018, differential expression gene (DEG) analysis was found to provide informative data on a common molecular characteristic of stages 2 and 4 bladder cancer [9]. Moreover, that same year, it was demonstrated that COL5A2 was significantly correlated with poor clinical outcomes and survival rates in bladder cancer patients, indicating its potential as a bladder cancer biomarker [10]. In 2021, previous studies reporting a range of mutations in T1 tumors intermediate between Ta and muscle-invasive tumors were challenged. Instead, it was found that some T1 tumors harbor similar mutations to both Ta and muscle-invasive tumors [4]. In 2022, another study examined adjacent non-malignant urothelium and bladder cancer samples ranging from Ta to T4 stages, and identified a gene fingerprint consisting of eight genes, namely AKAP7, ANLN, CBX7,

CDC14B, ENO1, GTPBP4, MED19, and ZFP2. The authors suggested that this eight-gene signature can aid in decision making for therapeutic options and surveillance programs. [7]

These and other studies indicate the importance of advancing the molecular and biological characterization of bladder cancer stages to improve diagnosis and treatment.

In advanced stages of bladder cancer, the prognosis is much more severe, and the survival rate decreases significantly. Therefore, an accurate tool is needed to diagnose tumors in the T2-T4 range and precisely classify the disease stage. The objective of this study is to identify a genetic signature consisting of a limited number of biomarkers, enabling the effective classification of different stages within the muscle-invasive spectrum of bladder cancer (T2, T3, and T4). To achieve this goal, we employ the KnowSeq tool, which implements a complete pipeline for gene expression data processing, including various machine learning techniques that are useful for characterizing tumor stages.

2 Data Description

The present study utilized a dataset obtained from the Genomic Data Commons (GDC) repository, consisting of 406 cases of bladder cancer and 431 files from the Cancer Genome Atlas Bladder Urothelial Carcinoma (TCGA-BLCA) project. The files contained information generated using RNA sequencing with the STAR-Counts workflow. Our investigation focused on three disease stages, namely Stage 2, Stage 3, and Stage 4, with Stage 1 being merged with Stage 2 due to its limited sample size (n = 2). Notably, Stages 3 and 4 had sufficient samples for each category. To ensure a balanced training dataset, we integrated the clinical data from the GDC dataset using the case code as unique identifier, resulting in 137 samples for Stage 2, 140 samples for Stage 3, and 138 samples for Stage 4 (1). This approach allowed us to explore the association between gene expression and bladder cancer disease stages comprehensively.

Table 1. Training-Test data partition table of the three classes

Clases	STAGE (1 - 2)	STAGE 3	STAGE 4
train	110	112	111
test	27	28	27
total	137	140	138

3 Metodology

3.1 Differential Expressed Genes Extraction and Visualization

The present study was developed using the R programming language and the KnowSeq library [2]. Specific functions from KnowSeq were used to transform the counts from the GDC files into expression data. Quality analysis to eliminate outliers and batch effect removal were aplied. Then DEG extraction was applied: for the current study, the LFC parameter was set at 0.7, as setting it higher would lead to a shortage of genes, potentially reducing representativeness; the p-value was set to 0.001 and the cov parameter to 1. Using this configuration, the resulting DEGMatrix contained 27 genes.

3.2   Development of Classification Models

The development of machine learning models for predicting bladder tumor stage involves two main steps: biomarker ranking using feature selection algorithms, and building a classification model.

Several feature selection algorithms have been proposed in the biomedical literature with the objective of selecting a subset of features based on criteria such as mutual information, distance measures, and their performance in classification. These algorithms typically measure the relationship between the features and the objective class, and rank the features accordingly. In this study, we compare the results of three ranking algorithms: minimum redundancy maximum relevance (mRMR), Random Forest (RF), and Disease Association (DA).

To evaluate the ranking algorithms, we use two conventional classifiers: K-nearest neighbors (KNN) and Support Vector Machines (SVM). The models are trained using the training set mentioned previously, and the average accuracy under cross-validation is used as the performance metric. We calculate the accuracy for each set of genes, adding one at a time according to the ranking established by each algorithm.

Once the feature extraction algorithm is selected, a final signature of the most relevant genes for classification. Finally, the resulting classifier and genetic signature are evaluated using cross-validation on the test set to obtain the accuracy of the model.

4   Results and Discussion

Figures 1 and 2 depict the outcomes of various biomarker ranking algorithms applied to different classifiers. The X-axis shows the number of genes in the ranking order utilized for classification, and the Y-axis represents the accuracy achieved. It is noteworthy that in both classifiers, the mRMR algorithm delivers the best accuracies for models with a small number of genes (up to 9), achieving approximately 90%. For more than nine genes, DA, RF and mRMR produce similar results, while all three algorithms yield very similar accuracy percentages for over 20 genes. The mRMR algorithm dominates the SVM comparisons up to 20 genes, after which it performs similarly to the KNN classifier. The objective is to strike a balance between model accuracy and the number of genes employed. The fewer genes required to achieve competent classification, the easier the stage detection process will be with actual patients. Consequently, we have decided to use mRMR, which attains the best results using few genes.

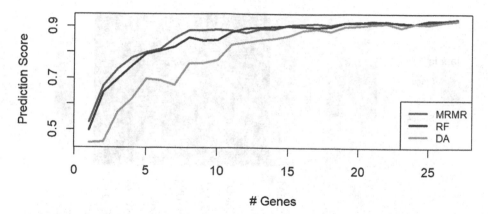

Fig. 1. Diagram of a KNN model performance on the train set using the genes of the ranking obtained by different feature selection algorithms. The x axis shows the number of genes from the top of the ranking used for the classification and the y axis shows the accuracy of the model

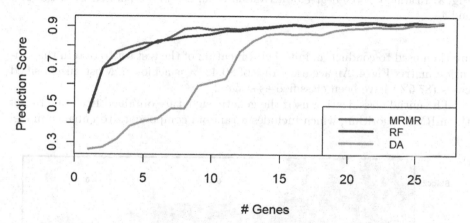

Fig. 2. Diagram of a SVM model performance on the train set using the genes of the ranking obtained by different feature selection algorithms. The x axis shows the number of genes from the top of the ranking used for the classification and the y axis shows the accuracy of the model

The genetic signature comprises the first 8 genes (the most important) of the DEGMatrix in mRMR order, since the models do not appear to improve beyond with more genes. A heatmap-like illustration, in Fig. 3, demonstrates the degree of activation of each gene in each case, separated by stage. Although subtle, color variations are visible in sections belonging to different stages for the same gene. This supports the idea that the selected genes are relevant for identifying the target class, as their activation depends on the corresponding tumor stage.

KNN has been chosen as the classifier between the two options since it provides better results with fewer genes than SVM. The selected genes and model

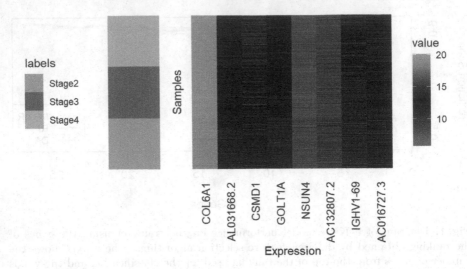

Fig. 3. Heatmap of the degree of activation of all genes in separated by stages 2, 3 and 4

are then used to conduct an initial classification of the test set to obtain the confusion matrix Fig. 4. An accuracy rate of 80.488% is achieved. Most misclassified cases (87.5%) have been classified as stage 2.

The initial classification used the genetic signature obtained from one run of the mRMR algorithm, which includes a random component. To enhance model

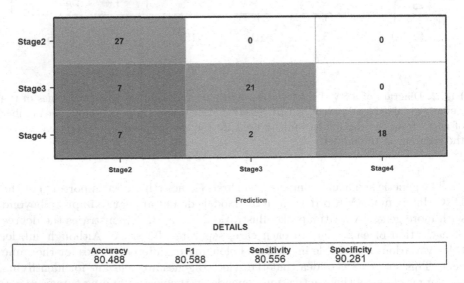

Fig. 4. Confusion matrix of the model performance on the test set with a signature of genes obtained trough mRMR algorithm

performance, we carried out ten runs of the mRMR algorithm and evaluated them using the train set. Ten rankings of 15 genes were obtained from the ten runs, and Table 2 displays all of them. Some are highlighted in color to indicate which genes occupied important positions in each run.

Table 2. Table with the 10 gene rankings obtained by the mRMR algorithm

		AC016727.3	CHMP1B2P		CSMD1	HMGB1P6	CCN4	THRA1/BTR	IGHV1-69	AC010618.2	KMT2E	AC132807.2	ACTN2
	CSMD1		AC010618.2	ACTN2	AC132807.2	CHMP1B2P	AC016727.3	THRA1/BTR	IGHV1-69	C9orf152	AC027277.2	LRWD1	
AC025370.2	AC016727.3	CHMP1B2P	AC132807.2	CSMD1	C9orf152		KMT2E	HMGB1P6	THRA1/BTR	IGHV1-69	LRWD1	LMNB1	
	CSMD1	AC132807.2	THRA1/BTR	IGHV1-69	AC016727.3	HMGB1P6		PLAT	CHMP1B2P	LRWD1	AC027287.2	C9orf152	
AC025370.2	AC132807.2	IGHV1-69	AC016727.3	NSUN4	AC132807.2	CSMD1		KMT2E	CHMP1B2P	HMGB1P6	APCDD1L	LRWD1	
	AC132807.2	IGHV1-69	AC016727.3	THRA1/BTR	AL021407.2	CHMP1B2P	CCN4	CSMD1	KMT2E	LINC00707	LRWD1	APCDD1L	
	DSG1	CHMP1B2P	TMEM242		CCN4	THRA1/BTR	KMT2E	CSMD1	AC016727.3	AC132807.2	LINC01602	C9orf152	
CSMD1	AC132807.2	THRA1/BTR	AC016727.3	IGHV1-69	CCN4	AC010618.2	CHMP1B2P		C9orf152	PLA2G3	KMT2E	PLAT	
AC025370.2	AC132807.2	LRRC14B	CHMP1B2P	AC016727.3	CSMD1	DSG1	THRA1/BTR	IGHV1-69	AC132807.2	HMGB1P6	IGHVII-60-1	C9orf152	
AC006982.2		AC132807.2	KMT2E	CSMD1		CHMP1B2P	PLAT	AC016727.3	THRA1/BTR	C9orf152	ACTN2	IGHV1-69	

By weighting gene positions using the accuracy attained in each ranking, we constructed a final signature consisting of 8 genes that we will use as the training base for the model. The footprint is made up of the following genes: COL6A1, AL031668.2, CSMD1, GOLT1A, NSUN4, AC132807.2, IGHV1-69 and AC016727.3. Figure 5 shows the gene activation distribution of this signature in every stage. The KNN model was trained with k = 23 using this new genetic signature and predicted the generated test sets via 5 k-fold cross-validation. The

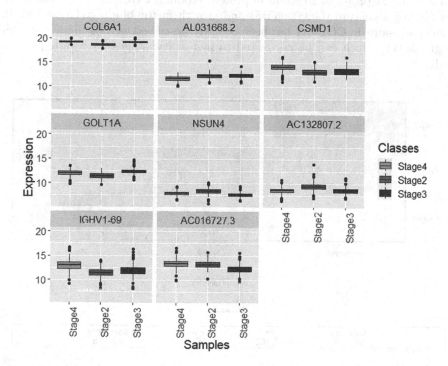

Fig. 5. Boxplot of the first eight genes obtained trough KnowSeq library functions

results of the confusion matrix are presented in Fig. 6, which demonstrates an improvement over the previous ranking. It is more accurate cases where it was previously incorrect, especially for cases of stage 4 classified as stage 2, which decreased from seven to zero. The final accuracy is 90.244%. Additionally, a 5-fold cross validation was performed with the entire dataset, reaching an accuracy of 87%.

4.1   Gene Enrichment

Gene enrichment of the genes belonging to this new end fingerprint (composed of COL6A1, AL031668.2, CSMD1, GOLT1A, NSUN4, AC132807.2, IGHV1-69 and AC016727.3) shows information about which biological processes each gene is involved. Regarding Gene Ontology, four of the genes (COL6A1, GOLT1A, NSUN4, and CSMD1) are all related to protein binding (GO:0005515). Some of these genes (COL6A1 and GOLT1A) are more specifically associated to the extracellular matrix (ECM) organization, which is a process of binding and maintaining the structural components of the extracellular matrix. The ECM is essential for cell adhesion, proliferation, and migration, all of which are critical processes in cancer [8]. COL6A1 gene encodes for collagen type VI alpha 1, which is a structural component of the ECM (GO:0030198) whereas GOLT1A is involved in Golgi organization (GO:0007030), which plays a key role in processing and secreting ECM proteins. In terms of protein binding, COL6A1 affects the collagen binding structure (GO:0005518) and growth factor binding (GO:0019838). High expressions in COL6A1 has been recently described as a biomarker of poor prognosis [11] which can be correlated to metastatic stages, as resulted in our study.

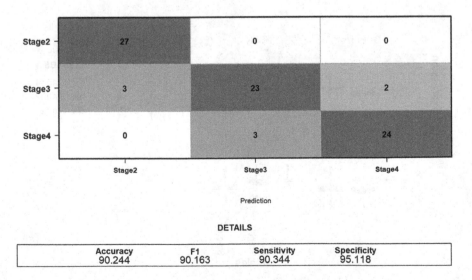

Fig. 6. Confusion matrix of the model performance on the test set with the signature of genes obtained through a 10-CV of a Knn model using de mRMR algorithm

Moreover, two of the selected genes are associated to RNA binding and modification: NSUN4 and CSMD1. NSUN4 is involved in RNA modification (GO:0009451), which includes processes such as ribosomal RNA methylation and binding (GO:0031167 and GO:0019843, respectively) that can regulate RNA stability and function. NSUN4 gene has been shown to aberrantly express in other carcinomas and adenocarcinomas (lung, hepatocellular, renal, etc.) and it has been described as a good prognosis predictor [5]. CSMD1 is involved in RNA splicing (GO:0008380), which is critical for generating diverse protein isoforms. Dysregulation of RNA processing and modification like these genes has been linked to various types of cancer.

Finally, COL6A1 and CSMD1 are involved in tissue development (GO:0009888), both including multicellular organism and anatomical structure developments (GO:0007275 and GO:0048856, respectively). Tissue development involves the coordinated growth, differentiation, and organization of cells to form functional tissues and organs, which could also be alternated by the effects of cancer.

Regarding pathways, only information about the gene COL6A1 was found. The gene COL6A1 is involved in various cellular pathways and processes. In the PI3K-Akt signaling pathway (map04151), the Akt protein controls important cellular processes, including apoptosis, protein synthesis, metabolism, and the cell cycle. COL6A1 is also involved in the formation of focal adhesions (map04510), which play an essential role in biological processes such as cell motility, cell proliferation, differentiation, gene expression regulation, and cell survival. The gene also participates when interaction between cells and the extracellular matrix (ECM) is mediated by transmembrane molecules, primarily integrins, and the resulting interaction directly or indirectly controls cellular activities such as adhesion, migration, differentiation, proliferation, and apoptosis (map04512). Regarding protein digestion and absorption (map04974), COL6A1 does not seem to be directly involved in the process, but the protein is an essential component for nutritional homeostasis and undergoes a complex series of degradative processes in the gastrointestinal tract. Finally, COL6A1 is also related to human papillomavirus infection (map05165), which causes pathologies such as cervical cancer. In particular, the viral oncogenes E5, E6, and E7 are involved in cellular transformation and carcinogenesis.

5   Conclusions

This study used a dataset obtained from the GDC repository for the detection of bladder cancer stages. Three classes were considered based on the stages of the disease: Stage 2, Stage 3, and Stage 4. The R programming language and the KnowSeq library were used for the development of the study. Differentially expressed genes (DEGs) were extracted to differentiate among the classes, resulting in a DEGMatrix containing 27 genes. The performance of three biomarker ranking algorithms, namely MRMR, RF, and DA, was compared, and it was concluded that the MRMR algorithm yielded the best results with a lower number of genes utilized. KNN and SVM-based classification models, using the genes selected by the MRMR algorithm, achieved high accuracy in classifying bladder cancer stages, with the KNN-based model exhibiting a train accuracy of 90% and test accuracy of 80% using only the top 9 genes from the DEGMatrix. Additionally, results using 8 genes were presented, and this gene new signature was proposed for bladder cancer stage classification, demonstrating a test accuracy of 90%, and showing its usability for real patients characterization.

Acknowledgements. This work was funded by the Spanish Ministry of Sciences, Innovation and Universities under Project PID2021-128317OB-I00 and the projects from Junta de Andalucia P20-00163.

References

1. Andersen, J.B., Aaboe, M., Borden, E.C., Goloubeva, O.G., Hassel, B.A., Ørntoft, T.F.: Stage-associated overexpression of the ubiquitin-like protein, ISG15, in bladder cancer. Br. J. Cancer **94**(10), 1465–1471 (2006). https://doi.org/10.1038/sj.bjc.6603099
2. Castillo-Secilla, D., et al.: Knowseq r-bioc package: the automatic smart gene expression tool for retrieving relevant biological knowledge. Comput. Biol. Med. **133**, 104387 (2021)
3. Fang, Z.Q., et al.: Gene expression profile and enrichment pathways in different stages of bladder cancer. Genet. Mol. Res. **12**(2), 1479–1489 (2013). https://doi.org/10.4238/2013.may.6.1
4. Hurst, C.D., et al.: Stage-stratified molecular profiling of non-muscle-invasive bladder cancer enhances biological, clinical, and therapeutic insight. Cell Rep. Med. **2**(12), 100472 (2021). https://doi.org/10.1016/j.xcrm.2021.100472
5. Li, M., et al.: 5-methylcytosine RNA methyltransferases and their potential roles in cancer. J. Transl. Med. **20**(1) (2022). https://doi.org/10.1186/s12967-022-03427-2
6. Siegel, R.L., Miller, K.D., Fuchs, H.E., Jemal, A.: Cancer statistics, 2022. CA Cancer J. Clin. **72**(1), 7–33 (2022). https://doi.org/10.3322/caac.21708
7. Stroggilos, R., et al.: Gene expression monotonicity across bladder cancer stages informs on the molecular pathogenesis and identifies a prognostic eight-gene signature. Cancers **14**(10), 2542 (2022). https://doi.org/10.3390/cancers14102542
8. Walker, C., Mojares, E., del Río Hernández, A.: Role of extracellular matrix in development and cancer progression. Int. J. Mol. Sci. **19**(10), 3028 (2018). https://doi.org/10.3390/ijms19103028
9. Zamanian Azodi, M., Rezaei-Tavirani, M., Rostami-Nejad, M., Rezaei-Tavirani, M.: Comparative bioinformatics characteristic of bladder cancer stage 2 from stage 4 expression profile: a network-based study. Galen Med. J. (Articles in Press), December 2018. https://doi.org/10.22086/gmj.v0i0.1279
10. Zeng, X.T., Liu, X.P., Liu, T.Z., Wang, X.H.: The clinical significance of COL5a2 in patients with bladder cancer. Medicine **97**(10), e0091 (2018). https://doi.org/10.1097/md.0000000000010091
11. Zhang, X., et al.: High expression of COL6a1 predicts poor prognosis and response to immunotherapy in bladder cancer. Cell Cycle **22**(5), 610–618 (2022). https://doi.org/10.1080/15384101.2022.2154551

Predicting Cancer Stage from Circulating microRNA: A Comparative Analysis of Machine Learning Algorithms

Sören Richard Stahlschmidt[1]([✉]), Benjamin Ulfenborg[1], and Jane Synnergren[1,2]

[1] Systems Biology Research Center, University of Skövde, Högskolevägen 1,
541 28 Skövde, Sweden
{soren.richard.stahlschmidt,benjamin.ulfenborg,jane.synnergren}@his.se
[2] Department of Molecular and Clinical Medicine, Institute of Medicine, Sahlgrenska
Academy at University of Gothenburg, Gothenburg, Sweden

Abstract. In recent years, serum-based tests for early detection and detection of tissue of origin are being developed. Circulating microRNA has been shown to be a potential source of diagnostic information that can be collected non-invasively. In this study, we investigate circulating microRNAs as predictors of cancer stage. Specifically, we predict whether a sample stems from a patient with early stage (0-II) or late stage cancer (III-IV). We trained five machine learning algorithms on a data set of cancers from twelve different primary sites. The results showed that cancer stage can be predicted from circulating microRNA with a sensitivity of 71.73%, specificity of 79.97%, as well as positive and negative predictive value of 54.81% and 89.29%, respectively. Furthermore, we compared the best pan-cancer model with models specialized on individual cancers and found no statistically significant difference. Finally, in the best performing pan-cancer model 185 microRNAs were significant. Comparing the five most relevant circulating microRNAs in the best performing model with the current literature showed some known associations to various cancers. In conclusion, the study showed the potential of circulating microRNA and machine learning algorithms to predict cancer stage and thus suggests that further research into its potential as a non-invasive clinical test is warranted.

Keywords: machine learning · circulating microRNA · cancer stage · liquid biopsy

1 Introduction

With almost 2 million estimated new cases and more than 600,000 estimated death during 2022 in the United States alone, cancer is the second leading cause of death. While overall incidences of cancer are slightly decreasing, this cannot be said for all cancer types [22]. Therefore, cancer remains a challenging group of diseases that burdens patients and society.

I. Rojas et al. (Eds.): IWBBIO 2023, LNBI 13919, pp. 103–115, 2023.
https://doi.org/10.1007/978-3-031-34953-9_8

The stage at diagnosis is a major predictor of mortality where early diagnosis is associated with an increased survival rate. While the impact of earlier detection on the excess mortality rate varies by cancer type, earlier detection generally improves survival chances [17]. For instance, the 5-year survival rate for cancers with the pancreas as the primary site is 44% if detected when still local and merely 3% when distant metastasis occur [23]. This association between survival and cancer stage makes reliable staging essential for informing treatment decisions.

An appealing approach for early detection would be to develop methods that enable screening and surveillance of large patient populations using cost effective and non-invasive strategies that are scalable and deliver results of high precision. In recent years, research into serum-based diagnostic tests for cancers has become prominent. One promising approach is to measure circulating microRNA (miRNA) in the serum or plasma. In normal tissue, miRNAs play a key role in maintaining the homeostasis of cells by regulating gene expression through post-transcriptional alterations [9]. For instance, miRNAs regulate apoptotic behavior of cells thereby mitigating tumorgenesis [13]. In cancerous tissue, the expression of miRNAs can be altered such that cells avoid, for example, apoptosis or detection by immune cells. miRNAs can act intracellularly but can also be exported from the cell packed into extracellular vesicles to act upon other cells. Alterations in the levels of circulating miRNA can be detected in serum, thus making these molecules potential biomarkers for cancer detection [9].

The potential of circulating miRNA for early detection of cancer has been investigated for various cancers by distinguishing samples of cancer patients from healthy controls [7,16,26]. Additionally, recent studies have shown that the tissue of origin of various cancers can be successfully predicted from circulating miRNA [16]. To our knowledge, predicting the stage of various cancers from circulating miRNAs has not yet been extensively investigated. A sensitive and precise serum-based test could serve as a non-invasive complimentary tool to current imaging modalities for cancer staging.

In this study, we therefore investigate whether (1) the cancer stage can be predicted from circulating miRNAs with machine learning (ML) methods, (2) it is beneficial to train such algorithms to detect pan-cancer patterns, and (3) which miRNAs are relevant to use as diagnostic biomarkers for cancer stage. To do so, we compare the performance of five common ML-algorithms to distinguish early stage (0-II) from late stage (III-IV) cancers based on their circulating miRNA profiles for a large cohort consisting of cancers from twelve different primary sites. The former class consists of carcinoma *in situ* (stage 0) as well as localized stages that have maximally spread to nearby lymph nodes (stage I-II). The latter class is made up of cancers that have grown significantly in size and spread (stage III-IV).

2   Materials and Methods

2.1   Study Cohort

To study the performance of ML algorithms for predicting cancer stage from circulating miRNAs, we obtained the recently published data by Matsuzaki et al. with GEO accession number *GSE211692* [16]. The original data set consists of serum samples from 10,547 patients collected at the National Cancer Center (NCC) Hospital in Japan between 2007 and 2016. Samples were stored in the NCC biobank until data generation. The data consists of miRNA levels that were assayed using a microarray platform (3D-Gene v21, Toray Industries, Tokyo, Japan). Details about sample collection and data generation can be found in [16].

After excluding patients with benign disease 10,234 patients remained. Excluding patients without stage information resulted in 8,753 patients, constituting the data set used in this study. The data set is comprised of serum samples from patients with a tumor from any of the twelve different primary sites. The distribution between these primary sites are highly imbalanced (see Table 1). The microarray assay provides expression data for 2,565 miRNAs.

Table 1. *GSE211692* Distribution of target classes by tumor primary site.

Cancer	Samples	Early Stage	Late Stage	Minority %
Pan-cancer	8,753	6,537	2,216	25.32%
Lung	1,676	1,409	267	15.90%
Colorectal	1,511	719	792	47.58%
Gastric	1,359	1,337	22	1.62%
Prostate	809	173	636	21.38%
Pancreatic	757	25	732	3.30%
Breast	598	396	202	33.78%
Esophageal	501	247	254	49.30%
Ovarian	382	131	251	34.29%
Hepatocellular	331	154	177	46.52%
Bone&Soft Tissue	296	29	267	10.86%
Biliary Tract	295	68	227	23.05%
Bladder	238	201	37	15.54%

As the target for the ML models, we operationalize cancer staging as a two-class classification task where early stage is defined as stage 0-II and late stage is defined as stage III-IV. With this aggregation into two classes the signal in the data set is potentially enhanced allowing the ML algorithms to identify more coarse patterns. In future research, the more complex task of fine-grained prediction for each stage should be studied.

The target classes are imbalanced with the early stage class consisting of 6,537 samples and the late stage class of 2,216 samples. This has important implications during training and for evaluating the predictions. Contributions to the loss functions during training are therefore weighed by,

$$class\_weight_i = 1 - p_i, \tag{1}$$

where p_i is the proportion of the ith class. Appropriate evaluation metrics for imbalanced data sets are discussed below.

We train the ML algorithms on a pan-cancer data set consisting of samples from the twelve described primary sites (Table 1). Many algorithms perform better with increased sample size and thus pooling samples into a pan-cancer data set might allow the algorithms to find patterns common to several cancers. To investigate the impact of pan-cancer training, we compare the performance between the best performing pan-cancer model and models trained on samples of single tissue of origin.

2.2 Algorithms

In this study, we are performing a comparative analysis of five common ML algorithms predicting early and late stage from circulating miRNA. In the following sections we will briefly discuss the chosen algorithms.

Logistic Regression. A logistic regression (LR) classifier models the relationship between input x and target y as a linear combination of x weighted by a vector of corresponding coefficients w and bias b, such that

$$P(Y = 1|X = x) = \frac{1}{1 + exp(b + w^T x)},\tag{2}$$

thus providing the probability of an instance being of the positive class [11]. The inductive bias is therefore a restriction of the hypothesis space to the linear function family. However, this low complexity allows a clear interpretation of the models' decisions.

Support Vector Machine. Support vector machines (SVMs) [5] learn to linearly separate two classes by finding a hyperplane maximally distant from the data points of each class, i.e. maximizing the margins to the decision boundary. SVMs have the potential to separate classes with nonlinear decision boundaries by mapping the data from the original input space to a feature space where they are linearly separable. This is done through kernels that transform the input data.

K-Nearest Neighbors. In a binary classification, the k-Nearest Neighbors (kNN) algorithms [8] predicts the label $\hat{y}$ of an instance by considering its nearest neighbors in the input space x. Formally this can be denoted as

$$\hat{y} = \frac{1}{k} \sum_{x_i \in N_k(x)} y_i,\tag{3}$$

where $\hat{y}$ is the predicted label, k is the number of neighbors to consider, and $N_x(x)$ is the neighborhood of x [11]. k is a tuneable hyperparameter and $N_x(x)$ is determined by a chosen distance metric which can have significant impact on performance [1]. With kNN nonlinear decision boundaries can be learned, thus relaxing the inductive bias of logistic regression. However, this lower bias might come at a trade off with higher variance [11].

Random Forest. Random Forest (RF) classifiers are ensembles of decision trees that are trained on respective data sets that are sampled (with replacement) from the original training set. At each node the samples are split based on a random subset of input features. RFs have been shown to perform generally well, even with significant noise present in the data [6]. It has been shown that when increasing the number of trees in the forest RFs relatively quickly converge on a generalization error and adding more estimators does not necessarily improve performance [19]. This in combination with an easy parallelization of the training makes RF efficient to train.

Deep Neural Network. Deep neural networks (DNNs) are made up of multiple layers of nodes, which each perform a linear combination of the output of the nodes of the previous layer and then transform the resulting score with a nonlinear activation function. Formally, this can be expressed as a composite function

$$\hat{y} = f_3(f_2(f_1(x))), \tag{4}$$

where f_i represents the ith hidden layer [10,14]. Given sufficiently many nodes, DNNs are universal function approximators and thus represent a large hypothesis space. Additionally, DNNs learn hierarchical representations that capture increasingly more simple relationships between underlying explanatory factors [3].

Implementation. We apply the *python* implementation from *scikit-learn* version 0.22 [20] of LR, SVM, kNN, and RF. The DNN is implemented with *PyTorch* version 1.7.1.

2.3   Evaluation Metrics

The data set used for training and evaluating the ML algorithms is imbalanced where the minority class, late stage, constitutes approximately 25% of the cohort. Therefore, common classification metrics such as the area under the receiver-operating characteristics curve (AUROC) and accuracy can be inflated, giving incorrectly optimistic views on the models' performance. Thus, additionally, we evaluate the models with the following metrics which are less sensitive to class imbalance. We define the *late stage* class as positive cases. This however does not imply that we care more about predicting this class. Rather we are interest in a balanced performance.

Sensitivity/recall is the probability that a true positive (TP) is classified as such by the model. Thus,

$$sensitivity = \frac{TP}{TP + FN} \tag{5}$$

where FN are the false negative samples. Specificity is the probability that a true negative, in our case an early stage sample, is classified as early stage by the model,

$$specificity = \frac{TN}{TN + FP} \tag{6}$$

where FP are the false positive samples. For imbalanced data sets, specificity can however be misleading since the FP samples could be large in relation to the TP and small in relation to TN. Thus, we also measure positive predictive value (PPV) and negative predictive value (NPV) defined as

$$PPV = \frac{TP}{TP + FP} \tag{7}$$

and

$$NPV = \frac{TN}{TN + FN}. \tag{8}$$

To measure the balance between sensitivity and PPV we also measure F1 as their harmonic mean,

$$F1_{late} = \frac{2 * sensitivity * PPV}{sensitivity + PPV} \tag{9}$$

and the equivalent for specificity and NPV,

$$F1_{early} = \frac{2 * specificity * NPV}{sensitivity + NPV}. \tag{10}$$

Finally, we measure the combination of sensitivity and specificity with balanced accuracy,

$$BAcc = \frac{sensitivity + specificity}{2}. \tag{11}$$

To measure this trade off at different thresholds we also report the AUROC.

2.4   Experimental Design

The aim of the experimental design is to gain an unbiased estimate of the generalization error of the trained ML models while at the same time optimize the hyperparameters of each algorithm. To that end we trained and evaluated the above described methods in a 5×3 nested cross-validation. Briefly, during cross-validation the data set is divided into k partitions. The ML algorithms repeatedly train on $k - 1$ folds and are evaluated on the remaining fold. By calculating the mean and standard deviation of the repeated evaluations a less biased estimation can be obtained.

During nested cross-validation, after the outer split into partitions the training set is further repeatedly partitioned into training and test set. On this inner cross-validation loop, hyperparameter optimization can be performed. This potentially results in a better selection of hyperparameters due to less biased estimates of the performance on the inner test sets. Figure 1 displays the experimental design applied in this study. On the inner cross-validation, we selected hyperparameters for the different algorithms by performing a grid search over specified values (hyperparameters for each algorithm can be found in the appendix). Also on the inner loop, we performed feature selection so not to introduce information leakage between train and test folds. miRNAs are selected based on their variance. The p most varying miRNAs in the training set are chosen as features. The hyperparameter p is chosen during the inner cross-validation.

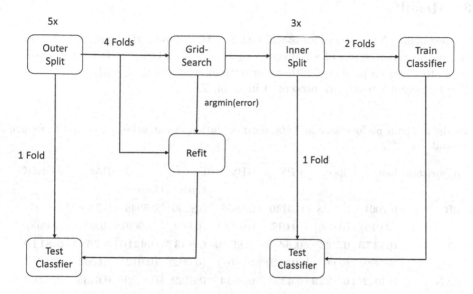

Fig. 1. Nested cross-validation with grid search

To investigate the effect of training models on the pan-cancer data set compared to training individual models on each cancer, the best performing algorithm is also applied to each single-cancer subset of *GSE211692*. Note that the models are again randomly initialized so that no transfer learning occurs. This approach enables the comparison of between pan-cancer training and cancer-specific training.

2.5   Statistical Analysis

We test the difference between the single-cancer approach and a pan-cancer approach with Wilcoxon signed-rank test due to outliers in the differences in performance. Our alternative hypothesis is that there is a difference between the pan-cancer model and the single cancer models. To test whether there is an association between model differences and sample size of single-cancer cohorts we apply Spearman's rank correlation test as a permutation test. Both tests are implemented with the *python* package *SciPy* version 1.10.1. To identify relevant miRNAs, we performed the permutation importance algorithm proposed by Altmann et al. [2] which repeatedly permutes the target variable and thereby extracts a null distribution of variable relevance. From this null distribution a hypothesis test for the actual variable importance can be performed. We implemented the algorithm in *python* ourselves according to the description in the original paper. We corrected for multiple hypothesis testing with the Benjamini-Hochberg method [4] with a false discovery rate of 5%.

3   Results

We performed 5×3 nested cross-validation with different ML algorithms training on a pan-cancer cohort of 8,753 patients. Serum-circulating miRNA was used as input features to predict whether the patient's cancer is at early or late stage. The pan-cancer result are presented in Table 2.

Table 2. Mean performance and standard deviation in parenthesis. Best performance in bold.

Algorithm	Sen.	Spec.	PPV	NPV	F1		BAcc	AUROC
					Early	Late		
LR	0.7061	0.7738	0.5140	0.8858	0.8260	0.5946	0.7404	0.7995
	(0.015)	(0.008)	(0.017)	(0.009)	(0.004)	(0.009)	(0.004)	(0.005)
SVM	**0.7173**	0.7997	**0.5481**	**0.8929**	**0.8437**	**0.6210**	**0.7584**	**0.8119**
	(0.009)	(0.008)	(0.022)	(0.007)	(0.002)	(0.012)	(0.002)	(0.008)
kNN	0.4042	**0.8229**	0.4360	0.8034	0.8129	0.4189	0.6135	0.6695
	(0.042)	(0.016)	(0.038)	(0.006)	(0.007)	(0.037)	(0.019)	(0.009)
RF	0.6375	0.7760	0.4908	0.8635	0.8173	0.5543	0.7066	0.7738
	(0.030)	(0.012)	(0.029)	(0.009)	(0.007)	(0.027)	(0.016)	(0.013)
DNN	0.6472	0.7760	0.5007	0.8669	0.8174	0.5594	0.7212	0.7925
	(0.060)	(0.053)	(0.052)	(0.018)	(0.022)	(0.017)	(0.014)	(0.007)
Baseline	0.4927	0.5035	0.2535	0.7468	0.6061	0.3336	0.4980	0.500
"uniform"	(0.009)	(0.01)	(0.020)	(0.008)	(0.009)	(0.012)	(0.005)	(0.000)

Comparing the performance of the ML models, all models show a higher specificity, PPV, NPV, F1 (early and late), BAcc, and AUROC than a baseline model that predicts labels randomly with uniform probability. Only kNN underperforms the baseline in regards to sensitivity, while LR, RF, SVM, and DNN show a clearly higher sensitivity than the random 0.4927. The F1 (late) score shows that even for the minority class, balanced performance between sensitivity and PPV can be observed. It is to note that specificity and NPV are higher than sensitivity and PPV, suggesting that the models perform better on the majority class (comparing F1 early and F1 late). This is primarily driven by the lower PPV compared to NPV. The weighting by class distribution in the loss functions was not able to entirely address the imbalance.

Except for specificity, SVM shows the best results for all metrics. The LR and DNN are the next best performing models, however, showing lower performance in all metrics. Therefore, we decided to address the second and third research question with SVM.

Table 3. F1-score for SVM models trained on a single-cancer data sets or on the pan-cancer data set. Difference (Diff) between the models is calculated by subtracting F1 of the single-cancer model from the pan-cancer result.

Cancer	F1 Early		Diff	F1 Late		Diff
	Single	Pan		Single	Pan	
Lung	0.766	0.955	0.1889	0.0869	0.079	−0.0131
Colorectal	0.841	0.868	0.0268	0.6626	0.677	−0.0447
Gastric	0.752	0.993	0.2407	0.0000	0.000	−0.3726
Prostate	0.579	0.707	0.1274	0.3377	0.290	−0.0106
Pancreatic	0.610	0.228	−0.3829	0.9034	0.875	−0.0301
Breast	0.950	0.909	−0.041	0.3951	0.135	−0.318
Esophageal	0.552	0.406	-0.146	0.3376	0.337	0.0626
Ovarian	0.359	0.237	-0.122	0.5892	0.679	0.1121
Hepatocellular	0.545	0.556	0.011	0.3215	0.445	0.1124
Bone&Soft Tis	0.500	0.358	−0.1422	0.6105	0.665	0.4657
Biliary Tract	0.439	0.682	0.2426	0.6408	0.811	0.1023
Bladder	0.874	0.892	0.0176	0.1733	0.138	0.0411
Mean	0.6475	**0.6492**	0.0017	0.4215	**0.4275**	0.0060
Std	(0.186)	(0.284)	(0.184)	(0.265)	(0.3060)	(0.1075)

To test whether there is a difference between single-cancer models and a pan-cancer model, we compared the performance for SVM models (see Table 3). Neither for predicting the early stage class (F1-score differences, p-value: 0.8501), nor the late stage cancer (F1-score differences, p-value: 0.7221), statistically significant differences could be observed. Therefore we fail to reject the null hypothesis. However, for individual cancers, performance between the models varies strongly (standard deviation of 0.3275 and 0.2041, respectively for the two classes). There seems to be no correlation between sample size of the individual cancer and difference in performance between single-cancer models and the pan-cancer model (early stage p-value:0.4914; late stage p-value:0.2954).

For the best performing pan-cancer model, SVM, we perform permutation importance for the 500 miRNAs with highest variance. For computational reasons, it was not possible to perform the analysis on all 1,500 chosen miRNAs. We identified 185 significant miRNAs with the permutation importance method by [2] using 30 repeats. Table 4 shows the five most important miRNAs according to their original importance.

Table 4. Five most relevant miRNAs in the SVM model.

miRNA	Adjusted p-value	Importance
hsa-miR-1290	0.000000	0.009657
hsa-miR-31-5p	0.002684	0.004783
hsa-miR-8058	0.016269	0.004513
hsa-miR-4528	0.000016	0.083917
hsa-miR-5701	0.000019	0.003610

4   Discussion

The results presented in this study show that cancer stage can be predicted from circulating miRNA with relatively high sensitivity and specificity. However, the performance varies significantly between cancers of different primary sites, as can be seen in Table 3. SVM outperforms all other models in almost all metrics, including a fully-connected DNN.

We could not detect any statistically significant difference between single-cancer and pan-cancer models. However, differences varied strongly between cancers of primary sites. This indicates that miRNAs which are shed into the blood stream at different stages of the disease might be similar between cancers of different primary sites. Alternatively, the algorithms might learn different decision boundaries for different cancers. However, the good performance of rather simple models such as LR speak against this.

With 185 statistically significant miRNAs, many molecules seem to be altered between early and advanced stages of cancers. Searching the existing literature for the five most relevant miRNAs, we find most of them having known associations to cancers. For instance, hsa-miR-1290 was shown to be upregulated in colorectal [12], pancreatic and gastric cancer patients' serum and associated with stage in at least pancreatic and colorectal cancer patients [25]. hsa-miR-31–5p has been shown to be differentially expressed in serum of oral cancer patients [15] and its expression in tissue is involved in the progression of colon cancer [18]. hsa-miR-4528 has been found upregulated in colon cancer [24] tissue but its role in disease progression remains unclear. hsa-miR-5701 was previously reported to inhibit cell proliferation of cervical cancer cells [21]. Cervical cancer was not included in the present study. However, this might indicate a role in progression of cancers of different primary site. We did not find evidence in the literature of hsa-miR-8058's role in cancer, which might justifies further investigation.

5   Conclusion

We have shown that it is possible to predict early and late cancer from a large panel of circulating miRNAs. Moreover, a pan-cancer model can predict this endpoint at least equally well as primary site-specific models of the same algorithm type. Furthermore, 185 circulating miRNAs were found significantly relevant in the final model. The five most relevant miRNAs were investigated with a literature search and most were found to be associated to different types of cancer.

While these results justify further research into predicting cancer stage from circulating miRNA, this study has the following limitations. On several of the primary sites, both the single-cancer and pan-cancer models do not perform better than random chance. Thus, in future research, ML models should be optimized for these cancers and more extensive data collection might further aid this development. Tissue of origin can be very reliable predicted from circulating miRNA [16]. Hence, stage prediction models might only be applied to samples where performance is sufficiently sensitive and precise. Furthermore, the predictive performance of the models presented in this study require further improvement to be clinically applicable. While stage prediction does not face the same challenge as screening tests (requirement of high PPV), the current PPV leaves room for improvement. Finally, the models should be further improved to predict individual stages (0-IV).

Despite these limitations, we have provided evidence that cancer stage can be predicted from circulating miRNA. Serum-based tests have the key advantage of being non-invasive and thus might bare less risks to patients than other test modalities. Additionally, they might become cost-effective alternatives.

6   Ethical Statement

Ethical approval was provided to the original study generating the data [16]. No additional approval of the present study is required.

Acknowledgments. This work was supported by the University of Skövde, Sweden under grants from the Knowledge Foundation (20170302, 20200014). The computations were enabled by resources provided by the National Academic Infrastructure for Supercomputing in Sweden (NAISS) at Chalmers University of Technology partially funded by the Swedish Research Council through grant agreement no. 2022–06725.

7   Appendix

Table 5.

Table 5. Hyperparameters used during grid search

Algorithm	Hyperparameters	Search Grid
LR	Penalty	["l1", "l2", "elasticnet"]
	Penalty Weight	[0.1, 0.5, 1.0]
	l1 Ratio (ElasticNet)	[0.5, 0.7]
	Max. iterations for optimization	[200]
	Number of Features	[100,500, 1000, 1500]
SVM	Penalty Weight	[0.5, 1.0, 1.5]
	Kernel	["linear", "rbf"]
	Number of Features	[100,500, 1000, 1500]
kNN	k	[5, 7, 11, 15, 31, 77]
	Weight	["uniform", "distance"]
	Distance Metric	["minkowski", "euclidean", "manhattan"]
	Number of Features	[50, 100, 500, 1000, 1500]
RF	Number of Trees	[100, 150, 200]
	Max Tree Depth	[6, 10, 15, 30, 50]
	Proportion of Samples	[0.3, 0.5, 0.8]
	Max. Features During Split	["sqrt"]
	Number of Features	[100, 500, 1000, 1500]
DNN	Hidden Dimensions	[32, 64]
	Number of Layers	[2, 3]
	Learning Rate	[1e-3, 1e-4, 1e-5]
	Weight Decay	[0.75, 1.25]
	Dropout	[0.5]
	Number of Epochs	[10, 30, 80]
	Batch Size	[32]

References

1. Abu Alfeilat, H.A., et al.: Effects of distance measure choice on k-nearest neighbor classifier performance: a review. Big Data **7**(4), 221–248 (2019)
2. Altmann, A., Toloşi, L., Sander, O., Lengauer, T.: Permutation importance: a corrected feature importance measure. Bioinformatics **26**(10), 1340–1347 (2010)
3. Bengio, Y., Courville, A.C., Vincent, P.: Unsupervised feature learning and deep learning: a review and new perspectives. CoRR, abs/1206.5538 (2012)
4. Benjamini, Y., Hochberg, Y.: Controlling the false discovery rate: a practical and powerful approach to multiple testing. J. R. Stat. Soc. Ser. B (Methodological) **57**(1), 289–300 (1995)
5. Boser, B.E., Guyon, I.M., Vapnik, V.N.: A training algorithm for optimal margin classifiers. In: Proceedings of the Fifth Annual Workshop on Computational Learning Theory, COLT 1992, New York, NY, USA, pp. 144–152. Association for Computing Machinery (1992)
6. Breiman, L.: Random forests. Mach. Learn. **45**(1), 5–32 (2001)

7. Elias, K.M., et al.: Diagnostic potential for a serum miRNA neural network for detection of ovarian cancer. Elife **6**, e28932 (2017)
8. Fix, E., Hodges, J.L.: Discriminatory analysis. nonparametric discrimination: consistency properties. Int. Stat. Rev. Revue Internationale de Statistique **57**(3), 238–247 (1989)
9. Galvão-Lima, L.J., Morais, A.H.F., Valentim, Ricardo A.M., Barreto, E.J.S.S.: mirnas as biomarkers for early cancer detection and their application in the development of new diagnostic tools. BioMedical Eng. OnLine **20**(1), 21 (2021)
10. Goodfellow, I., Bengio, Y., Courville, A.: Deep Learning. MIT Press, Cambridge (2016)
11. Hastie, T., Tibshirani, R., Friedman, J.: Data Mining, Inference, and Prediction. Springer, The Elements of Statistical Learning (2009). https://doi.org/10.1007/978-0-387-21606-5
12. Imaoka, H., et al.: Circulating microrna-1290 as a novel diagnostic and prognostic biomarker in human colorectal cancer. Ann. Oncol. **27**(10), 1879–1886 (2016)
13. Iorio, M.V., Croce, C.M.: Microrna dysregulation in cancer: diagnostics, monitoring and therapeutics. a comprehensive review. EMBO Mol. Med. **4**(3), 143–159 (2012)
14. LeCun, Y., Bengio, Y., Hinton, G.: Deep learning. Nature **521**(7553), 436–444 (2015)
15. Zhiyuan, L., et al.: MiR-31-5p is a potential circulating biomarker and therapeutic target for oral cancer. Mol. Ther. Nucleic Acids **16**, 471–480 (2019)
16. Matsuzaki, J., Kato, K., Oono, K., et al.: Prediction of tissue-of-origin of early stage cancers using serum miRNomes. JNCI Can. Spectrum **7**(1), (2022). pkac080
17. McPhail, S., Johnson, S., Greenberg, D., Peake, M., Rous, B.: Stage at diagnosis and early mortality from cancer in England. Br. J. Can. **112**(1), S108–S115 (2015)
18. Mi, B., Li, Q., Li, T., Liu, G., Sai, J.: High mir-31-5p expression promotes colon adenocarcinoma progression by targeting TNS1. Aging (Albany NY) **12**(8), 7480–7490 (2020)
19. Oshiro, T.M., Perez, P.S., Baranauskas, J.A.: How many trees in a random forest? In: Perner, P. (ed.) MLDM 2012. LNCS (LNAI), vol. 7376, pp. 154–168. Springer, Heidelberg (2012). https://doi.org/10.1007/978-3-642-31537-4_13
20. Pedregosa, F., et al.: Scikit-learn: machine learning in python. J. Mach. Learn. Res. **12**, 2825–2830 (2011)
21. Pulati, N., Zhang, Z., Gulimilamu, A., Qi, X., Yang, J.: HPV16+ -miRNAs in cervical cancer and the anti-tumor role played by mir-5701. J. Gene Med. **21**(11), e3126 (2019)
22. Siegel, R.L., Miller, K.D., Fuchs, H.E., Jemal, A.: Cancer statistics, 2022. CA Can. J. Clin. **72**(1), 7–33 (2022)
23. American cancer society. Survival rates for pancreatic cancer. https://www.cancer.org/cancer/pancreatic-cancer/detection-diagnosis-staging/survival-rates.html. Accessed 01 Apr 2023
24. Wang, Y.-N., Chen, Z.-H., Chen, W.-C.: Novel circulating microRNAs expression profile in colon cancer: a pilot study. Eur. J. Med. Res. **22**(1), 51 (2017)
25. Liyi, X., Cai, Y., Chen, X., Zhu, Y., Cai, J.: Circulating mir-1290 as a potential diagnostic and disease monitoring biomarker of human gastrointestinal tumors. BMC Cancer **21**(1), 989 (2021)
26. Yokoi, A., et al.: Integrated extracellular microrna profiling for ovarian cancer screening. Nat. Commun. **9**(1), 4319 (2018)

Gait Asymmetry Evaluation Using FMCW Radar in Daily Life Environments

Shahzad Ahmed⑩, Yudam Seo, and Sung Ho Cho$^{(\boxtimes)}$⑩

Department of Electronic Engineering, Hanyang University, Seoul 04763, Korea
{shahzad1,skyway21c,dragon}@hanyang.ac.kr

Abstract. Gait analysis plays a crucial role in medical diagnostics due to its ability to determine and quantify the patient's physical abilities and limitations. Unlike the other competing sensors, radar is capable of measuring human gait in non-contact fashion. In this paper, we present the extraction of six different gait parameters using Frequency Modulated Continuous Wave (FMCW) radar within five walking steps. The range-time and Doppler-time information are used to extract the parameters. The range-time information of FMCW radar yields the walking duration, walking time, and average walking velocity whereas, the velocity-time information yields pause time during walking, inter-step distance variation, and inter-step time variations. An Inertial Measurement Unit (IMU) is deployed as a ground-truth reference sensor to track the gait movement and a high correlation is found between radar and the reference sensor. Finally, as a use case example, gait parameters analysis is performed to detect asymmetric gait movement. Symmetric and asymmetric walking data is collected with radar and features analysis is performed which suggests that inter-step time and velocity variations contributes greatly in asymmetry detection.

Keywords: Gait Analysis · Wireless Sensor · FMCW Radar · Gait Speed

1 Introduction

Gait analysis deals with the measurement and evaluation of parameters associated with human walk [1]. Analysis of human walking pattern (gait) can provide useful information about the one's health and well-being. Several industries such as biomedical (therapeutics) and sport sciences have utilized gait analysis for a vast range of applications [2,3]. It serves as a diagnostic tool for several movement and neurodegenerative disorders such as musculoskeletal disorders and dementia. Particularly, an asymmetric gait, which is referred to as a variability occurring between the left and right step, is a huge indicator of such disorders.

This research was supported by National Research Foundation (NRF) of Korea. (NRF-2022R1A2C2008783).

S. Ahmed and Y. Seo—Contributed equally as co-first authors.

I. Rojas et al. (Eds.): IWBBIO 2023, LNBI 13919, pp. 116–127, 2023.
https://doi.org/10.1007/978-3-031-34953-9_9

For instance, individuals suffering from stroke appears to have a slow walking speed and an asymmetric gait due to lack of coordination between the two limbs [4].

Gait analysis consist of extracting several parameters [5] while walking. These parameters includes quantifying step-size, step-time, gait-speed total distance and number of steps. Gait analysis serves as a diagnostic tool for fall-risk analysis and cognitive decline [6]. It has also shown its effectiveness in measuring several other neurological disorders.

To record human movement with an aim of gait analysis, several sensors exist. These sensors can be classified into two groups which are wearable and wireless sensors. Wearable sensors include Inertial Measurement Unit (IMU) and dedicated gait-marking hardware systems. However, users are required to wear these sensors while moving around which additionally adds constrains. Also, the sensors are normally attached at different joints and wrongly placed sensor will affect the overall accuracy of gait analysis. Wireless sensors on the other hand, can extract gait parameters without being in contact with the human under consideration. Few studies have used Red Green Blue depth (RGBD) cameras for wireless gait analysis. However, these sensors are largely dependent on the environmental conditions such as lightning and mist. Unlike camera, radar sensor provides a non-contact and non-invasive way for gait analysis. For this purpose, recently several studies have recently exploited radar signals. Radar sensors can quantify minute vibrations such as human vital signs [7].

One of the earliest attempt for gait analysis using radar was presented in early 2000 s where a Continuous Wave (CW) radar was used to record the gait related Doppler signatures [8]. Recently, Wang et al. [9] proposed Frequency Modulated Continuous Wave (FMCW) radar-based gait analysis framework. However, the parameters were extracted using treadmill. The participants were not allowed to move freely, hence the system does not matches the real-world requirements. Similarly, gait abnormalities in unconstrained environment were discussed in [10]. However, the overall gait patterns are being considered by the authors and pre-defined classes are used as normal and abnormal classification instead of detecting the asymmetry. Few studies have recently considered gait symmetry analysis with radar. For instance, Seifert et al. [11,12] used Doppler radar for asymmetry analysis which considered velocity (Doppler) analysis. On the other hand, MIMO FMCW radar can provide the information about the range, velocity and angle simultaneously [13]. Despite several works, FMCW radar-based gait symmetry analysis in living environment is not discussed in details.

In this work, we used FMCW radar to extract several gait parameters in a real world environment. As a use-case example, these parameters are further used to measure gait symmetry. Unlike the previous studies using treadmills or other control factors, no restrictions were imposed in this work. We extracted gait parameters within six steps of walking which greatly resembles the real world scenarios. Consequently, the proposed system can be installed inside a room at a home or healthcare facility to (automatically) observe gait of elderly person. The experimental design is shown in Fig. 1 where the participants were asked to walk for six meters (round-trip) in front of radar.

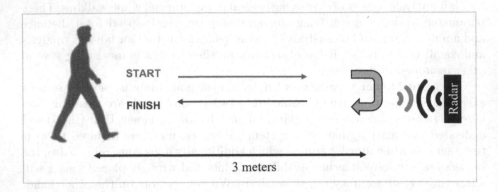

Fig. 1. Data capturing conditions for our gait analysis Experiment.

2   Methodology

2.1   Gait Parameters Definition

Figure 2 shows a normal gait cycle of a healthy human. The left and right step distance and deviation from the walking base must be similar for a symmetric gait. In most of the cases, a single step might not be enough to extract all gait parameter. A series of consecutive steps is often used to extract several different gait parameters.

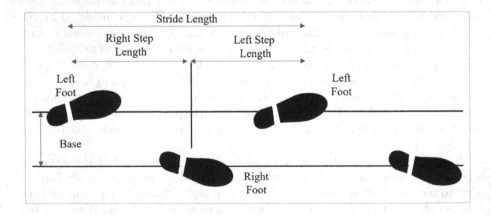

Fig. 2. Gait Parameters Definition

Gait parameters extraction systems based on radar have few limitations. For instance the time when the heel touches or leaves the ground may not be easy to detect via radar. As a result, finding the double support time with radar is not an easy task. Considering the available gait information available from the raw radar-returns, we (define and) extract below gait parameters:

1. Round trip distance: The starting and ending point of journey
2. Round trip time: Time taken to make the journey
3. Average walking velocity: For several consecutive steps
4. Pause between steps: The average pause between two consecutive steps.
5. Inter-step distance difference/variation: The distance traveled by several consecutive step is supposed to be consistent for symmetric gait.
6. Inter-step time difference/variation: The flight time for consecutive steps should also be same for symmetric gait.

We aim to intuitively quantify the aforementioned parameters for normal and abnormal walking patterns. As a use-case scenario, we calculated these parameters for symmetric and asymmetric walking patterns. Note that, not all of the parameters mentioned above contributed in symmetry analysis. Perhaps, inter-step distance and time variations are being used for asymmetry analysis. Next we define the methodology.

2.2   FMCW Radar Signal Processing

The FMCW radar transmits a signal whose frequency increases linearly with time and the signal consequently spans over a finite bandwidth (B). The signal transmitted by radar is reflected back from the target present within the radar-cross-section (RCS) and collected at the receiver [14]. The time delay between transmitted and received signal can be expressed as:

$$\tau = \frac{2(R + v_r t)}{c}, \tag{1}$$

where τ represents the time delay, R represents the distance, v_r represents the target (legs) velocity, t represents the instantaneous time, and c represents speed of light respectively. The received signal is mixed with the copy of transmitted signal to get the down-converted (low-frequency) signal known as the Intermediate-Frequency (IF) signal which can be expressed as [15]:

$$x_{IF}(t) = exp(j2\pi(f_c\tau + \frac{B}{2T}\tau^2)), \tag{2}$$

where f_c and T represents the carrier frequency and pulse duration respectively. Figure 3 represents the process of extracting the range and velocity of gait based on the IF signal presented above. As shown in Fig. 3, the received down-converted

IF signal is gathered for few seconds and saved as a 2-D matrix (termed as radar data-matrix) where the horizontal axis represents the time and vertical axis represents the distance respectively. After that, the Fast Fourier Transform (FFT) is taken along the horizontal axis and the peak in the FFT will represent the distance of the target. Another FFT is performed across the vertical direction which resolves the velocity of (gait) target. Both the distance and velocity are used to extract different gait parameters.

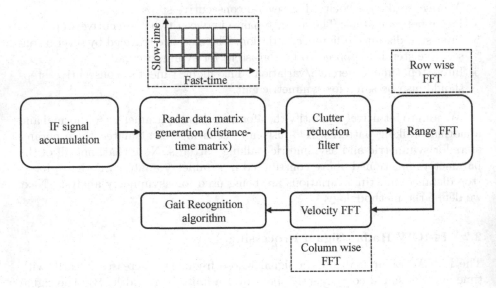

Fig. 3. Target range and velocity extraction.

2.3   Extraction of Radar-Based Gait Parameters

Figure 4 presents the gait recognition algorithm for the six gait parameters introduced in section one. The first three parameters are related to the overall movement instead of individual limb movement. We utilized range-FFT plot to extract these parameters. Peaks are collected from the range-FFT plot of each frame to form a vector containing peaks of each frame which is then passed through the outlier removal filter. This vector shows the range point detection from each frame. As shown in Fig. 4 (a), distance variations quantification can be used to extract first three parameters which are starting and the ending point of human walking, the average walking time, and the average walking velocity. Similarly, Fig. 4 (b) shows the velocity based features being extracted from human walking. When the velocity of gait is zero, step distance and step time can be visualized.

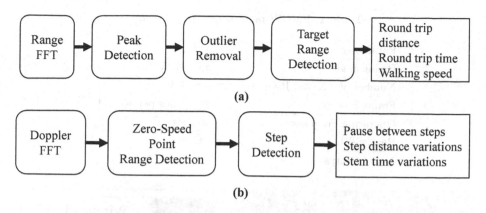

Fig. 4. Radar-based gait parameters extraction (a) Range (distance) based parameters and (b) Doppler (velocity) based parameters.

3   Experimental Setup and Equipment

The experimental setup to capture data and the reference IMU sensor installed on the shoes of human participant is shown in Fig. 5. The participants were asked to move in front of radar for a distance of 6 m. The participant moved three meters towards radar and turned around and moved back to the starting position. The core objective of this experimental environment is to imitate the real-world bedroom or hospital environment.

As shown in Fig. 5, the radar was installed at a height of 30 cm above the ground in order to reduce the torso movement as much as possible. In addition to that, as shown in Fig. 5, a pair of IMU sensors was installed at each shoes of the participant to capture the ground truth velocity information of each foot independently. Total six participants were invited for data capturing purpose to confirm that the proposed system can work independently on different individuals.

The technical specification of FMCW radar and IMU sensor are discussed in Table 1 and 2 respectively. We used IWR-6843 FMCW radar designed by Texas Instrument, USA. The radar sensor has previously been used in several studies related to human behavior quantification [13]. Note that we used only one transmitter and receiver pair in this study since one angle of arrival is being considered. The reference IMU sensor is a wireless sensor operating at a frequency of 2.4 GHz, having a frame rate of 100 frames per second.

Table 1. FMCW radar sensor parameters.

Parameter	Value
Radar model	TI - IWR683
Number of TX and RX antennas	2 and 4
Frame rate	20 frames per second
Operating Frequency	60 GHz
Bandwidth	4 GHz
No. of Chirps	64

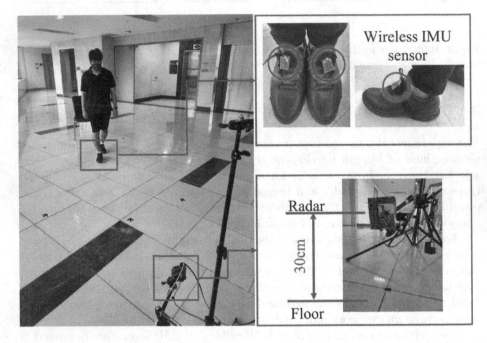

Fig. 5. Experimental setup showing the reference (IMU) sensor and FMCW radar used in this study.

Table 2. Reference IMU sensor parameters.

Parameter	value
IMU sensor chip	EBIMU24GV5
Frequency	2.4 GHz
Frames	100 per second
Velocity frames	100 per second
Absolute Orientation frames	100

4   Experimental Results

4.1   Raw Data Visualization

Figure 6 shows the radar data cube and extracted range range-time map for the data capturing scenario represented in Sect. 1. The horizontal axis represents the time in seconds whereas the vertical axis representing range in meters. We extracted gait parameters for both the cases when participant moves towards and away from radar. An outline rejection lowpass filter was also applied to create a smooth trajectory shown in Fig. 6(b).

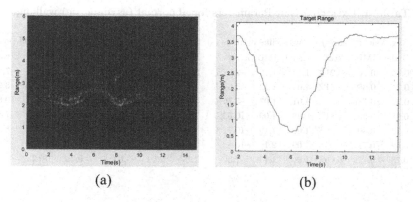

(a)   (b)

Fig. 6. Range-Time map of person moving towards and away from radar:(a) Radar data cube and (b) extracted range information.

4.2   Gait Parameters Extraction Results

Table 3 and 4 respectively represents the extracted gait parameters for normal (symmetric) and abnormal (asymmetric) walking cases. For each parameter, the parameter extracted using radar and IMU is presented along with their difference in terms of percentage. As shown in Table 3 and 4, the error between radar and ground truth is always within 10% range which suggests that radar is capable of gait parameters extraction.

Note that the total walking distance was fixed in both the scenarios since all the participants followed same trajectory. In addition to that, walking time does not contributed significantly in asymmetry analysis. Consequently, the analysis of remaining four parameters is discussed Table 3 and 4.

Table 3. Extracted Gait Parameters for normal (symmetric) walking.

Parti	Avg. Velocity (m/s)			Pause Time in Steps (s)			Step Distance Diff.(mm)			Step Time Diff.(ms)		
No.	Radar	IMU	Error	Radar	IMU	Error	Radar	IMU	Error	Radar	IMU	Error
P-1	1.10	1.05	4.65%	0.62	0.59	4.95%	0.81	0.86	−5.9%	9.02	8.84	2.01%
P-2	1.05	1.03	1.92%	0.65	0.66	−1.52%	0.94	0.89	5.46%	8.97	8.85	1.34%
P-3	0.94	0.98	−4.16%	0.71	0.68	4.31%	0.74	0.81	−9.03%	8.89	9.04	−1.67%
P-4	1.12	1.07	4.56%	0.59	0.62	−4.95%	0.76	0.82	−7.59%	9.11	9.17	−0.65%
P-5	1.08	1.11	−2.73%	0.57	0.55	3.57%	0.89	0.85	4.59%	9.04	9.15	−1.20%
P-6	1.11	1.01	9.43%	0.58	0.59	−1.70%	0.91	0.89	2.22%	9.31	9.22	0.97%
Avg.	1.07	1.04	2.28%	0.62	0.615	0.78%	0.84	0.85	−1.72%	9.06	9.05	0.13%

Table 4. Extracted Gait Parameters for abnormal (asymmetric) walking.

Parti	Avg. Velocity (m/s)			Pause Time in Steps (s)			Step Distance Diff.(mm)			Step Time Diff.(ms)		
No.	Radar	IMU	Error	Radar	IMU	Error	Radar	IMU	Error	Radar	IMU	Error
P-1	0.70	0.76	−8.21%	1.13	1.16	−2.62%	2.13	2.26	−5.92%	9.57	10.23	−7.92%
P-2	0.71	0.68	4.31%	1.08	1.12	−3.63%	1.88	2.01	−6.68%	9.61	10.51	−8.94%
P-3	0.72	0.79	−9.27%	1.01	1.07	−5.76%	1.85	1.96	−5.77%	9.11	10.16	−10.88%
P-4	0.81	0.85	−4.81%	0.98	1.09	−10.62%	1.92	1.21	−8.95%	9.36	10.12	−7.80%
P-5	0.73	0.80	−9.15%	1.06	1.18	−10.71%	1.96	1.90	3.10%	9.16	9.98	−8.56%
P-6	0.69	0.74	−6.99%	1.09	2.1	−10.43%	2.06	2.01	2.45%	9.51	10.26	−7.53%
Avg.	0.73	0.77	−5.69%	1.06	1.14	−7.3%	1.97	2.01	−3.63	9.36	10.23	−8.62%

Figure 7(a) and (b) respectively represents extracted gait parameters for normal (symmetric) and abnormal (asymmetric) walking pattern. The radar measured range information for each step is plotted alongside the reference sensor velocity data to show the distance variations over time. The brown sine alike waveform represents the velocity extracted using IMU sensor with each cycle representing one step whereas the blue line shows the corresponding range (distance) variations with time. As shown in Fig. 7, the inter-step distance and the inter-step time variation contributed greatly in asymmetry analysis. While wearing the knee orthosis and walking in asymmetric fashion, the velocity was lower and the pause time between steps was higher in comparison to the normal walking.

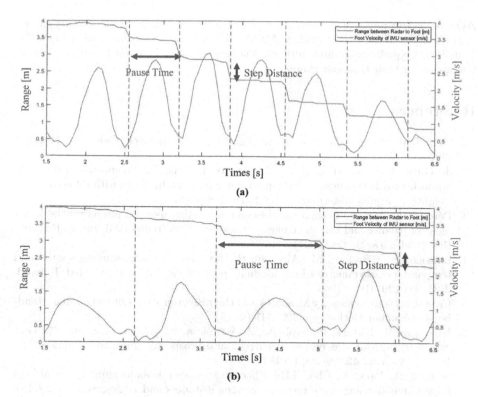

Fig. 7. Gait parameters extraction using range information for (a) normal case and (b) abnormal walking case

5   Conclusion and Further Work

This study presents the gait parameters extraction framework based on FMCW radar in an unconstrained environment. Multiple participants were involved in the study in the interest of robustness. The error between radar and ground truth values was within a range of 10% which confirms that radar is capable of extracting gait parameters. In addition to that, an example of symmetric and asymmetric walking pattern detection is also presented.

In future, the pre-clinical trial must be performed to verify the effectiveness of proposed algorithm on real patients and elderly individuals. In addition to that, the contribution of each feature individually in asymmetry analysis must be analyzed.

Institutional Review Board (IRB) Statement: This study was approved by the local ethics committee of Hanyang University Hospital under IRB number HYU-IRB-202101-015-2.

Acknowledgements. This research was supported by National Research Foundation (NRF) of Korea. (NRF-2022R1A2C2008783). Shahzad Ahmed and Yudam Seo contributed equally as co-first authors. The authors are thankfull to all the human volunteers for their time and effort.

References

1. Silva, L.M., Stergiou, N.: The basics of gait analysis. Biomech. Gait Anal **164**, 231 (2020)
2. de Oliveira, K., Clayton, H.M., dos Santos Harada, É.: Gymnastic training of hippotherapy horses benefits gait quality when ridden by riders with different body weights. J. Equine Veterinary Sci. **94**, 103248 (2020)
3. Pérez-de la Cruz, S.: Comparison between three therapeutic options for the treatment of balance and gait in stroke: a randomized controlled trial. Int. J. Environ. Res. Health **18**(2), 426 (2021)
4. Lauziere, S., Betschart, M., Aissaoui, R., Nadeau, S.: Understanding spatial and temporal gait asymmetries in individuals post stroke. Int. J. Phys. Med. Rehabil. **2**(3), 201 (2014)
5. Cabral, S.: Gait Symmetry Measures and their Relevance to Gait Retraining. Handbook of Human Motion, pp. 429–447 (2018)
6. Viteckova, S., Kutilek, P., Svoboda, Z., Krupicka, R., Kauler, J., Szabo, Z.: Gait symmetry measures: a review of current and prospective methods. Biomed. Sig. Process. Control **42**, 89–100 (2018)
7. Ahmed, S., Park, J., Cho, S.H.: Effects of receiver beamforming for vital sign measurements using fmcw radar at various distances and angles. Sensors **22**(18), 6877 (2022)
8. Geisheimer, J.L., Marshall, W.S., Greneker, E.: A continuous-wave (cw) radar for gait analysis. In: Conference Record of Thirty-Fifth Asilomar Conference on Signals, Systems and Computers (Cat. No. 01CH37256), vol. 1, pp. 834–838. IEEE (2001)
9. Wang, D., Park, J., Kim, H.-J., Lee, K., Cho, S.H.: Noncontact extraction of biomechanical parameters in gait analysis using a multi-input and multi-output radar sensor. IEEE Access **9**, 138496–138508 (2021)
10. Seifert, A.K., Amin, M.G., Zoubir, A.M.: Toward unobtrusive in-home gait analysis based on radar micro-doppler signatures. IEEE Trans. Biomed. Eng. **66**(9), 2629–2640 (2019)
11. Seifert, A.-K., Zoubir, A.M., Amin, M.G.: Detection of gait asymmetry using indoor doppler radar. In: 2019 IEEE Radar Conference (RadarConf), pp. 1–6. IEEE (2019)
12. Seifert, A.-K., Reinhard, D., Zoubir, A.M., Amin, M.G.: A robust and sequential approach for detecting gait asymmetry based on radar micro-doppler signatures. In: 2019 27th European Signal Processing Conference (EUSIPCO), pp. 1–5. IEEE (2019)
13. Ahmed, S., Park, J., Cho, S.H.: Fmcw radar sensor based human activity recognition using deep learning. In: 2022 International Conference on Electronics, Information, and Communication (ICEIC), pp. 1–5. IEEE (2022)

14. Ahmed, S., Kallu, K.D., Ahmed, S., Cho, S.H.: Hand gestures recognition using radar sensors for human-computer-interaction: a review. Remote Sens. **13**(3), 527 (2021)
15. Ahmed, S., Kim, W., Park, J., Cho, S.H.: Radar-based air-writing gesture recognition using a novel multistream CNN approach. IEEE Internet Things J. **9**(23), 23869–23880 (2022)

Biomedical Computing

Whole Tumor Area Estimation in Incremental Brain MRI Using Dilation and Erosion-Based Binary Morphing

Orcan Alpar[1] and Ondrej Krejcar[1,2(✉)]

[1] Faculty of Informatics and Management, Center for Basic and Applied Research, University of Hradec Kralove, Rokitanskeho 62, Hradec Kralove 500 03, Czech Republic
{orcan.alpar,ondrej.krejcar}@uhk.cz
[2] Malaysia Japan International Institute of Technology (MJIIT), Universiti Teknologi Malaysia, Kuala Lumpur, Malaysia

Abstract. Magnetic resonance imaging (MRI) technology is rapidly advancing and three-dimensional (3D) scanners started to play an important role on diagnosis. However, not every medical center has access to 3D magnetic resonance imaging (MRI) devices; therefore, it is safe to state that the majority of MRI scans are still two-dimensional. According to the setup values adjusted before any scan, there might be consistent gaps between the MRI slices, especially when the increment value exceeds the thickness. The gap causes miscalculation of the lesion volumes and misjudgments when the lesions are reconstructed in three-dimensional space due to excessive interpolation. Therefore, in this paper, we present the details of three types of conventional morphing methods, one dilation-based and two erosion-based, and compare them to figure out which one provides better solution for filling up the gaps in incremental brain MRI. Among three types of morphing methods, the highest average dice score coefficient (DSC) is calculated as %91.95, which is obtained by the multiplicative dilation morphing method for HG/0004 set of BraTS 2012.

Keywords: morphing · brain MRI · erosion · dilation · increment

1 Introduction

Emerging 3D scanning devices provide better visualization of the lesions in 3D space and thus better volumetric computation; while it is not possible to claim that every center employed these devices. Despite being the most common method, 2D scans have some drawbacks indeed, which could be collected in two groups; pre-scan and post scan. The pre-scan involves the setup step, which is executed by a radiologist who predetermines the required slice thickness and increment per scan sequence; while the post-scan issues emerge due to 2D representation and built-in device operations. In setup, the radiologists or the technologists need to determine several parameters affecting the scan results; while the increment and thickness values are more important than the other image specifications.

© The Author(s), under exclusive license to Springer Nature Switzerland AG 2023
I. Rojas et al. (Eds.): IWBBIO 2023, LNBI 13919, pp. 131–142, 2023.
https://doi.org/10.1007/978-3-031-34953-9_10

The increment value represents the physical motion of the MRI devices per slice which has a rational interval; while this value might be drastically increased if the total scan time is needed to be minimum. The slice thickness is the height of the 3D prisms scanned by the device; while the voxels in the prisms are averaged and represented by pixels in the post-scan to generate 2D MR images. Theoretically, as the thickness value increases, the tissues captured inside of the prisms per increment will also increase; therefore, after averaging, the detail level and the spatial resolution would decrease due to blurriness. Sometimes, the increment value $(t + g)$, deliberately or not, exceeds the thickness (t), causing gap (g) between the slices, as presented in Fig. 1.

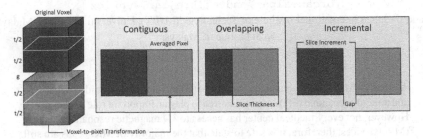

Fig. 1. MRI properties and built-in averaging process

Given that a voxel is averaged while creating the corresponding pixel, the averaged pixel needs to be at the very center of the voxel; therefore, the distance between two pixels is always equal to the increment value $(t/2 + g + t/2 = t + g)$, not the gap. Therefore, the loss of information is not only due to the gap value; but also, due to the thickness, itself. What we propose in this research paper is filling the distance between two genuine ground truth (GT) images by morphing to generate three imaginary slices between. The methods, we will be presenting and comparing, are one common dilation morphing against the two erosion-based variants in this paper using synthetic images in the methods section and real examples in the experiments section.

Morphing basically is a morphological operation between two shapes, mostly for estimating the shapes between. As a branch of geometrics, the source and the target images are usually synthetically produced or generic shapes, like: triangles, letters and trees in [1] and animals, continent and country shapes in [2]. However, what we focus on here are the real amorphic images seen in the MRI slices for filling the gaps between the slices. Provided that two MRI slices have two identical filled circles of a perfect sphere of a tumor at the same coordinates, the tumor regions would be linearly interpolated that would form a perfect rectangle in lateral cross-section without any morphing operation. Conversely, if two images are morphed with the morphing size 1, the resulting image would look like as a hectogon instead. If the morphing size increases, the image will start to approach the perfect sphere as seen in Fig. 2.

In Fig. 2, the circle represents the lateral cross-section of a perfect sphere tumor. Assuming that two MRI slices, red and blue, consist of same portions of the 3D tumor that would appear as filled circles in the 2D slices. Without morphing, the volumetric estimations would be held by the 3D reconstruction of these two circles which will

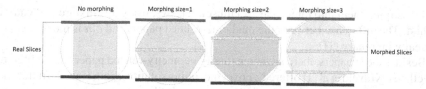

Fig. 2. Illustration of the cross-sections of a tumor changing by morphing

form a cylinder having a cross section of a rectangle. If a morphed image is placed in the middle of these circles, the 3D reconstruction will be smoother. For reaching the perfect sphere, the morphing size could be increased to form octagons and decagons and more-sided polygons.

Despite the lack of a proper image set consisting of incremental MRI slices and corresponding GT images to compare the superiority, we borrowed the genuine GTs from BraTS 2012 [3, 4] and placed n^{th} and $n + 4^{th}$ GT images as the source and the target while estimating $n + 1^{th}$, $n + 2^{th}$ and $n + 3^{th}$ GT slices to calculate the dice score coefficients (DSC), as presented in Fig. 3. The DSC, or the F1 score, is one of the main performance area-wise metrics which is calculated by 2x the intersection area between GT and segmented image over the sum of GT and the segmented image areas.

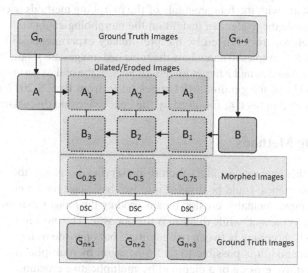

Fig. 3. The basics of the morphing protocols

There basically are a few important papers published in the recent years on geometrical morphing methods. One of the medicine-related papers belong to Cheng et al [5] on 3D reconstruction of livers using morphed 2D computed tomography (CT) images. They extracted the features by a curvature method from the 2D CTs and mapped the key-points in the 3D space using a unit sphere method. They executed the voxelization process by reconstructing the morphed images with the genuine ones using Hermite interpolation. Moreover, Chavez et al [6] dealt with boundary-based morphing, disregarding the inside of the shapes. Their purpose is evaluation the breast lesions found in terahertz imaging and classifying them into tumor, fat and muscle classes, using a t-mixture and gaussian method. The difference between the medicine-related papers and our is the focus; since we deal with binary GTs.

Besides these papers, there are some totally geometry-related papers in the literature as well. van Kraveld et al [1] offered a novel morphing solution using Hausdorff distances between two abstract shapes. They also employed dilation algorithm in a section of their paper to show the dilation morphing using Hausdorff distances. Furthermore, [7] focused on binary morphing between two shapes, like we did. However, they presented a geometrical Delaunay triangulation algorithm to generated morphed images; while their main performance criterion is the similarity between the morphed images with the source and target images. Although, both of these papers might be considered as working with binary images; these images are abstract; while we morph two genuine GT images to estimate the real GTs between.

Despite the limited relevancy, there are some state-of-art papers published in the recent years on segmentation with various kernels. All papers might be classified by means of the investigation areas and diseases in general; such as: brain tumors [8–11], breast tumors [12–15], multiple sclerosis [16–19], skin cancer [20] [21] and kidney/liver masses [22, 23]. There is no instance of segmentation in this paper; while the performance evaluation steps using the dice score coefficients for benchmarking the methods proposed in this paper are totally similar.

The paper starts with the fundamentals of the morphing methods used in this paper in Sect. 2 to provide the readers an insight on the morphological operations. The mathematical foundations are followed by the preliminary experiments with the synthetic geometrical shape examples to show how the morphed images are produced during each process between a source and a target image. In Sect. 3, the real experiment is presented using the real GTs of the genuine MRI slices taken from BraTS 2012 database. The results are presented in Sect. 4, followed by the conclusion and discussion in Sect. 5.

2   Morphing Methods

Among many other alternatives, like structure-preserving morphing, the topic of morphing between two shapes, could be divided into two basic categories: erosion-based and dilation-based. Fundamentally, erosion means eroding the images making them smaller within a ratio or a constant; while dilation is the opposite. The most intriguing and interesting example is morphing between two identical but opposite triangles. According to the morphing method, it is possible to find a hexagon by multiplicative dilation or an hourglass by additive erosion or a diamond by multiplicative erosion.

The morphed images are generated by union or intersection of the imaginary slices created by erosion or dilation using Hausdorff distance. The directional Hausdorff distance from set A to B is defined as

$$d_{\vec{H}}(A, B) = \sup_{a \in A, b \in B} \inf d(a, b) \qquad (1)$$

and B to A, as:

$$d_{\vec{H}}(B, A) = \sup_{a \in A, b \in B} \inf d(b, a) \qquad (2)$$

where d represents the Euclidean distance. The undirected distance is the maximum distance between two directional distances, namely:

$$d_H(A, B) = \max\left(d_{\overrightarrow{H}}(A, B), d_{\overrightarrow{H}}(B, A)\right) \tag{3}$$

The morphed images are generated by:

$$C_\delta = (A \oplus D_\delta) \cap (B \oplus D_{1-\delta}) \tag{4}$$

for multiplicative morphing systems, and by:

$$C_\delta = (A \oplus D_\delta) \cup (B \oplus D_{1-\delta}) \tag{5}$$

for additive systems, where $\oplus$ represents Minkowski sum and D_δ is the resulting area by dilation/erosion defined by the $d_H(A, B)$, in a very basic and comprehensible manner. The examples with two opposite triangles are presented in Table 1 to show the differences of these morphing methods while morphing two synthetic images.

Table 1. Synthetic Morphing Examples

	Multiplicative Dilation	Additive Erosion	Multiplicative Erosion
A			
$C_{0.25}$			
$C_{0.5}$			
$C_{0.75}$			
B			

These preliminary experiments in Table 1 are conducted using the triangles inspired from the previous shape-based morphing research papers to show the possible outcomes. All morphing operations are executed by the square morphing element in parallel with these papers to reach their outputs; therefore, no rounding took place. Given that a binary image is consisting of zeros and ones in the matrix form, the resulting areas D_δ are generated by adding or subtracting the matrices with a morphological structuring element. The morphed images are generated by a single square dilation/erosion element, which is explained in Fig. 4.

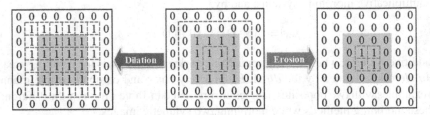

Fig. 4. Dilation/Erosion with square morphological structuring element

The middle image is the original GT with ones representing the whole tumors; while the left-hand side is the dilated and the right-hand side is the eroded image by square morphing element. The gray background section represents the whole tumor in the original GT; while the dotted regions are the results of applying dilation/erosion by $D_{0.25} = 1$, given the morphing size is 3. As seen from the images, the overall D value represent the whole morphing while the D_δ is only one step further from the source image.

Although we used square element in the triangle example for better comprehension in Table 1 and in Fig. 4, the main experiment in this paper is conducted with spherical element, which would fit the tumor shapes better. The idea beneath the spherical or circular morphological structuring element is similar indeed; while a spherical shape with the origin at the center of gravity is created instead of a square.

3   Experiments

Until this point, the preliminary experiments using two synthetic opposite triangles are conducted by square element disregarding the Hausdorff distance to show the readers how two triangles could be morphed and what kind of output images could be generated. Conversely, in this section, we only focus on the real GT images and morphed them by taking two genuine GTs, G_n and G_{n+4}, to estimate G_{n+1}, G_{n+2} and G_{n+3} with the exact Hausdorff distances. The GT data is taken from the BraTS 2012 dataset, more specifically from HG/0004, for the preliminary experiments. The image #86 is taken as the source and the #90 as the target to estimate the #87, #88 and #89 with the highest DSC possible, as presented in Table 2.

However, the morphing algorithms don't involve any adjustable parameters; since they strictly depend on the Hausdorff distance. As small as expected, the Hausdorff distance $d_H(A, B) = 7.6158$ is found prior to any further process and the morphed images are created according to this value; while $d_H(A, B) = 7.6158$ was calculated for the triangles However, this non-integer value cannot be used in morphing; therefore, rounded in each step so that:

$$d_H(A, C_{0.25}) = d_H(C_{0.25}, C_{0.5}) = d_H(C_{0.5}, C_{0.75}) = d_H(C_{0.75}, B) = 2 \qquad (6)$$

Table 2. Preliminary experiment using HG/0004 set between the images #86 and #90

As clearly seen in the Table 2, the multiplicative dilation (dilation with intersection) algorithm tends to enlarge the images; conversely, multiplicative erosion (erosion with intersection) shrinks them. The most structure preserving one seems to be additive erosion (erosion with union) in the middle column. From the preliminary experiments, it is clear that the multiplicative erosion shown on the right column is the least precise one due to the omnidirectional proliferation nature of the tumors in general. The performance metrics, which hare also used in segmentation, also prove this mismatch; while the other methods seem to have a potential, as summarized in Table 3.

Table 3. Performance evaluation of the preliminary experiment HG/0004, #86 - #90

%	Sin.	SE	SP	FPR	FNR	PPV	ACC	JAC	DSC	Ave. DSC
Multiplicative Dilation	$C_{0.25}$	99.85	99.63	0.37	0.15	91.55	99.64	91.43	95.52	
	$C_{0.5}$	100	99	1	0	80.56	99.04	80.56	89.23	**91.52**
	$C_{0.75}$	99.51	99.06	0.94	0.49	81.82	99.07	81.49	89.8	
Additive Erosion	$C_{0.25}$	76.52	99.99	0.01	23.48	99.81	99.08	76.41	86.63	
	$C_{0.5}$	72.61	99.98	0.02	27.39	99.21	98.88	72.19	83.85	**87.87**
	$C_{0.75}$	87.15	100	0	12.85	100	99.47	87.15	93.13	
Multiplicative Erosion	$C_{0.25}$	55.42	100	0	44.58	100	98.26	55.42	71.32	
	$C_{0.5}$	55.51	100	0	44.49	100	98.22	55.51	71.39	**67.22**
	$C_{0.75}$	41.81	100	0	58.19	100	97.62	41.81	58.96	

In the Table 3, the column titles are: SE is sensitivity (TPR), SP is specificity (TNR), TPR is true positive rate, TNR is true negative rate, FPR is false positive rate, FNR is false negative rate, PPV is positive predictive value (Precision), ACC is accuracy, JAC is Jaccard index and DSC is the dice score coefficient. Since our main purpose is finding the highest DSC possible, only the DSC columns will be our concern, though we presented every metrics for the readers. Each first row in any title colored in different colors represents the single results for $C_{0.25}$, the second, $C_{0.5}$ and the third, $C_{0.75}$.

Theocratically, there might be more than one performance criteria, even if only the DSC is taken into consideration, which are the average DSC and the DSC of farthest morph, $C_{0.5}$. Actually, both criteria could be meaningful based on the context; while we are trying to fill the gaps with the morphed slices; therefore, we will take the average DSC into the account. According to the preliminary results, the dilation-based morphing is apparently the dominant strategy among others with the highest average DSC = 91.52.

The other columns consist of only some complementary metrics, which are automatically computed by our framework. For instance, the ACC has no meaning in segmentation-based systems which indeed represents the overall true pixels in the segmented image with respect to the whole image. Therefore, it cannot be stated the accuracy of these methods are similar; since the accuracy is a special metric; not the accuracy of the segmentation. This experiment is repeated for the whole set to reveal the superior strategy and the results are presented in the following section.

4 Results

For the rest of the experiment, all images are in the same set are morphed by same methods to understand the overall capability of the methods while morphing all genuine slices. The GT images in the set all are is 160 px × 216 px grayscale images; while all of them are binarized for binary morphing. A total number of 70 images, containing at least one lesion, from #81 to #150 are morphed; yet in algorithmic perspective, the loop size is only 66; since the last morph is between #146 and #150. In parallel with the preliminary experiment, the following results are obtained, in Table 4.

Table 4. Performance evaluation of the whole experiment HG/0004, #81 - #150

%	Ave.	SE	SP	FPR	FNR	PPV	ACC	JAC	DSC	Ave. DSC
Multiplicative Dilation	$C_{0.25}$	98.52	99.35	0.65	1.48	87.71	99.34	86.71	92.46	
	$C_{0.5}$	98.2	99.03	0.97	1.8	84.04	99.03	82.99	90.15	**91.95**
	$C_{0.75}$	99	99.33	0.67	1	88.59	99.34	87.95	93.25	
Additive Erosion	$C_{0.25}$	84.28	99.99	0.01	15.72	98.47	99.36	84.06	89.76	
	$C_{0.5}$	80.06	99.98	0.02	19.94	99.09	99.07	79.71	87.81	**89.71**
	$C_{0.75}$	85.83	99.98	0.02	14.17	99.18	99.34	85.27	91.56	
Multiplicative Erosion	$C_{0.25}$	59.03	100	0	40.97	96.96	98.13	59.03	70.15	
	$C_{0.5}$	63.65	100	0	36.35	95.45	98.34	63.65	74.12	**72.31**
	$C_{0.75}$	61.92	100	0	38.08	92.42	98.15	61.92	72.67	

Based on the overall results, the multiplicative dilation, which actually is the most common method, has the superiority over others; despite the unavoidable expansion of the tumor areas in each morph. The multiplicative erosion, totally oppositely, generates smaller images in each step lowering the DSC very significantly. In more detail, the overall average DSC are calculated as 91.95%, 89.71% and 72.31% for the methods respectively; while the overall average standard deviations are 7.51%, 12.17% and %25.61; which also support the reliability of the multiplicative dilation method.

Generally speaking, each method could be very beneficial in different circumstances based on the expected shapes of lesions. A tumor is expected to grow omnidirectionally; so that a convexity, strict or not, might emerge in each cross section for each axis; therefore dilation-based morphing might be beneficial. On the other hand, MS is a demyelination disease that attacks the myelin covers of the nerves; therefore, structure-preserving morphing, like additive erosion, by might be so useful for these cases.

The framework is fully optimized and automated after the preliminary experiments that we also extracted all the morphed images. All codes are written in MATLAB 2019a syntax and executed on Dell Inspiron 15 with 4 processors and Intel Core i7-8750 CPU @ 2.20 Ghz and 1 6GB RAM. The average process time is calculated as 9.75 s including Hausdorff distance calculations and morphing and storing all 594 slices as 300 DPI TIFF files without showing the 9 morphed images per step. If the morphed slices is necessary to be shown on the screen, the operation time drastically increases up to 433 s for the whole set of images.

5 Conclusions and Discussions

In this paper, we introduced three binary morphing methods to offer solutions for recovery of information loss in incremental brain MRI. These methods basically are dilation and erosion using intersection and union of imaginary areas while morphing between two GTs. The purpose of this research is finding the optimal morphing strategy that will reach the highest DSC possible after comparing the original GTs with the corresponding morphed images.

The GTs are selected from the BraTS 2012 dataset; while they are grayscale images consisting of edema, enhancing and non-enhancing tumors and necrosis. However, we firstly binarized them to form binary GT images representing the whole tumor areas. The binary GTs are selected from a subset with a distance of three GTs, as the source and target images; while the images between these images are estimated by multiplicative dilation, multiplication erosion and additive erosion methods. Consequently, the highest DSC among all methods is calculated as %91.95 using the multiplicative dilation (dilation with intersection) morphing method for HG/0004 set of BraTS 2012.

The outcomes of this research will mainly be very beneficial for filling the gaps of incremental brain MRIs. On the other hand, the morphing strategies are very useful in 3D reconstruction of the lesions from 2D MRI slices, if more precision is needed. Moreover, the morphed images would lead to better volumetric estimation of the lesions. The 3D reconstruction and volume estimations of the lesions might be unrealistic due to excessive interpolation; while the morphed images would be placed between the genuine slices for smoothing the interpolation. These are the basic advantages of using a morphing algorithm for increasing the number of slices by morphed ones.

We are presenting and comparing three different solutions; therefore, it is not possible to state any overall weakness. If one of the morphing methods doesn't fit to the GT images of a dataset, other method could be tried for better average DSC. However, according to the image size and selection of morphological structuring element of dilation/erosion algorithm, the operation time might drastically increase. The element we used in Table 1 is square, which takes significantly shorter time, approximately 5 s for one morphing sequence; however, the spherical element takes at least 25 s for the same sequence.

If we consider the strengths and weaknesses of each method separately, there are some findings worth-mentioning. As seen from the experiments, the multiplicative dilation methods, which is the most common one in the literature, fits the tumor formation and proliferations better; while it distorts and deteriorates the GTs by expanding the areas. Given this kernel, this method perfectly fits the low-grade tumors rather than high-grades, which have smoother and expanding 3D shapes. Conversely, the multiplicative erosion method always shrinks the GTs, which is not a proper way to morph the tumors. The one and only structure-preserving algorithm, the additive erosion method, could be very beneficial for some datasets indeed; since this method doesn't deteriorate the original GT shapes.

This paper is solely based on the existing and conventional methods for binary morphing; therefore, no relevant and comparable studies could be found in the literature for the dataset we analyzed. In addition, this kind of comparison hasn't been studied for the same images; which could be considered as the novelty of the paper. The clinical relevance of this paper could only be of concern, when the morphing is needed between the MRI slices with gaps. Given the differences of the mathematical basis of three morphing methods, only "additive erosion" is structure-preserving; while the others are eventually distorting the binary GTs.

As the future research ideas, these methods could be very useful if employed as benchmarking algorithms, when a new morphing framework is presented. The application area is brain MRIs in this paper; yet could be easily applicable to any other areas and diseases, if a morphing algorithm is needed for 3D reconstruction. It is always possible to achieve better DSC results by changing the distance calculation method, if the Hausdorff distance is not suitable for any set. On the other hand, since the slices are the planes averaged from the prisms, the distance of the lesion would be very small regardless of the algorithm.

Acknowledgment. The work and the contribution were also supported by the SPEV project, University of Hradec Kralove, Faculty of Informatics and Management, Czech Republic (ID: 2102–2023), "Smart Solutions in Ubiquitous Computing Environments". We are also grateful for the support of student Michal Dobrovolny in consultations regarding application aspects.

References

1. van Kreveld, M., Miltzow, T., Ophelders, T., Sonke, W., Vermeulen, J.L.: Between shapes, using the Hausdorff distance. Comput. Geom. **100**, 101817 (2022)
2. Bouts, Q.W., Kostitsyna, I., van Kreveld, M., Meulemans, W., Sonke, W., Verbeek, K.: Mapping polygons to the grid with small Hausdorff and Fréchet distance. In: 24th Annual European Symposium on Algorithms (ESA 2016), pp. 22:1–22:16 (2016)
3. Kistler, M., Bonaretti, S., Pfahrer, M., Niklaus, R., Buchler, P.: The virtual skeleton database: an open access repository for biomedical research and collaboration. J. Med. Internet Res. **15**(11), e245 (2013)
4. Menze, B.H., et al.: The multimodal brain tumor image segmentation benchmark (BRATS). IEEE Trans. Med. Imaging **34**(10), 1993–2024 (2014)
5. Cheng, Q., Sun, P., Yang, C., Yang, Y., Liu, P.X.: A morphing-Based 3D point cloud reconstruction framework for medical image processing. Comput. Meth. Program. Biomed. **193**, 105495 (2020)

6. Chavez, T., Bowman, T., Wu, J., Bailey, K., El-Shenawee, M.: Assessment of terahertz imaging for excised breast cancer tumors with image morphing. J. Infrared Millimeter Terahertz Waves **39**(12), 1283–1302 (2018)
7. Cheddad, A.: Structure preserving binary image morphing using Delaunay triangulation. Pattern Recogn. Lett. **85**, 8–14 (2017). https://doi.org/10.1016/j.patrec.2016.11.010
8. Alpar, O., Dolezal, R., Ryska, P., Krejcar, O.: Nakagami-Fuzzy imaging framework for precise lesion segmentation in MRI. Pattern Recogn. **128**, 108675 (2022)
9. Ayadi, W., Elhamzi, W., Charfi, I., Atri, M.: Deep CNN for brain tumor classification. Neural Process. Lett. **53**(1), 671–700 (2021)
10. Alpar, O., Dolezal, R., Ryska, P., Krejcar, O.: Low-contrast lesion segmentation in advanced MRI experiments by time-domain Ricker-type wavelets and fuzzy 2-means. Appl. Intell. (2002a). https://doi.org/10.1007/s10489-022-03184-1
11. Alpar, O.: A mathematical fuzzy fusion framework for whole tumor segmentation in multimodal MRI using Nakagami imaging. Expert Syst. Appl. **216**, 119462 (2023)
12. Singh, V.K., et al.: Breast tumor segmentation and shape classification in mammograms using generative adversarial and convolutional neural network. Expert Syst. Appl. **139**, 112855 (2020)
13. Alpar, O.: Nakagami imaging with related distributions for advanced thermogram pseudo-colorization. J. Therm. Biol **93**, 102704 (2020)
14. Pramanik, S., Banik, D., Bhattacharjee, D., Nasipuri, M., Bhowmik, M.K., Majumdar, G.: Suspicious-region segmentation from breast thermogram using DLPE-based level set method. IEEE Trans. Med. Imaging **38**(2), 572–584 (2018)
15. Kumar, V., et al.: Automated and real-time segmentation of suspicious breast masses using convolutional neural network. PLoS ONE **13**(5), e0195816 (2018)
16. Nair, T., Precup, D., Arnold, D.L., Arbel, T.: Exploring uncertainty measures in deep networks for multiple sclerosis lesion detection and segmentation. Med. Image Anal. **59**, 101557 (2020)
17. Alpar, O., Krejcar, O., Dolezal, R.: Distribution-based imaging for multiple sclerosis lesion segmentation using specialized fuzzy 2-means powered by Nakagami transmutations. Appl. Soft Comput. **108**, 107481 (2021)
18. Aslani, S., et al.: Multi-branch convolutional neural network for multiple sclerosis lesion segmentation. Neuroimage **196**, 1–15 (2019)
19. Billast, M., Meyer, M.I., Sima, D.M., Robben, D.: Improved inter-scanner MS lesion segmentation by adversarial training on longitudinal data. In: International MICCAI Brainlesion Workshop (2019)
20. Khouloud, S., Ahlem, M., Fadel, T., Amel, S.: W-net and inception residual network for skin lesion segmentation and classification. Appl. Intell. 1–19 (2021). https://doi.org/10.1007/s10489-021-02652-4
21. Tan, T.Y., Zhang, L., Lim, C.P.: Adaptive melanoma diagnosis using evolving clustering, ensemble and deep neural networks. Knowl. Based Syst. **187**, 104807 (2020)
22. Corbat, L., Nauval, M., Henriet, J., Lapayre, J.C.: A fusion method based on deep learning and case-based reasoning which improves the resulting medical image segmentations.. Expert Syst. Appl. 113200 (2020)
23. Song, L.I., Geoffrey, K.F., Kaijian, H.E.: Bottleneck feature supervised U-Net for pixel-wise liver and tumor segmentation. Expert Syst. Appl. **145**, 113131 (2020)

Three-Dimensional Representation and Visualization of High-Grade and Low-Grade Glioma by Nakagami Imaging

Orcan Alpar[1] and Ondrej Krejcar[1,2(✉)]

[1] Faculty of Informatics and Management, Center for Basic and Applied Research, University of Hradec Kralove, Rokitanskeho 62, Hradec Kralove 500 03, Czech Republic
{orcan.alpar,ondrej.krejcar}@uhk.cz
[2] Malaysia Japan International Institute of Technology (MJIIT), Universiti Teknologi Malaysia, Kuala Lumpur, Malaysia

Abstract. Three-dimensional (3D) visualization of the brain tumors reconstructed from the two-dimensional (2D) magnetic resonance imaging (MRI) sequences plays an important role in volumetric calculations. The reconstructions are usually executed using the fluid attenuated inversion recovery (FLAIR) sequences, where the whole tumors appear brighter than the healthy surrounding tissues. Without any processing; however, reconstruction results might be inconclusive; therefore, we propose a mathematical m-parametric Nakagami imaging for highlighting the lesions. The raw 2D FLAIR MRI images are taken from BraTS 2012 dataset and the highlighted images are generated by the Nakagami imaging. The information on the MRI slices is compiled in three-layered Nakagami images for better visualization of the high-grade and low-grade glioma in 3D space. By the flexible m-parametric design, on the other hand, the reconstructed images might easily be adjusted according to the GT images for precise representation.

Keywords: three-dimensional · brain MRI · visualization · nakagami imaging · reconstruction

1 Introduction

Three-dimensional (3D) MRI devices have become widely-used instruments day by day; yet the majority of the MRI scans are still two-dimensional (2D), worldwide. Despite the better representation of the brain tumors obtained by 3D devices; the 3D visualization of the tumors could be generated by reconstructing 2D MRI slices, when necessary. However, 3D visualization, if executed with the raw images solely, could lead to inconclusive and inadequate results; unless the images are processed and the lesions are highlighted. Therefore, we proposed a Nakagami imaging protocol for highlighting the lesions in brain MRI slices and for visualization of the whole tumors in 3D space.

As a promising method in ultrasonography and actually as an applied mode in ultrasound devices to analyze B-mode images, the ultrasonic Nakagami imaging is an intriguing protocol. The core of the imaging method is the Nakagami distribution [1], a kind of

© The Author(s), under exclusive license to Springer Nature Switzerland AG 2023
I. Rojas et al. (Eds.): IWBBIO 2023, LNBI 13919, pp. 143–154, 2023.
https://doi.org/10.1007/978-3-031-34953-9_11

parametric Rayleigh distribution, which became an imaging protocol with the researches of Shankar [2, 3]. The basic idea beneath this method is tissue characterization by analyzing the back-scattered signals inevitably caused by reflection of ultrasound waves in B-mode images. The ultrasonographic Nakagami imaging was used in various areas such as: monitoring heart regeneration of zebrafish [4], classification of hepatic steatosis [5], assessment of thermal lesions [6], characterization of breast tumors [7] and cavitation erosion [8], evaluation of liver fibrosis [9, 10] and diagnosis of fatty liver [11].

The conventional Nakagami method very basically consists of two distinctive steps, while only the first step is slightly similar with our imaging technique. In the first step, the Nakagami distribution is applied to all pixels for differentiation by changing the m-parameter inside of an interval and collecting the images generated with each parameter. In the second, the images are concatenated with a very unique visualization algorithm than shows all images in one output image by representing the output pixels by the color dictated by the m-parameter value. We, however, find the optimal m-parameter and generate one output image per raw image and show the results with one single image where the lesions are highlighted.

Our Nakagami and related imaging methods, when combined with a specialized fuzzy c-means, provided very promising and competitive results as seen in [12, 13] for brain tumor segmentation and in [14] for multiple sclerosis segmentation using FLAIR images. We also tried our method to segment the tumor areas from T1, T1c and T2 sequences to fuse with the areas segmented from FLAIR in [15]. With a different viewpoint, we focused on pseudocolorization of the thermograms to show the retroareolar tumors better than the built-in thermograms in [16] which was our very first research on Nakagami imaging and indeed the very first research of Nakagami outside of ultrasonography in medicine.

Briefly, what we propose in this paper is application of our single and double Nakagami imaging protocol to the brain FLAIR-MRIs taken from BraTS 2012 [17, 18] to show its potential in highlighting the whole tumor areas when visualizing and reconstructing in 3D space. According to our previous studies, the raw images solely are not adequate for highlighting the tumor areas when visualized in 3D space. Neglecting the segmentation step, we investigated the highlighting capability of our Nakagami imaging procedure and compared with the raw images.

We firstly introduced the mathematical fundamentals of Nakagami distribution and m-parametric imaging process in Sect. 2. In this section we also provided several synthetic examples to support our claim on the highlighting capacity and the flexibility of our design. Afterwards, we presented several experiments in Sect. 3 using the data from the BraTS 2012 public dataset to show the output images of our imaging system. The experimental results section consists of two subsections: Imaging and visualization. In imaging subsection, we provided several outputs by altering the m-parameter; while in visualization subsection, we highlighted all lesions and reconstructed the whole set in 3D space. The paper ends with the conclusion and discussion section in Sect. 4.

2 Nakagami Imaging

The Nakagami imitating is based on the Nakagami distribution, which indeed is an m-parametric version of Rayleigh, as mentioned before. The Nakagami distribution has the following probability density function:

$$f(r; m, \Omega) = \frac{2m^m r^{2m-1}}{\Gamma(m)\Omega^m} exp\left(-\frac{m}{\Omega}r^2\right) \tag{1}$$

where m is the shape, Ω is the scaling parameter: while $\Gamma(m)$ is the gamma function with the equation of $\Gamma(m) = (m-1)!$. The distribution could be converted to several distributions; yet automatically turns into Rayleigh if $m = 1$. In our method, r represents a pixel value on an image R; while in ultrasonographic Nakagami imaging, it is totally different. Among nearly all distributions, the Nakagami distribution has a unique feature that it has no independent parameters to be manually adjusted, for the ultrasonographic case. The shaping parameter Ω totally depends on the image itself:

$$\Omega = E(R^2) \tag{2}$$

where E is the statistical mean; while the m-parameter is strictly dictated by Ω and the image values, namely:

$$m = \frac{\Omega^2}{E\left[(R^2 - \Omega)^2\right]} \tag{3}$$

The Nakagami distribution generates a probability density function for different m and Ω parameters as in Fig. 1.

Against what we see in the conventional ultrasonic Nakagami imaging, we do control the m-parameter for achieving better results, in general. Therefore, a novel imaging protocol is studied by us to alter the conventional protocol a bit so that our new protocol would produce 2D images using an independent m-parameter that will be trainable. Provided that $r_{x,y}$ represents a pixel at (x, y) coordinates on a grayscale FLAIR MRI image R, which could be mathematically denoted as:

$$r_{x,y} \in R(x = [1:h], y = [1:w]) \tag{4}$$

where h is height, w is width of the image and $0 \leq r_{x,y} \leq 1$. A grayscale image indeed is a matrix consisting of values, which could be represented in matrix form as:

$$R = \begin{bmatrix} r_{1,1} & r_{1,2} & \cdots & r_{1,y} \\ r_{2,1} & \cdot & \cdot & \vdots \\ \vdots & & \cdot & \vdots \\ r_{x,1} & \cdots & \cdots & r_{h,w} \end{bmatrix} \tag{5}$$

Any matrix consists of arrays, so that all arrays could be extracted and appended as:

$$A = \begin{bmatrix} r_{1,1} & r_{1,2} & .. & r_{h,w} \end{bmatrix} \rightarrow \begin{bmatrix} a_1 & a_2 & .. & a_z \end{bmatrix} \tag{6}$$

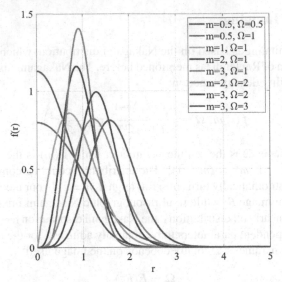

Fig. 1. Nakagami probability distribution function

where $z =$ hw. If we take the transpose of R, we get R^T with the new transposed array:

$$\overline{A} = \left[r_{1,1} \; r_{2,1} \; .. \; r_{h,w} \right] \rightarrow \left[\overline{a}_1 \; \overline{a}_2 \; .. \; \overline{a}_z \right] \tag{7}$$

If we let the system calculate all dependent parameters automatically, it should use the following steps.

$$\Omega = \frac{1}{z} \sum_{i=1}^{z} a_i \tag{8}$$

$$m = \Omega^2 / \sum_{i=1}^{z} \left(a_i^2 - \Omega \right)^2 \tag{9}$$

while the second step could be omitted and the optimal m-parameter could be searched within a plausible interval. The Nakagami distribution is applied to each pixel by:

$$\dot{a}_i \in \dot{A}(m) = \frac{2m^m a_i^{2m-1}}{\Gamma(m)\Omega^m} e^{\left(-\frac{m}{\Omega}a_i^2\right)} \tag{10}$$

$$\ddot{a}_i \in \ddot{A}(m) = \frac{2m^m (\overline{a}_i)^{(2m-1)}}{\Gamma(m)\Omega^m} e^{\left(-\frac{m}{\Omega}\overline{a}_i^2\right)} \tag{11}$$

to obtain the synthetic Nakagami arrays by reversely reshaping the vectors into matrices $\dot{R}(m)$ and $\ddot{R}(m)$, as presented above. The ultimate output of the imaging process is the dot product of these matrices by $H(m) = \dot{R}(m) \cdot \ddot{R}(m)$, what we name "dual" or "dual-scan"; while the matrices $\dot{R}(m)$ and $\ddot{R}(m)$ are named after "single" or "single-scan" Nakagami images. There are two basic superiorities of this method: revealing near-zero contrast lesions and flexibility of highlighting. Initially, this method applies a distribution to all images so that; if any pixel has a different value than the neighbors, this pixel will be highlighted; unlike thresholding techniques. In Fig. 2, a low-contrast example is presented to show the readers how our framework highlights the lesion areas.

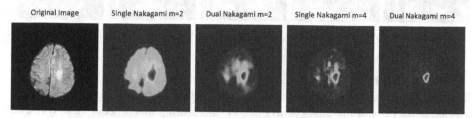

Fig. 2. Low-contrast experiment

As seen in Fig. 2, the low contrasted tumor is highlighted without deteriorating the original image and the original lesion area. Based on the m-parameter, the highlighted area might expand or shrink; while it is not possible to highlight greater area than the original. In addition to this real example, a synthetic example is given in Fig. 3 to show the differentiation of the near-zero pixels by single and dual Nakagami protocol.

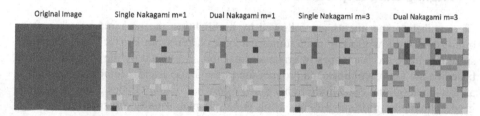

Fig. 3. Near-zero contrast experiment with synthetic images

The original image in Fig. 3 is deliberately prepared to include various very-low contrast squares as lesions, in a main square. Although the squares are barely visible; since all of them are gray, all pixel values are changed according to their neighbors and initial contrast by Nakagami imaging. Dual scan doubles the precision for the squares, as seen in the Dual Nakagami m = 3 image; therefore, some red-like squares are tuned back to yellow after dual Nakagami. On the other hand, the flexibility comes from the m-parameter as seen in the Fig. 2; while it is not so meaningful for this image set. Therefore, we show in Fig. 4 how the m-parameter might change the highlighted trajectories.

The experiment shown in Fig. 4 is about the highlighting capability and flexibility of our Nakagami imaging method. The filled gradient circles have very high contrast indeed; while they are the perfect imitations of the real tumors which could be seen in 2D MRI slices considering the gradient intensity. The highlighted area could easily be enlarged by decreasing the m-parameter and vice versa, as shown clearly in the Fig. 2. The combination of the features presented in these examples is the key to providing better 3D visualization in real experiments.

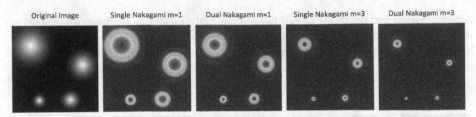

Fig. 4. Visible contrast experiment with synthetic images

On the other hand, it is safe to state that no deterioration could occur when our farmwork is applied to the images. The system is based on the Nakagami distribution applied to the cells of the image arrays one by one. During this process, new synthetic RGB images are generated by applying the m-parametric imaging algorithm. Therefore, the new images are not fully identical with the originals which are grayscale; while these images are created using the data of the originals, only. Additionally, the highlighting level could easily be adjusted by finding the proper m-parameter; while it is not possible to highlight the lesions bigger that they are. Given these facts, there is no possibility to deteriorate the original images and generate unreal synthetic images.

3 Highlighting Experiments

The synthetic experiments are followed by highlighting the real images taken from high-grade and low-grade glioma tumor databases taken from BraTS 2012. Generally speaking, there are four types of grades from 1 to 4, evaluated by the medical doctors based on the tumor volumes and progression rates. Grade 1 and 2 tumors are classified as low-grade glioma which are rather slowly growing; while Grades 3 and 4 as high-grade glioma are more aggressive. High-grade gliomas usually are supposed to have larger areas of necrosis, edema and tumor cores, which makes the whole tumor regions rather larger and more amorphous with high intensity shapes. Therefore, we divided the experiments section into two to highlight and reconstruct high-grade and low-grade glioma sets separately.

3.1 High-Grade Glioma Experiment

As presented in the previous synthetic examples, our Nakagami imaging protocol could find low-contrast lesions and highlight them based on the m-parameter. In addition, it is possible to enhance the precision of the highlighting by changing the protocol from single to double scan. When the m-parameter is decreased, the highlighted are automatically expands and vice versa. In Table, 1, an experiment is presented using the raw image HG/0001 #95 from BraTS 2012, to show the differences when the scan protocol changed from single to dual for various m-parameters. Since we are dealing with visualization and reconstruction of the whole tumor areas, the parameter intervals are selected accordingly.

Table 1. Nakagami imaging experiment for high-grade glioma, HG/0001, #95

Raw Image	m	Single Nakagami	Dual Nakagami
	0.7		
	0.75		
	0.8		
	0.85		
	0.9		
	0.95		
	1		

As seen in the Table 1, that for one raw image, infinite number of highlighted images could be provided; yet we presented the single and dual Nakagami outputs for the interval of $0.7 \leq m \leq 1$ by changing m with $\Delta m = 0.05$. Strictly depending on the m-parameter, unless the scaling parameter Ω is also independent, the imaging method provides the highlighted regions by differentiating all pixels from the neighbor pixels based on the intensities. However, the boundary values of the chosen interval don't provide sufficient highlighting; since $m = 0.7$ of Single Nakagami process lacks the precise boundary of the whole tumor regions; while $m = 1$ of Dual Nakagami process highlights a very narrow region. Given this deficit, a more plausible value $m = 0.9$ is selected for the 3D visualization experiment in Sect. 4.

3.2 Low-Grade Glioma Experiment

The low-grade glioma case is taken from LG/0006 set of BraTS 2012 and the single and dual Nakagami imaging is applied to the image #127. The intensity seems lower than HG, while the tumor region is still not so small. The highlighting results are presented in Table 2 with single and dual Nakagami method for various m-parameters.

Table 2. Nakagami imaging experiment for low-grade glioma, LG/0006, #127

Raw Image	m	Single Nakagami	Dual Nakagami
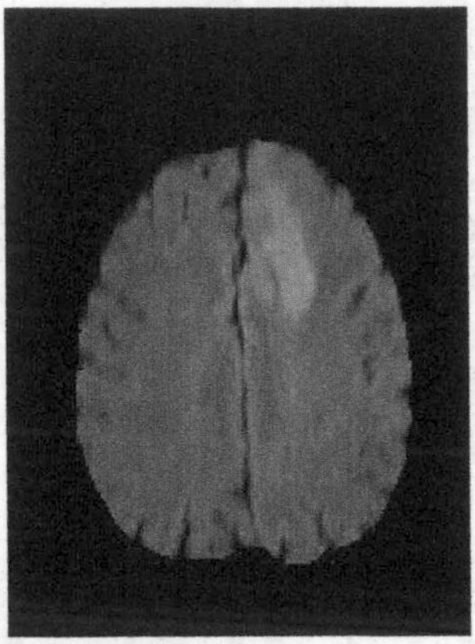	0.85	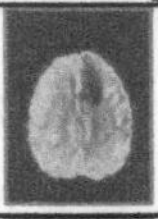	
	0.9		
	0.95		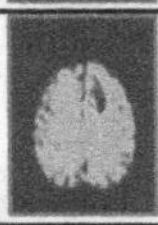
	1		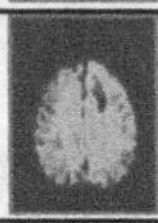
	1.05		
	1.10		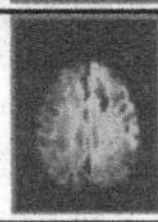
	1.15		

Similar to the high-grade glioma experiment, the highlighting level is increasing based on the m-parameter; while again the boundary values don't provide sufficient differentiation. The visualization experiment will be held by $m = 1.1$ in the following section.

4   Results

In this section, we present examples of visualization protocol for the high-grade and low-grade gliomas in 3D space. Above all, we reconstructed the original raw images from HG/0001 and LG/0006 sets by linear maximum intensity rendering to show the tumor areas as visible as possible. Afterwards, we executed the whole framework for the sets separately to generate the single and dual Nakagami images using $m = 0.9$ for the HG/0001 and $m = 1.1$ for the LG/0006 sets. We obtained the following images presented in Fig. 5, below, for the high-grade set.

ORIGINAL 3D
REPRESENTATION

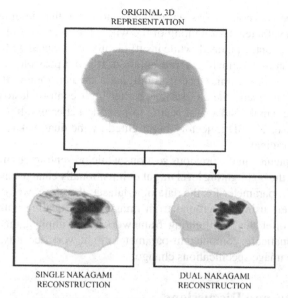

SINGLE NAKAGAMI
RECONSTRUCTION

DUAL NAKAGAMI
RECONSTRUCTION

Fig. 5. 3D Visualization of high-grade glioma

The same visualization technique is applied to the low-grade set and the following images are reconstructed in Fig. 6. All visualizations are executed by basic linear maximum intensity projection without any other modes or adjustments to see the real differences.

ORIGINAL 3D
REPRESENTATION

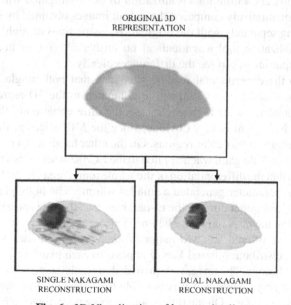

SINGLE NAKAGAMI
RECONSTRUCTION

DUAL NAKAGAMI
RECONSTRUCTION

Fig. 6. 3D Visualization of low-grade glioma

In a qualitative viewpoint, it is seen in Fig. 5 and Fig. 6 that, both protocols, single or dual, provide a sufficient highlighting of the whole tumor areas. In Fig. 5, the whole lesion has very low contrast indeed; while it sufficiently highlighted with dual Nakagami imaging better than the original 3D representation. It is not possible to claim the same thing for the single Nakagami, since it highlights a greater volume than the original representation. Conversely, the LG set has rather high contrast lesions compared to the HG, therefore, single Nakagami seems to provide better highlighting; while it is not always the case. The 3D trajectory highlighted by the dual Nakagami seems to be smaller than the original.

As clearly explained in the previous sections while presenting our method using the synthetic images, the highlighting level might unintentionally change based on the image itself. Not every m-parameter fits the data in a dataset; therefore, we occasionally train our imaging system using the ground truth images when dealing with segmentation. However, as we offer this highlighting framework as an implantable mode for MRI devices, the automatically computed m-parameter might work decently, while it should be adjusted if the image specifications change.

5   Conclusions and Discussions

In this paper, we proposed a novel utilization area of our Nakagami imaging method that we used in segmentation and pseudocoloring before. The idea behind this research is actually found in the conventional Nakagami imaging procedure applied to the B-mode images in ultrasonography. The ultrasonic Nakagami imaging is a tool for analyzing the back-scattered signals with a unique 2D representation of the 3D output images; while we changed the process a little for visualization of the highlighted whole tumor areas in 2D space. We qualitatively compared the output images obtained by single and dual Nakagami imaging separately with the original reconstructions of high- and low-grade lesions. The visualization tool was identical no further filtering or image processing applied to the output images to see the differences clearly.

According to the experimental results, it is seen that both single and dual Nakagami imaging provided better visualization compared to the 3D representation with raw images. Especially, for the high-grade case, the differentiation of the whole tumor volumes by dual Nakagami is very obvious; since the MRI slices of the HG/0001 set have very low contrast in particular regions. On the other hand, it is a bit opposite given the results of the dual Nakagami when applied to the LG/0006 set. This set contains more visible lesions; so that the differentiation of the whole tumor areas from the surroundings for the selected m-parameter generated a smaller volume. The highlighted areas might be expanded or shrunk depending on the m-parameter, so that, it is possible to reach GT volumes by training the m-parameter when needed.

The protocol we offered has two major advantages: the m-parametric design for flexibility and the distribution-based kernel applied to each pixel for highlighting even near-zero contrast lesions. The images we provided in results section are only a few brief outputs of our imaging system; yet it is possible to change the m-parameter for better highlighting when necessary. On the other hand, despite numerous CNN-like black-box solutions, we proposed a fully mathematical framework, which is easy to implement.

There is no disadvantage worth-mentioning; since this paper is an output of more like a qualitative research progress that has no numeric performance criteria. Despite being a qualitative paper; the quantitative results of our Nakagami imaging method could be found in our previous research papers [13–15] for BraTS datasets. This paper is not related with segmentation; but more highlighting lesion areas; while it is possible to check the segmentation results by our protocol in [13].

As the future research, we are using our Nakagami imaging method in 3D reconstruction and progression rate estimation of MS lesions. On the other hand, we also working on novel morphing and fusion algorithms using this method. Like the ultrasonic Nakagami imaging protocol was applied to the ultrasound devices years ago, it is probable to see similar implementations to the MRI devices as an alternative visualization mode. It is also possible to see likes of these mathematical imaging modes in 3D MRI devices, in the future.

Acknowledgment. The work and the contribution were also supported by the SPEV project, University of Hradec Kralove, Faculty of Informatics and Management, Czech Republic (ID: 2102–2023), "Smart Solutions in Ubiquitous Computing Environments". We are also grateful for the support of student Michal Dobrovolny in consultations regarding application aspects.

References

1. Nakagami, M.: The m distribution—a general formula of intensity. In: Statistical Methods in Radio Wave Propagation, Pergamon, pp. 3–36 (1960)
2. Shankar, P.M.: A general statistical model for ultrasonic backscattering from tissues. IEEE Trans. Ultrason. Ferroelectr. Freq. Control **47**(3), 727–736 (2000)
3. Shankar, P.M.: Ultrasonic tissue characterization using a generalized Nakagami model. IEEE Trans. Ultrason. Ferroelectr. Freq. Control **48**(6), 1716–1720 (2001)
4. Yeo, S., Yoon, C., Lien, C.L., Song, T.K., Shung, K.K.: Monitoring of adult zebrafish heart regeneration using high-frequency ultrasound spectral doppler and nakagami imaging. Sensors **19**(19), 4094 (2019)
5. Zhou, Z., et al.: Hepatic steatosis assessment with ultrasound small-window entropy imaging. Ultrasound Med. Biol. **44**(7), 1327–1340 (2018)
6. Zhang, S., et al.: Ex Vivo and In Vivo monitoring and characterization of thermal lesions by high-intensity focused ultrasound and microwave ablation using ultrasonic Nakagami imaging. IEEE Trans. Med. Imaging **37**(7), 1701–1710 (2018)
7. Tsui, P.H., et al.: Small-window parametric imaging based on information entropy for ultrasound tissue characterization. Sci. Rep. **7**, 41004 (2017)
8. Han, M., Wang, N., Guo, S., Chang, N., Lu, S., Wan, M.: Nakagami-m parametric imaging for characterization of thermal coagulation and cavitation erosion induced by HIFU. Ultrason. Sonochem. **45**, 78–85 (2018)
9. Tsui, P.H., Ho, M.C., Tai, D.I., Lin, Y.H., Wang, C.Y., Ma, H.Y.: Acoustic structure quantification by using ultrasound Nakagami imaging for assessing liver fibrosis. Sci. Rep. **6**, 33075 (2016)
10. Ma, H.Y., Lin, Y.H., Wang, C.Y., Chen, C.N., Ho, M.C., Tsui, P.H.: Ultrasound window-modulated compounding Nakagami imaging: resolution improvement and computational acceleration for liver characterization. Ultrasonics **70**, 18–28 (2016)

11. Tsui, P.H., Wan, Y.L.: Application of ultrasound nakagami imaging for the diagnosis of fatty liver. J. Med. Ultrasound **24**(2), 47–49 (2016)
12. Alpar, O., Dolezal, R., Ryska, P., Krejcar, O.: Low-contrast lesion segmentation in advanced MRI experiments by time-domain Ricker-type wavelets and fuzzy 2-means. Appl. Intell. (2022a). https://doi.org/10.1007/s10489-022-03184-1
13. Alpar, O., Dolezal, R., Ryska, P., Krejcar, O.: Nakagami-Fuzzy imaging framework for precise lesion segmentation in MRI. Pattern Recogn. **128**, 108675 (2022)
14. Alpar, O., Krejcar, O., Dolezal, R.: Distribution-based imaging for multiple sclerosis lesion segmentation using specialized fuzzy 2-means powered by Nakagami transmutations. Appl. Soft Comput. **108**, 107481 (2021)
15. Alpar, O.: A mathematical fuzzy fusion framework for whole tumor segmentation in multimodal MRI using Nakagami imaging. Expert Syst. Appl. **216**, 119462 (2023)
16. Alpar, O.: Nakagami imaging with related distributions for advanced thermogram pseudo-colorization. J. Therm. Biol **93**, 102704 (2020)
17. Kistler, M., Bonaretti, S., Pfahrer, M., Niklaus, R., Büchler, P.: The virtual skeleton database: an open access repository for biomedical research and collaboration. J. Med. Internet Res. **15**(11), e245 (2013)
18. Menze, B.H., et al.: The multimodal brain tumor image segmentation benchmark (BRATS). IEEE Trans. Med. Imaging **34**(10), 1993–2024 (2014)

Speeding up Simulations for Radiotherapy Research by Means of Machine Learning

I. Fernández[1], C. Ovejero[2], L.J. Herrera[1], I. Rojas[1], F. Carrillo-Perez[1], and A. Guillén[1(✉)]

[1] Computer Engineering, Automatics and Robotics Department,
University of Granada, Granada, Spain
aguillen@ugr.es
[2] Kerma S.L., Granada, Spain

Abstract. Radiotherapy is one of the most widely used treatments for cancer by irradiating the tumor volume. However, one of its disadvantages is that healthy tissue is also affected, producing various side effects. For this reason, preliminary studies are required beforehand to determine the dose to be administered in each case, to avoid possible damage and to make sure that the dose received by the tumor is the correct one. These studies are carried out both using simulations and with routine machinery procedures using a mannequin that simulates the area to be treated. In this work a way of speeding up the previous study process is tackled, starting from simulated data whose optimized obtaining will be the objective of this work. The PENELOPE Monte Carlo simulation software is used to recreate the process and obtain the necessary previous data. Subsequently, regression models are applied to obtain the values of interest and accelerate the procedure, reducing, in addition, the energy consumption and storage required while obtaining very accurate approximations.

Keywords: Radiotherapy · Regression · Machine Learning · Simulation · PENELOPE

1 Introduction

Radiotherapy involves using ionizing radiation to destroy cancer cells, while minimizing damage to healthy tissue. This is achieved by delivering the radiation from a machine, typically a linear accelerator (LINAC) that will accelarate the electrons that will generate X-ray beams. Before treatment is delivered to the patient, the quality of the treatment is verified through dosimetric verification, which ensures that the absorbed dose is distributed correctly.

During dosimetric verification, a mannequin is irradiated to simulate the treatment area, and the distribution of the absorbed dose is measured. This ensures that the tumor receives the correct dose while minimizing damage to surrounding healthy tissue and ensuring the QUANTEC protocol.

I. Rojas et al. (Eds.): IWBBIO 2023, LNBI 13919, pp. 155–164, 2023.
https://doi.org/10.1007/978-3-031-34953-9_12

A mannequin is a device that mimics the human body or a part of it. There are three main types of dummies: physical, physiological and computational. Mannequins are used in various applications such as quality control or dosimetry which consists on the test performed for the purpose of planning a radiotherapy treatment. It calculates the doses absorbed in tissues and matters as a result of exposure to radiation, both direct and indirect [1]. The dummy with which the basic tests and quality controls are usually carried out is a basin filled with liquid water.

A detector is placed inside the dummy. This is a device for detecting and characterizing ionizing radiation. For this purpose, the detectors interact with the radiation and generate a signal that can be modified to a quantity representative of the characteristic to be determined.

In this work, we are going to simulate a semiconductor detector formed by silicon multistrips in order to generate dose data by varying different parameters and thus, obtain a series of data faithful to the real ones in a simpler way. The simulations' results are then used to compare several regression techniques to determine if they are able to reproduce the simulator output with the same precision but in less time and energy usage. The rest of the paper is organized as follows: Sect. 2 describes in detail the problem. Afterwards, Sect. 3 depicts the algorithms and paradimgs considered as well as the data generation process and Sect. 4 shows the approximation results for both multi-output and single-output cases. Finally, Sect. 5 draws conclusions.

2   Problem Description

The main objective of this work is to optimize the process to obtain reliable solutions in a shorter time requiring less energy consumption. This task was to be performed by simulating a silicon detector consisting of 16 2 mm strips for different spacings. This detector is placed at a depth of 1.5 cm in a basin of water and a beam of radiation is incident from the air at a distance of 100 cm from the surface of the basin. The idea is to simulate this process as close to reality as possible and verify that it resembles real data. Generally, the reference conditions are SSD (Source-to-Surface Distance) of 100 cm, square field of (10×10) cm^2, energy of 6 MeV and a depth of 1.5 cm.

In conventional radiotherapy, photons are usually used, although electrons are used in some cases. These particles undergo various attenuation and scattering mechanisms when interacting with matter and when moving. Therefore, when the incident photon beam passes through the photon phantom/patient it is affected, complicating dose deposition on them and making direct measurement essentially impossible.

Nevertheless, it is necessary to know, as precisely and accurately as possible, the dose distribution in the irradiated volume in order to achieve a successful treatment outcome. For this purpose, various functions are applied that relate the dose at any point on the patient (or phantom) to the dose at the reference point in a water phantom. This point is determined for a specific set of reference conditions described previously.

The graphical representation of dose versus depth is called the depth yield curve and is related to how radiation attenuates as it penetrates a medium. The value of the maximum absorbed dose varies with energy, as does its depth. For the same energy, the dose decreases exponentially with depth. Therefore, the dose absorbed by a patient varies with it, [2]. The depth of the maximum dose below the surface also depends on the beam field size.

When dealing with real patients (instead of mannequins), the low surface dose compared to the maximum dose is called the skin-sparing effect and represents an important advantage in the treatment of deep tumors. Surface and orthovoltage beams do not exhibit the skin-sparing effect, since their maximum dose occurs at the surface, i.e., the surface dose is equal to the maximum dose.

Thus, if we could approximate the dose curve without requiring simulation repeatedly, it could be possible to save time and energy. The fact of speeding up this process could also reduce the time required to plan the several sessions a patient has to go through. Therefore, in this paper, we will characterize the dose curve with three points: first charge, maximum charge, and last charge as a first approximation to resolve this problem.

2.1 PENELOPE Simulator

PENELOPE (*PENetration and EnergyLoss of Positrons and Electrons*, [3], is a Monte Carlo code that simulates the coupled transport of photons, electrons and positrons in arbitrary materials in the energy range from 50 eV to 1 GeV. It is a free and open source code developed at UPC and distributed through the NEA (Nuclear Energy Agency). This code is often used in the field of medical physics, since the energy range corresponds to that used in this field.

It is a package of subroutines written in FORTRAN that simulates photon transport in a simple and detailed way. It simulates all interactions in chronological order. For photon transport, the processes of photon interaction with matter are taken into account: Rayleigh scattering, Compton scattering, pair production and photoelectric absorption.

For the case of electrons and positrons it employs a mixed procedure, i.e., single elastic collisions, inelastic collisions and *Bremsstrahlung* emission, the three main interactions of the electron with matter, are simulated in detail, while the trajectory of a particle between successive interactions or between an interaction and the crossing of an interface is generated as a series of steps of limited length, [3].

Furthermore, for the case of these two particles one can distinguish between hard and soft events. Hard events are those that are simulated in detail and are characterized by an energy loss or a polar angular deviation greater than that set by the user. The soft ones correspond to events where these values are below the mentioned threshold.

The basic sources of PENELOPE are:

- penelope.f: subroutine package for Monte Carlo simulation of coupled photon-electron transport in a homogeneous medium.

- pengeom.f: allows the generation of electron-photon showers in any material system formed by homogeneous bodies bounded by quadratic surfaces.
- material.f: allows to create the material file with all the necessary physical information.

Some of the programs available in PENELOPE:

- GVIEW: program to visualize the geometry.
- SHOWER: program to visualize the particle trajectory.

Today, it is still one of the most widely used codes for research in the field of radiotherapy. An example is [4], where the PENELOPE code is used to perform Monte Carlo simulations of Low Dose Radiation Therapy (LDRT) with kilo voltage beams delivering doses of 0.3, 0.5 and 1.0 Gy. It is used as a unit of absorbed dose of ionizing radiation to the lungs to determine dosimetric adequacy and to provide an alternative method of treating patients with COVID-19 pneumonia.

Penmain. The PENMAIN program simulates particle transport in regular geometries. By default, this program assumes that primary particles of a particular type are emitted from a point or source with a given energy or energy spectrum.

The program can read the initial state variables of the primary particles from a precomputed phase space. This option is useful for splitting the simulation of complex problems into several consecutive stages.

The PENMAIN program provides global simulation results such as the energy and angular distributions of the particles emerging from the material system, the energy deposited in each body, etc. To generate more specific information, impact detectors, external angular detectors and energy deposition detectors can be defined. Each detector consists of a set of active bodies that must be defined as part of the geometry beforehand.

3   Methodology: Data Generation with PENELOPE and Modeling the Output by Means of Machine Learning Algorithms

This section describes, first, how data with PENELOPE was generated, defining the parameters tuned. Then, it presents the alternatives to approximate the three points commented previously and will provide the implementation details of those models.

3.1   Data Generation

The relevant input data is written to the file named "penmain.dat". The output file enmain-res.dat contains a report on the overall simulation and some partial results. The calculated continuous distributions are written to separate files whose names have the extension penmain_.dat.

The input file contains all the parameters that PENMAIN and PENGEOM will need to perform the simulation. This file consists of several parts:

- **Source definition:** This part defines the primary particle type, initial energy (monoenergetic source), source position and beam shape, among others.
- **Material information and simulation parameters:** The files of the materials that make up the geometry are mentioned per row and the parameters of *scattering* and shear energies are associated. The files corresponding to the materials are created using the auxiliary program MATERIAL. The simulation parameters are: **EABS**: value from which the energy will be considered absorbed by the material. *C1*: average angular deviation due to the elastic dispersions suffered along the path between two strong elastic interaction events. It is better to consider a small value. *C2*: maximum average energy loss between events. It is better to consider a large value. **WCC**: shear energy for strong inelastic collisions. **WCR**: cutoff energy for strong *Bremsstrahlung* emission.
- **Geometry.** The geometry file defined in a separate file is associated.
- **Phase space and impact detector:** A phase space is defined on a particular body belonging to the geometry. A phase space contains all the corresponding information of the particles arriving at that surface. The body in which the phase space is defined corresponds to the impact detector to which one can define which particle to detect (electrons, photons, positrons or all), the energy range and the number of bins in which the range is subdivided.
- **Dose distribution:** The space where the dose distribution is to be measured is defined by the maximum and minimum values of the X, Y and Z coordinates and the number of bins for each coordinate is set.
- **Working properties:** The number of primary particles to be simulated, the maximum simulation time, the seeds to generate the random numbers, the time period to renew the information stored in the dump file, among others, are defined.

With these definitions, three files corresponding to that detector can be obtained: one where all the information of the particles arriving to the volume is stored, which is used to perform a second simulation from that point, another one containing the energy distribution with respect to the density probability function and the last one containing the integrated fluence over the volume.

The dump file saves the necessary information to be able to continue running the simulation at another time. In this way, if the simulation is interrupted unexpectedly, the information already obtained is not lost.

3.2   Models Evaluated and Hyperparameter Tunning

The models used to approximate the set of input-ouput pairs were: lineal regression, K-Nearest Neighbors, Support Vector Regressors, Decision Trees, Random Forest, Extra Trees, and boosting techniques like Gradient Boosting, Histogram-based Gradient Boosting and XGBoost.

3.3   Implementation Details and Model Deployment

All the implementations have been performed using Python 3.6. As a Deep Learning framework for the coding of the FFNN and CNN architectures, Pytorch was chosen [5]. It was chosen based on the utilities it provides for easy and fast prototyping. For SVM implementation the library Scikit-Learn [6] was chosen, which SVM implementation is based on the famous library LIBSVM [7]. For data preprocessing and treatment the Python libraries Pandas [8] and Numpy [9] were used.

4   Experiments and Results

This section is subdivided in two subsections: one considering the regression over the three points at the same time using multi-output models, then, considering regression of each point isolatedly using single output models. To evaluate the performance of the regression models two metrics will be considered: R^2 and Root Mean Squared Error.

Then the models are entered with the default hyperparameters, which can be seeded with the variable *random state* and given the same value. They are going to be trained first for the csv of 120 data, then of 240 and finally of 400 data. In this way we will see if the models improve as the dataset is enlarged. Cross-validation using 5 splits was carried out to define the train/test data sets (Tables 1, 2 and 3).

4.1   Multi-Output Regression

– **120 samples**

Table 1. 5 split average for RMSE and R^2 for both train and test using 120 samples.

Model	RMSE train	RMSE test	R^2 train	R^2 test	t tr	t tst
GB	53.68992	841.2112	0.99996	0.9898	0.055	0.001
HGB	7417.8002	8145.5160	0.4624	−0.6174	0.126	0.009
KNN	7260.2621	9195.2436	0.4910	0.0590	0.001	0.002
ET	$1.6850 \cdot 10^{-11}$	938.6923	1	0.9846	0.100	0.013
RF	604.79654	1472.2111	0.9948	0.9948	0.080	0.011
DT	0	958.2350	1	0.9879	0.001	0.001
LR	9136.3715	9673.0802	0.2080	−1.0372	0.000	0.000
SVR	10998.0863	10483.8120	−0.1245	−0.1418	0.003	0.004
XGBR	0.5213	936.2736	0.999999994	0.999999994	0.075	0.003

– 240 samples

Table 2. 5 split average for RMSE and R^2 for both train and test using 240 samples.

Modelo	RMSE train	RMSE test	R^2 train	R^2 test	t tr	t tst
GB	63.8497	462.5502	0.99995	0.9961	0.063	0.003
HGB	5082.4651	6458.1112	0.7520	0.4892	0.235	0.012
KNN	7215.7550	7868.041	0.5250	0.3816	0.001	0.002
ET	$1.5688 \cdot 10^{-11}$	393.3563	1	0.9971	0.106	0.013
RF	189.4465	559.6805	0.9995	0.9995	0.098	0.014
DT	0	524.0483	1	0.995	0.001	0.000
LR	9497.9205	9162.5308	0.1981	0.1603	0.002	0.000
SVR	11325.9281	12842.9102	−0.1190	−0.1303	0.005	0.024
XGBR	1.5807	596.7499	0.99999996	0.99999996	0.080	0.004

– 400 samples

Table 3. 5 split average for RMSE and R^2 for both train and test using 400 samples.

Modelo	RMSE train	RMSE test	R^2 train	R^2 test	t tr	t tst
GB	61.1578	142.7611	0.99995	0.99995	0.074	0.003
HGB	3607.4946	3918.7482	0.8720	0.8088	0.345	0.010
KNN	6156.6455	6856.9788	0.6455	0.4940	0.001	0.002
ET	$1.6504 \cdot 10^{-11}$	155.1992	1	0.9995	0.118	0.015
RF	118.4126	224.4564	0.9997	0.9997	0.106	0.016
DT	0	198.9078	1	0.998	0.001	0.001
LR	9397.4700	8957.6901	0.1979	0.15926	0.005	0.000
SVR	11210.9819	10572.1514	−0.1193	−0.1180	0.023	0.039
XGBR	3.7440	245.6224	0.9999998	0.9999998	0.100	0.004

When comparing the errors obtained in Tables 1, 2 and 3 a decrease in the RMSE value (especially for the test part) can be observed as the number of data increases. In general, an improvement of the results is obtained for all models as the number of data increases. We are going to work with the csv of 400 data from now on to try to obtain the best possible result.

4.2 Single Output Regression

In this section the same process of the previous section will be performed but for single output models. The same preprocessing (normalization) and the same 5 training and test splits obtained before will be applied. The difference is that in this case the output variable will have a single value instead of an array of 3 values. We will also consider the csv of 400 data, because as it was previously proved, better results are obtained for a larger number of data (Tables 4, 5 and 6).

– **First dose**

Table 4. 5 split average for RMSE and R^2 for both train and test using 400 samples for the first dose.

Modelo	RMSE train	RMSE test	R^2 train	R^2 test	t tr	t tst
GB	17.1819	35.3079	0.99998	0.99992	0.057	0.002
HGB	1973.8978	1956.0436	0.8465	0.7717	0.134	0.003
KNN	3150.5698	3398.6809	0.6096	0.4175	0.001	0.002
ET	$8.42082 \cdot 10^{-12}$	37.7751	1	0.99992	0.116	0.016
RF	33.5727	69.8232	0.99995	0.99995	0.145	0.015
DT	0	39.2752	1	0.99991	0.001	0.000
LR	4597.7326	4203.8007	0.1712	0.0988	0.001	0.000
SVR	5309.5756	4760.8679	−0.1048	−0.1079	0.005	0.009
XGBR	1.5234	55.2711	0.99999990	0.99999990	0.047	0.005

– **Second dose**

Table 5. 5 split average for RMSE and R^2 for both train and test using 400 samples for the second dose .

Modelo	RMSE train	RMSE test	R^2 train	R^2 test	t tr	t tst
GB	114.9507	254.3694	0.99996	0.9998	0.037	0.001
HGB	6899.1941	7484.5738	0.8924	0.8473	0.153	0.005
KNN	12049.0245	13428.7005	0.6724	0.5455	0.001	0.002
ET	$3.1955 \cdot 10^{-11}$	229.5695	1	0.9998	0.121	0.035
RF	203.0966	384.1774	0.99990	0.99990	0.261	0.031
DT	0	295.3186	1	0.9995	0.002	0.000
LR	18627.7176	17793.0404	0.2186	0.1964	0.005	0.001
SVR	22428.6852	21239.3449	−0.1324	−0.1319	0.009	0.024
XGBR	6.7531	413.0400	0.9999998	0.9999998	0.409	0.007

– **Third dose**

Table 6. 5 split average for RMSE and R^2 for both train and test using 400 samples for the third dose.

Modelo	RMSE train	RMSE test	R^2 train	R^2 test	t tr	t tst
GB	51.34084	138.6061	0.99991	0.9992	0.041	0.002
HGB	1949.3918	2315.6271	0.8770	0.8074	0.162	0.004
KNN	3270.3423	3743.5550	0.6544	0.5190	0.001	0.002
ET	$9.1386 \cdot 10^{-12}$	198.3798	1	0.9986	0.121	0.025
RF	125.3441	227.8772	0.9994	0.9994	0.210	0.030
DT	0	263.8044	1	0.997	0.003	0.001
LR	4966.9597	4876.2291	0.2040	0.1825	0.005	0.001
SVR	5894.6849	5716.2415	−0.1207	−0.1142	0.011	0.26
XGBR	2.9557	268.5561	0.9999997	0.9999997	0.491	0.005

5   Conclusions

One of the most widely used mechanisms in the fight against cancer is external radiation therapy. This uses a linear electron accelerator (LINAC) to irradiate the tumor volume. The two most important objectives for this task are to know what dose to deliver and to avoid irradiating the healthy tissue around the tumor.

For this purpose, several preliminary studies are carried out both telematically and with routine procedures of the machinery, using a mannequin with the precise shape and materials to simulate the area to be treated. The realization of these studies is very time consuming. In this work we have sought a way to optimize the telematic part of the preliminary analysis required before treatment, specifically we have applied Machine Learning techniques to speed up the process of Monte Carlo simulation PENELOPE code.

First, the geometry and the necessary execution files have been generated: a beam of monoenergetic photons from a point source is incident on a $40 \times 40 \times 40$ cm^3 square water dummy containing a 5 cm^3 thick silicon detector, formed by 2 mm strips up to a width of 64 mm. The energy, aperture/field and SSD have been modified until 400 different samples have been obtained.

Different regression models were trained to predict these three dose values and the $RMSE$ and R^2 for training and test were calculated.

When training the multi-output models, it was obtained that the best model to predict is *XGBRegressor* with the hyperparameter values obtained with *GridSearchCV*, since it is the one that suffers less *overfitting*, its *RMSE* for test is the best and the training RMSE is the one that obtains the lowest value after *Decision Tree*.

When training a proper model for each point and comparing the results individually, it is obtained that the best model is still *XGBoost* for all three.

Then, a comparison was made between the results of the single and multiple models and it was observed that for the third point better results were obtained with the single model, for the second and first point with the multiple model (lower *overfitting*). Therefore, it was decided to choose this model. In addition, the multi-output model is better if in the future it is intended to increase the number of characteristic points because it speeds up the whole process of study and final predictions.

The development of this work has shown that generating a regression model trained with data obtained from PENELOPE simulations is a good substitute option when it comes to obtaining the dose deposited on the axis for a case study similar to the one created. In this way, both radiophysicists and students/researchers can perform some simple studies in much less time and consuming less energy.

In addition, this same process can be applied for other types of simulations with different geometries that are more complex and even for the other types of data provided by PENELOPE. In this way, anyone who wanted to use these results could do so without having to learn a new tool and saving both learning and execution time.

Acknowledgements. This work was funded by the Spanish Ministry of Sciences, Innovation and Universities under Project PID2021-128317OB-I00 and the projects from Junta de Andalucia P20-00163.

Conflict of interest. The authors declare that they have no conflict of interest.

References

1. QuironSalud. https://initiaoncologia.com/glosario/dosimetria/
2. Brosed, A., Lizuain, M.C., Radioterapia externa I: bases físicas, equipos, determinación de la dosis absorbida y programa de garantía de calidad. Sociedad Española de Física Médica (2012)
3. Salvat, F.: Penelope. a code system for monte carlo simulation of electron and photon transport. In: Workshop Proceedings Barcelona, January 2019
4. Gonzales Ccoscco, A.E.: "Simulación monte carlo de radioterapia de dosis bajas con haces de kilo voltaje, para tratar neumonía inducida por covid 19," Master's thesis (2022)
5. Paszke, A., et al.: Automatic differentiation in pytorch (2017)
6. Pedregosa, F., et al.: Scikit-learn: machine learning in python. J. Mach. Learn. Res. **12**, 2825–2830 (2011)
7. Chang, C.-C., Lin, C.-J.: LIBSVM: a library for support vector machines. ACM Trans. Intell. Syst. Technol. **2**, 27:1–27:27 (2011). http://www.csie.ntu.edu.tw/cjlin/libsvm
8. McKinney, W.: Data structures for statistical computing in python. In: Proceedings of the 9th Python in Science Conference (S. van der Walt and J. Millman, eds.), pp. 51–56 (2010)
9. Walt, S.V.d., Colbert, S.C., Varoquaux, G.: The numpy array: a structure for efficient numerical computation. Comput. Sci. Eng. **13**(2), 22–30 (2011)

A Meta-Graph for the Construction of an RNA-Centered Knowledge Graph

Emanuele Cavalleri[ID], Sara Bonfitto[ID], Alberto Cabri[ID], Jessica Gliozzo[ID], Paolo Perlasca[ID], Mauricio Soto-Gomez[ID], Gabriella Trucco[ID], Elena Casiraghi[ID], Giorgio Valentini[✉][ID], and Marco Mesiti[ID]

AnacletoLab - Dipartimento di Informatica, Università degli Studi di Milano, Via Celoria 18, Milano, Italy
{emanuele.cavalleri,sara.bonfitto,alberto.cabri,jessica.gliozzo, paolo.perlasca,mauricio.soto-gomez,gabriella.trucco,elena.casiraghi, marco.mesiti}@unimi.it, valentini@di.unimi.it

Abstract. The COVID-19 pandemic highlighted the importance of RNA-based technologies for the development of new vaccines. Besides vaccines, a world of RNA-based drugs, including small non-coding RNA, could open new avenues for the development of novel therapies covering the full spectrum of the main human diseases. In the context of the "National Center for Gene Therapy and Drugs based on RNA Technology" funded by the Italian PNRR and the NextGenerationEU program, our lab will contribute to the construction of a Knowledge Graph (KG) for RNA-drug analysis and the development of innovative algorithms to support RNA-drug discovery. In this paper, we describe the initial steps for the identification of public data sources from which information about different kinds of non-coding RNA sequences (and their relationships with other molecules) can be collected and used for feeding the KG. An in-depth analysis of the characteristics of these sources is provided, along with a meta-graph we developed to guide the RNA-KG construction by exploiting and integrating biomedical ontologies and relevant data from public databases.

Keywords: RNA molecules · RNA data sources · Biological KGs

1 Introduction

RNA-based drugs represent one of the most promising advances in therapeutics, as evidenced by the recent success of mRNA-based vaccines for the COVID-19 pandemic [2]. More generally, coding and non-coding RNA molecules can potentially lead to a significant breakthrough in the therapy of cancer, genetic and neurodegenerative disorders, cardiovascular and infectious diseases [17].

Conventional drugs show relevant limitations in their druggable targets, because they usually consist of small molecules targeting proteins and only about 10% of proteins have druggable binding sites and no more than 2% of the human genome is protein-coding. On the contrary, RNA drugs can target

I. Rojas et al. (Eds.): IWBBIO 2023, LNBI 13919, pp. 165–180, 2023.
https://doi.org/10.1007/978-3-031-34953-9_13

both proteins and mRNA, as well as other non-coding RNA. Moreover, they can encode missing or defective proteins, regulate the transcriptome, and mediate DNA or RNA editing. Thus, RNA technology significantly broadens the set of druggable targets, and is also less expensive than other technologies (e.g. drug synthesis based on recombinant proteins), due to the relatively simple structure or RNA molecules that facilitates their biochemical synthesis and chemical modifications [54]. Guided by these considerations, the scientific community is now focusing on the study and development of novel RNA drugs, ranging from mRNA vaccines to small non-coding RNA, such as microRNA (miRNA), small interfering RNA (siRNA) and antisense oligonucleotides (ASO), to use their therapeutic potential for a large range of relevant human diseases [17,54].

In the framework of the NextGenerationEU funded "National Center for Gene Therapy and Drugs based on RNA Technology", we aim to support the discovery of novel RNA-based drugs by developing a RNA-centered KG (RNA-KG).[1] RNA-KG will collect and organize data and knowledge about RNA molecules, retrieved from public databases and/or generated from the results of the research groups involved in the national center. Moreover, it will provide a comprehensive description of the relationships among the various kind of RNAs, diseases, drugs, phenotypes, and other bio-medical entities. RNA-KG will be the basis for the development of novel cutting-edge AI methods specifically tailored for the analysis of various biological processes involving RNA. These methods could also open the way to RNA-drug prioritization, RNA drug-target prediction, and other prediction tasks for discovering RNA drugs for specific human diseases.

In this paper we summarize our initial research efforts and results toward the construction of RNA-KG. In particular, we present our thorough analysis of the literature about RNA databases (RNA-DBs) in order to both collect RNA data and select the relevant relationships between different RNA molecules and other biomedical entities that will constitute the core of RNA-KG. We then describe the meta-graph we developed to guide the RNA-KG construction by exploiting and integrating biomedical ontologies and relevant data from public databases. The meta-graph uses multiple types of nodes and edges to detail and represent the "RNA-world", i.e. the main types of RNA molecules and their interactions, as well as their relationships with other relevant biomolecules and medical concepts.

The paper is organized as follows. Section 2 introduces approaches developed for the construction of RNA-KG starting from different heterogeneous databases and highlights the issues that should be faced. Then, Sect. 3 describes the main characteristics of coding and non-coding RNA sequences and their classification. Section 4 provides the methodology for the identification of the RNA-DBs and the main characteristics of the more than 50 data sources currently selected. Section 5 highlights the main types of relationships that can be extracted from the sources and introduces a meta-graph that shows the potential relationships that will be available among the RNA molecules. Finally, Sect. 6 reports our concluding remarks and next research activities.

---

[1] Initial results available at https://github.com/AnacletoLAB/RNA-KG.

2   KG Construction from Bio-Medical Data Sources

The data integration issue is a well-known problem in the area of data management and many approaches have been devised to deal with relational data [24]. However, the explosion of data formats (like CSV, JSON, XML) and the variability in the representation of the same types of information [50] has pushed the need to exploit ontologies as global common models both for accessing (OBDA - Object-Based Data Access) and integrating (OBDI - Object-Based Data Integration) data sources [57]. In OBDA, queries are expressed in terms of an ontology, and the mappings between the ontology and the data sources' schema are described in the form of declarative rules. Two approaches are usually proposed to enable access and integration to different data sources: *materialization*, where data are converted from the local data format according to the ontology concepts and relationships (i.e. data are converted into an RDF KG and locally stored in a data-warehouse of triples that can be queried through SPARQL); *virtualization*, where the transformation is executed on the fly during the evaluation of queries by exploiting the mapping rules and the ontology. In this case, only the data from the original sources involved in the query are accessed. Materialization can provide fast and accurate access to data because already organized in a centralized repository. However, data freshness can be compromised when data sources frequently change. On the other hand, virtualization allows access to fresh data but requires the application of transformations during query evaluation and can cause delay, and inconsistency when the structures of the local sources change. Many approaches are available for the specification of mapping rules like R2RML [18] (a W3C standard for relational to RDF mapping), and RML [20] that extends the standard for dealing with other formats. Moreover, SPARQL-Generate [40], YARRRML [25] and ShExML [23] were also proposed for dealing with data heterogeneity.

In the biological context, many efforts are nowadays devoted to the construction of KGs by integrating different public sources that exploit the materialization and virtualization approaches previously described. An approach for integrating different biological data into a biological KG was proposed in [71]. The approach designs a Connecting Ontology CO to integrate all the external ontologies describing the involved data sources. By exploiting algorithms for fusing and integrating annotations, an enriched KG is obtained that spans multiple data sources and is annotated by the integrated biological ontology. The effectiveness of this approach is shown by integrating `rice gene-phenotype` and `lactobacillus` data sources by gluing together the GO, Trait, Disease, and Plant Ontologies. In [9], the Precision Medicine KG (PrimeKG) was developed to represent holistic and multimodal views of diseases. PrimeKG integrates more than 20 high-quality resources with more than 4 million relations that capture information like disease-associated perturbations in the proteome, biological processes, and molecular pathways. The considered data were collected and annotated using diverse ontologies such as Disease Gene Network, Mayo Clinical knowledgebase, Mondo, Bgee, and DrugBank. A virtualization approach based on an ontology-based federation of three data sources (Bgee, OMA, and

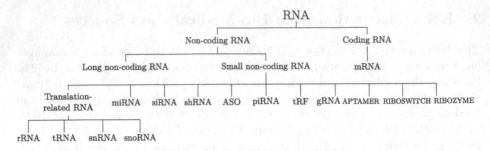

Fig. 1. RNA classification.

UNIProtKB) was presented in [59]. Starting from a semantic model for gene expression named GenEx, the authors propose using mapping rules for dealing with the different formats of the three sources and allow the issue of joint queries across the sources by exploiting SPARQL endpoints. A fully automated Python 3 library named PheKnowLator (Phenotype Knowledge Translator) was proposed in [8] for the construction of semantically rich, large-scale biomedical KGs that are Semantic Web compliant and amenable to automatic OWL reasoning, conform to contemporary property graph standards. The library offers tools to download data, transform and/or pre-processing of resources into edge lists, construct knowledge graphs, and generate a wide-range of outputs.

All these papers point out the difficulties that arise when trying to integrate different data sources that exploit different data models, formats, and ontologies. Specifically, data redundancies, data duplicates, and lack of common identifier mechanisms must be properly addressed. In the case of RNA data integration, we also have to consider the lack of specific ontologies for the description of all possible non-coding RNA sequences, and the presence of ontologies that are not well-recognized by the community because still in their infancy. All these aspects need to be properly addressed in the generation of RNA-KG.

3   RNA Molecules

The wide variety of RNA molecules, which can be classified as sketched in Fig. 1, are translated into proteins, regulate gene expression, have enzymatic activity, and modify other RNAs.

Coding RNA. RNA molecules that encode proteins are named messenger RNAs (mRNAs). In Eukaryotes, mRNA primary transcripts follow a cascade of biological processes [63] for transforming them into mature, functional mRNA molecules that are read by ribosomes, transformed into amino acids, and assembled in proteins (through peptide bonds) during the translation process.

Non-coding RNA. Transcripts that are not translated into proteins are named non-coding RNAs (ncRNAs). They can be further classified into long non-coding RNAs (lncRNAs – with more than 200 nucleotides) and small non-coding RNAs (sncRNAs – with less than 200 nucleotides). lncRNAs are the majority of transcription products and play a pivotal role in disease development and progression. Circular RNAs (circRNAs) are lncRNAs produced from alternative splicing events, and may play a role as splicing event regulators. circRNAs have been involved in many human illnesses, including cancer and neurodegenerative disorders such as Alzheimer's and Parkinson's disease, due to their aberrant expression in different pathological conditions [51].

Small Non-coding RNA (sncRNA). sncRNAs are involved in several cellular biological processes, including: translation processes; RNA interference (RNAi) pathways; splicing and self-cleavage processes; catalysis of biochemical reactions; and, targeted gene editing.

sncRNAs Involved in the Translation Process. Several sncRNA are involved in this process, including ribosomal RNA (rRNA), transfer RNA (tRNA), small nuclear RNA (snRNA), and small nucleolar RNA (snoRNA). While rRNA constitutes the core structural and enzymatic framework of the ribosome, tRNAs are characterized by a structure consisting of an acceptor stem that links to a particular amino acid and of a specific anticodon sequence of 3 bases complementary to the corresponding mRNA codon, thus assuring the translation from the mRNA codon triplets to the corresponding sequence of amino acids. snRNAs and snoRNAs primarily guide chemical modifications of other RNAs, mainly rRNAs, tRNAs, and control chromatin compaction and accessibility.

sncRNAs Associated with RNA Interference Pathways. RNA interference pathways play a central role in gene expression and their misregulation is associated with several diseases. sncRNAs associated with RNAi pathways include: microRNAs (miRNAs), short interfering RNAs (siRNAs), short hairpin RNAs (shRNAs), ribozymes, antisense oligonucleotides (ASOs), piwi-interacting RNAs (piRNAs), tRNA-derived fragments (tRFs), and aptamers. Mature miRNAs, siRNAs, shRNAs, and ASOs regulate the mRNA expression by blocking translation or promoting degradation of the target mRNA (complementary base pairings). Unlike siRNAs, each miRNA can simultaneously regulate the expression and the activity of hundreds of protein-coding genes and transcription factors (TFs). miRNAs from various exogenous sources, which are present in human circulation, are named xeno-miRNAs [60]. By contrast, ASOs are more effective to knock down nuclear targets, whereas siRNAs are superior at suppressing cytoplasmic targets due to their interaction with specific enzymes forming different editing-machinery complexes. Similarly, throughout RNAi, piRNAs and tRFs promote genome integrity, avoiding potential threats to cellular homeostasis, by silencing transposons, retrotransposons and repeat sequences.

Aptamers, Riboswitches, Ribozymes and Guide-RNA. The tertiary structure of RNA sequences can also be investigated to identify interferences. Aptamers are short single-stranded nucleic acids that can bind to a variety of targets (e.g. proteins, peptides, carbohydrates, DNA, and RNA) thanks to their 3D conformation. Riboswitches are small non-coding RNAs involved in alternative splicing and self-cleavage processes that cause gene expression control and mRNA degradation, critical for survival of the cell [46]. Some RNAs, such as ribozymes, even possess enzymatic activity therefore catalyzing biochemical reactions (e.g. mRNA and protein cleavage). Synthetic ribozymes can and have already been designed to target viral RNA. Synthetic guide RNAs (gRNAs) are usually involved in the application of CRISPR-Cas9 technique, used for gene editing and gene therapy [61].

4   RNA-Based Data Sources

In order to construct RNA-KG, we examined public online repositories for ncRNA sequences and annotations, and for the interactions among RNAs and between RNAs and different types of molecules. An extensive literature review was carried out to identify repositories developed by well-reputed organizations and published in top journals of the sector. Moreover, we have considered only recent repositories that are periodically updated, and contain significant amounts of molecules and relationships. Furthermore, repositories that are included in other bigger repositories have been highlighted, as well as those that are used as collectors of other repositories. We have also identified the presence of controlled vocabularies, thesaurus, ontologies that formally describe the repository content, and the presence of well-recognized identification schemes.

Our research ended up identifying more than 50 public repositories. The papers presenting these data sources were published in top journals (like NAR, BMC Bioninformatics, Science, RNA Journal, IEEE/ACM TCBB) between 2008 and 2023 but the majority were published in the last 5 years. The main characteristics of the identified repositories are reported in Table 1, whose entries are organized according to the main type of ncRNA made available. Moreover, for each repository we report the number of sequences and species (note that the HS and MM tags are used when the involved species are only Homo Sapiens and Mus Musculus). No species is present for aptamers because they are synthetic. Note that, none of the databases contains piRNA molecules related to Homo Sapiens. Finally, the ontologies used for describing the sequences and their relationships are reported.

Some data sources are included in larger ones. For example, the data sources Epimir, HMDD and miR2Disease (containing miRNA) are included in miRNet. The data source TAM (containing relationships of different miRNA molecules with diseases in which they are under/over-expressed) is included in HMDD which in turn is included in miRNet. The data source NONCODE is a collector of other data sources (like LncRNADisease, RNADisease, Lnc2Cancer and LncRNAWiki). miRandola collects the miRNA entries that are also made available in Vesciclepedia along with the relationships with circRNA and lncRNA. RNAcentral is a big collector containing the same data of miRBase, NONCODE, snoDB, LncBook, LNCipedia, lncRNAdb, and TarBase.

Several standard ontologies are used in the considered data sources (their specifications are reported in ebi.ac.uk/ols, bioportal.bioontology.org). The "Gene Ontology" (GO) for representing the name of ncRNA molecules,

Table 1. Main RNA-based data sources with the adopted ontology (when available)

Type	DB name	species	size	DO	GO	NCRO	ChEBI	MeSH	Others
				Ontology					
miRNA	miRBase [37]	271	38,589		x	x			
	miRDB [12]	5	7,086		x	x			
	miRNet [21]	10	7,928	x	x	x			OBI
	EpimiR [16]	7	617			x			
	HMDD [29]	HS	1,206	x		x			OBI
	miR2Disease [31]	HS	299	x		x			
	TargetScan [49]	5	5,168		x	x			
	SomamiR DB [3]	HS	1,078			x			EDAM
	TarBase [35]	18	2,156		x	x			
	miRTarBase [26]	28	5,818		x	x			
	SM2miR [45]	21	1,658		x	x	x		
	TransmiR [45]	19	785		x	x			
	PolymiRTS [4]	HS	11,182		x	x			SNPO
	dbDEMC [70]	HS	3,268		x	x			
	TAM [69]	HS	1,209	x		x			
	PuTmiR [1]	HS	1,296			x			
	miRPathDB [36]	HS, MM	29,430		x	x			
	miRCancer [68]	HS	57,984			x			
	miRdSNP [7]	HS	249		x	x			
s(i/h)RNA	ICBP siRNA [30]	HS, MM	147			x			
Aptamers	Apta-Index [43]		7,770			x			
ASO	eSkip-Finder [14]	4	2,196			x			
	DrugBank [65]	HS	12				x		
lncRNA	LncBook [41]	HS	323,950		x	x		x	EDAM
	LNCipedia [62]	60	127,802			x			
	LncRNADisease [10]	4	19,166	x	x	x		x	
	LncExpDB [42]	HS	101,293			x			
	dbEssLnc [72]	HS, MM	207			x			
	lncATLAS [48]	HS	6,768						
	NONCODE [73]	39	644,510			x			
	Lnc2Cancer [22]	HS	3,402	x	x	x		x	
	LncRNAWiki [44]	HS	106,063			x			
gRNA	Addgene [33]	29	296			x			
Ribozyme	Ribocentre [19]	1,195	21,084						SO
Viral RNA	ViroidDB [39]	9	9,891						
Riboswitch	TBDB [47]	3,621	23,497						
	RSwitch [55]	50	215						
tRF	tRFdb [38]	7	863						
	tsRFun [64]	HS	3,940						
	MINTbase [56]	HS	28,824						
snoRNA	snoDB [6]	HS	751			x			
tRNA	tRNAdb [32]	681	9,758			x			
all RNA	RNAInter [34]	156	455,887			x	x		
	RNALocate [15]	101	123,592			x			
	RNADisease [11]	117	91,245	x		x		x	
	ncRDeathDB [67]	12	648			x			
	cncRNADB [28]	21	2,002			x			BTO, EFO
	ViRBase [13]	152	56,614			x			
	miRandola [58]	14	1,002			x			
	Vesiclepedia [53]	41	20,490			x			

whereas the "Disease Ontology" (DO) for the description of associated diseases and CheBI (Chemical Entities of Biological Interest) for the description of the biologically relevant 'small' chemical compounds. Finally, data formats specifically developed for biological pathways (like Panther, Reactome or Wikipathways) are used for semantically annotating the RNA molecules. Non-Coding RNA Ontology (NCRO) provides a systematically structured and precisely defined controlled vocabulary for the structures of ncRNAs, their molecular and cellular functions, and their impacts on phenotypes [27]. Ontology for Biomedical Investigations (OBI) serves as a resource for annotating biomedical investigations, including the study design, protocols and instrumentation used, the data generated and the types of data analysis. Single-Nucleotide Polymorphism Ontology (SNPO) is a domain ontology that provides a formal and unambiguous representation of genomic variations. Despite its name, SNPO is not limited to the representation of SNPs but can be used for the representation of variations observed in the genome of various species. EMBRACE Data And Methods (EDAM) is an ontology of bioinformatics operations (tool, application, or workflow functions), data types and formats, and topics (application domains). EDAM supports semantic annotation of diverse entities such as Web services, databases, programmatic libraries, standalone tools, interactive applications, data schemas, datasets and publications within bioinformatics. MeSH (Medical Subject Headings) is the National Library of Medicine (NLM) controlled vocabulary thesaurus used for indexing PubMed citations. Sequence Ontology (SO) is a structured controlled vocabulary for sequence annotation, for the exchange of annotations, and for the description of sequences in databases. The BRENDA Tissue Ontology (BTO) is a structured controlled vocabulary for the source of an enzyme comprising tissues, cell lines, cell types and cell cultures from uni- and multicellular organisms. The Experimental Factor Ontology (EFO) is an application-focused ontology modeling the experimental variables in multiple resources at the EBI and Open Targets. The ontology pulls together classes from reference ontologies such as disease, cell line, cell type and anatomy and adds axiomatization to connect areas such as disease to phenotype.

Some data sources, even though obtained or quantified through experimental protocols and widely used and accepted sequencing techniques (e.g. RNA-Seq or CLIP-Seq), do not exploit any ontology (see Table 1), being the focus of recent studies, for which a semantic conceptualization has not yet been defined. These experimental data can be associated with Homo Sapiens ncRNA molecules (like tRFs in tRFdb, tsRFun, MINTbase) or with molecules of viruses and bacteria that are involved in alterations of the human phenotype (as in the case of viroidDB, and TBDB), or, finally, contain information on antibacterial drug targets, such as riboswitches in the RSwitch database.

In general, a well-recognized and globally accepted ontology for the representation of any kind of ncRNA molecule is still lacking. Often for referring to lncRNA molecules, the name of the gene encoding a protein that is physically closest is used. Moreover, lncRNA genes with no known function are named pragmatically based on their genomic context; if there is a proximal (genomically adjacent close in physical proximity) protein coding gene (PCG) then

Table 2. Main relations among bio-entities involving RNA with the RO identifier

Relation ID	Name	Abbreviation
RO:0002429	involved in positive regulation of	regulates$^+$
RO:0002430	involved in negative regulation of	regulates$^-$
RO:0002434	interacts with	interacts
RO:0002436	molecularly interacts with	interacts^{MOL}
RO:0010002	is carrier of	carries
RO:0002204	gene product of	gene product
RO:0002526	overlaps sequence of	overlaps
RO:0002528	is upstream of sequence of	upstream
RO:0002529	is downstream of sequence of	downstream
RO:0002202	develops from	develops from
RO:0002203	develops into	develops into

the lncRNA genes are given a gene symbol beginning with the PCG symbol [66]. The identification scheme used for miRNA (which are the majority of data sources) is always borrowed from miRBase. This makes all the other data sources associated with miRNAs, miRBase "compliant". Furthermore, the identification scheme associated with miRNAs is partially included in NCRO, which includes miRNA transcripts from Homo Sapiens cells [27].

5   A Meta-Graph Representing RNA-Centered Relationships

Besides the sequences, these data sources also contain different kinds of relationships that can be exploited for the construction of the KG. These relationships can be represented by exploiting the Relation Ontology (RO) [52] which provides precise semantics. Table 2 reports the main relationships that have been identified in the data sources discussed in Sect. 4. For each relation, Table 2 reports the RO identifier, the corresponding meaning, and an abbreviated form that is used in our paper. Whenever feasible, we have used bidirectional relationships (e.g. interacts with) and introduced the inverse relationships in case only a unidirectional relationship is available in the data source (e.g. develops from and develops into). The general relationships "interacts with" available in RO with the meaning "A relationship that holds between two entities in which the processes executed by the two entities are causally connected" have been specified in the most specific relationships "molecularly interacts with" in our classification to represent the situation in which the two partners are molecular entities that directly physically interact with each other (e.g. via a stable binding interaction or a brief interaction during which one modifies the other). We use this relationship to represent a specific interaction process at the molecular level (e.g.

Table 3. Available relations involving RNA molecules

entity_1	relation	entity_2	DBname
miRNA	develops into develops from	miRNA	miRBase (48,860)
miRNA	interacts^{MOL}	mRNA	miRDB (3,519,884); SomamiR DB(2,313,416); tsRFun(10,829); miRNet (1,106,242); TargetScan (2,850,014); TarBase (665,843); miRTarBase (5,173,924)
xeno-miRNA	interacts^{MOL}	mRNA	miRNet(1,919,245)
miRNA	interacts^{MOL}	pseudogene	miRNet(59,417)
protein (TF)	regulates$^{+/-}$	miRNA	miRNet(3,311); TransmiR (3,730); TAM (665)
miRNA	interacts	lncRNA	SomamiR DB (127,025); LncBook (146,092,274)
miRNA	interacts	circRNA	SomamiR DB (428,237)
miRNA	interacts	miRNA	TAM (1,218)
siRNA	interacts^{MOL}	mRNA	ICBP siRNA (94)
shRNA	interacts^{MOL}	mRNA	ICBP siRNA (53)
protein	interacts	aptamer	Apta-Index (7,770)
ASO	interacts^{MOL}	mRNA	eSkip-Finder (11,778); DrugBank (7)
ASO	interacts	protein	DrugBank (11)
ASO	carries	protein	DrugBank (1)
lncRNA	interacts	mRNA	LncExpDB(28,443,865); LncRNADisease (478)
lncRNA	interacts	protein	LncBook (772,745)
small protein	gene product	lncRNA	LncRNAWiki (9,387)
gRNA	interacts	gene	Addgene(321)
ribozyme	interacts	gene	Ribocentre(48)
viral RNA	overlaps	ribozyme	ViroidDB(17,460)
riboswitch	downstream	protein	TBDB(23,535)
protein	upstream	riboswitch	TBDB(23,535)
tRF	interacts	miRNA	tsRFun(45,165)
tRF	develops from	tRNA	MINTbase(125,285); tRFdb(792); tsRFun(46,798)
snoRNA	interacts	gene	snoDB(763)
snoRNA	interacts	lncRNA	snoDB(45)
snoRNA	interacts	miRNA	snoDB(17)
snoRNA	interacts	mRNA	snoDB(276)
snoRNA	interacts	pseudogene	snoDB(10)
snoRNA	interacts	rRNA	snoDB(735)
snoRNA	interacts	snoRNA	snoDB(670)
snoRNA	interacts	snRNA	snoDB(164)
snoRNA	interacts	tRNA	snoDB(164)
tRNA	interacts^{MOL}	amino acid	tRNAdb(8,872)

complementary base pairing occurring in RNAi in miRNA-mRNA interaction or tRNA molecule charged with a specific amino acid).

Table 3 summarizes the relationships among RNA molecules that we have identified in the different data sources. For each type of relationship, the table reports the data sources containing it and the number of occurrences. The most representative relationships, with more than one million occurrences, are those

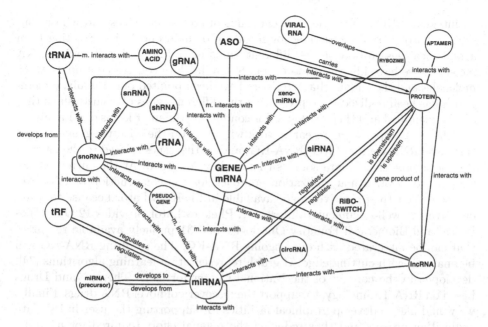

Fig. 2. RNA-centered meta-graph aka the schema for RNA-KG

involving miRNA or lncRNA molecules. These categories of molecules often interact with mRNA molecules and proteins involved in diseases.

The content of the two tables along with the bio-molecules described in Sect. 3 have been used for the generation of the meta-graph reported in Fig. 2. The graphical representation provides a global overview of the richness of information that is currently provided. Moreover, the meta-graph points out the presence of a central hub, named "GENE/mRNA", that is bound to many kinds of ncRNA. This characteristic might have a deep impact on the discovery of new unconsidered interactions among ncRNA molecules. To simplify the visualization of the meta-graph, we omitted most of the non-RNA biomolecular and medical entities that are known to play an important role to study the biology and support the discovery of novel RNA drugs. Indeed the meta-graph in Fig. 2 can be further extended with other nodes representing other biological entities (e.g. diseases, epigenetic modifications, small molecules, tissues, biological pathways, cellular compartments) and relationships relevant to the analysis of RNA-KG.

6   Concluding Remarks

This paper reports the initial results of an ongoing project for the creation of a biomedical knowledge graph for the representation of non-coding RNA molecules and their relationships made available in different publicly available data sources. We have described different approaches and techniques that have been used in the biomedical context for the integration of heterogeneous data sources into an integrated KG presenting different kinds of nodes and edges and whose content and structure are described by means of domain ontologies. Starting from a biological description of the different kinds of coding and non-coding RNA molecules we have identified more than 50 public data sources containing these molecules and their main characteristics have been pointed out. Finally, a metagraph has been realized for representing the different types of relationships that can be identified and then exploited for conducting different kinds of AI analysis.

The next steps of our research work will focus on the effective generation of this novel RNA-centered KG (RNA-KG) and on the issues related to the integration of heterogeneous data sources that we have outlined. At the current stage, we plan to follow a materialization approach for the construction of RNA-KG and to exploit the primitives made available in PheKnowLator because they are effective and well-documented. Moreover, PheknowLator provides 12 Open Biological and Biomedical Foundry Ontologies and 31 publicly available resources that can be integrated with our ongoing RNA-KG. The resulting RNA-KG will be analyzed with cutting-edge AI graph representation learning algorithms [74], developed in the context of the National Center for Gene Therapy and Drugs based on RNA Technology, to support the discovery of novel RNA drugs. Finally, we would like to develop graphical facilities for supporting the user in the data acquisition process and thus reducing the manual effort required for mapping the data available in the different data sources into RNA-KG [5].

Acknowledgments. This research was supported by the "National Center for Gene Therapy and Drugs based on RNA Technology", PNRR-NextGenerationEU program [G43C22001320007]. The authors wish to thank Tiffany J. Callahan, Justin T. Reese, and Peter N. Robinson for useful discussions on the topics of this paper.

References

1. Bandyopadhyay, S., et al.: PuTmiR: a database for extracting neighboring transcription factors of human microRNAs. BMC Bioinf. **11**(190) (2010). http://isical.ac.in/~bioinfo_miu/TF-miRNA1.php
2. Barbier, A., et al.: The clinical progress of mRNA vaccines and immunotherapies. Nat. Biotechnol. **40**, 840–865 (2022)
3. Bhattacharya, A., Cui, Y.: SomamiR 2.0: a database of cancer somatic mutations altering microRNA–ceRNA interactions. Nucleic Acids Res. **44**, D1005–D1010 (2015). http://compbio.uthsc.edu/SomamiR/home.php
4. Bhattacharya, A., Ziebarth, J.D., Cui, Y.: PolymiRTS database 3.0: linking polymorphisms in microRNAs and their target sites with human diseases and biological pathways. Nucleic Acids Res. **42**(D1), D86–D91 (2013). http://compbio.uthsc.edu/miRSNP
5. Bonfitto, S., Perlasca, P., Mesiti, M.: Easy-to-use interfaces for supporting the semantic annotation of web tables. In: International Workshop on Data Platforms Design, Management, and Optimization (2023)
6. Bouchard-Bourelle, P., et al.: snoDB: an interactive database of human snoRNA sequences, abundance and interactions. Nucleic Acids Res. **48**(D1), D220–D225 (2020). http://bioinfo-scottgroup.med.usherbrooke.ca/snoDB

7. Bruno, A., et al.: miRdSNP: a database of disease-associated SNPs and microRNA target sites on 3'UTRs of human genes. BMC Genomics **13**(5) (2012). http://mirdsnp.ccr.buffalo.edu

8. Callahan, T.J., et al.: A framework for automated construction of heterogeneous large-scale biomedical knowledge graphs. bioRxiv (2020)

9. Chandak, P., et al.: Building a knowledge graph to enable precision medicine. Sci. Data **10**(1), 67 (2023)

10. Chen, G., et al.: LncRNADisease: a database for long-non-coding RNA-associated diseases. Nucleic Acids Res. **41**(D1), D983–D986 (2012). http://rnanut.net/lncrnadisease

11. Chen, J., et al.: RNADisease v4.0: an updated resource of RNA-associated diseases, providing RNA-disease analysis, enrichment and prediction. Nucleic Acids Res. **51**(D1), D1397–D1404 (2023). http://rnadisease.org/download

12. Chen, Y., Wang, X.: miRDB: an online database for prediction of functional microRNA targets. Nucleic Acids Res. **48**(D1), D127–D131 (2019). http://mirdb.org

13. Cheng, J., et al.: ViRBase v3.0: a virus and host ncRNA-associated interaction repository with increased coverage and annotation. Nucleic Acids Res. **50**(D1), D928–D933 (2022). http://rna-society.org/virbase

14. Chiba, S., et al.: eSkip-Finder: a machine learning-based web application and database to identify the optimal sequences of antisense oligonucleotides for exon skipping. Nucleic Acids Res. **49**(W1), 193–198 (2021). http://eskip-finder.org/cgi-bin/input.cgi

15. Cui, T., et al.: RNALocate v2.0: an updated resource for RNA subcellular localization with increased coverage and annotation. Nucleic Acids Res. **50**(D1), 333–339 (2022). http://rna-society.org/rnalocate/

16. Dai, E., et al.: EpimiR: a database of curated mutual regulation between miR-NAs and epigenetic modifications. Database (Oxford), 6 (2014). http://jianglab.cn/EpimiR

17. Damase, T.R., et al.: The limitless future of RNA therapeutics. Front. Bioeng. Biotechnolo. **9** (2021). http://frontiersin.org/articles/10.3389/fbioe.2021.628137

18. Das, S., et al.: R2rml: Rdb to RDF mapping language. In: W3C (2012). http://www.w3.org/TR/r2rml/

19. Deng, J., et al.: Ribocentre: a database of ribozymes. Nucleic Acids Res. **51**(D1), D262–D268 (2023). http://ribocentre.org

20. Dimou, A.: RML: a generic language for integrated RDF mappings of heterogeneous data. In: Proceedings of Workshop on Linked Data on the Web. CEUR Workshop Proceedings, vol. 1184 (2014)

21. Fan, Y., et al.: Xeno-miRNet: a comprehensive database and analytics platform to explore xeno-miRNAs and their potential targets. PeerJ **6** 12 (2018). http://mirnet.ca/miRNet

22. Gao, Y.: Lnc2Cancer 3.0: an updated resource for experimentally supported lncRNA/circRNA cancer associations and web tools based on RNA seq and scRNA seq data. Nucleic Acids Res. **49**(D1), 1251–1258 (2021). http://bio-bigdata.hrbmu.edu.cn/lnc2cancer

23. García-González, H., et al.: ShExML: improving the usability of heterogeneous data mapping languages for first-time users. PeerJ Comput. Sci. **6**, 27 (2020). http://hal.science/hal-03110745

24. Halevy, A.: Information Integration, pp. 1490–1496. Springer, Cham (2009)

25. Heyvaert, P., De Meester, B., Dimou, A., Verborgh, R.: Declarative rules for linked data generation at your fingertips! In: Gangemi, A., et al. (eds.) ESWC 2018. LNCS, vol. 11155, pp. 213–217. Springer, Cham (2018). https://doi.org/10.1007/978-3-319-98192-5_40

26. Huang, H.Y., et al.: miRTarBase update 2022: an informative resource for experimentally validated miRNA-target interactions. Nucleic Acids Res. **50**(D1), 222–230 (2021). http://mirtarbase.cuhk.edu.cn

27. Huang, J., et al.: The non-coding RNA Ontology (NCRO): a comprehensive resource for the unification of non-coding RNA biology. J. Biomed. Semant. **7**(1), 24 (2016)

28. Huang, Y., et al.: cncRNAdb: a manually curated resource of experimentally supported RNAs with both protein-coding and noncoding function. Nucleic Acids Res. **49**(D1), 65–70 (2021). http://rna-society.org/cncrnadb

29. Huang, Z., et al.: HMDD v3.0: a database for experimentally supported human microRNA-disease associations. Nucleic Acids Res. **47**, 1013–1017 (2018). http://www.cuilab.cn/hmdd

30. ICB Program: siRNA (2010). http://web.mit.edu/sirna/

31. Jiang, Q., et al.: miR2Disease: a manually curated database for microRNA deregulation in human disease. Nucleic Acids Res. **37**, 98–104 (2008). http://www.mir2disease.org

32. Jühling, F., et al.: tRNAdb 2009: compilation of tRNA sequences and tRNA genes. Nucleic Acids Res. **37**(suppl_1), 159–162 (2009). http://trna.bioinf.uni-leipzig.de/DataOutput/

33. Kamens, J.: The Addgene repository: an international nonprofit plasmid and data resource. Nucleic Acids Res. **43**(D1), 1152–1157 (2015). http://addgene.org

34. Kang, J., et al.: RNAInter v4. 0: RNA interactome repository with redefined confidence scoring system and improved accessibility. Nucleic Acids Res. **50**(D1), 326–332 (2022). http://rnainter.org

35. Karagkouni, D., et al.: DIANA-TarBase v8: a decade-long collection of experimentally supported miRNA-gene interactions. Nucleic Acids Res. **46**, 239–245 (2017). http://dianalab.e-ce.uth.gr/tools

36. Kehl, T., et al.: miRPathDB 2.0: a novel release of the miRNA Pathway Dictionary Database. Nucleic Acids Res. **48**(D1), 142–147 (2019). http://mpd.bioinf.uni-sb.de

37. Kozomara, A., et al.: miRBase: from microRNA sequences to function. Nucleic Acids Res. **47**(D1), 155–162 (2018). http://mirbase.org

38. Kumar, P., et al.: tRFdb: a database for transfer RNA fragments. Nucleic Acids Res. **43**(D1), 141–145 (2015), http://genome.bioch.virginia.edu/trfdb

39. Lee, B.D., et al.: ViroidDB: a database of viroids and viroid-like circular RNAs. Nucleic Acids Res. **50**(D1), 432–438 (2022). http://viroids.org

40. Lefrançois, M., Zimmermann, A., Bakerally, N.: A SPARQL extension for generating RDF from heterogeneous formats. In: Blomqvist, E., Maynard, D., Gangemi, A., Hoekstra, R., Hitzler, P., Hartig, O. (eds.) ESWC 2017. LNCS, vol. 10249, pp. 35–50. Springer, Cham (2017). https://doi.org/10.1007/978-3-319-58068-5_3

41. Li, Z., et al.: LncBook 2.0: integrating human long non-coding RNAs with multi-omics annotations. Nucleic Acids Res. **51**(D1), 186–191 (2023). http://ngdc.cncb.ac.cn/lncbook

42. Li, Z., et al.: LncExpDB: an expression database of human long non-coding RNAs. Nucleic Acids Res. **49**(D1), 962–968 (2021). http://ngdc.cncb.ac.cn/lncexpdb

43. Liao, A.M., et al.: Aptamer-based target detection facilitated by a 3-stage G-quadruplex isothermal exponential amplification reaction. Bioengineering **188** (2022). https://doi.org/10.3791/64342. http://aptagen.com/apta-index

44. Liu, L., et al.: LncRNAWiki 2.0: a knowledgebase of human long non-coding RNAs with enhanced curation model and database system. Nucleic Acids Res. **50**(D1), 190–195 (2022). http://ngdc.cncb.ac.cn/lncrnawiki1

45. Liu, X., et al.: SM2miR: a database of the experimentally validated small molecules' effects on microRNA expression. Bioinformatics **29**(3), 409–411 (2012). http://jianglab.cn/SM2miR

46. Machtel, P., et al.: Emerging applications of riboswitches - from antibacterial targets to molecular tools. J. Appl. Genet. **57**(4), 531–541 (2016)

47. Marchand, J.A., et al.: TBDB: a database of structurally annotated T-box riboswitch: tRNA pairs. Nucleic Acids Res. **49**(D1), 229–235 (2021). http://tbdb.io

48. Mas-Ponte, D., et al.: LncATLAS database for subcellular localization of long noncoding RNAs. RNA **23**(7), 1080–1087 (2017). http://lncatlas.crg.eu

49. McGeary, S.E., et al.: The biochemical basis of microRNA targeting efficacy. Science **366** (2019). http://targetscan.org

50. Mesiti, M., et al.: XML-based approaches for the integration of heterogeneous biomolecular data. BMC Bioinf. **10**(Suppl 12), S7 (2009)

51. Nisar, S., et al.: Insights into the role of circRNAs: Biogenesis, characterization, functional, and clinical impact in human malignancies. Front. Cell Dev. Biol. **9** (2021)

52. Ong, E., et al.: Ontobee: a linked ontology data server to support ontology term dereferencing, linkage, query and integration. Nucleic Acids Res. **45**(D1), 347–352 (2016)

53. Pathan, M., et al.: Vesiclepedia 2019: a compendium of RNA, proteins, lipids and metabolites in extracellular vesicles. Nucleic Acids Res. **47**(D1), 516–519 (2019). http://microvesicles.org

54. Paunovska, K., et al.: Drug delivery systems for RNA therapeutics. Nat. Rev. Genet. **23**, 265–280 (2022)

55. Penchovsky, R., et al.: RSwitch: a novel bioinformatics database on riboswitches as antibacterial drug targets. IEEE/ACM Trans. Comput. Biol. Bioinf. **18**(2), 804–808 (2020). http://penchovsky.atwebpages.com

56. Pliatsika, V., et al.: MINTbase: a framework for the interactive exploration of mitochondrial and nuclear tRNA fragments. Bioinformatics **32**(16), 2481–2489 (2016). http://cm.jefferson.edu/MINTbase

57. Poggi, A., Lembo, D., Calvanese, D., De Giacomo, G., Lenzerini, M., Rosati, R.: Linking data to ontologies. In: Spaccapietra, S. (ed.) Journal on Data Semantics X. LNCS, vol. 4900, pp. 133–173. Springer, Heidelberg (2008). https://doi.org/10.1007/978-3-540-77688-8_5

58. Russo, F., et al.: miRandola 2017: a curated knowledge base of non-invasive biomarkers. Nucleic Acids Res. **46**(D1), 354–359 (2018). http://mirandola.iit.cnr.it

59. Sima, A.C., et al.: Enabling semantic queries across federated bioinformatics databases. Database 2019 (2019)

60. Stephen, B.J., et al.: Xeno-miRNA in maternal-infant immune crosstalk: an aid to disease alleviation. Front. Immunol. **11**(404) (2020)

61. Sun, L., et al.: The CRISPR/Cas9 system for gene editing and its potential application in pain research. Transl. Perioperative Pain Med. **1**(3) (2016)

62. Volders, P.J., et al.: LNCipedia 5: towards a reference set of human long non-coding RNAs. Nucleic Acids Res. **47**(D1), 135–139 (2018). http://lncipedia.org

63. Vorländer, M.K., et al.: Structural basis of mRNA maturation: time to put it together. ScienceDirect **75**, 102431 (2022)

64. Wang, J.H., et al.: tsRFun: a comprehensive platform for decoding human tsRNA expression, functions and prognostic value by high-throughput small RNA-Seq and CLIP-Seq data. Nucleic Acids Res. **50**(D1), 421–431 (2022). http://rna.sysu.edu.cn/tsRFun

65. Wishart, D., et al.: DrugBank 5.0: a major update to the DrugBank database for 2018. Nucleic Acids Res. **46**(D1), 1074–1082 (2018). http://go.drugbank.com/categories/DBCAT001709

66. Wright, M.W.: A short guide to long non-coding RNA gene nomenclature. Hum. Genomics **8**(1), 7 (2014)

67. Wu, D., et al.: ncRDeathDB: a comprehensive bioinformatics resource for deciphering network organization of the ncRNA-mediated cell death system. Autophagy **11**(10), 1917–1926 (2015). http://rna-society.org/ncrdeathdb

68. Xie, B., et al.: miRCancer: a microRNA-cancer association database constructed by text mining on literature. Bioinformatics **29**(5), 638–644 (2013). http://mircancer.ecu.edu

69. Xu, F., et al.: TAM: a method for enrichment and depletion analysis of a microRNA category in a list of microRNAs. BMC Bioinf. **11**(419) (2010). http://lirmed.com/tam2

70. Xu, F., et al.: dbDEMC 3.0: Functional exploration of differentially expressed miR-NAs in cancers of human and model organisms. Genomics Proteomics Bioinf. **20**(3), 446–454 (2022). http://biosino.org/dbDEMC

71. Zhang, S., et al.: A graph-based approach for integrating biological heterogeneous data based on connecting ontology. In: IEEE International Conference on Bioinformatics and Biomedicine, pp. 600–607 (2021)

72. Zhang, Y.Y., et al.: dbEssLnc: a manually curated database of human and mouse essential lncRNA genes. Comput. Struct. Biotechnol. J. **20**, 2657–2663 (2022). http://esslnc.pufengdu.org

73. Zhao, L., et al.: NONCODEV6: an updated database dedicated to long non-coding RNA annotation in both animals and plants. Nucleic Acids Res. **49**(D1), 165–171 (2021). http://noncode.org

74. Xia, F., et al.: Graph learning: a survey. IEEE Trans. Artif. Intell. **2**(2), 109–127 (2021)

Structural Analysis of RNA-Binding Protein EWSR1 Involved in Ewing's Sarcoma Through Domain Assembly and Conformational Molecular Dynamics Studies

Saba Shahzadi[1] (iD), Mubashir Hassan[1,1] (iD), and Andrzej Kloczkowski[1,2(✉)] (iD)

[1] The Steve and Cindy Rasmussen Institute for Genomic Medicine at Nationwide Children's Hospital, Columbus, OH 43205, USA
{Saba.Shahzadi,Andrzej.Kloczkowski}@nationwidechildrens.org
[2] Department of Pediatrics, The Ohio State University, Columbus, OH 43205, USA

Abstract. Proteins are active players in different sarcomas through actively participating in downstream signaling pathways. EWS RNA-Binding Protein 1 (EWSR1) is considered as good therapeutic target for treatment of Ewing's sarcoma. In current study, comparative modeling approach was employed to predict the structural models of three different domains of EWSR1 and Domain Enhanced MOdeling (DEMO) server was used to compose the predicted models of three domains. Furthermore, RNA motifs binding to EWSR1 were predicted and the 3D model of RNA bound to EWSR1 was built by using MC-Fold. The conformational interactions between EWSR1 and RNA were studied with HNADOCK. Moreover, MD simulations were performed to check the stability of EWSR1-RNA complex by computing RMSD, RMSF, R_g, and SASA graphs. Based on computational assessments, it has been concluded that certain structural and dynamical features of EWSR1-RNA complex could be used as a target in future development of drugs against Ewing's sarcoma.

Keywords: Ewing's sarcoma · EWSR1 · Computational Modeling · Domain assembly · Molecular Docking · Molecular Dynamics simulations

1 Introduction

Ewing's sarcoma (ES) is a cancerous tumor that occurs in long bones and soft tissues like cartilages and nerve tissues [1, 2]. Ewing's sarcoma family of tumors includes extraskeletal, and bone tumors, Askin's tumors and peripheral primitive neuroectodermal tumors [3]. The EWSR1 protein contains 656 amino acids and belongs to erythroblast transformation specific (ETS) family involved in the cellular transcriptional regulations [4]. The EWSR1 expresses in the nucleus and comprises couple of binding domains such as ETS DNA-binding and pointed (PNT) domains, respectively [5]. The genetic studies showed that EWSR1 is associated with chromosomal translocation (chromosome breaks and a portion of it reattaches to a different chromosome) event and as a result forms

© The Author(s), under exclusive license to Springer Nature Switzerland AG 2023
I. Rojas et al. (Eds.): IWBBIO 2023, LNBI 13919, pp. 181–190, 2023.
https://doi.org/10.1007/978-3-031-34953-9_14

multiple fusion gene products, such as transmembrane protease serine 2 (TMPSSR2) and N-Myc downstream-regulated Gene 1 (NDRG1) fused to ERG in in prostate cancer [6, 7], and to EWS in Ewing sarcoma, respectively [8]. The structural data reported that N-terminus of EWS/FLI1 retains the prion-like transactivation domain which allows EWS/FLI1 to bind with RNA polymerase II and recruit the barrier-to-autointegration factor (BAF) complex [9]. These interactions change heterochromatin to euchromatin at EWS/FLI1 DNA-binding sites effectively and generate de novo enhancers. The C-terminus of EWS/FLI1 retains the DNA-binding domain of FLI1 and particularly binds to ACCGGAAG core sequence. The EWS/FLI1 preferentially binds to GGAA-repetitive regions and there is a positive correlation between the GGAA microsatellites, EWS/FLI1 binding, and target gene expression [10, 11]. There is an urgent demand for latest targeted therapies that might offer a greater efficiency and less unfavorable side effects than the conventional chemo and radiotherapy that are used in present days. In this regard, understanding of EWS pathogenesis at molecular level provides key information that might help to generate new targeted biological therapies. Therefore, present study has been designed to explore the structural basis of EWSR1 protein domains through the interaction with RNA to check its involvement in insurgence of ES.

2   Computational Methodology

The current study is based on in silico methodology which utilized multiple different online tools and servers to explore the structural features of EWSR1 involved in Ewing Sarcoma through interactions with RNA motifs.

2.1   Retrieval of Human EWSR1 Sequence and Alignment

Human EWSR1 with accession no. Q01844 was accessed from the UniProt database (http://www.uniprot.org/). Furthermore, BLAST was run to check the sequence similarity in different species and results were visualized by AliView [12].

2.2   EWSR1 Conserved Sequence Analysis

The human EWSR1 sequence was analyzed through multiple alignments (MSA) to check the conservation pattern of amino acids in different species. The mature protein EWSR1 consists of 656 amino acids, having major region EAD composed of 1–256 AA. In EWSR1, there are two others major domains: IQ (256–285: 29 AA) and RRM (361–447: 86 AA), respectively. Here, 36 different species were analyzed through MSA and it has been observed that protein region with residues 1–70 AA was conserved, however, in rest of sequence there is also matching/mismatching behavior among species. This conservation pattern may ensure the significance of EWSR1 domains in the regulation of transcriptional activity and in downstream signaling pathway (see Fig. 1).

The prior results also ensured that common residue pattern among multiple protein sequences depicts the core significant behavior of conserved amino acids that might be involved in downstream signaling pathways.

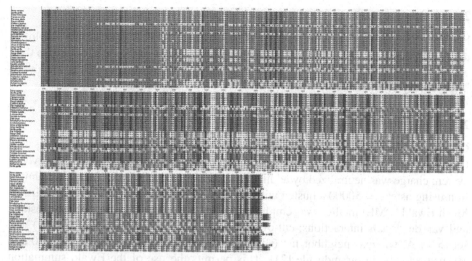

Fig. 1. Residue conservation pattern of EWSR1

2.3 Structure Prediction of Human EWSR1

The EWSR1 is a large protein composed of 656 residues and presently its crystal structure is unavailable in the Protein Data Bank (PDB) (http://www.rcsb.org). Therefore, a
homology-based modeling approach was employed to predict the structure of EWSR1
using Swiss modeling approach [13]. In Swiss modeling approach three EWSR1 domain
models were predicted using three different suitable templates (5W3N, 6GBM and
2CPE) for all three domains, respectively. The templates for domains have been selected
from features of the target-template alignment and highest quality models have then
been selected for EWSR1 domain building. The constructed target protein models were
minimized by employing the conjugate gradient algorithm and the Amber force field
with UCSF Chimera 1.10.1 [14]. Furthermore, the MolProbity server [15] and ProSA-
web [16] were utilized to assess the stereo-chemical properties of the predicted domain
structures. Finally, Domain Enhanced MOdeling (DEMO) was used to assemble all three
domains based on already known protein architectures available in the database.

2.4 Prediction of RNA Motifs Interacting with EWSR1

The hPDI, an online tool was used to predict the genomic motifs interacting with EWSR1
[17]. The best genomic motif, having best interaction with EWSR1 was selected. Furthermore, a conversion of DNA to RNA sequence was done by using an online Transcription and Translation Tool (http://biomodel.uah.es/en/lab/cybertory/analysis/trans.
htm). Finally, MC-Fold, a web-based service was employed to predict RNA secondary
and tertiary structures. The MC-Fold uses the input sequence and predicts secondary
structures that are uses as input to MC-Sym online server [18]. The hPDI, an online tool
was used to predict the genomic motifs having interactions with EWSR1.

2.5 Molecular Docking

HNADOCK docking (http://huanglab.phys.hust.edu.cn/hnadock/) server was employed to check binding pattern of EWSR1 with predicted genomic motifs and to observe their conformational behavior. The top ten docking complexes were downloaded and analyzed in Discovery Studio 2.1 Client and by Chimera 1.10.1 respectively.

2.6 Molecular Dynamics (MD) Simulations

Groningen Machine for Chemicals Simulations (GROMACS 4.5.4 package, [19] with GROMOS 96 force field [20] was employed to check the protein stability. The overall system charge was neutralized by adding ions and further underwent energy minimization using nsteps = 50000 adjusted with energy step size (emstep) 0.01 value. Particle Mesh Ewald (PME) method was employed for energy calculation and for electrostatic and van der Waals interactions; cut-off distance for the short-range vdW (rvdw) was set to 14 Å, whereas neighbor list (rlist) and nstlist values were adjusted as 1.0 and 10, respectively, in em.mdp file [21]. This permits the use of the Ewald summation at a computational cost comparable with that of a simple truncation method of 10 Å or less, and the linear constraint solver (LINCS) [22] algorithm was used for covalent bond constraints and the time step was set to 0.002 ps. Finally, the molecular dynamics simulations were carried out for 100 ns with nsteps 50,000,000 in md.mdp file.

3 Results

3.1 Structural Analysis of Predicted EWSR1

The VADAR 1.8 analysis shows that EWSR1 contains 25% helices, 30% β-sheets, 43% coils and 20% turns, respectively. The predicted model exhibited good ERRAT score value 61.01. Furthermore, the Ramachandran plots indicated that 97.1% residues were present in allowed regions which showed the good accuracy of phi (φ) and psi (ψ) angles among the coordinates of receptor molecule. Moreover, MolProbity and ProSA-web servers, also confirmed good stability of the predicted models. The predicted model of EWSR1 is shown in Fig. 2.

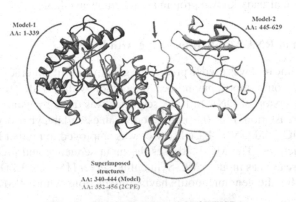

Fig. 2. The predicted model of EWSR1

3.2 Docking Energy Evaluation

In generated docking results, the energy values range from −363 to −284. The first three docking poses possessing −363, −322 and −315 scores showed higher variations, whereas, in these complexes RNA binds EWSR1 at different binding regions (Fig. 3).

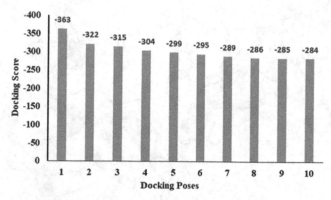

Fig. 3. Docking scores of best docked complexes.

3.3 Binding Pocket Conformational and Interaction Analysis

Top ten complexes were observed to find the binding pattern between EWSR1 and genomic motifs. The overall results showed that most of docked complexes showed similar binding pattern and conformations around the third domain. In docked complex 1 and 2 the binding pattern of genomic motifs showed little variations against target protein, however, the binding regions were same in both complexes. The most out backed complex among all docked complexes is Complex 3 bound at different binding site. The rest of all docked complexes (Complex 4 to Complex 10) were quite similar and bound at same position in EWSR1 with little varying conformations. The docking results showed that third domain of target protein could be used a RNA receptor binding site which may have different cellular activities (Figs. 4 and 5).

The detail interaction analysis showed that, different core residues showed interaction with RNA strand at different nucleotides. The Cys508 showed interaction with Uracil-15 (U15), Arg434 with Uracil-20 (U20) and Gly467-with Uracil-22 (U22), respectively. The other binding residues such as Arg548, Arg502, Trp520, GLy467, His387, Tyr389, Arg434, respectively actively take part against RNA counter conformation in EWSR1 protein (Fig. 6).

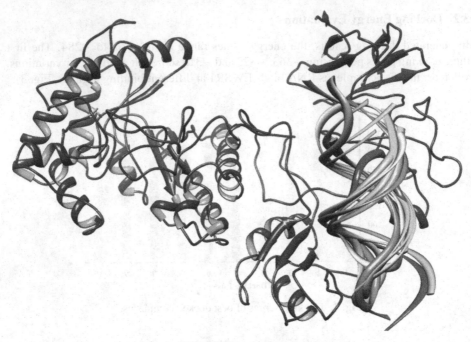

Fig. 4. RNA-EWSR1 docked complexes

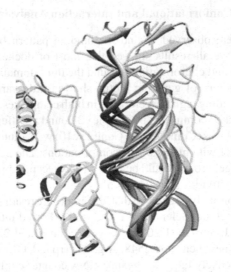

Fig. 5. Binding conformation and superimposition against target protein

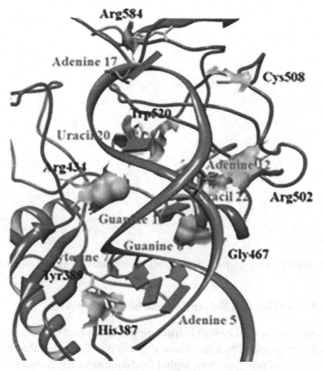

Fig. 6. RNA-EWSR1 docked complexes.

3.4 Root Mean Square Deviation Analysis

The HDOCK results of the best docked complexes were analyzed using Molecular Dynamics simulations to check conformational changes and backbone residue flexibility. The generated graphs show small fluctuations in the equilibrium phase of MD simulations with RMSD values ranging from 0 to 0.8 nm during 20000 ns simulations. In static phase the graph line remained constant, and no fluctuations were observed in protein backbone structure. The RMSD simulation results gave insight on the binding behavior of predicted RNA structure with EWSR1. The RMSD graph shows that RNA model firmly binds inside the active site of EWSR1. From 40000 to 100000 ns the RMSD value remained stable with steady graph line (Fig. 7).

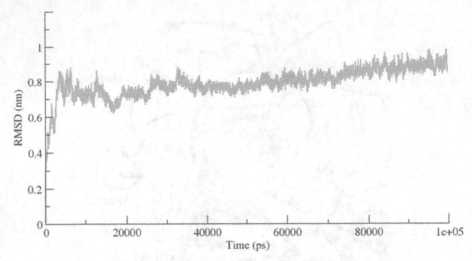

Fig. 7. RMSD graph of docked structures during 100,000 ns MD simulations.

3.5 Radius of Gyration and Solvent Accessible Surface Area

The structural compactness of EWSR1 structure was evaluated by radius of gyration Rg. The stable folded regions of proteins show the relatively steady value of Rg, whereas the disturbed regions of proteins show higher fluctuations of Rg during simulations. Our results show that Rg values for all docked proteins showed small variations of Rg from 2.7 to 3.1 nm. Initially, the graph line was slightly fluctuating and after gaining stability it remained stable through the simulation time period with average Rg value 2.55 nm (Fig. 8). Finally, the SASA analysis was performed to observe the solvent accessible surface area during simulations. The SASA results showed that docking complex SASA is centered at 135 nm^2 during 100000 ns simulations (Fig. 9).

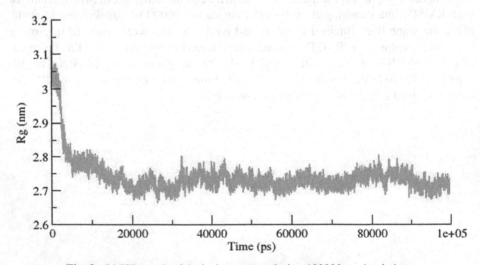

Fig. 8. RMSD graph of docked structures during 100000 ns simulations.

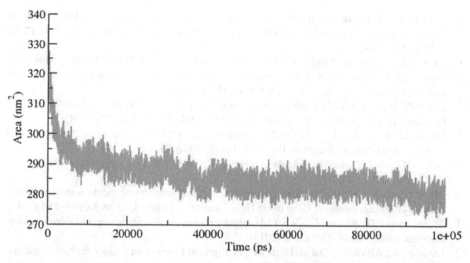

Fig. 9. RMSD graph of docked structures during 100000 ns simulations.

4 Conclusion

Current work is shedding some insights on the structural features and molecular interactions of EWSR1 and RNA in relation to ES. Domain assembly results showed the conformation and architecture of EWSR1 and molecular interaction profile explored the most significant amino acids that could be possible targets for drug design. Furthermore, stable behavior of predicted assembled EWSR1 structure has been confirmed by MD simulation. Taken together, the computational structural assessments shows that EWSR1 could be used as a possible therapeutic target for treatment of ES.

Acknowledgment. AK acknowledges the financial support from NSF grant DBI 1661391, and NIH grants R01GM127701, and R01HG012117. M.H. acknowledges The Ohio State University for "President's Postdoctoral Scholars Program (PPSP)" fellowship award.

References

1. Ordóñez, J.L., Osuna, D., Herrero, D., de Alava, E., Madoz-Gúrpide, J.: Advances in Ewing's sarcoma research: where are we now and what lies ahead? Cancer Res. **69**(18), 7140–7150 (2009). https://doi.org/10.1158/0008-5472.CAN-08-4041
2. Kauer, M., et al.: A molecular function map of Ewing's sarcoma. PLoS ONE **4**, e5415 (2009)
3. Tsokos, M., Alaggio, R.D., Dehner, L.P., Dickman, P.S.: Ewing sarcoma/peripheral primitive neuroectodermal tumor and related tumors. Pediatr Dev Pathol. **15**(1 Suppl), 108–126 (2012)
4. Mavrothalassitis, G., Ghysdael, J.: Proteins of the ETS family with transcriptional repressor activity. Oncogene **19**(55), 6524–6532 (2000)

5. Kedage, V., et al.: An interaction with Ewing's sarcoma breakpoint protein EWS defines a specific oncogenic mechanism of ETS factors rearranged in prostate cancer. Cell Rep. **17**(5), 1289–1301 (2016)

6. John, S.J., Powell, K., Conley-LaComb, M.K., Chinni, S.R.: TMPRSS2-ERG fusion gene expression in prostate tumor cells and its clinical and biological significance in prostate cancer progression. J. Cancer Sci. Ther. **4**(4), 94–101 (2012)

7. Lin, P.P., et al.: EWS-FLI1 induces developmental abnormalities and accelerates sarcoma formation in a transgenic mouse model. Cancer Res. **68**(21), 8968–8975 (2008)

8. Lessnick, S.L., Ladanyi, M.: Molecular pathogenesis of Ewing sarcoma: new therapeutic and transcriptional targets. Annu Rev Pathol. **7**, 145–159 (2012)

9. Boulay, G., et al.: Cancer-specific retargeting of BAF complexes by a prion-like domain. Cell **171**(1), 163-178.e19 (2017)

10. Riggi, N., et al.: EWS-FLI1 utilizes divergent chromatin remodeling mechanisms to directly activate or repress enhancer elements in Ewing sarcoma. Cancer Cell **26**(5), 668–681 (2014)

11. Cidre-Aranaz, F., Alonso, J.: EWS/FLI1 target genes and therapeutic opportunities in ewing sarcoma. Front. Oncol. **20**(5), 162 (2015)

12. Larsson, A.: AliView: a fast and lightweight alignment viewer and editor for large datasets. Bioinformatics **30**(22), 3276–3278 (2014)

13. Arnold, K., Bordoli, L., Kopp, J., Schwede, T.: The SWISS-MODEL workspace: a web-based environment for protein structure homology modelling. Bioinformatics **22**(2), 195–201 (2006)

14. Pettersen, E.F., et al.: UCSF Chimera–a visualization system for exploratory research and analysis. J. Comput. Chem. **25**(13), 1605–1612 (2004)

15. Chen, V.B., et al.: MolProbity: all-atom structure validation for macromolecular crystallography. Acta Crystallogr. D Biol. Crystallogr. **66**, 12–21 (2010)

16. Wiederstein, M., Sippl, M.J.: ProSA-web: interactive web service for the recognition of errors in three-dimensional structures of proteins. Nucleic Acids Res. **35**, 407–410 (2007)

17. Xie, Z., Hu, S., Blackshaw, S., Zhu, H., Qian, J.: hPDI: a database of experimental human protein-DNA interactions. Bioinformatics **26**(2), 287–289 (2010)

18. Parisien, M., Major, F.: The MC-Fold and MC-Sym pipeline infers RNA structure from sequence data. Nature **452**(7183), 51–55 (2008)

19. Pronk S, et al.: GROMACS 4.5: a high-throughput and highly parallel open source molecular simulation toolkit. Bioinformatics **29**(7), 845–54 (2013)

20. Chiu, S.W., Pandit, S.A., Scott, H.L., Jakobsson, E.: An improved united atom force field for simulation of mixed lipid bilayers. J. Phys. Chem. B. **113**(9), 2748–2763 (2009)

21. Wang, H., Dommert, F., Holm, C.: Optimizing working parameters of the smooth particle mesh Ewald algorithm in terms of accuracy and efficiency. J. Chem. Phys. **133**(3), 034117 (2010)

22. Amiri, S., Sansom, M.S., Biggin, P.C.: Molecular dynamics studies of AChBP with nicotine and carbamylcholine: the role of water in the binding pocket. Protein Eng. Des. Sel. **20**(7), 353–359 (2007)

Pharmacoinformatic Analysis of Drug Leads for Alzheimer's Disease from FDA-Approved Dataset Through Drug Repositioning Studies

Mubashir Hassan[1,1] (iD), Saba Shahzadi[1] (iD), and Andrzej Kloczkowski[1,2](✉) (iD)

[1] The Steve and Cindy Rasmussen Institute for Genomic Medicine at Nationwide Children's Hospital, Columbus, OH 43205, USA
{Mubasher.Hassan,Saba.Shahzadi,
Andrzej.Kloczkowski}@nationwidechildrens.org
[2] Department of Pediatrics, The Ohio State University, Columbus, OH 43205, USA

Abstract. Drug design is highly priced and time-consuming procedure. Therefore, to overcome this problem, *in silico* drug design methods, particularly drug repositioning have been developed to assess therapeutic effects of known drugs against other diseases. In this study, computational drug repositioning method is used to explore the alternative therapeutic effects of FDA approved drugs to treat Alzheimer's disease. The chemical shape-based screening was employed to fetch some potential new drugs based on the structure of standard drugs. The screened drugs were further evaluated through pharmacogenomics, molecular docking, and molecular dynamics simulation studies. The best lead drugs, such as darifenacin, astemizole, tubocurarine, elacridar, sertindole and tariquidar displayed promising repositioned effects in the treatment of Alzheimer's disease and may be used as potential medicines after thorough experimental and clinical studies.

Keywords: Alzheimer's disease · Drug repositioning · Protein modelling · Shape based screening · Pharmacogenomics · MADD protein

1 Introduction

Alzheimer's disease (AD) is brain disorder characterized by dementia in older age [1, 2]. There are over 55 million people worldwide living with dementia in 2020. This number will almost double every 20 years, reaching 78 million in 2030 and 139 million in 2050 [3]. The basic mechanism of AD is complex however, low levels of acetylcholine, β-amyloid (Aβ) deposits, tau-protein aggregation, neurofibrillary tangles (NFTs), and oxidative stress are considered as biomarkers of AD [2, 4]. These biomarkers are partially effective in detecting AD, however effective therapeutic agents are still needed to treat AD [5].

Drug design time consuming, highly priced with low success rate process. Therefore, multiple computational approaches such as drug repositioning are being used to predict novel is drugs to treat AD [6–9]. However, it has been observed that drug design for AD has been largely ineffective in the last decade. Currently, medical research is focused

I. Rojas et al. (Eds.): IWBBIO 2023, LNBI 13919, pp. 191–201, 2023.
https://doi.org/10.1007/978-3-031-34953-9_15

on the factors that are thought to contribute to AD development, such as tau proteins and Aβ deposits [10]. However, new protein target could also play a significant role in the development of drugs against AD. MAP kinase-activating death domain (MADD) is tetra-domain (uDENN, cDENN, dDENN and death domain) protein with 1647 amino acids. MADD plays a substantial role in regulating cell proliferation, survival, and death through alternative mRNA splicing in various diseases such as Deeah syndrome and Alzheimer's disease [11]. MADD protein also links TNF Receptor Superfamily Member 1A (TNFRSF1A) with MAP kinase activation and plays important regulatory role in physiological cell death [12]. In the present study, a computational drug repositioning approach is employed to screen the Food and Drug Administration (FDA)-approved drugs against AD by employing multiple computational approaches.

2 Computational Methodology

2.1 Retrieval of MADD Sequence and Protein Modelling

The amino acid sequence of MADD protein having accession number Q8WXG6 was accessed from the UniProt database [13]. MADD contains four domains: uDENN, cDENN, dDENN and Death. The three-dimensional (3D) structure of MADD is not available in Protein Data Bank (PDB); therefore, trRosetta a web-based platform (https://yanglab.nankai.edu.cn/trRosetta/) was employed to predict MADD structure. For prediction, all four domains were modelled separately and assembled through Domain Enhanced MOdeling (DEMO) server [14].

2.2 Structure Analysis of MADD Protein

The assemble model structure of MADD was analyzed using different computational approaches to check its stability. To do this, initially Ramachandran Plot Server (https://zlab.umassmed.edu/bu/rama/index.pl) and MolProbity server (http://molprobity.biochem.duke.edu/) were used to check protein conformation and its phi (φ) and psi (ψ) angles by generating Ramachandran graph. The stability of protein is dependent upon the distribution of residues in allowed and disallowed regions.

2.3 Virtual Screening of Chemical Scaffolds

The SwissSimilarity web tool was employed you to identify the similar chemical hits from FDA-approved list based on standards drugs; donepezil and galantamine [15–17]. The chemical structures of donepezil and galantamine were retrieved from DrugBank (https://go.drugbank.com/) having accession numbers (DB00843 & DB00674), respectively. All the screened drugs were ranked according to their predicted similarity scoring values.

2.4 Active Binding Site of MADD Protein and Docking

The Prankweb (http://prankweb.cz/), is a template-free machine learning method based on the prediction of local chemical neighborhood ligandability centered on points placed on a solvent-accessible protein surface [18]. Before conducting our docking experiments, all the screened drugs were sketched in ACD/ChemSketch tool and accessed in mol format. In PyRx docking experiment, all screened drugs were docked with death domain of MADD using default procedure [19]. In docking experiments, the grid box dimension values were adjusted as center_X = -0.8961, Y = -1.6716 and Z = 0.3732 with default exhaustiveness = 8. Furthermore, the generated docked complexes were thoroughly analyzed to check their binding conformational poses at active binding site of MADD protein. Furthermore, another docking experiment was employed on best screened drugs against MADD protein using AutoDock 4.2 tool [20]. In brief, for receptor protein, the polar hydrogen atoms and Kollman charges were assigned. For ligand, Gasteiger partial charges were designated, and non-polar hydrogen atoms were merged. A grid map of $80 \times 80 \times 80$ Å was adjusted on the binding pocket of MADD to generate the grid map and to get the best conformational state of docking. A total of 100 number of runs were examined using docking experiments. The Lamarckian genetic algorithm (LGA) and empirical free energy function were applied by using default docking parameters [20]. All the docked complexes were further evaluated by analyzing lowest binding energy (kcal/mol) values and hydrogen and hydrophobic interactions by using Discovery Studio (2.1.0) and UCSF Chimera 1.10.1.

2.5 Designing of Pharmacogenomics Networks

Drug Gene Interaction Databases (DGIdb) (https://www.dgidb.org/) and Drug Signatures Database (DSigDB) (http://dsigdb.tanlab.org/DSigDBv1.0/) were employed to obtain the possible list of different disease-associated genes. Furthermore, a detailed literature survey was performed against all predicted genes to identify its involvement in AD.

2.6 Molecular Dynamics (MD) Simulations

The best screened drugs-complexes having good energy values were selected to understand the residual backbone flexibility of protein structure; MD simulations were carried out by Groningen Machine for Chemicals Simulations (GROMACS 4.5.4 package [21], with GROMOS 96 force field [22]. The protein topology was designed by pdb2gmx command by employing GROMOS 96 force field. Moreover, simulation box with a minimum distance to any wall of 10 Å (1.0 nm) was generated by editconf command. Additionally, box was filled with solvent molecules using gmx solvate command by employing spc216.gro water model. The overall system charge was neutralized by adding ions. The steepest descent approach (1000 ps) for protein structure was applied for energy minimization. For energy minimization the nsteps = 50,000 were employed with energy step size (emstep) 0.01 value. Particle Mesh Ewald (PME) method was employed for energy calculation and for electrostatic and van der Waals interactions; cut-off distance for the short-range vdW (rvdw) was set to 14 Å, whereas neighbor list (rlist) and nstlist values were adjusted to 1.0 and 10, respectively, in em.mdp file [23]. This method permits the use of the Ewald summation at a computational cost comparable with that of a simple truncation method of 10 Å or less, and the linear constraint solver (LINCS) [24] algorithm was used for covalent bond constraints and the time step was set to 0.002 ps. Finally, the molecular dynamics simulation was carried out for 100 ns with nsteps 50,000,000 in md.mdp file. Different structural evaluations such as root mean square deviations and fluctuations (RMSD/RMSF), solvent accessible surface areas (SASA) and radii of gyration (Rg) of backbone residues were analyzed through Xmgrace software (http://plasma-gate.weizmann.ac.il/Grace/).

3   Results and Discussion

3.1   Predicted Model of MADD Protein

MADD is a bulky protein structure comprised of 1647 amino acids with molecular mass 183.303 kDa. Detail structural analysis showed that MADD contains four major domains such as uDENN (14–268 AA), cDENN (289–429 AA), dDENN (431–565 AA) and death domain (1340–1415 AA), respectively with various disordered regions with different lengths (Fig. 1).

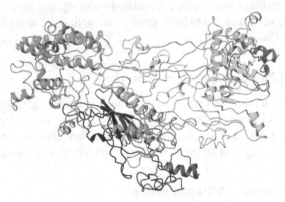

Fig. 1. Predicted model of MADD protein

3.2   VADAR and Galaxy Refine Analysis

VADAR online server predicts the volume, area, dihedral angle and statistical evaluation of protein architecture. The MADD results show that, predicted model consists of 25% α-helices, 19% β-sheets, 55% coils and 20% turns, respectively. The Ramachandran plots indicated that 92.68% residues are placed in favored region having proper values of phi (φ) and psi (ψ) angles. Moreover, the coordinates of MADD residues were also placed in the acceptable region. Galaxy-refine, an online tool to rectify the modelled structure of MADD protein improved Ramachandran graph from 72.4 to 85.8 and overall value was increased from 83.45 to 92.68%. The online Ramachandran graph server (https://www.umassmed.edu/zlab/) graphs shown in Fig. 2.

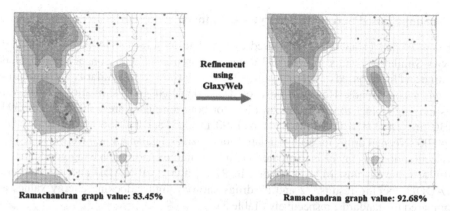

Ramachandran graph value: 83.45%   Ramachandran graph value: 92.68%

Fig. 2. Ramachandran graphs of predicted MADD protein structure

3.3 Binding Pocket Analysis of MADD Protein

Prankweb tool explores the binding pocket of target protein with different pocket sizes and position inside the target protein. The predicted results showed five different binding pockets having different scoring values 15.4, 4.42, 3.68, 2.44 and 1.41, respectively. The highest pocket score value is 15.4. The pocket 1 showed good SAS value 152 as compared to other binding pockets values (Fig. 3).

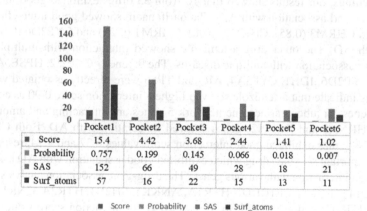

	Pocket1	Pocket2	Pocket3	Pocket4	Pocket5	Pocket6
■ Score	15.4	4.42	3.68	2.44	1.41	1.02
■ Probability	0.757	0.199	0.145	0.066	0.018	0.007
■ SAS	152	66	49	28	18	21
■ Surf_atoms	57	16	22	15	13	11

■ Score ■ Probability ■ SAS ■ Surf_atoms

Fig. 3. Binding pocket evaluation of targeted protein

3.4 Shape Based Screening and PyRx Docking

Donepezil and galantamine were used as template to screen FDA approved dataset. Swiss-Similarity results showed 633 drugs were retrieved with donepezil and galantamine and listed based on scoring values 0–1 which showed dissimilarity and similarity between compounds. However, most of drugs depicted good similarity scoring values ranging from 1 to 0.111. The best hits for both templates are DB07701, DB3393, DB07561, DB00318, DB01466, and DB11490, DB00318, DB01466, and DB11490. The top 100 screened drugs for each standard (donepezil and galantamine) were selected for molecular docking analysis to predict the most suitable candidate having good binding affinity values against MADD protein. In PyRx, donepezil-MADD and galantamine-MADDD docking results, 72 and 67 drugs showed higher docking energy values as compared to standards, respectively (Table 3).

3.5 Pharmacogenomics Analysis

Based on SwissSimilarity and virtual screening result, the selected drugs underwent further pharmacogenomics analysis to check their possible interaction with putative genes encoded proteins and their association with AD. Multiple reported studies justify this approach to predict the drug having repositioned functionality [5, 12]. All the screened drugs using both templates (donepezil and galantamine) exhibited good interaction score values with predicted genes which showed good correlation with AD. From donezepil-based sceenings, our results showed that 10 from 35 drugs exhibited good interaction score values and association with AD. The darifenacin showed good interaction score values with CHRM3 (0.85), CHRM2 (0.38), CHRM1 (0.29) and CYP2D6 (0.04) associated with AD. The other drug astemizole showed interaction with multiple genes (14) having association with multiple diseases. The 9 genes - CYP2J2, HPSE, ABCB1, PPARD, CYP2D6, IDH1, CYP3A4, AR, and TP53 were directly associated with AD. Our results indicate that astemizole showed highest interaction score 0.90 as compared to other genes. In tubocurarine gene network 5 genes are interacting and among these four 2 (CHRNA2 and KCNN2) genes showed association with AD. Both CHRNA2 and KCNN2 genes exhibited good interaction score values 6.37 and 4.24, respectively. Another important screened drug which exhibited good association with AD is aripiprazole through interaction with various genes. The drug-gene association analysis showed that out of 14 genes, eleven (DRD2, HTR1A, ANKK1, SH2B1, HTR2A, CNR1, FAAH, HTR1B, HTR2C, ABCB1, CYP2D6) possessed good interaction score values and are directly involved in AD.

Fluspirilene gene network analysis also showed involvment in AD by targeting multiple genes. The predicted results indicate that six genes: DRD2, HTR1E, XBP1, HTR2A, HTR1A, and PPARD showed their association with AD with different interaction score values. In elacridar-genetic network, two genes ABCG2 (1.54) and ABCB1 (0.27) showed association with AD. Another important screened drug was sertindole which showed correlations with multiple genes involved in AD and other diseases. A total of 17 genes showed interaction with sertindole and from 17, eight genes (DRD2 (0.15), HTR2A (0.14), HTR2C (0.17), HTR1E (0.1), HTR1A (0.03), HTR1B (0.05),

CYP2D6 (0.01), CYP3A4 (0.01)) showed association with AD with different interaction score values. Tariquidar interacted with ABCB1 and ABCG2 with association values 0.66 and 1.03, respectively. Sarizotan is another screened drug showing interactions with HTR1A (0.91) and DRD2 (0.36) having good association with AD. Similarly, fipexide interacted with CYP2D6 (0.03) and CYP3A4 (0.02) showing association with AD. The galantamine was another drug template employed to improve the accuracy and reliability of our prediction results. Forty drugs having good similarity score with galatamine were screened in the pharmacogenomics analysis. Our analysis showed that multiple drugs showed association with AD and were further used in docking analysis. Eight screened drugs having good interaction score values and good association with AD were analyzed. Morphine shows interactions with 24 genes having association with different diseases including seven genes associated with AD. These seven genes PDYN (4.11), OPRK1 (0.23), PER1 (2.05), HMOX2 (2.05), ABCB1 (0.09), CYP2D6 (0.01), DRD2 (0.02) may suggest good therapeutic behavior of morphine as an anti-AD drug. Codeine is another drug based on galantamine template that could be repositioned against AD. Our drug-gene predictions showed that eight genes OPRD1 (1.22), OPRM1 (0.58), OPRK1 (0.63), UGT2B7 (0.76), ABCB1 (0.13), CYP3A4 (0.02), CYP2D6 (0.01), and AR (0.01) interacted having different interaction values. The literature mining showed that four genes among eight were involved in AD. Quinine-gene interaction analysis showed the involvement of the following genes: CYP3A7 (1.27), SLC29A4 (0.64), COP1 (1.27), IL2 (0.6), G6PD (0.18), ABCB1 (0.03), CYP3A4 (0.01), and CYP2D6 (0.01). Moreover, from literature mining it has been found that 5 genes were involved in AD. Similarly, atropine, nalbuphine, quinidine, volinanserin and vernakalant were also selected as potential drugs interacting with various genes linked with AD.

4 Molecular Docking Using Autodock

4.1 Docking Energy Analysis Against Death Domain of MADD

The selected drugs-death domain docked complexes were analyzed based on lowest binding energy values (kcal/mol) and hydrogen/hydrophobic interaction analyses. Results showed that all drugs showed good binding energy values and binds within the active site with appropriate conformational poses (Table 1). Docking energy is most significant parameter to screen and evaluate the drugs for binding with target proteins [5].

$$\Delta G_{binding} = \Delta G_{vdW} + \Delta G_{elec} + \Delta G_{Hbond} + \Delta G_{desolv} + \Delta G_{torsional} \quad (1)$$

Here, ΔG_{vdW} is energy of van der Waals interactions, ΔG_{elec} is energy of electrostatic interactions, ΔG_{Hbond} is energy of hydrogen bonds, ΔG_{desolv} refers to energy of desolvation, and $\Delta G_{torsional}$ is energy of torsional interactions, proportional to the number of rotatable bonds. Top six drug having docking energy greater than -7 kcal/mol were selected to check the binding conformation behavior and draggability behavior of MADD. The best six drug such as darifenacin (-7.59), astemizole (-7.19), tubocurarine (-8.26), elacridar (-7.72), sertindole (-7.58) and tariquidar (-8.42) exhibited good docking energy values and showed good binding interaction profiles.

Table 1. The predicted docking energy values of selected drugs

Docking complexes	Docking energy (kcal/mol)
Darifenacin	−7.59
Astemizole	−7.19
Tubocurarine	−8.26
Elacridar	−7.72
Sertindole	−7.58
Tariquidar	−8.42

4.2 Binding Interaction Profile

In darifenacin docking, single hydrogen bond was observed with appropriate binding distance. Similarly, elacridar and tariquidar also formed hydrogen bonds with Glu1023 having appropriate bond lengths. The nitrogen atom of NH2 elacridar formed hydrogen bond with Glu1023 having bond length 3.00 Å, whereas tariquidar formed two hydrogen bonds with Glu1023 having bond lenghts 2.50 and 2.80 Å, respectively (Fig. 4).

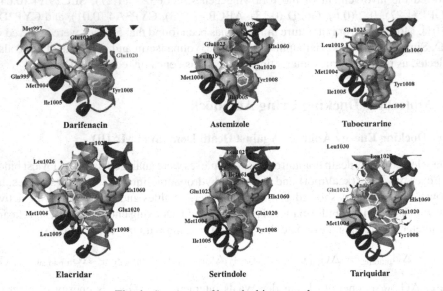

Fig. 4. Structures of best docking complexes

5 Molecular Dynamic Simulations

5.1 Root Mean Square Deviation and Rg Analysis

The RMSD graph results of six docking complexes showed steady and little fluctuating behavior throughout the simulation time. The RMSD graphs displayed an increasing trend with RMSD values ranging from 0 to 0.1 nm during 100ns simulations (Fig. 5A). The Rg values of all docked structures showed small variations from 1.25 to 2.25 nm (Fig. 5B).

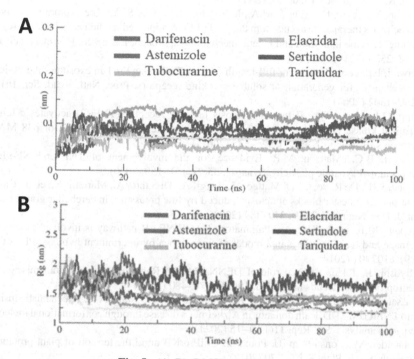

Fig. 5. (A, B). RMSD and Rg graphs

6 Conclusions

In the present research repositioning profiles of known drugs against AD have been explored using shape-based screening, molecular docking pharmacogenomics and MD simulation approaches. Our results show that darifenacin, astemizole, tubocurarine, elacridar, sertindole and tariquidar drugs exhibited good lead like behavior against AD. The comparative analysis shows that tariquidar exhibited the best repositioning profile in the treatment of AD.

Acknowledgment. AK acknowledges the financial support from NSF grant DBI 1661391, and NIH grants R01GM127701, and R01HG012117.

References

1. Acarin, L., González, B., Castellano, B.: Stat3 and NFκB glial expression after excitotoxic damage to the postnatal brain. NeuroReport 9(12), 2869–2873 (1998)
2. Moustafa, A.A., et al.: Genetic underpinnings in Alzheimer's disease–a review. Rev. Neurosci. 29(1), 21–38 (2018)
3. Cutsuridis, V., Moustafa, A.A.: Neurocomputational models of Alzheimer's disease. Scholarpedia (2017)
4. Ahn, K.S., Aggarwal, B.B.: Transcription factor NF-κB: a sensor for smoke and stress signals. Ann. N. Y. Acad. Sci. 1056(1), 218–233 (2005)
5. Hassan, M., Raza, H., Abbasi, M.A., Moustafa, A.A., Seo, S.-Y.: The exploration of novel Alzheimer's therapeutic agents from the pool of FDA approved medicines using drug repositioning, enzyme inhibition and kinetic mechanism approaches. Biomed. Pharmacother. 109, 2513–2526 (2019)
6. Hawari, F.I., et al.: Release of full-length 55-kDa TNF receptor 1 in exosome-like vesicles: a mechanism for generation of soluble cytokine receptors. Proc. Natl. Acad. Sci. 101(5), 1297–1302 (2004)
7. Jupp, O.J., et al.: Type II tumour necrosis factor-α receptor (TNFR2) activates c-Jun N-terminal kinase (JNK) but not mitogen-activated protein kinase (MAPK) or p38 MAPK pathways. Biochem. J. 359(3), 525–535 (2001)
8. Albensi, B.C., Mattson, M.P.: Evidence for the involvement of TNF and NF-κB in hippocampal synaptic plasticity. Synapse. 35(2), 151–159 (2000)
9. Amodio, R., De Ruvo, C., Di Matteo, V., Poggi, A., Di Santo, A., Martelli, N., et al.: Caffeic acid phenethyl ester blocks apoptosis induced by low potassium in cerebellar granule cells. Int. J. Dev. Neurosci. 21(7), 379–389 (2003)
10. Angelo, M.F., et al.: The proinflammatory RAGE/NF-κB pathway is involved in neuronal damage and reactive gliosis in a model of sleep apnea by intermittent hypoxia. PLoS ONE 9(9), e107901 (2014)
11. Miyoshi, J., Takai, Y.: Dual role of DENN/MADD (Rab3GEP) in neurotransmission and neuroprotection. Trends Mol. Med. 10(10), 476–480 (2004)
12. Hassan, M., Zahid, S., Alashwal, H., Kloczkowski, A., Moustafa, A.A.: Mechanistic insights into TNFR1/MADD death domains in Alzheimer's disease through conformational molecular dynamic analysis. Sci. Rep. 11(1), 1–15 (2021)
13. Schneider, M., Consortium tU, Poux, S.: UniProtKB amid the turmoil of plant proteomics research. Front. Plant Sci. 3, 270 (2012)
14. Zhou, X., Hu, J., Zhang, C., Zhang, G., Zhang, Y.: Assembling multidomain protein structures through analogous global structural alignments. Proc. Natl. Acad. Sci. 116(32), 15930–15938 (2019)
15. Bragina, M.E., Daina, A., Perez, M.A., Michielin, O., Zoete, V.: The SwissSimilarity 2021 web tool: novel chemical libraries and additional methods for an enhanced ligand-based virtual screening experience. Int. J. Mol. Sci. 23(2), 811 (2022)
16. Cacabelos, R.: Donepezil in Alzheimer's disease: from conventional trials to pharmacogenetics. Neuropsychiatric disease and treatment (2007)
17. Razay, G., Wilcock, G.K.: Galantamine in Alzheimer's disease. Expert Rev. Neurother. 8(1), 9–17 (2008)
18. Jendele, L., Krivak, R., Skoda, P., Novotny, M., Hoksza, D.: PrankWeb: a web server for ligand binding site prediction and visualization. Nucleic Acids Res. 47(W1), W345–W349 (2019)
19. Dallakyan, S., Olson, A.J.: Small-molecule library screening by docking with PyRx. In: Hempel, J.E., Williams, C.H., Hong, C.C. (eds.) Chemical biology. MMB, vol. 1263, pp. 243–250. Springer, New York (2015). https://doi.org/10.1007/978-1-4939-2269-7_19

20. Morris, G.M., et al.: AutoDock4 and AutoDockTools4: automated docking with selective receptor flexibility. J. Comput. Chem. **30**(16), 2785–2791 (2009)
21. Pronk, S., et al.: GROMACS 4.5: a high-throughput and highly parallel open source molecular simulation toolkit. Bioinformatics. **29**(7), 845–54 (2013)
22. Chiu, S.-W., Pandit, S.A., Scott, H., Jakobsson, E.: An improved united atom force field for simulation of mixed lipid bilayers. J. Phys. Chem. B **113**(9), 2748–2763 (2009)
23. Wang, H., Dommert, F., Holm, C.: Optimizing working parameters of the smooth particle mesh Ewald algorithm in terms of accuracy and efficiency. J. Chem. Phys. **133**(3), 034117 (2010)
24. Amiri, S., Sansom, M.S., Biggin, P.C.: Molecular dynamics studies of AChBP with nicotine and carbamylcholine: the role of water in the binding pocket. Protein Eng. Des. Sel. **20**(7), 353–359 (2007)

Biomedical Engineering

Motion Control of a Robotic Lumbar Spine Model

Thuanne Paixão[1]([✉]) [iD], Ana Beatriz Alvarez[1] [iD], Ruben Florez[2] [iD],
and Facundo Palomino-Quispe[2] [iD]

[1] PAVIC Laboratory, University of Acre, Rio Branco, Brazil
thuanne.paixao@sou.ufac.br, ana.alvarez@ufac.br
[2] LIECAR Laboratory, School of Electronic Engineering, University of San Antonio
Abad del Cusco, Cuzco, Peru
ruben.florez@ieee.org, facundo.palomino@unsaac.edu.pe

Abstract. The study of the movement of the vertebrae of the lumbar
spine is classified as a relevant theme for research, considering the possibility of exploring the pathological dysfunctions of this region. This paper
presents the development of a trajectory motion control for a lumbar
spine model. The spine model is being represented by a 2 DOF (Degrees
of Freedom) manipulator robot, which represents the motion of two lumbar vertebrae. For the computational simulations of the controlled spine
behavior the mathematical dynamic model of the manipulator based
on the Lagrange approach is being considered. Preliminary simulation
results show that the implemented conventional controller robustly follows the references given for the angles of the vertebrae, guaranteeing
the planned movement.

Keywords: Biomechanics · Motion control · Robotic lumbar spine

1 Introduction

In the study of biomechanics, models that aim to represent anatomical structures
are developed to promote the simulation and analysis of characteristic biological
behaviors of human beings. These models, can be elaborated through computational tools and by developing humanoid robots with the ability to assimilate the
behavior of human functions [10]. The construction of experimental models of
the human spine has as main motivation the design of computational and physical models with structural characteristics present in the development, in order
to provide the study, evaluation and demonstration of the specific structural
behavior of these elements [3].

In particular, the lumbar region of the spine has important characteristics
for the investigation of pathological dysfunctions, and is considered a fundamental factor segment. In this region of the spine that the greatest abnormalities
of this bone structure are caused, so much exploratory research is carried out.

I. Rojas et al. (Eds.): IWBBIO 2023, LNBI 13919, pp. 205–216, 2023.
https://doi.org/10.1007/978-3-031-34953-9_16

The lumbar spine can present anomalies and diseases resulting from the deformation of its vertebral structure. Low back pain, for example, can be characterized by several factors that can lead to injury, inflammation and degeneration of the lumbar spine structure, such as sedentariness, excessive physical effort when lifting heavy objects, habit of positioning oneself with an incorrect posture, among others, considering that in certain cases, some factors become more pertinent [17].

The procedures related to cadaveric specimens, used for demonstration and practical teaching of the spine and other structures, have limitations and costly characteristics. At present, there are notable efforts dedicated to the possibility of developing educational modules that have computational and tangible structures that accurately represent models of the spine for simulation, with an interest in the training of students. In this sense, applications of biomechanical models, about the lumbar spine, are extremely important for the training of students and for the assimilation of the behavior of structural and functional movement, considering the fact that it represents a vital region for the composition of the dynamic stability of movement of the human being [7,18].

The representation of the lumbar spine through a robotic structure, with the ability to perform controlled movement becomes a challenge. Among the various control techniques that could be used to realize the movement, the PID (Proportional Integral Derivative) controller can be implemented to control the expected movement. Based on differential equations and feedback control, the PID controller has extensive applicability for robotic manipulator control systems and in contemporary control systems [4].

In this context, this paper presents a controller designed for the motion of a robotic lumbar spine model. A PID controller enables the controlled movement of each joint of the lumbar vertebrae of the robotic structure. For the representation of the lumbar spine, the model of a two Degree of Freedom (2 DOF) manipulator is used, representing the motion of two lumbar vertebrae. The dynamic mathematical model to represent the motion in the computer simulations, is based on the Lagrange approach, the simulations and experimental tests were developed in MATLAB/Simulink software.

2   Related Work

In the literature, there are approaches and compositions of research that report the construction of models of the spine, such as the work of Gao et al. [10] who developed a virtual model of the spine, based on multi-body theory, to contribute to surgeons' understanding of spinal biomechanics. From another perspective, Okyar et al. [15] built a model of the lumbar spine in CAD Model by using CT scans in order to expedite the development of biomechanical physical models. In addition, Akinci et al. [1] performed a systematic review of papers related to the development of finite element modeling for the construction of biomechanical structures of the spine.

Tjessova et al. [18] carried out a biomechanical research of the lumbar spine, focusing on the geometry of the model, using Bézier curves as a manipulation tool. On the other hand, Kakehashi et al. [12] built a robotic mechanism capable of replicating the spine movement, applying a continuous robot technique with 9 motors and 24 wires distributed among the cervical, thoracic and lumbar segments, performing the movements of flexion, extension, lateral flexion and rotation.

In the work by De Zee et al. [6], a detailed 3D biomechanical model of the lumbar spine was developed, with 7 segments, 18 degrees of freedom and 154 muscles, for research and study of this region. In view of this, to analyze movement behaviors, computer simulations become necessary, and can be implemented even by using a robust mathematical model applied to a robotic structure that represents a lumbar spine model. Among the works that used mathematical models, we have Karadogan et al. [13] who produced a dynamic robotic lumbar spine model with 15 degrees of freedom, driven by 20 cables, where the dynamic structure was established, based on a mathematical model and a controller with feedback linearization, the structure comprised the lumbar vertebrae and the sacrum, connected by spherical joints, which allowed the simulation of the behavior of the dynamic spine.

Similarly, in the work of Clifton et al. [5], a 3D-printed model of the spine was created using multiple structures to simulate the composition of the cancellous bone in the vertebrae, for educational and biomechanical use, considering the limitations presented for the use of cadaveric models. In the same way, Bohl et al. [3] and his research group Barrow Biomimetic Spine, report the elaboration of a spine synthetic model, developed by 3D printing, with biomechanical quality specifications similar to human tissue, with the aim of representing cadaveric models in the educational sphere of surgical planning, development and testing of medical devices.

Urrea et al. [19] developed a new approach for the fuzzy controller model, established with the possibility of inserting the linguistic variable acceleration, in order to perform the movement of the joints of a robotic manipulator with 2 DOF, the PID controller was also implemented to the system, in order to compare the response output of the PID controller with the fuzzy controller.

Angst [2] performed the construction of an experimental model of the spine, with musculoskeletal anatomical correlation of flexion and extension movement, with emphasis on the vertebrae of the lumbar region and the pelvic base. The construction occurred through the use of the 3D Slicer program, using tomographic images for the physical object reconstruction, resulting in STL files, which represent vertebral models in formats available in a database, which were applied to SolidWorks software for subsequent 3D printing, with the Ultimaker Cura software.

In this paper, the lumbar spine mathematical model was considered to be non-linearized, due to the interest of evaluating the controller behavior dealing with the modeled structure inherent characteristics.

3   Kinematics

The spine has thirty-three vertebrae, and has the cervical, thoracic, lumbar, sacral, and coccygeal segments. Each segment refers to specific features of the spine consisting of independent bony sections (Fig. 1). In this sense, they are classified into seven cervical vertebrae (C7), twelve thoracic vertebrae (T12), five lumbar vertebrae (L5), and the sacrum (S1-5) which has five vertebrae that are static and directly interconnected to each other, these regions can suffer distortions in their formation, except for the sacrum. The bony structure of the spine provides the foundation for the composition of the spinal cord, as well as providing the necessary balance for movement of the other limbs of the human body [8].

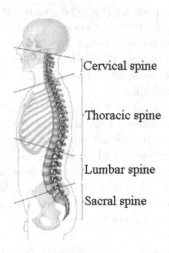

Fig. 1. Spinal column. [12]

According to White et al. [20] the study of the basic anatomy of the spine consists initially of identifying the anatomical planes of this structure in relation to the coronal plane, sagittal plane, and the axial plane (Fig. 2a). The coronal plane is related to frontal and posterior visibility, highlighting the proportionality of the spine. In sequence, the sagittal plane represents the proportions of the aspects on the left and right sides, where in this case, the curvatures of the cervical, thoracic convex posterior, lumbar convex anterior, and sacral positions are highlighted, which promote efficiency in the structure's ability to balance and control. Finally, the axial plane provides the segmentation of the upper and lower parts with respect to the observed physical constitution.

The behavior of the movements performed with respect to the bodies and elements identified by mechanics comprise the study of kinematics, which is concerned with describing the movement occurred, but does not detail the force-related factors responsible for the movement of the bodies. This approach can

be applied to the spine model, starting with the selection of two adjacent vertebrae through translation and rotation movements. For each vertebral unit, the
vertebrae can perform three translations and three rotations, corresponding to
six degrees of freedom [9, 16].

The degrees of freedom represent the possible conditions of movement of a
rigid body, in this sense, considering the three-dimensional coordinate system
with x, y and z axes, a rigid body can perform three translational movements
and three rotational movements. These references are used to characterize the
movements of the spine, considering the x, y and z axes with respect to the
anatomical planes of movement of the spine, coronal, sagittal and axial, as shown
in Fig. 2a [21].

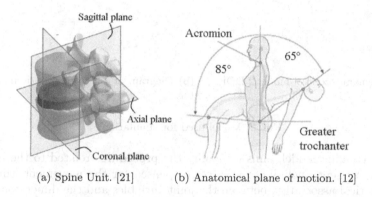

(a) Spine Unit. [21]   (b) Anatomical plane of motion. [12]

Fig. 2. Spinal motion composition.

The movements that can be performed are applied to a unit or set of vertebrae, for the x-axis respective to the coronal plane of right lateral flexion and
left lateral flexion, referring to the y-axis and the sagittal plane of flexion and
extension, and for the z-axis related to the axial plane of rotation comprising
the entire plane [9, 16]. Figure 2b shows the motion relative to the sagittal plane,
which is the base motion for the mathematical model presented in this paper.

4   Dynamic Model

The sagittal symmetry described by Karadogan et al. [13], can be considered
from a facet plane of motion and a facet angle, which represent the plane of
connection of the centers of the four facets of a vertebra and the angle between
the facet plane and the posterior wall of the vertebral body. In this sense, a
dynamic representation of a structure formed by two vertebrae can be seen
in Fig. 3a where m_1 and m_2 represent the masses, l_1 and l_2 characterize the
height of each vertebra, θ_1 and θ_2 comprise the angles of motion rotation, and g
refers to the gravitational force. In addition, θ_1 and θ_2 determine the respective
torques, which determine the forces applied to the joints of each structure. This

representation is being correlated with the modeling performed by Lagrange, where the problems presented for the robotic structure motions formulation with 2 DOF, are elucidated.

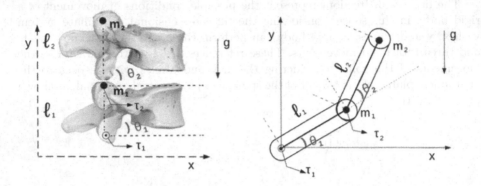

(a) Dynamic control model 2 DOF.   (b) Diagram of a 2 DOF robotic manipulator.

Fig. 3. Model used for simulation.

The dynamic model, aims to qualify the procedures related to the manipulator's movement, including the forces dispensed for the movement, through a mathematical association between the joint variables and the dimensional identifications of the manipulator.

The mathematical modeling used in this work, is based on [11,14]. Thus, the geometric construction scheme of a robotic manipulator with 2 DOF, necessary for the movement of the joints of a structure, with its specific characteristics, is represented in Fig. 3b. This diagram defines the values of the variables, where m_1 and m_2 refer to the masses, l_1 and l_2 comprise the height or body length of each connecting artifact of the manipulator, θ_1 and θ_2 characterize the angles of rotation, and g represents the gravitational force. The torques are determined by τ_1 and τ_2.

The formulations of the dynamic model refer to the partial derivatives of the kinetic and potential energies, in relation to the calculation of the mechanical systems motion equations, and are composed by means of a non-linear equation, commonly based on the Lagrangean methodology, presented in equation (1), where q indicates the position values, $\dot{q}$ represents the velocity and $\ddot{q}$ refers to the acceleration.

$$\tau = M(q)\ddot{q} + C(q,\dot{q}) + G(q) \tag{1}$$

The mathematical modeling established for the robotic model is represented by Eqs. (2–12). Equations (2–6) show the composition of the inertia matrix (M), Eqs. (7–9) refer to the coriolis/centrifugal force vector (C) and Eqs. (10–12) identify the gravity force vector (G).

$$\boldsymbol{M} = \begin{bmatrix} M_{11} & M_{12} \\ M_{21} & M_{22} \end{bmatrix} \tag{2}$$

$$M_{11} = (m_1 + m_2)l_1^2 + m_2 l_2^2 + 2m_2 l_1 l_2 \cos\theta_2 \tag{3}$$

$$M_{12} = m_2 l_2^2 + m_2 l_1 l_2 \cos\theta_2 \tag{4}$$

$$M_{21} = m_2 l_2^2 + m_2 l_1 l_2 \cos\theta_2 \tag{5}$$

$$M_{22} = m_2 l_2^2 \tag{6}$$

$$\boldsymbol{C} = \begin{bmatrix} C_{11} \\ C_{21} \end{bmatrix} \tag{7}$$

$$C_{11} = -m_2 l_1 l_2 (2\dot{\theta}_1 \dot{\theta}_2 + \dot{\theta}_1^2) \sin(\theta_2) \tag{8}$$

$$C_{21} = -m_2 l_1 l_2 \dot{\theta}_1 \dot{\theta}_2 \sin(\theta_2) \tag{9}$$

$$\boldsymbol{G} = \begin{bmatrix} G_{11} \\ G_{21} \end{bmatrix} \tag{10}$$

$$G_{11} = -(m_1 + m_2)gl_1 \sin(\theta_1) - m_2 gl_2 \sin(\theta_1 + \theta_2) \tag{11}$$

$$G_{21} = -m_2 gl_2 \sin(\theta_1 + \theta_2) \tag{12}$$

The mathematical modeling occurs through the association of the robotic structure's characteristics, where the values of the manipulator's dimensions, which refer to the values of the masses and length of each artifact, are associated with the forces of inertia, coriolis/centrifugal, and gravitational matrices, in order to obtain the torques, related to the angles of rotation θ_1 and θ_2 of movement of the dynamic systems.

5   PID Controller

The PID (Proportional Integral Derivative) controller performs the parameter aggregation required to control the system with respect to the response error of the control output. The parameter P refers to the gain result of the proportional control, which can be reflected in a higher response speed. The integral control I regulates the error signal, according to the system response. Finally, the derivative control gain D produces overshoot minimization, increasing the ability of the response signal to achieve the objective. All these gains are received through the error, which travels between the desired reference and the output of the controller. Given this, the mathematical behavior of PID control can be described through Eq. (13).

$$\boldsymbol{\tau} = \boldsymbol{K_p}\mathbf{e} + \boldsymbol{K_v}\dot{\mathbf{e}} + \boldsymbol{K_i} \int_0^t \mathbf{e}(\sigma)d\sigma \tag{13}$$

where K_p characterizes the proportional gain, K_v indicates the derivative gain and K_i represents the integral gain. For the control model of the robotic manipulator system, the values determined for each controller gain, referring to the controllers of each joint of the structure, were stipulated at $K_p= 9$, $K_v= 0.7$ and $K_i= 0.4$.

6   Simulation and Results

To simulate the lumbar spine dynamic model, the physical parameters shown in Table 1 [19] are used.

Table 1. Manipulator robot parameters.

Joint 1	Joint 2	Units
$m_1 = 0.392924$	$m_2 = 0.094403$	[kg]
$l_1 = 0.2032$	$l_2 = 0.1524$	[m]

The computer simulations were performed using the MATLAB/Simulink software. The control scheme developed using the block diagram tool is shown in Fig. 4, where the PID control blocks of the system are represented according to the data in Sect. 5. The input reference signals to the system are the θ_1 and θ_2 angles for the movement of the L1 and L2 vertebrae, respectively. The dynamic model block involves characterizing the behavior of the lumbar spine structure by delivering at the output the current angular displacements θ_1 and θ_2, signals that serve to feed back to the control system (Fig. 5). The equations and parameters follow the descriptions of Sect. 4 and Table 1, respectively.

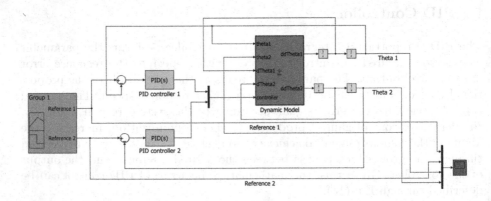

Fig. 4. Control System for a Robotic Manipulator 2 DOF.

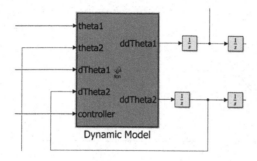

Fig. 5. Block of the dynamic model.

The modeling of the system considered only the first two vertebrae of the robotic lumbar spine model, in order to perform the first experimental tests with the developed system. The dynamic model is based on the control of a manipulator with 2 DOF, considering the flexion and extension movement performed within the sagittal plane of motion of the vertebrae. The input references considered for the computer simulation are of two types: a pulse train with acceleration and deceleration of 0.5 s and a signal composed of three steps of different amplitudes. For the tests, flexion and extension movements are considered, so four reference signals were generated. Figure 6 show the result of the output signals from the simulation performed in Matlab/Simulink.

Analytical results of the references and responses of the controlled system are shown in Table 2.

Table 2. Analytical results of the references and responses of the controlled system.

	θ_1	%PO1	θ_2	%PO2
Signal 1	0° to 6°	3.3%	0° to 3°	3.3%
Signal 2	0° to −6°	3.3%	0° to −3°	3.3%
Signal 3	0°, 2°, 4°, 6°	50%	0°, 4°, 8°, 12°	50%
Signal 4	0°, −2°, −4°, −6°	50%	0°, −4°, −8°, −12°	50%

The planning of the initial movements, indicating the movement for the θ_1 and θ_2 angles, occurred from 0° to 6° and from 0° to 3° for Signal 1, (Fig. 6a). For Signal 2 the references for were determined from 0° to −6° and from 0° to −3°, (Fig. 6b). For Signal 3 the references for were stipulated from 0°, 2°, 4° and 6°, (Fig. 6c). Finally, the references given for Signal 4 comprised the angles 0°, −4°, −8°, −12°, (Fig. 6d). All these values, represent the angular test displacements of the simulations of the flexion and extension motion of the robotic lumbar spine model.

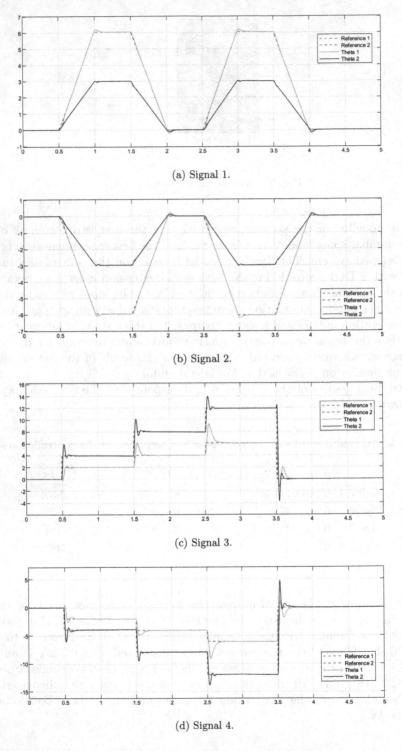

(a) Signal 1.

(b) Signal 2.

(c) Signal 3.

(d) Signal 4.

Fig. 6. References and simulation of controlled behavior.

7 Conclusions

This paper presents the mathematical model to represent a robotic lumbar spine structure and its motion control. Computational simulation was performed using reference input signals. The analysis of the results shows a fast response of at most 0.2 s, a zero steady state error in all cases and also a maximum overshoot peak of 3.3% for the Signal 1 and Signal 2 references and 50% for the Signal 3 and Signal 4 references. From the results obtained, the responses for references with acceleration and deceleration in time (Signal 1 and Signal 2) stand out. Thus, the simulations show that the implementation of the control system returns acceptable responses for the execution of the intended movement for each robotic lumbar vertebra.

This movement control of the lumbar spine model structure, can also be performed using computational intelligence techniques, such as fuzzy controllers, which when well designed, can improve the response time and decrease the maximum overshoot of the system, this technique is currently being studied by the authors for development. Finally, as future work the controller intends to be implemented in the 3D structure of the lumbar spine idealized by Angst [2]. From the implemented 3D structure a behavior database can also be generated to analyze and validate the mathematical model used.

Acknowledgment. The work presented in this paper was supported by the PAVIC Laboratory (*Pesquisa Aplicada em Visão e Inteligência Computacional*) at University of Acre , Brazil.

References

1. Akinci, S.Z., Arslan, Y.Z.: Finite element spine models and spinal instruments: a review. J. Mech. Med. Biol. **22**(04), 2230001 (2022)
2. Angst, L.R.: Construção e validação de um modelo experimental da cinesiologia da coluna lombar humana. Dissertação de Mestrado do Programa de Pós-Graduação em Ciências da Saúde da Amazônia Ocidental, UFAC, Brazil (2022)
3. Bohl, M.A.: Range of motion testing of a novel 3d-printed synthetic spine model. Global Spine J. **10**(4), 419–424 (2020)
4. Chevalier, M., Gómez-Schiavon, M., Ng, A.H., El-Samad, H.: Design and analysis of a proportional-integral-derivative controller with biological molecules. Cell Syst. **9**(4), 338–353 (2019)
5. Clifton, W., Nottmeier, E., Damon, A., Dove, C., Chen, S.G., Pichelmann, M.: A feasibility study for the production of three-dimensional-printed spine models using simultaneously extruded thermoplastic polymers. Cureus 11(4) (2019)
6. De Zee, M., Hansen, L., Wong, C., Rasmussen, J., Simonsen, E.B.: A generic detailed rigid-body lumbar spine model. J. Biomech. **40**(6), 1219–1227 (2007)
7. Eremina, G., Smolin, A., Martyshina, I.: Convergence analysis and validation of a discrete element model of the human lumbar spine. Rep. Mech. Eng. **3**(1), 62–70 (2022)
8. Frost, B.A., Camarero-Espinosa, S., Foster, E.J.: Materials for the spine: anatomy, problems, and solutions. Materials **12**(2), 253 (2019)

9. Galbusera, F., Wilke, H.J.: Biomechanics of the Spine: Basic Concepts, Spinal Disorders and Treatments. Academic Press, Cambridge (2018)
10. Gao, Z., Gibson, I., Ding, C., Wang, J., Wang, J.: Virtual lumbar spine of multi-body model based on simbody. Procedia Technol. **20**, 26–31 (2015)
11. Ghaleb, N.M., Aly, A.A.: Modeling and control of 2-DOF robot arm. Int. J. Emerg. Eng. Res. Technol. **6**(11), 24–31 (2018)
12. Kakehashi, Y., Okada, K., Inaba, M.: Development of continuum spine mechanism for humanoid robot: biomimetic supple and curvilinear spine driven by tendon. In: 2020 3rd IEEE International Conference on Soft Robotics (RoboSoft), pp. 312–317. IEEE (2020)
13. Karadogan, E., Williams, R.L.: The robotic lumbar spine: Dynamics and feedback linearization control. Comput. Math. Methods Med. 2013 (2013)
14. Okubanjo, A., Oyetola, O., Osifeko, M., Olaluwoye, O., Alao, P.: Modeling of 2-DOF robot arm and control. Fed Univ. Technol. Owerri, J. Ser. (Futojnls) **3**(2), 80–32 (2017)
15. Okyar, F., Guldeniz, O., Atalay, B.: A holistic parametric design attempt towards geometric modeling of the lumbar spine. Comput. Methods Biomech. Biomed. Eng.: Imaging Visual. (2019)
16. Panjabi, M.M., White, A.A., III.: Basic biomechanics of the spine. Neurosurgery **7**(1), 76–93 (1980)
17. Swain, C.T., Pan, F., Owen, P.J., Schmidt, H., Belavy, D.L.: No consensus on causality of spine postures or physical exposure and low back pain: a systematic review of systematic reviews. J. Biomech. **102**, 109312 (2020)
18. Tjessova, M., Minarova, M.: Precising of the vertebral body geometry by using bézier curves. In: Proceedings of Algoritmy, pp. 55–61 (2016)
19. Urrea, C., Kern, J., Alvarado, J.: Design and evaluation of a new fuzzy control algorithm applied to a manipulator robot. Appl. Sci. **10**(21), 7482 (2020)
20. White 3rd, A., Panjabi, M.M.: The basic kinematics of the human spine. a review of past and current knowledge. Spine **3**(1), 12–20 (1978)
21. Zhou, C.: Multi-objective design optimization of a mobile-bearing total disc arthroplasty considering spinal kinematics, facet joint loads, and metal-on-polyethylene contact mechanics. Ph.D. thesis, State University of New York at Binghamton (2018)

Annotation-Free Identification of Potential Synteny Anchors

Karl Käther[1]([✉])[iD], Steffen Lemke[2][iD], and Peter F. Stadler[1,3,4,5,6][iD]

[1] Bioinformatics Group, Department of Computer Science, and Interdisciplinary Center for Bioinformatics, Universität Leipzig, Härtelstrasse 16-18, 04107 Leipzig, Germany
{karl,studla}@bioinf.uni-leipzig.de
[2] Center for Organismal Studies, Universität Heidelberg, Heidelberg, Germany
[3] Max Planck Institute for Mathematics in the Sciences, Leipzig, Germany
[4] Institute for Theoretical Chemistry, University of Vienna, Vienna, Austria
[5] Facultad de Ciencias, Universidad Nacional de Colombia, Bogotá, Colombia
[6] Santa Fe Institute, Santa Fe, NM, USA

Abstract. Orthology assignment between genetic elements lies at the heart of comparative genomics. Current methods primarily rely on sequence and structural similarity. Both low sequence similarity and the presence of multiple copies limit similarity-based methods. Synteny, i.e., conservation of (relative) genomic location, can help resolve many such cases. The mapping of synteny is based on "synteny anchors", defined as intervals of genomic locations for which unique orthologs in related genomes can be determined unambiguously. Usually, annotated elements such as protein-coding genes are utilized for this purpose. Here we describe an annotation-free approach and devise a k-mer-based heuristic to identify synteny anchors. To demonstrate the practicality of the approach, we compute and analyze a set of synteny anchors for 25 *Drosophila* species.

Keywords: Synteny · anchor · k-mer

1   Background

Orthology inference between genetic elements is the central problem in comparative genomics and the basis for many downstream analyses. Sequence similarity is the primary mean of establishing orthology, often by means of identifying pairwise best matches. Even in two closely related genomes, such as human and chimpanzee, not the entire sequence is composed of one-to-one orthologs. Deletions and duplications of stretches of genomic DNA already are relevant contributions to genetic differences. Comparing more distant genomes, furthermore, one usually encounters regions where the sequence has diverged so far that

---

This work was funded by the German Research Foundation as part of SPP 2349 *"Genomic Basis of Evolutionary Innovations (GEvol)"*.

I. Rojas et al. (Eds.): IWBBIO 2023, LNBI 13919, pp. 217–230, 2023.
https://doi.org/10.1007/978-3-031-34953-9_17

no significantly similar sequences can be recognized. Failure to identify orthologs cause underestimates of gene age [11] and overestimates of the number of taxon-restricted genes [19], which in turn biases the study of mechanisms of evolutionary innovations. Moreover, lineage-specific differences in evolutionary rates complicate the distinction of orthologs and paralogs based on sequence similarity alone [2].

The conservation of genomic position can help in either scenario. Sequences that are below the detection limit in genome-wide searches can become readily recognizable when the search space is restricted to a region between two well-conserved anchor genes. Paralogous genes, even if subject to concerted evolution such as tRNAs, can be disambiguated by their location between unambiguous orthologous regions [20]. Computational tools to map syntenic regions between genomes such as DAGchainer [7] and MCScanX [21] rely on genome annotations. This can be problematic in situations where annotations are unavailable or are of limited quality. While there are several efficient and accurate pipelines to (re-)annotate genomes, considerable issues remain: Contamination by foreign DNA may cause many false positive annotations [16]. Annotation accuracy is limited for fragmented assemblies. The distances between annotation items may be very extensive, particularly in large genomes; this is bound to limit the accuracy of the synteny information even in the presence of high-quality genome annotations. It is therefore desirable to address synteny in an annotation-free setting. In principle, the problem is solved by multiple genome alignments [3]. The construction of genome-alignments, however, is computationally very demanding. Synteny anchors, furthermore, may serve as a useful intermediate for genome-alignments.

Thus we consider the problem of determining synteny anchors in an annotation-free setting. Starting from a concise presentation of the theoretical background, we describe an efficient heuristic. As proof of concept we apply our prototype to 25 fruit-fly genomes.

2   Theory and Algorithmic Considerations

We consider a genome G as a string over the DNA alphabet $\{A, C, G, T\}$ and a "line break" character to accommodate chromosome ends of assembly fragments. We denote by $S(G)$ the set of all substrings of G and its reverse complement that do not contain a "line break". Furthermore, let d be a metric distance function on $\{A, C, G, T\}^*$. In practice $d(w, w')$ is realized e.g. by a pairwise sequence alignment, blast bit scores or similar. Consider a string $w \in S(G)$ with length $|w|$. Clearly w will be similar to any $w' \in S(G)$ that overlaps w in the same reading direction for a large fraction of its length.

We therefore say that w is d_0-unique in G if $d(w, w') > d_0$ for all $w' \in S(G \setminus \{w\})$, where $G \setminus \{w\}$ is obtained from G by replacing w by a "line break". Note that $w' \in S(G \setminus \{w\})$ implies that the reverse complement $\bar{w}'$ is also in $S(G \setminus \{w\})$. Now consider two genomes G and H with d_0-unique substrings $w \in S(G)$ and $y \in S(H)$ such that $d(w, y) \leq d_0/2$. Then the triangle inequality implies that

$d(w,y') > d_0/2 > d(w,y)$ for all $y' \in S(H \setminus \{y\})$ and $d(y,w') > d_0/2 > d(y,w)$ for all $w' \in S(G \setminus \{w\})$. That is, $w \in S(G)$ and $y \in S(H)$ form a unique match and thus can safely serve as an *anchor*.

This observation suggests that synteny anchors can be computed by first identifying a set $C(G, d_0^G)$ of d_0-unique sequences in each genome G. Then anchors are identified as pairs (w, y) such that $w \in C(G, d_0^G)$, $y \in C(H, d_0^H)$ and $d(w, y) < d_1$, where the distance cutoff is determined by the thresholds d_0^G and d_0^H. Following the argument above, $d_1 = \min(d_0^G, d_0^H)/2$ ensure that (w, y) is the unique best match. Pragmatically, it seems sufficient to ensure that y closer to w than to any $y' \in S(H \setminus \{y\})$ and w is closer to y than to any $w' \in S(H \setminus \{w\})$ by some safety margin ε. This amount to choosing a threshold of $d_1 = \min(d_0^G, d_0^H) - \varepsilon$.

The practical challenges are (1) to determine the candidate sets $C(G, d_0^G)$ and the anchor pairs without explicitly comparing each substring of a fixed length ℓ against the rest of the genome, and (2) to avoid an explicit all-against-all comparison of the candidate sets $C(G, d_0^G)$ and $C(H, d_0^H)$.

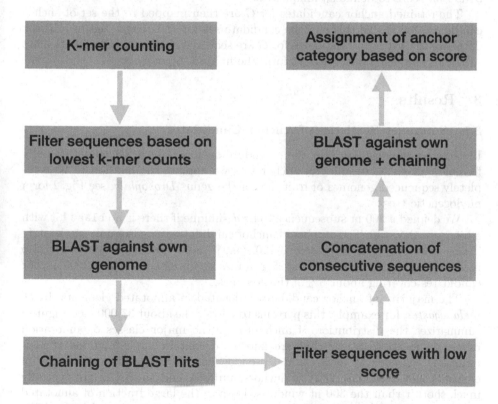

Fig. 1. Workflow for determining candidate synteny anchors for a single genome.

Since similar strings are in particular composed of many similar substrings it seems plausible to measure the similarity of a substring w with the rest of its genome by the total frequency of k-mers contained in w. In other words, fixing a length ℓ, the sum of the k-mer abundances for all k-mers in a given sequence window $G[i, i + \ell - 1]$ is a predictor for $w = G[i, i + \ell - 1]$ being a good anchor candidate. GenMap [14] was used to count k-mer frequencies. In the showcase study listed below we do not allow mismatches and use a k-mer length of $k = 13$. We settled on substring of length 300 that overlap by 150 nt to reduce the computational effort, and retained a tenth of the windows with the lowest aggregate k-mer count.

The resulting anchor candidates are then compared against the same genome G with blast. Since blast does not necessarily return a single alignment for each locus but may instead produce multiple local alignments, such fragmented hits were chained to produce a single full-length match [1]. We use clasp [12] for this purpose. Candidates for which chained blast hits are too similar to the query were discarded. The workflow to generate sets of potential synteny anchors is summarizes in Fig. 1. We tested different values between 40 and 120 of the blast bit score to define d_0-uniqueness.

The retained anchor candidates for G are then mapped to the set of anchor candidates for H. To this end the candidate set for H is converted to a blast database against with candidates for G are searched as queries. Candidate pairs satisfying the distance constraint form the final set of anchors.

3   Results

3.1   Summary Statistics of Anchor Candidates

In order to demonstrate that the procedure outlined above is computationally feasible and yields meaningful anchor sets we applied the method to 25 completely sequenced genomes of fruit flies of the genus *Drosophila*, see Fig. 2 for a phylogenetic tree.

We defined a 300 nt subsequence to be d_0-unique if there is no blast hit with a bit-score above 40. The number of anchor candidates increases only moderately if this score cut-off is increased to 120, i.e., if more similar sequences are also included, see Fig. 3. In each of the 25 genomes we obtained at least 30000 anchor candidates covering about 5% of the genomes.

The majority of anchor candidates is located in annotated elements. In *D. melanogaster*, for example, this pertains to 70% of the about 37,000 loci. Figure 4 summarizes the distribution of anchors over the major classes of annotation items. Annotation classes that are known to contain large fractions of non-unique DNA, such as lncRNA and pseudogenes, have the highest fraction of elements which cannot serve as anchors. miRNA and snoRNAs are typically much shorter than the 300 nt windows. Despite the large fraction of annotated anchor candidates, about 30% are un-annotated and thus extend beyond the reach of annotation-based methods. Even though tRNAs are among the best-conserved genes, there are almost no anchor candidates among them. The reason is that tRNA genes are multi-copy genes that evolve under concerted evolution, see [20] and the references therein.

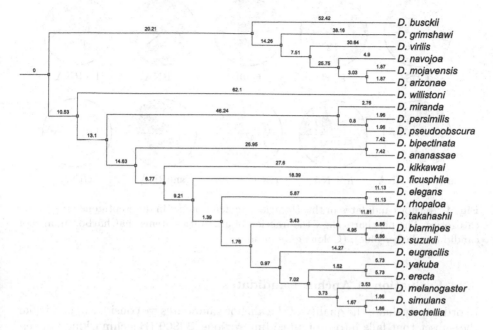

Fig. 2. Phylogenetic tree of the *Drosophila* species used in this contribution. Estimated divergence times (in Myr) are annotated at all inner nodes. Accession numbers for all genomes are listed in the Appendix. The *melanogaster* and *obscura* subgroups are indicated by blue and red species names, respectively. Figure adapted from [18].

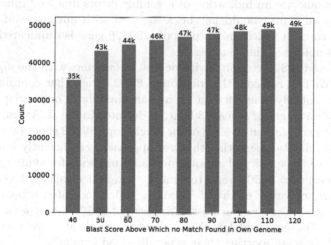

Fig. 3. Number of anchor candidates as a function of the bit-score cutoff. Note increasing bit-scores imply decreasing distance and thus correspond to a less stringent definition of d_0-uniqueness.

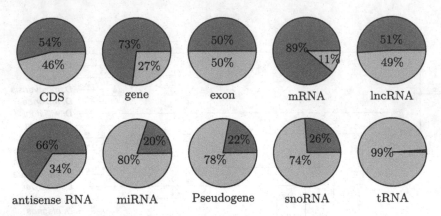

Fig. 4. Anchor candidates in the *D. melanogaster* genome in different gene annotation categories. The pie charts show the fraction of annotation items that harbor an anchor candidate (darker blue). (Color figure online)

3.2   Evaluation of Anchor Candidates

In order to evaluate the quality of the anchor candidates we consider in particular the subset that falls into protein-coding regions. BUSCO (Benchmarking Universal Single-Copy Orthologs) [17] has become *the* tool to assess the completeness and accuracy of genome assemblies. The tool is based on curated collections of single-copy genes. These serve as excellent anchor candidates. The sensitivity of our approach thus can be assessed by considering the fraction of BUSCO genes containing an anchor. Conversely, BUSCO genes that appear in more than one copy in a genome are an indication of assembly errors due to duplicated DNA. In general, such genes should not be associated with anchor candidates. The exception is cases in which only a part of a BUSCO gene is duplicated, while the remainder is unique in the genome.

Here we used BUSCO version 5 with default parameters with the *diptera_odb10* dataset [9]. We first consider the fraction of BUSCO genes that contain an anchor candidate, is entirely contained in an anchor candidate, or overlap an anchor candidate over a length of at least 300 nt, as depicted in Fig. 5. Averaged over the 25 fruit fly genomes, about 93±4% of the single-copy BUSCO genes overlap at least one anchor candidate, suggesting that our approach is sufficiently sensitive to at least reproduce the quality of annotation-based methods for synteny detection.

For *D. miranda*, BUSCO reported more than 500 duplicated genes, by far more than in any other of the 25 genomes which indicates a poor quality of the genome assembly. Comparing the duplicated BUSCO genes with our anchor candidates, we observe that less than 4 % of the duplicates are associated with anchor candidates, supporting their generally good quality.

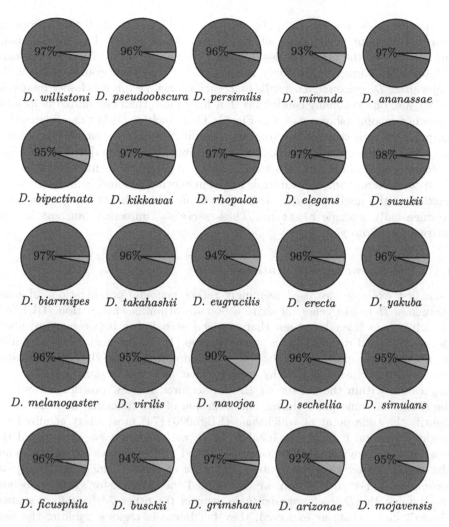

Fig. 5. Recall of single-copy `BUSCO` genes containing anchor candidates. As expected, the overwhelming majority of single-copy `busco` genes contains one or more anchor candidate (darker blue). (Color figure online)

3.3 Mapping of Anchors Candidates Across Genomes

Figure 6 displays the average number of anchor candidates from a given genome that have `blast` matches in a certain number of other genomes. Here, all matches with bit-scores after chaining are counted that exceed the bit-score threshold used to define d_0-uniqueness with in the candidate's own genome plus a tolerance of this bit score's square root. Error bars indicate the variation between genomes. Not surprisingly, the number of anchor candidates matching very few other genomes is highly variable and depends on the number of very closely related genomes. For example, *D. willistoni*, which is most distance from its nearest neighbour in our data set, shows over 20000 potential anchors without any matches in the other species. On the other hand, most of the anchor candidates in *D. novoja's* with only a few matches find their counterparts in *D. mojavensis*, *D. arizonae*, *D. virilis* or *D. grimshawi*, i.e., in its closest relatives.

We also observe many candidates that have more than one match in at least one other genome (shown as −1 in Fig. 6). These include anchor candidates that happen to match sequences erroneously duplicated due to assembly problems, but also true gene duplications. The inset shows that the majority of multiple matches occur in only one or a small number of genomes. In more than 80% of these cases there is only one multiple match in another genome. Still, a significant fraction of the anchor candidates is present in all or almost all genomes and produces only a single `blast` hit. These serve as "universal" anchors for the entire genus *Drosophila*.

3.4   Two Showcase Applications

As a first example, we used seemingly recent genomic organizations of *halo*-like genes. *Halo*-like genes all share a domain of unknown function (DUF733) [6], while *halo* itself has been characterized as a linker between microtubule motors and lipid droplets in the early embryo [4]. While most DUF733 family members appear clustered to a single locus on chromosome 2L, *halo* is located outside of this cluster [6]. We hypothesize that *halo* was initially born by gene duplication within the cluster on 2L and acquired its new position away from the cluster at some point during the radiation of the subgenus *Drosophila*. To identify the time point at which *halo* (FBgn0001174) most likely acquired the genomic location it currently inhabits in *D. melanogaster*, we determined the local genomic environment of *halo* and asked which species in the subgenus *Drosophila* shared a similar synteny signature at the *halo* locus as *Drosophila melanogaster*. For comparison, we determined the matches for synteny anchors surrounding the *D. melanogaster halo* gene in the other 24 fruit fly genomes. Figure 7 shows that, as expected, they fall into the regions flanking the *halo* orthologs (as annotated in `OrthoDB` [8]). We found at least one anchor in all species and observed that the number of anchors sharply decreases outside of the *melanogaster* group. This is consistent with synteny assessment by direct pairwise sequence comparison.

Our results indicate that *halo* acquired its current genomic location in *D. melanogaster* in the stem group of the *melanogaster* group, and they hint to the existence of an evolutionary period during which such a gene jump could have been facilitated. This interpretation is supported by independently acquired *halo* synteny in *D. pseudobscura* and *D. persimilis*, which both are part of the *obscura* group and represent the sister taxon of the *melanogaster* group. In relation to the synteny anchors shared with *D. melanogaster*, we observe a local translocation of *halo* in *D. persimilis*. A similar translocation is observed for *D. bipectinata*. Even within the *melanogaster* group relative locations of the anchors indicate several small local genome rearrangements, indicating the flexibility of this genomic locus.

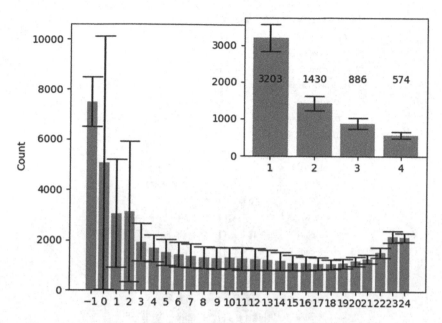

Fig. 6. Distribution of the number of other genomes with unique match anchor candidates from a given genome. The number of anchor candidates with more than one match in at least one other genome is show as -1. The inset shows the number of anchor candidates that have multiple matches as function of the number of genomes in which multiple matches are detected. Error bars show the variation for query anchor candidates across the 25 different fruit fly genomes. Note that anchor candidates with no or very few other matching genomes have a very high variance, owing to the differences in phylogenetic distance to the closest relatives.

As a second example, we revisited the use of anchors to determine orthology among multi-copy genes. As noted above, tRNAs usually appear in multiple copies and thus cannot serve as anchors. As an example demonstrating the usefulness of our anchor construction, we searched for the orthologs of tRNA:Gly-GCC (`FlyBase` ID *FBgn0011859*). A `blast` search using the tRNA as query yields on average 16 potential orthologs with an E-value $< 10^{-10}$. The anchor candidates within ± 30 kb of *D. melanogaster* tRNA:Gly-GCC yield on average 21 anchor matches in 12 out 24 genomes that in each case identify a single `blast` hit. In the remaining genomes, no anchor match is associated with any of the tRNA homologs. This pattern persists even if the neighborhood of *FBgn0011859* is increased to 1 Mb. Together with the phylogeny in Fig. 2, these data imply that this particular tRNA:Gly-GCC originated about 49 Myr ago in the last common ancestor of *D. melanogaster* and *D. pseudoobscura*. The corresponding anchors are also exclusive to this clade.

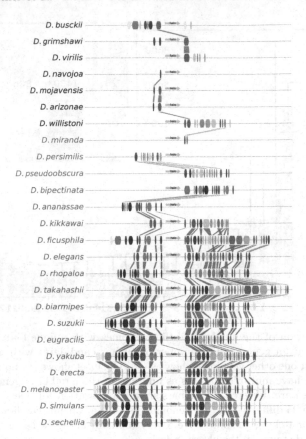

Fig. 7. Anchors around the gene *halo* (blue arrow) in a region of 100000 nucleotides up- and downstream in *D. melanogaster* and its orthologs in other *Drosophila* species. For readability anchors are enlarged by a factor of 5 and spaces without anchors are reduced to a maximum length of 4000 nucleotides. Species are ordered according to the phylogenetic tree in Fig. 2 and the reading direction is given relative to the orientation of *halo*. As in Fig. 2, the *melanogaster* and *obscura* groups are indicated by blue and red, respectively. Colors indicate anchors globally. Grey lines connect corresponding anchors in adjacent genomes to highlight local rearrangements. (Color figure online)

3.5   Computational Effort

We performed the computation for the 25 *Drosophila* genomes on a workstation with a 12-core *Intel(R) Core(TM) i7-8700 CPU @ 3.20GHz* processor and 16 GB RAM. The running time of approximately 12h suggests that the approach is also feasible for larger genomes, in particular given that we have not yet invested efforts to optimize the implementation.

4   Discussion and Outlook

We have introduced here an annotation-independent approach for the construction of synteny anchors. Instead of starting with sequence comparisons between pairs of genomes as in the case of genome-wide alignment procedures, we first identify subsequences that are "sufficiently unique" within their own genome. These sequences likely have a unique significant match in other, related genomes. To demonstrate the feasibility of this approach, we computed synteny anchors for a data set of 25 fruit fly genomes.

Our preliminary results show that the proposed algorithm indeed yields tens of thousands of potential synteny anchors for each individual genome. While the initial set contains some false positives, i.e., anchor candidates that yield multiple significant blast hits in the other genomes, this is not the case for the majority of the candidates. A comparison with BUSCO single-copy orthologs, furthermore, shows that these typically overlap with our anchor candidates, suggesting that anchor candidates typically achieve a very good coverage. Of course, the method will require much more extensive benchmarking on data sets comprising genomes of different sizes and phylogenetic scopes. We expect that parameters can also be optimized further in the wake of such a more extensive evaluation. While the use of BUSCOs as references was convenient, it will be interesting to consider commonly used orthology detectors such as OrthoFinder [5] that are not restricted to single-copy genes and consider also paralogs as long as they are sufficiently distinct.

We conclude that the construction of synteny anchors from "sufficiently unique" genomic sequences is practically feasible. It is a promising alternative in particular for novel, un-annotated genomes. Furthermore, it altogether avoids biases that may arise from incomplete annotations as well as inconsistent annotations obtained with different tools. It also has the advantage that anchor candidates can be pre-computed independently for each genome and stored as simple annotation tracks. This enables an efficient re-computation of anchor sets for different collections of genomes and allows the incremental inclusion of additional genomes. The example of the *halo* region also shows that our anchors provide a convenient method to detect local genome rearrangements.

Despite the encouraging computational results, several open problems remain. First, here we have used fixed length sequence windows with a pitch of half the window the size. It seems worthwhile to investigate whether these restrictions can be alleviated without prohibitive computational costs. Although the aggregate k-mer count proved to be an efficient filter, it is well conceivable that the performance can be improved by utilizing more sophisticated measures borrowed from alignment-free sequence comparison [22]. It should also be possible to use overlapping anchor matches to obtain consistent anchor sets for multiple genomes, using in particular ideas from much earlier work on consistent local alignment segments [10, 13, 15].

Furthermore, it will be interesting to explore the theory of synteny anchors in some more detail, in particular regarding overlapping anchor candidates (which are excluded here) and with respect to their relation with reciprocal best matches. While our simple considerations have resulted in a meaningful notion of unambiguity, there is no guarantee that the best match of an anchor candidate in one genome is also an anchor candidate in another genome: a sequence that is unique in the query genome may be duplicated in the target genome, whence it does not appear as an anchor candidate. Moreover, an insertion into an anchor candidate may lead to a split match in the target. In either case, such anchor candidates have to be removed from the final set of anchors. A related issue that is d_0-uniqueness for given level d_0 pertains to a specific phylogenetic scope. It would be interesting to see how the uniqueness-level d_0 needs to vary with phylogenetic distance to ensure consistency of anchors across large sets of genomes.

A   Genomes Used in this Contribution

Organism	GenBank Accession	Assembly Level
D. willistoni	GCA_018902025.2	chromosome
D. pseudoobscura	GCA_009870125.2	chromosome
D. persimilis	GCA_003286085.2	contig
D. miranda	GCA_003369915.2	chromosome
D. ananassae	GCA_017639315.2	chromosome
D. bipectinata	GCA_018153845.1	contig
D. kikkawai	GCA_018152535.1	contig
D. rhopaloa	GCA_018152115.1	contig
D. elegans	GCA_018152505.1	contig
D. suzukii	GCA_013340165.1	chromosome
D. biarmipes	GCA_025231255.1	chromosome
D. takahashii	GCA_018152695.1	contig
D. eugracilis	GCA_018153835.1	contig
D. erecta	GCA_003286155.2	contig
D. yakuba	GCA_016746365.2	chromosome
D. melanogaster	GCA_000001215.4	chromosome
D. sechellia	GCA_004382195.2	chromosome
D. simulans	GCA_016746395.2	chromosome
D. ficusphila	GCA_018152265.1	contig
D. busckii	GCA_011750605.1	chromosome
D. grimshawi	GCA_018153295.1	contig
D. arizonae	GCA_001654025.1	scaffold
D. mojavensis	GCA_018153725.1	contig
D. navojoa	GCA_001654015.2	scaffold
D. virilis	GCA_003285735.2	contig

References

1. Abouelhoda, M.I., Ohlebusch, E.: Multiple genome alignment: chaining algorithms revisited. In: Baeza-Yates, R., Chávez, E., Crochemore, M. (eds.) CPM 2003. LNCS, vol. 2676, pp. 1–16. Springer, Heidelberg (2003). https://doi.org/10.1007/3-540-44888-8_1

2. Altenhoff, A.M., et al.: Standardized benchmarking in the quest for orthologs. Nat. Methods **13**(5), 425–430 (2016). https://doi.org/10.1038/nmeth.3830

3. Armstrong, J., et al.: Progressive Cactus is a multiple-genome aligner for the thousand-genome era. Nature **587**, 246–251 (2020). https://doi.org/10.1038/s41586-020-2871-y

4. Cohen, R.S.: Halo: a guiding light for transport. Curr. Biol. **13**(22), R869–R870 (2003). https://doi.org/10.1016/j.cub.2003.10.046

5. Emms, D.M., Kelly, S.: OrthoFinder: phylogenetic orthology inference for comparative genomics. Genome Biol. **20**(1), 238 (2009). https://doi.org/10.1186/s13059-019-1832-y

6. Gramates, L.S., et al.: FlyBase: a guided tour of highlighted features. Genetics **220**(4), iyac035 (2022). https://doi.org/10.1093/genetics/iyac035

7. Haas, B.J., Delcher, A.L., Wortman, J.R., Salzberg, S.L.: DAGchainer: a tool for mining segmental genome duplications and synteny. Bioinformatics **20**(18), 3643–3646 (2004). https://doi.org/10.1093/bioinformatics/bth397

8. Kuznetsov, D., et al.: OrthoDB v11: annotation of orthologs in the widest sampling of organismal diversity. Nucleic Acids Res. **51**, D445–D451 (2023). https://doi.org/10.1093/nar/gkac996

9. Manni, M., Berkeley, M.R., Seppey, M., Simão, F.A., Zdobnov, E.M.: BUSCO update: novel and streamlined workflows along with broader and deeper phylogenetic coverage for scoring of eukaryotic, prokaryotic, and viral genomes. Mol. Biol. Evol. **38**(10), 4647–4654 (2021). https://doi.org/10.1093/molbev/msab199

10. Morgenstern, B.: DIALIGN 2: improvement of the segment-to-segment approach to multiple sequence alignment. Bioinformatics **15**, 211–218 (1999). https://doi.org/10.1093/bioinformatics/15.3.211

11. Moyers, B.A., Zhang, J.: Further simulations and analyses demonstrate open problems of phylostratigraphy. Genome Biol. Evol. **9**(6), 1519–1527 (2017). https://doi.org/10.1093/gbe/evx109

12. Otto, C., Hoffmann, S., Gorodkin, J., Stadler, P.F.: Fast local fragment chaining using sum-of-pair gap costs. Alg. Mol. Biol. **6**, 4 (2011). https://doi.org/10.1186/1748-7188-6-4

13. Otto, W., Stadler, P.F., Prohaska, S.J.: Phylogenetic footprinting and consistent sets of local aligments. In: Giancarlo, R., Manzini, G. (eds.) CPM 2011. LNCS, vol. 6661, pp. 118–131. Springer, Heidelberg (2011). https://doi.org/10.1007/978-3-642-21458-5_12

14. Pockrandt, C., Alzamel, M., Iliopoulos, C.S., Reinert, K.: GenMap: ultra-fast computation of genome mappability. Bioinformatics **36**(12), 3687–3692 (2020). https://doi.org/10.1093/bioinformatics/btaa222

15. Prohaska, S., Fried, C., Flamm, C., Wagner, G., Stadler, P.F.: Surveying phylogenetic footprints in large gene clusters: applications to Hox cluster duplications. Mol. Phyl. Evol. **31**, 581–604 (2004). https://doi.org/10.1002/jez.b.20007

16. Salzberg, S.L.: Next-generation genome annotation: we still struggle to get it right. Genome Biol. **20**, 92 (2019). https://doi.org/10.1186/s13059-019-1715-2

17. Simão, F.A., Waterhouse, R.M., Ioannidis, P., Kriventseva, E.V., Zdobnov, E.M.: BUSCO: assessing genome assembly and annotation completeness with single-copy orthologs. Bioinformatics **31**(19), 3210–3212 (2015). https://doi.org/10.1093/bioinformatics/btv351

18. Thomas, G., Hahn, M.: Drosophila 25 species phylogeny. Figshare **10**, m6089 (2017). https://doi.org/10.6084/m9.figshare.5450602.v1

19. Vakirlis, N., Carvunis, A.R., McLysaght, A.: Synteny-based analyses indicate that sequence divergence is not the main source of orphan genes. eLife **9**, e53500 (2000). https://doi.org/10.7554/eLife.53500
20. Velandia-Huerto, C.A., et al.: Orthologs, turn-over, and remolding of tRNAs in primates and fruit flies. BMC Genomics **17**, 617 (2016). https://doi.org/10.1186/s12864-016-2927-4
21. Wang, Y., et al.: MCScanX: a toolkit for detection and evolutionary analysis of gene synteny and collinearity. Nucleic Acids Res. **40**(7), e49 (2012). https://doi.org/10.1093/nar/gkr1293
22. Zielezinski, A., Vinga, S., Almeida, J., Karlowski, W.M.: Alignment-free sequence comparison: benefits, applications, and tools. Genome Biol. **18**, 186 (2017). https://doi.org/10.1186/s13059-017-1319-7

Analysing Dose Parameters of Radiation Therapy Treatment Planning and Estimation of Their Influence on Secondary Cancer Risks

Lily Petriashvili[1]([✉]) [ID], Irine Khomeriki[1] [ID], Maia Topeshashvili[3] [ID],
Tamar Lominadze[1] [ID], Revaz Shanidze[2] [ID], and Mariam Osepashvili[2] [ID]

[1] Faculty of Informatics and Control Systems, Georgian Technical University, Tbilisi, Georgia
l.petriashvili@gtu.ge
[2] Kutaisi International University, Youth Avenue, 5th Lane, K Building, 4600 Kutaisi, Georgia
[3] Radiation Oncology Department, Todua Clinic, 13 Thevdore Mghvdeli Street, 0112 Tbilisi, Georgia

Abstract. Breast Cancer is no longer considered a death sentence and the number of patients receiving radiotherapy is increasing every year. Advancements in RT technology allow more accurate and accumulated delivery of radiation significantly reducing the risk of side effects. The use of these techniques has led to better clinical outcomes and significantly increased long-term survival rates. Therefore, secondary cancer risk after breast conserving therapy is becoming more important. In this study, we estimate the risks of developing a solid second cancer after radiotherapy of breast cancer using the concept of organ equivalent dose (OED).

Eight breast cancer patients were retrospectively selected for this study. Three-dimensional conformal radiotherapy (3DCRT), intensity modulated radiotherapy (IMRT), and volumetric modulated arc therapy (VMAT) were planned to deliver a prescribed dose. Differential dose volume histograms (dDVHs) were created and the OEDs calculated. Secondary cancer risks of ipsilateral, contralateral lung and thyroid gland were estimated using linear, linear-exponential and plateau models.

The highest interest of our study was the evaluation of secondary cancer risk for the organs at risk (OAR), which are located far from the treatment region and are very sensitive to radiation exposure. Our results showed very high secondary cancer excess absolute risk (EAR) values for IMRT and VMAT compared with 3DCRT. It has to be noted, that a significant reduction of the EARs for the contralateral lung, ipsilateral lung and thyroid gland was observed in all dose–response models.

Keywords: Intensity modulated radiotherapy · secondary cancer risk · organ equivalent dose (OED)

1 Introduction

Currently, radiation therapy presents as an integral component of early-stage, localized breast cancer treatment after breast conserving surgery as it helps to reduce recurrence rates [15]. The early breast cancer trialists' collaborative group (EBCTCG) meta-analysis

© The Author(s), under exclusive license to Springer Nature Switzerland AG 2023
I. Rojas et al. (Eds.): IWBBIO 2023, LNBI 13919, pp. 231–243, 2023.
https://doi.org/10.1007/978-3-031-34953-9_18

has shown an overall survival benefit in favor of adjuvant radiotherapy after breast cancer surgery [1, 6]. Although the risk for radiotherapy treated patients regarding the induction of secondary cancer is small, it remains a relevant consideration particularly among younger patients [2, 28].

Secondary cancer risk (SCR) after radiotherapy has been analyzed by international organizations, including the United Nations Scientific Committee on the Effects of Atomic Radiation (UNSCEAR), the International Commission on Radiological Protection (ICRP), the National Council on Radiation Protection and Measurement (NCRP), and the American Association of Physicists in Medicine (AAPM) and epidemiological studies as well [1, 25, 27]. In the ongoing search to improve the treatment quality and disease control rates and reduce the treatment related morbidity, IMRT/VMAT is becoming an increasingly popular method of breast cancer radiotherapy. As documented in the literature, several institutions and radiation therapy departments utilize varying treatment techniques and fractionation schedules for breast radiation treatments [7, 13] and consensus on the optimal technique to be employed is still lacking both in literature and clinically, hence the full extent of their effectiveness on breast cancer treatment remains only partially understood [5].

The clinical consequences regarding SCR are still under investigation and need deep analysis, although from the technical properties of IMRT/VMAT technologies, it can be supposed, that SCR may be increased relative to 3DCRT as:

(1) They involve more beams and therefore a larger volume of normal tissue is exposed to a low dose;
(2) They require a longer beam-on time because typically higher monitor units (MU) are used, and subsequently the integral dose may increase because of head leakage and collimator scatter [1, 9, 11].

It is obvious, that radiation exposure is a strong risk factor for cancer, particularly when exposure occurs in childhood and adolescence. The relative risk of radiation-induced cancer in men and women exposed at different ages varies among different organs [12, 16, 26]. The organs most susceptible to radiation-induced cancer in those exposed to external radiation at a young age are the female breast [8], followed by the female lung and the female thyroid [3, 10, 14, 19]. The attributable risk of secondary thyroid cancer in females is 5.5 times higher than of males. The vulnerability of girls relative to boys remains high through puberty and after age 50, when most women reach menopause, the relative risk is decreased to the point where there is no longer a meaningful difference between men and women, indicating some modulation by hormonal status. This pattern appears to be unique to breast and thyroid and likely explains the female-specific shift in the likelihood (rank order) of various cancers as a function of age/hormonal status.

Female radiation-attributable lung cancer rates are also higher than those of males, although the ratio of female to male remains constant and independent suggesting that, unlike breast and thyroid, the risk of radiation-induced lung cancer is modulated by age and sex, but not by hormonal status.

In this study, we have considered the above-mentioned issues and aim to: a) perform treatment planning with different available modalities: three-dimensional conformal radiotherapy (3DCRT), intensity modulated radiotherapy (IMRT), and volumetric modulated arc therapy (VMAT) with the same total prescribed dose; b) using achieved dosimetry parameters of each plan for defining OED for thyroid and ipsilateral and con-tralateral lung regarding secondary risk assessment; c) estimation of secondary cancer risk with Biological Effects of Ionizing Radiation (BEIR) VII cancer incidence models, which means defining excess absolute risks (EAR) and excess relative risks (ERR) based on OED data; d) making conclusion as a consensus on the priority among disease control rate, secondary cancer risk and toxicity.

It is important to mention, that OED concept encounters several problems and uncer-tainties and the shape of the dose–response curve for radiation induced cancer is currently under much debate [20, 24]. It is not known whether cancer risk as a function of dose continues to be linear or decreases at high doses due to cell killing or levels off due to, for example, a balance between cell killing and repopulation effects [23]. However, since very little is currently known about the shape of dose–response relationships for radiation-induced cancer in the radiotherapy dose range, this approach could be regarded as a first attempt to acquire more information on this area.

As it is reported in the literature [17, 22, 24, 30] at low doses the OED is the average organ dose because for low doses the risks are proportional to dose. At doses higher than 2 Gy radiation induced cancer incidence rates are not necessarily a linear function and the following three possibilities are reported for the shape of dose response relationship as the dose increases [1]:

1. The curve continues to be linear
2. Decreases linear-exponentially
3. May reach a plateau [21, 22, 24].

2 Materials and Methods

2.1 Patient Selection

This retrospective study utilized 8 planning computed tomography (CT) datasets of early-stage female breast cancer patients. Planning CT datasets were acquired using Siemens CT (Somatom Definition AS, Siemens Medical Solutions, USA) with patients lying supine on a breastboard and arms raised above the head. Each slice of acquired CT datasets was 2 mm in thickness. Image registration and delineation of gross tumor volume (GTV), planning target volume (PTV) and OARs were performed using Eclipse Treatment Planning System (TPS) (version 13.7; Varian Medical Systems, Inc, USA). Contoured OARs included ipsilateral lung, contralateral breast and lung, heart, spinal cord, oesophagus and thyroid gland.

2.2 Treatment Planning

For each dataset, 3 different plans were generated and compared against each other. These included 3DCRT, IMRT and VMAT. A prescription dose of 46.2 Gy in 21 fractions to the whole breast and regional lymph nodes with a simultaneously integrated boost 55.86 Gy was applied for all planning techniques. Each treatment plan was created to meet the planning goals defined in Table 1.

Table 1. Planning goals

Structure	Planning Goal
PTV	Max < 110% D95% > 95%
Ipsilateral Lung	V20 Gy < 20% V5 Gy < 50%
Heart	Dmean < 2 Gy
Contralateral Breast	V5 Gy < 15%
Spinal Cord	Dmax < 39 Gy
Thyroid	Dmean < 30 Gy
Oesophagus	V100% of prescription < 0,03 cc

3DCRT Planning. 3DCRT plans involved a pair of half-beam blocked tangential fields with anterior-posterior and posterior-anterior slightly oblique fields. The choices of gantry angles for 3DCRT plans were selected to provide the best PTV dose coverage while minimizing as much exposure as possible to adjacent OARs and to minimize low-dose areas contributing to the contralateral side of the body. Dynamic wedge angles and weightings were selected to give the best PTV dose coverage and homogeneity. For all 3DCRT plans, 6 MV was used for tangential fields and 15 MV for posterior-anterior slightly oblique fields if the separation was > 23 cm. The field in Field technique was incorporated and shaped using multi-leaf collimators (MLCs) to remove any hot spots and improve dose homogeneity.

IMRT Planning. For IMRT planning, 7 coplanar intensity modulated fields were used, 3 for supra-infra clavicular lymph nodes and 4 for the breast region. Half-beam blocked and fixed jaw options were used to have gantry angles individually selected for each treatment area. The couch and collimator were kept at zero degrees. The gantry angle ranged between 155–300°. The IMRT plans were optimized to cover at least 97% of the planned target volume (PTV) with 95% of the prescribed dose while reducing the exposure to OARs to the greatest extent possible. To expand the fluence outside the body contour, an auto flash margin of 1.5 cm to PTV was applied. The optimization process was used with as many iterations as were required to meet the planning objectives. Post-optimization improvements such as the removal of hot spots and smoothing were performed through manual fluence editing.

VMAT Planning. VMAT plan utilized 195° dual arcs. Objective functions were specified accordingly to achieve plan objectives and dose constraints as illustrated in Table 1.

It is important to mention, that while the study is based on dose calculations, made on commercially available treatment planning systems, some inaccuracies in the estimates corresponding to deficiencies within each calculation algorithm can be considered. Monte Carlo methods are recognized as the most accurate dose calculation algorithms for both in-field and out-of-field dose calculations [18, 29]. However, because of the lack of Monte Carlo calculations, Anisotropic Analytical Algorithm (AAA) was used in this technique which is a well-reported accurate algorithm [2, 28] and there are some reports demonstrating that for all the scenarios, AAA and Monte-Carlo results are within ± 1% in the preinhomogeneity region and up to 10% higher accuracy of Monte-Carlo algorithm in high contrast inhomogeneous region [29].

Organ Equivalent Dose (OED) and Excess Absolute Risk (EAR) Calculations. For the calculation of secondary malignancy risk, Schneider's concept of OED was used [22]. According to the OED concept, two different RT plans that result in the same risk of secondary malignancy have the same OED. Therefore, the OED of one plan relative to another gives the relative risk of secondary malignancy for those two techniques. The OEDs for thyroid gland, contralateral lung and ipsilateral lung were calculated on the basis of the dDVHs for the linear, linear–exponential and plateau dose–response models [20–24, 30] according to:

$$OED_{T,linear} = \frac{1}{V_T} \sum_i \{DVH(D_i) \cdot D_i\} \tag{1}$$

$$OED_{T,linear-exp} = \frac{1}{V_T} \sum_i \left\{ DVH(D_i) \cdot D_i \cdot e^{-\alpha D_i} \right\} \tag{2}$$

$$OED_{T.plateau} = \frac{1}{V_T} \sum_i \left\{ DVH(D_i) \cdot \left(1 - e^{-\delta D_i}\right)/\delta \right\} \tag{3}$$

where DVH(D_i) is the volume receiving dose D_i and the summation runs over all voxels of organ T with volume V_T. The model parameters α and δ were estimated from the fits to the Hodgkin data for the different dose-response models for carcinoma induction. In the results of these fits esophagus and thyroid were excluded from the analysis, since these organs were covered by a limited dose range of 30–55 Gy and 44–46 Gy. Therefore, for the linear-exponential model the organ-specific parameter $\alpha = 0.065$ Gy^{-1} was taken for thyroid as was assigned to all solid tumors, and $\alpha = 0.022$ Gy^{-1} was taken for the lungs. Similarly, $\delta = 0.317$ Gy^{-1} was taken for thyroid in the plateau model as assigned to all solid tumors and $\delta = 0.056$ Gy^{-1} was taken for lungs. Table 2 shows the parameters used for OED calculations.

Table 2. Parameters for second malignancy risk calculation. In addition the coefficient of variation (CV) is given.

Organ	α (Gy^{-1})	δ (Gy^{-1})
Lung	0.022 (CV 1.2×10^{-2})	0.056 (CV 1.7×10^{-3})
Thyroid	0.065 (CV 4.8×10^{-3})	0.317 (CV 8.7×10^{-1})

The risk of developing a solid secondary cancer after RT is usually represented by EAR.

The EAR describes the absolute difference in cancer rates of persons exposed to a dose d and those not exposed to a dose beyond the natural dose exposition per 10,000 person-years (PYs). The EAR can be calculated by using the concept of OED according to:

$$EAR = EAR_0 \times OED \tag{4}$$

where EAR_0 is the initial slope, which is the slope of the dose–response curve at a low dose. All population-related parameters, such as attained age (a), age at exposure (e), and sex (s), are included in EAR_0 (e, s, a).

In the calculation of the EAR, the difference between the baseline risks of developing cancer without exposition to radiation has been accounted for. As was already mentioned, radiation-attributable lung and thyroid cancer rates for females are much higher than those of males [4]. In our calculation, the fit parameters for females are taken and centered at an age at exposure of 30 years and an attained age of 70 years for a Japanese population (Table 3). The initial slopes and the sex and age modifying parameters for different sites are taken from [17].

Table 3. Cancer Radiation-Risk-Model Parameter Estimates (Excess cases per 10,000 per PY Gy, 90% confidence interval).

Organ	Male	Female
Lung	6.0 (2.3; 11)	9.1 (6.4; 12)
Thyroid	0.5 (0.3; 1.5)	1.9 (1.3; 4.2)

3 Results and Discussion

Figure 1 shows absolute EARs in organs of interests, thyroid gland, ipsilateral and contralateral lungs for 3DCRT, IMRT and VMAT plans using the linear, plateau and linear-exponential dose–response model.

Figure 2 shows relative EARs in organs of interests, thyroid gland, ipsilateral and contralateral lungs for 3DCRT, IMRT and VMAT plans using the linear, plateau and linear-exponential dose–response model. Relative EAR was depicted from 3DCRT plans. Figure 1 a) and b) represent the relative EAR for the organs of interest, close to the primary beam, while Fig. 1 c) represents the relative EAR for the organ of interest, located in the opposing side of the primary beam.

Figure 3 shows isodose distribution for different dose levels on planning CT of one of the representative patient in 3DCRT, IMRT and VMAT technique.

The mean OED values for 8 patients using the three treatment modalities and three different calculation models are presented in Table 4.

Obtained results indicated, when organ doses are in the range of approximately 2 Gy and dose distribution in this organ is more or less homogenous there is no difference in OED values for three different models: linear, linear-exponential and plateau [24].

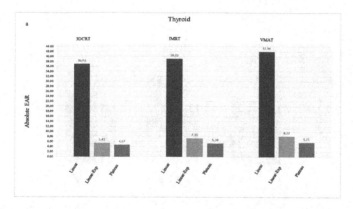

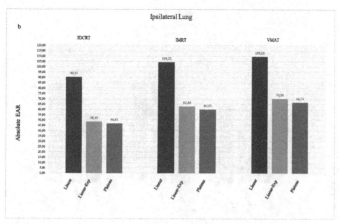

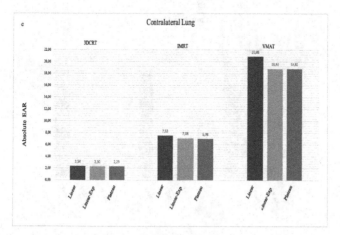

Fig. 1. Absolute Excess Absolute Risk (EAR) of developing a solid secondary cancer after RT for (a) thyroid, (b) ipsilateral lung, and (c) contralateral lung as a function of linear, plateau, and linear-exponential dose–response model using three-dimensional conformal radiotherapy (3DCRT), intensity-modulated radiotherapy (IMRT), Volumetric-Modulated Radiotherapy (VMAT) techniques.

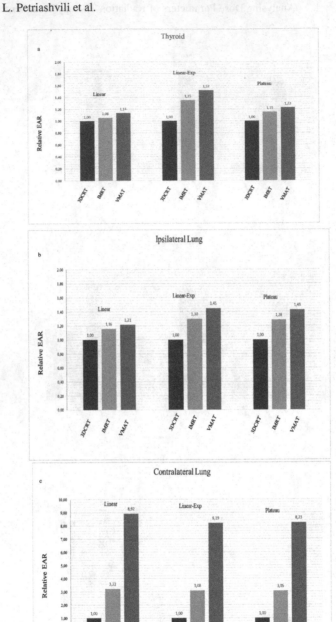

Fig. 2. Relative Excess Absolute Risk (EAR) of developing a solid secondary cancer after RT for (a) thyroid, (b) ipsilateral lung, and (c) contralateral lung as a function of three-dimensional conformal radiotherapy (3DCRT), intensity-modulated radiotherapy (IMRT), Volumetric-Modulated Radiotherapy (VMAT) techniques using linear, plateau, and linear-exponential

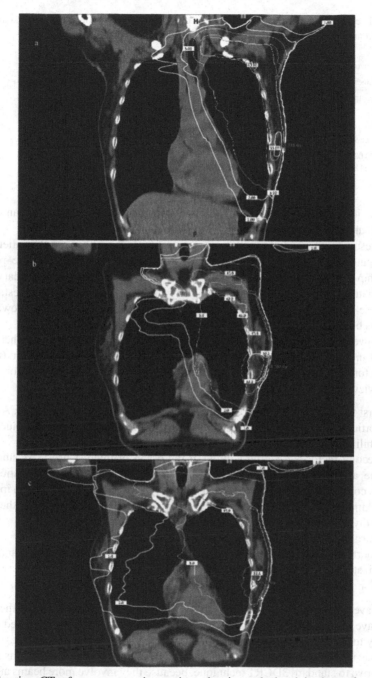

Fig. 3. Planning CT of a representative patient showing calculated isodoses for a) three-dimensional conformal radiotherapy (3DCRT), b) intensity-modulated radiotherapy (IMRT), c) volumetric-Modulated Radiotherapy (VMAT) techniques.

Table 4. The mean values of organ equivalent dose for the organs of interest in linear, linear-exponential and plateau models, including standard deviation.

	Thyroid			Ipsilateral Lung			Contralateral Lung		
	Linear	Linear-Exp	Plateau	Linear	Linear-Exp	Plateau	Linear	Linear-Exp	Plateau
3DCRT	19,42 (±6,87)	2,84 (±0,40)	2,45 (±0,42)	9,92 (±1,54)	5,33 (±0,59)	5,14 (±0,57)	0,26 (±0,06)	0,25 (±0,06)	0,25 (±0,06)
IMRT	20,53 (±6,99)	3,84 (±0,51)	2,82 (±0,41)	11,46 (±1,64)	6,90 (±0,87)	6,59 (±0,82)	0,83 (±0,44)	0,78 (±0,40)	0,77 (±0,39)
VMAT	22,08 (±4,88)	4,32 (±0,19)	3,00 (±0,2)	12,03 (±0,87)	7,71 (±0,5)	7,33 (±,48)	2,29 (±0,77)	2,07 (±0,60)	2,07 (±0,64)

Therefore, it can be presumed, that for low doses, the cancer incidence of an organ is proportional to the average dose applied to that organ.

Our results confirmed, that when the dose is above 2 Gy (e.g., in RT patients) and there is quite an inhomogeneous dose distribution in the organ [2], the dose–response relationship for carcinogenesis is no longer linear and its applicable in secondary cancer risk estimation is no more recommended. OED values of the organs of interest, located in the close vicinity to the treatment region (thyroid and ipsilateral lung) showed large deviations between linear and non-linear models.

Our investigation indicated, the standard deviation of OED data is higher for the organs of interest, that are located in close proximity to the treatment region (thyroid) and therefore are characterized by sharply inhomogeneous dose distribution. To our point of view, there should be several factors that have an influence on this.

a. The first factor is the fraction of the organ volume in the treatment region. As higher the portion of the volume in the projection of the treatment field, as much is the probability, that OED parameters will have a large standard deviation;
b. The second factor can be the volume of the organ of interest itself. As smaller the volume, as higher the deviation in dose distribution parameters between patients. Our results confirmed the highest discrepancy for thyroid gland compared with ipsilateral lung. Although both organs are in the close vicinity to the treatment region, the volume of the thyroid gland is 200-times smaller than the ipsilateral lung;
c. The third factor is anatomical differences between the patients;
d. The fourth factor can be human related. Although all parameters for planning and optimization are well reported and identical, the results achieved are never the same even for the same planner and there can always be some kind of variations.

We have discussed the results separately, according to the categories of the organs. As we have already noted, we have 2 different categories: a) organs, located in close proximity to the treatment region and b) organs, located at a distance from it.

As was mentioned in the introductory section, intensity modulated beams increase SCR relative to standard 3DCRT technique, because they involve more beams and therefore a larger volume of normal tissue is exposed. Also, they need much more treatment time, so exposure time is higher for modulated beams than for standard 3DCRT. All these results in the larger low dose region volume in the body and in surrounding sensitive organs compared with 3DCRT.

According to our results (Fig. 1 and Fig. 2) the difference in the organs of interest, which are located far from the treatment region is very high, about 3-times higher for IMRT compared with 3DCRT and about 9-times higher compared with VMAT.

The difference between EAR data for standard 3DCRT and modulated techniques is not so high for the organs located close to the irradiated region, the obtained results are quite noticeable. According to non-linear models, recommended for such an inhomogeneous dose distribution, there was about 30% higher EAR for IMRT compared with 3DCRT and about 50% higher compared with VMAT.

4 Conclusion

It is obvious, that medicine is always an individual approach and the final decision about the treatment is done only after a comprehensive judgement of patient's clinical anamnesis. Our results indicated, that there is a significant reduction of the radiation induced secondary cancer risk for the contralateral lung, ipsilateral lung and thyroid gland in 3DCRT technique relative to those for IMRT and VMAT in all dose–response models. It has to be mentioned, that the choice of the optimal treatment method must be chosen for every patient individually optimizing between the benefits of the treatment and all relevant risks, e.g., cardiac complications, deterministic acute normal tissue damage and risks for stochastic long-term effects.

There still does not exist any evidence based clinical analyses reporting on secondary cancer induction after intensity-modulated RT techniques. This is because of the limited time interval. Additional theoretical and clinical studies are needed to determine the correlations between treatment techniques and secondary cancer risk. Once sufficient data are collected, treatment plans in the future may be evaluated on the basis of not only conventional factors from a DVH but also on the secondary cancer risk.

Aknowledgement. This work was supported by Georgian Technical University.

References

1. Abo-Madyan, Y., et al.: Second cancer risk after 3D-CRT, IMRT and VMAT for breast cancer. Radiother. Oncol. **110**(3), 471–476 (2014). https://doi.org/10.1016/j.radonc.2013.12.002
2. Bartkowiak, D., et al.: Second cancer after radiotherapy, 1981–2007. Radiother. Oncol. **105**(1), 122–126 (2011). https://doi.org/10.1016/j.radonc.2011.09.013
3. Bhatti, P., et al.: Risk of second primary thyroid cancer after radiotherapy for a childhood cancer in a large cohort study: an update from the childhood cancer survivor study. Radiat. Res. **174**(6), 741–752. (2010). https://doi.org/10.1667/RR2240.1
4. Biogon, A., Cohen, S., Franceschi, D.: Modulation of secondary cancer risks from radiation exposure by sex, age and gonadal hormone status: progress, opportunities and challenges. J. Pers. Med. **12**(5), 725 (2022). https://doi.org/10.3390/jpm12050725
5. Chen, S.N., Ramachandran, P., Deb, P.: Dosimetric comparative study of 3DCRT, IMRT, VMAT, Ecomp, and hybrid techniques for breast radiation therapy. Radiat. Oncol. J. **38**(4), 270–281 (2022). https://doi.org/10.3857/roj.2020.00619

6. Clarke, M., et al.: Effects of radiotherapy and of differences in the extent of surgery for early breast cancer on local recurrence and 15-year survival: an overview of the randomised trials. Lancet **366**(9503), 2087–2106 (2005). https://doi.org/10.1016/S0140-6736(05)67887-7

7. Dayes, I.S., et al.: Cross-border referral for early breast cancer: an analysis of radiation fractionation patterns. Curr. Oncol. **13**(4), 124–129 (2006)

8. Elezaby, M., et al.: BRCA mutation carriers: breast and ovarian cancer screening guidelines and imaging considerations. Radiology **291**(3), 554–569 (2019). https://doi.org/10.1148/rad iol.2019181814

9. Fogliata, A., Nicolini, G., Alber, M., et al.: IMRT for breast: a planning study. Radiother. Oncol. **76**(3), 300–310 (2005). https://doi.org/10.1016/j.radonc.2005.08.004

10. Grantzau, T., Thomsen, M.S., Væth, M., Overgaard, J.: Risk of second primary lung cancer in women after radiotherapy for breast cancer. Radiother. Oncol. **111**(3), 366–373 (2014). https://doi.org/10.1016/j.radonc.2014.05.004

11. Hall, E.J.: Intensity-modulated radiation therapy, protons, and the risk of second cancers. Int. J. Radiat. Oncol. Biol. Phys. **65**(1), 1–7 (2006). https://doi.org/10.1016/j.ijrobp.2006.01.027

12. Inskip, P.D., et al.: Radiation-related new primary solid cancers in the childhood cancer survivor study: comparative radiation dose response and modification of treatment effects. Int. J. Radiat. Oncol. Biol. Phys. **94**(4), 800–807 (2016). https://doi.org/10.1016/j.ijrobp.2015. 11.046

13. Kara, F.G., Haydaroglu, A., Eren, H., Kitapcioglu, G.: Comparison of different techniques in breast cancer radiotherapy planning. J. Breast Health **10**(2), 83–87 (2014). https://doi.org/10. 5152/tjbh.2014.1772

14. Long, Q., Wang, Y., Che, G.: Primary lung cancer after treatment for breast cancer. Int. J. Women's Health **13**, 1217–1225 (2021). https://doi.org/10.2147/IJWH.S338910

15. Moran, M.S., Schnitt, S.J., Giuliano, A.E., et al.: Society of surgical oncology-American society for radiation oncology consensus guideline on margins for breast-conserving surgery with whole-breast irradiation in stages I and II invasive breast cancer. Int. J. Radiat, Oncol, Biol. Phys. **88**(3), 553–564 (2014). https://doi.org/10.1016/j.ijrobp.2013.11.012

16. National Academy of Sciences: Committee on the Biological Effects Radiation: BEIR VII Phase 2. National Academies Press, Washington, DC (2005)

17. Preston, D.L., et al.: Solid cancer incidence in atomic bomb survivors: 1958–1998. Radiat. Res. **168**(1), 1–64 (2007). https://doi.org/10.1667/RR0763.1

18. Rebecca, M.H., Sarah, B.S., Kry, S.F., Derek, Z.Y.: Accuracy of out-of-field dose calculations by a commercial treatment planning system. Phys. Med. Biol. **55**(23), 6999–7008 (2010). https://doi.org/10.1088/0031-9155/55/23/S03

19. Ronckers, C.M., et al.: Thyroid cancer in childhood cancer survivors: a detailed evaluation of radiation dose response and its modifiers. Radiat. Res. **166**(4), 618–28 (2006). https://doi. org/10.1667/RR3605.1

20. Schneider, U., Kaser-Hotz, B.: Radiation risk estimates after radiotherapy: application of the organ equivalent dose concept to plateau dose–response relationships. Radiat. Environ. Biophys. **44**(3), 235–239 (2005). https://doi.org/10.1007/s00411-005-0016-1

21. Schneider, U., Lomax, A., Timmermann, B.: Second cancers in children treated with modern radiotherapy techniques. Radiother. Oncol. **89**(2), 135–140 (2008). https://doi.org/10.1016/j. radonc.2008.07.017

22. Schneider, U., Sumila, M., Robotka, J.: Site-specific dose-response relationships for cancer induction from the combined Japanese A-bomb and Hodgkin cohorts for doses relevant to radiotherapy. Theor. Biol. Med. Model. **8**, 27 (2011). https://doi.org/10.1186/1742-4682-8-27

23. Schneider, U., Walsh, L.: Cancer risk estimates from the combined Japanese A-bomb and Hodgkin cohorts for doses relevant to radiotherapy. Radiat. Environ. Biophys. **47**(2), 253–263 (2008). https://doi.org/10.1007/s00411-007-0151-y

24. Schneider, U., Zwahlen, D., Ross, D., Kaser-Hotz, B.: Estimation of radiation induced cancer from three-dimensional dose distributions: concept of organ equivalent dose. Int. J. Radiat. Oncol. Biol. Phys. **61**(5), 1510–1515 (2005). https://doi.org/10.1016/j.ijrobp.2004.12.040

25. ICRP: The 2007 Recommendations of the International Commission on Radiological Protection. ICRP Publication 103. Ann. ICRP 37(2–4) (2007)

26. Thompson, D.E., et al.: Cancer incidence in atomic bomb survivors. Part II: Solid Tumors, 1958–1987. Radiat. Res. **137**(2), S17–S67 (1994). https://doi.org/10.2307/3578892

27. Effects of ionizing radiation. UNSCEAR 2006 Report to the General Assembly with Scientific Annex United Nations, New York (2006)

28. Van Leeuwen, F.E., Klokman, W.J., Veer, M.B.V.T., et al.: Long-term risk of second malignancy in survivors of Hodgkin's disease treated during adolescence or young adulthood. J. Clin. Oncol. **18**(3), 487–497 (2000). https://doi.org/10.1200/JCO.2000.18.3.487

29. Verhaegen, F., Seuntjens, J.: Monte Carlo modelling of external radiotherapy photon beams. Phys. Med. Biol. **48**(21), R107 (2003). https://doi.org/10.1088/0031-9155/48/21/r01

30. Zwahlen, D.R., Ruben, J.D., Jones, P., Gagliardi, F., Millar, J.L., Schneider, U.: Effect of intensity modulated pelvic radiotherapy on second cancer risk in the postoperative treatment of endometrial and cervical cancer. Int. J. Radiat. Oncol. Biol. Phys. **74**, 539–545 (2009). https://doi.org/10.1016/j.ijrobp.2009.01.051

A Platform for the Radiomic Analysis of Brain FDG PET Images: Detecting Alzheimer's Disease

Ramin Rasi(✉) ⓘ and Albert Guvenis ⓘ

Institute of Biomedical Engineering, Boğaziçi University, Istanbul, Türkiye
Ramin.Rasi@boun.edu.tr

Abstract. The objective of this work is to present a radiomics-based platform (RAB-PET) and method to detect Alzheimer's disease (AD) non-invasively using 18FDG-PET images. Radiomic analysis allows the identification of regional features that serve to predict the presence or characteristics of diseases using images as data. We first, used the FastSurfer a deep learning-based toolbox to segment the whole brain into 95 classes by the utilization of the DKT-atlas. Then the PyRadiomics toolbox was used to extract features from 18FDG-PET scans. After preprocessing, the features were subject to a selection process by making use of eight different methods, namely, ANOVA, PCA, Chi-square, LASSO, Recursive Feature Elimination (RFE), Feature Importance (FI), Mutual Information (MI), and Recursive Feature Addition (RFA). Finally, in order to classify the selected features by feature selection methods, we implemented nine different classifier methods, namely, GradientBoosting (GB), RandomForest (RF), DecisionTree (DT), GaussianNB (GNB), GaussianProcess (GP), MLP, QuadraticDiscriminantAnalysis (QDA), AdaBoost (AB), and KNeighbors (KNN) on selected feature subsets. All data (scans and clinical examination results) were obtained from the AD Neuroimaging Initiative (ADNI) database. The RF classifier with 100 iterations on features obtained with the LASSO algorithm yielded an area under the curve of AUC = 0.976 with a 95% confidence interval of 0.93–0.98 based on 30% independent test data. We conclude that a platform for radiomic analysis can serve as a potential method for deducing accurate information on brain diseases such as Alzheimer's disease non-invasively using 18FDG-PET images. Further studies are underway to extend this work by studying the association between the set of features and several characteristics of the Alzheimer's disease.

Keywords: Alzheimer's Disease · Radiomics · 18FDG-PET

1 Introduction

Alzheimer's disease (AD), the most common progressive neurological disease among the elderly, is a neuropsychiatric disorder that causes many economic and psychological difficulties for the patient's community and family [1]. Preliminary symptoms of AD starts 20 to 30 years before dementia [2, 3] so it is possible to detect this pathology

© The Author(s), under exclusive license to Springer Nature Switzerland AG 2023
I. Rojas et al. (Eds.): IWBBIO 2023, LNBI 13919, pp. 244–255, 2023.
https://doi.org/10.1007/978-3-031-34953-9_19

in vivo, using biomarkers such as molecular imaging techniques in the early stage of formation [4]. On the other, hand early detection is one of the leading problems in AD treatment and due to the complication of AD as an insidious, multifactorial disease, medical experts are using various clinical tests and neuroimaging techniques to detect dementia in the early stage [5, 6]. Imaging techniques can have a significant impact on the early diagnosis and definition of new treatments based on precision medicine [7]. In the present work, we hypothesize that we can predict different stages of AD non-invasively from 18FDG-PET [8] images using a radiomic analysis platform. There have been studies to predict AD from various biomarkers which can give early information on the pathology of the disease before even symptoms occur [9]. In recent years, radiomics through machine learning (ML) and deep learning (DL) have been advanced to analyze medical images as data. In the ML, approach a large number of phenotypic hand-crafted features are extracted to investigate their relationship with patients' prognosis and other characteristics [10]. Radiomic methods try to extract visible and invisible information from medical imaging using various computations on different regions. It has been introduced as an effective solution to deduce a number of disease characteristics using a high number of low-level imaging features as an extension of the computer-assisted medical decision-making (CMD) systems [11, 12].

The overall process of the radiomics approach consists of a series of sequential steps including image acquisition, image processing, ROIs/VOIs delineations, feature extraction, feature selection, and building a classification or regression model. These methods can use manual, semi-automatic or automatic delineation algorithms [40]. Due to the dependence of manual and semi-automatic segmentation on the individual user, features related to the ROI/ VOI region may not sufficiently resistant to intra and inter-observer changes. As a result, the use of artificial intelligence and deep learning for automatic segmentation to reduce the effects of subjective bias is one of the essential areas of radiomics research [42]. In order to automate the process, we use the FastSurfer [44] for volumetric analysis, which replicate the FreeSurfer's anatomical segmentation [43] to segment the whole brain into 95 classes using the DKT-atlas.

The objective of this study is to develop a platform for the radiomic analysis of brain FDG-PET images which can be used to derive quantitative information non-invasively for the diagnosis and personalized treatment of patients. We believe that this is the first radiomics platform that can be used to derive multiple results using a specific hardware. The system is first tested on AD patient's data obtained from an online database. The following section describe the methods used, the results obtained, their interpretations and the conclusions reached.

2 Methods

2.1 Participants and Data

In this study we have used the online database. Alzheimer's disease Neuroimaging Initiative (ADNI) is a multisite longitudinal study, with the aim to validate biomarkers for use in AD clinical treatment trials track the progression of AD in the human brain by clinical, imaging, genetic and biospecimen biomarkers through the process of normal aging, and mild cognitive impairment to dementia or AD (www.adni.loni.usc.edu). We extracted the 18FDG-PET scans of 350 individuals in the ADNI database as shown in Table 1.

Table 1. Clinical characteristic of participants

Clinical Diagnosis	No. Cases	Sex (M/F)	Age (mean ± SD)
AD	163	91/72	74.6 ± 8.12
CN	187	91/96	73.6 ± 6.37
Total	350	182/168	74.1 ± 7.02

2.2 PET Acquisition

The 18-Fluoro-DeoxyGlucose PET (185 MBq (5 mCi), dynamic 3D scan of six 5-min frames 30–60 min post-injection) imaging data from the baseline visit of the ADNI database were available as raw and preprocessed data and categorized into four different groups according to preprocessing steps (adni.loni.usc.edu). We used the third type of preprocessed 18FDG-PET image data (Co-Reg, AVG, Standardized Image, and Voxel Size). In this set, 18FDG-PET images were reoriented into a standard $160 \times 160 \times 96$ voxel image grid, having 1.5 mm cubic voxels [13].

2.3 ROI Extraction

Replacing time-consuming methods with new approaches has resulted in the development of more powerful toolboxes to process big datasets in a short time. We used the FastSurfer [14] for volumetric analysis, which replicate the FreeSurfer's anatomical segmentation [15] method to segment the whole brain into 95 classes using the DKT-atlas.

2.4 Feature Extraction

To extract the features from 18FDG-PET we used an open-source python package: PyRadiomics [16]. We calculated 120 feature classes: First Order Statistics (19), Gray Level Dependence Matrix (14), Shape-based (2D) (10), Gray Level Co-occurrence Matrix (24), Gray Level Run Length Matrix (16), Neighboring Gray Tone Difference Matrix (5), Gray Level Size Zone Matrix (16 features) and Shape-based (3D) (16). All features were computed from the extracted 95 ROIs in 18FDG-PET images.

2.5 Feature Selection

Radiomics studies in general include a feature selection step to reduce the enormous number of extracted features from available information and datasets [17, 18]. In order to find the best subset of features we compared the performance of various feature selection methods, namely, filtered methods, embedded methods, wrapper, and hybrid methods [19]. We chose the most common and widely used algorithms from each category [20, 21, and 22]. We used the ANOVA, PCA, Chi-square, LASSO, MI, RFA, FI, and RFE methods [21].

2.6 Classification and Tuning

Various classification methods have been developed by researchers [22]. Finding an optimal machine-learning method for radiomics applications is a crucial step and classifiers with high reliability and accuracy play an effective role in improving the quality of clinical applications based on radiomics. We examined the performance of nine classification methods arising from various classifier families (GradientBoosting (GB), RandomForest (RF), DecisionTree (DT), GaussianNB (GNB), GaussianProcess (GP), MLP, QuadraticDiscriminantAnalysis (QDA), AdaBoost (AB), and KNeighbors (KNN)) on eight feature selection methods. According to the obtained results shown in Table 2 from 72 combination of classification methods and feature selection methods we selected the classifiers with the best area under curve ROC to use in the proposed solution. In this step we evaluated the performance of all nine classifiers by the default parameters defined in the Scikit-learn library. In addition, selecting the appropriate hyper-parameters is one of the main steps in machine learning that should be considered. To optimize the hyper parameters various methods can be used, namely, Grid Search, Random Search, Bayesian Optimization, Particle Swarm Optimization (PSO), and Genetic Algorithm (GA). We performed the Randomized Search cross-validation (CV = 5) tuning using 70% of the data, excluding 30% of the independent test data. We implemented 100 replications in completely random conditions [15, 23].

2.7 Computational Hardware and Software

The hardware that was used for the RAB-PET platform was the Corei7 Gen10th processor, RTX2060 VGA card (1920 CUDA cores with 240 tensors), and 16 GB DDR5 RAM. We used Python as a programming language and PyTorch to utilize the CUDA technology. We used FastSurfer [44] components to segment the brain using a deep learning approach and PyRadiomics [16] components to extract features under IBSI [47, 48]. Phyton libraries were used to come up with a reduced set of features that can be used as input to our prediction model. The code for the prediction model can be found in [49].

3 Results

In this study, all 120 radiomics features of the PyRadiomic tool were calculated for all 95 regions obtained from 18FDG-PET scans. Given the number of features and regions of interest, we extracted 120*95 features for each scan. Over-fitting is an undesirable problem in supervised machine learning which can occur for various reasons such as the model complexity. Although various strategies have been proposed to reduce the effects of over-fitting, we used fine-tuning and dimension-reduction methods [13, 24]. Choosing the appropriate algorithm to reduce the features of the problem is an important step in machine learning [17]. Before performing the feature reduction methods on the extracted features, we preprocessed the features by removing the constant, quasi-constant, and duplicated features. After preprocessing the extracted features, we performed the eight different methods on the preprocessed features (n = 5351) to reduce the number of features and find the best subset of features for the predictions of AD. Finally, we calculated the accuracy for each subset of the most important features using the rank-based method. During comparison, we have used the default parameters defined in the Scikit-learn library for all classifiers without tuning the hyper-parameters.

Table 2. Results of performing nine classifiers GB, RF, DT, GNB, GP, MLP, QDA, AB, and KNN on the top 20 features selected by eight dimension-reduction methods ANOVA, PCA, Chi-Square, LASSO, MI, RFA, FI, and RFE.

	GB	RF	DT	GNB	GP	MLP	QDA	AB	KNN
ANOVA	0.957	0.962	0.850	0.961	0.910	0.935	0.958	0.951	0.912
PCA	0.920	0.929	0.784	0.919	0.677	0.853	0.918	0.904	0.790
Chi-Square	0.936	0.941	0.825	0.947	0.934	0.928	0.942	0.918	0.919
LASSO	0.971	0.976	0.874	0.971	0.914	0.919	0.972	0.967	0.909
MI	0.952	0.951	0.838	0.947	0.906	0.916	0.940	0.939	0.903
RFA	0.968	0.968	0.872	0.970	0.942	0.897	0.966	0.955	0.919
FI	0.949	0.955	0.838	0.955	0.920	0.941	0.952	0.933	0.921
RFE	0.959	0.964	0.853	0.965	0.899	0.939	0.955	0.943	0.927

According to the results shown in Table 2, the RF classifier had the highest accuracy (AUC = 0.976) on the selected features by LASSO method which is higher than the accuracy obtained using other combinations of classifiers and feature selection methods.

First, we performed the LASSO on the preprocessed features (n = 5351). Next, we used the rank-based method using the co-efficiency of features to calculate the average accuracy for all the subsets of the top 20 features. We calculated the average accuracy with 100 iterations in each step on the independent test data.

The RF classifier with 100 iterations on features obtained with the LASSO algorithm yielded an area under the curve of AUC = 0.976 with a 95% confidence interval 0.93–0.98 based on a 30% independent test data.

Figure 1 Shows that there is practically no improvement after four features. Four ROI's, namely, isthmuscingulate_(lh), inferiorparietal_(lh), Hippocampus_(lh) and entorhinal_(rh) and four features, namely, firstorder_90Percentile, firstorder Median, glrlm_LongRunEmphasis and gldm_DependenceEntropy have been chosen for the classification model. After selecting the RF classification method and tuning its hyperparameters using Randomized-Search-CV (CV = 5), the AUC was found as 0.961 for 100 iterations with a 95% confidence interval: 0.925–0.981.

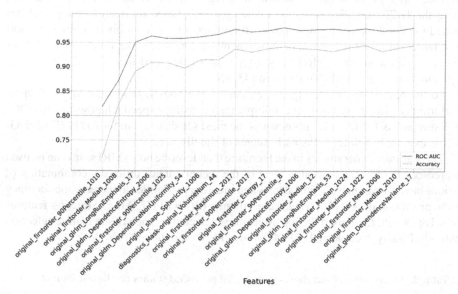

Fig. 1. Rank-based results for each subset of top 20 most important features selected by LASSO method.

4 Discussion

In this study we have developed a radiomics based platform (RAB-PET) and method for the analysis of brain FDG-PET images of AD patients. We have tested our platform for the Diagnosis of AD. Diagnosing AD in the early stages is one of the most important strategies for the prevention and treatment of AD. This is feasible by measuring specific features on specific regions of the human brain [35]. The platform allows us to find a satisfactory radiomics solution, by examining nine classification methods on eight different feature selection methods using 18FDG-PET images. We have achieved a relatively high accuracy level of AUC = 0.976 in comparison with other recent studies shown in Table 3. Please note that the comparison is limited by the fact that different datasets have been used in these studies. We have used the same database as the study in [46] although a slightly larger set of patients. We see that our results in our study compare favorably with respect to the results reported in this reference. In [46], they report an AUC of 0.96 whereas our results indicate an AUC of 0.98. Our method is also computationally fast (about one minute for segmentation and less than one minute for feature extraction and prediction). Another advantage is the interpretability of the prediction because of the low number of features and clinically meaningful ROI's. It would be desirable to experiment all reported methods using the same database as a future study. Our results seem to be better than the other reported studies probably due to completely automated segmentation, extensive analysis and optimization of features and prediction models. We used the FastSurfer for volumetric analysis, which replicates the FreeSurfer's anatomical segmentation to segment the whole brain into 95 classes using the DKT-atlas. FastSurfer is a new approach based on deep learning that has caused the development of more powerful toolboxes to process big datasets in a short time replacing time-consuming brain segmentation methods like FreeSurfer, SPM, or FSL [20]. We also used PyRadiomic [44] which calculated the 2D and 3D features separately under the IBSI [50] standard [43, 45].

To the best of our knowledge, as expressed previously, this is the first radiomics platform that can be used to derive multiple results using a specific hardware and FDG-PET studies. As 18FDG-PET is known as the most sensitive exam for AD [14]. 18FDG-PET is extremely helpful in the early stages of the disease.

There are also other studies in the literature that describe how sMRI scans can be used to predict AD [36, 37]. Although most of the recent studies utilized a combination of various biomarkers such as clinical tests and imaging modalities to increase the accuracy in the prediction of AD [38], we have used only 18FDG-PET. After extracting features from 18FDG-PET, we used the top 20 features selected by each of the feature selection methods. Finally, we selected only 4 features for the prediction of AD.

Table 3. Comparison of our results with related published studies on classification of AD

Dataset Size		Accuracy		Year	Ref
CN	AD	Acc (%)	AUC		
132	125	-	0.96	2022	[46]
302	209	89.24	0.95	2021	[25]
429	358	90.3	0.95	2020	[26]
119	97	88.9	0.93	2020	[27]
429	358	91.09	0.96	2018	[28]
430	358	83.7	-	2017	[29]
61	50	79	0.88	2017	[30]
207	154	88.3	0.94	2017	[31]
229	192	88.6	0.9	2017	[32]
52	51	87.26	0.93	2016	[33]
77	65	87.76	-	2014	[34]
188	163	**94.8**	**0.98**	Proposed Method	

It is worth to note that specific ROIs are early tau-deposited regions which can assist in the early diagnosis of AD disease [39]. In addition, according to the report of the National Institute on Aging (NIH), Alzheimer's disease typically destroys neurons and their connections in parts of the brain involved in memory. It later affects areas in the cerebral cortex responsible for language, reasoning, and social behavior [40]. Work is underway to associate these regions with the regions pinpointed by the radiomic analysis.

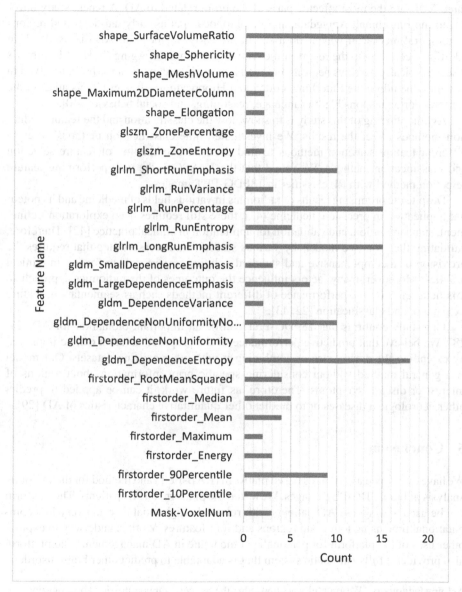

Fig. 2. Common feature types selected by different feature selection methods

An important finding of the study is the introduction of the four frequently used features (gldm_DependenceEntropy, shape_SurfaceVolumeRatio, glrlm_RunPercentage, glrlm_LongRunEmphasis) by the eight different feature selection methods as the most important features in the prediction of AD on three ROI's, namely, Amygdala, Entorhinal, and Hippocampus (Figs. 1, 2).

Considering recent clinical studies on patients with AD, it can be concluded that these three ROIs are the most affected parts of the brain related to AD. A recent study introduced the Entorhinal, Amygdala, and Parahippocampal as early tau-deposited regions of temporal meta-ROI which can assist in the early diagnosis of AD disease [39]. In addition, according to the report of the National Institute on Aging (NIH), Alzheimer's disease typically destroys neurons and their connections in parts of the brain involved in memory, including the Entorhinal cortex and Hippocampus. It later affects areas in the cerebral cortex responsible for language, reasoning, and social behavior [40].

Another finding of this study is to show how the classification and the feature reduction methods affect the results. We investigated nine classification methods on eight different feature selection methods to find the best combination of feature selection and classification methods. We found that the LASSO is the well-performing feature selection method with RF classifier in 18FDG-PET images.

Despite the promising results of radiomics in various fields of medicine and its potential application in precision medicine [41], there still requires deep exploration, refinement, standardization, and validation for application in clinical practice [42]. Therefore, validating the proposed models by various studies is a necessary step that requires the provision of a comprehensive and standard solution to verify them for use in clinical practice. Moreover, many factors influence the outcomes of the radiomics approach so this field relies on the performance of different methods such as segmentation, feature extraction, and classification [22, 43].

Our study confirms that 18FDG-PET can be an important biomarker for AD [27, 28]. We believe that predictive performance of the proposed solution can be improved by extending the dataset size or combining it with other existing datasets. Our model is a general method without considering specific prior information about regions of interest or disease symptoms. Therefore, as future work, it can be applied to predict other neurological diseases or to predict other quantitative characteristics of AD [29].

5   Conclusion

We have developed a radiomics based platform (RAB-PET) and method for the radiomic analysis of brain FDG-PET images. We tested the platform on AD patients. The platform can be used to diagnose AD patients with relatively high reliability and very low computational time using four brain regions and four features. Work is underway to explore other uses of the platform for personalized medicine in AD management. The platform also provides a fully automatic system that is adaptable to predict other brain disorders.

Acknowledgments. We gratefully acknowledge the financial support provided by Bogazici University Research Fund (BAP) for this research under project code 19774. Their support was instrumental in carrying out the experiments and analyzing the data.

References

1. Kumar, A., et al.: Amyloid and Tau in Alzheimer's disease: biomarkers or molecular targets for therapy? Are we shooting the messenger? Am. J. Psychiatry **178**, 1014–1025 (2021). https://doi.org/10.1176/appi.ajp.2021.19080873
2. Bateman, R.J., et al.: Clinical and biomarker changes in dominantly inherited Alzheimer's disease. N. Engl. J. Med. **367**, 795–804 (2012). https://doi.org/10.1056/NEJMoa1202753
3. Villemagne, V.L., et al.: Amyloid beta deposition, neurodegeneration, and cognitive decline in sporadic Alzheimer's disease: a prospective cohort study. Lancet. Neurol. **12**, 357–367 (2013). https://doi.org/10.1016/S1474-4422(13)70044-9
4. Grill, J.D., Cox, C.G., Harkins, K., Karlawish, J.: Reactions to learning a "not elevated" amyloid PET result in a preclinical Alzheimer's disease trial. Alzheimer's Res. Ther. **10**, 125 (2018). https://doi.org/10.1186/s13195-018-0452-1
5. Verma, R.K., et al.: An insight into the role of artificial intelligence in the early diagnosis of Alzheimer's disease. CNS Neurol. Disord. Drug Targets (2021). https://doi.org/10.2174/1871527320666210512014505
6. Revathi, A., et al.: Early detection of cognitive decline using machine learning algorithm and cognitive ability test. Secur. Commun. Netw. **2022**, 4190023 (2022). https://doi.org/10.1155/2022/4190023
7. Guiot, J., et al.: A review in radiomics: making personalized medicine a reality via routine imaging. Med. Res. Rev. **42**, 426–440 (2022). https://doi.org/10.1002/med.21846
8. Arbizu, J., Bastidas, J.F. (eds.): Clinical Nuclear Medicine in Neurology: An Atlas of Challenging Cases, pp. 9–13. Springer International Publishing (2022)
9. Chételat, G., et al.: Amyloid-PET and 18F-FDG-PET in the diagnostic investigation of Alzheimer's disease and other dementias. The Lancet Neurology **19**(11), 951–962 (2020)
10. Mannil, M., von Spiczak, J., Manka, R., Alkadhi, H.: Texture analysis and machine learning for detecting myocardial infarction in noncontrast low-dose computed tomography: unveiling the invisible. Invest. Radiol. **53**, 338–343 (2018). https://doi.org/10.1097/rli.0000000000000448
11. Li, Y., et al.: Radiomics: a novel feature extraction method for brain neuron degeneration disease using (18)F-FDG-PET imaging and its implementation for Alzheimer's disease and mild cognitive impairment. Ther. Adv. Neurol. Disord. **12**, 1756286419838682 (2019). https://doi.org/10.1177/1756286419838682
12. Zanfardino, M., et al.: Bringing radiomics into a multi-omics framework for a comprehensive genotype–phenotype characterization of oncological diseases. J. Transl. Med. **17**, 337 (2019). https://doi.org/10.1186/s12967-019-2073-2
13. Staartjes, V.E., et al. (eds.): Machine Learning in Clinical Neuroscience, pp. 51–57. Springer International Publishing (2022). https://doi.org/10.1007/978-3-030-85292-4
14. Wabik, A., et al.: Comparison of dynamic susceptibility contrast enhanced MR and FDG-PET brain studies in patients with Alzheimer's disease and amnestic mild cognitive impairment. J. Transl. Med. **20**, 1–14 (2022)
15. Kim, J.P., et al.: Predicting amyloid positivity in patients with mild cognitive impairment using a radiomics approach. Sci. Rep. **11**, 6954 (2021). https://doi.org/10.1038/s41598-021-86114-4
16. Sun, P., Wang, D., Mok, V.C., Shi, L.: Comparison of feature selection methods and machine learning classifiers for radiomics analysis in glioma grading. IEEE Access **7**, 102010–102020 (2019). https://doi.org/10.1109/ACCESS.2019.2928975
17. Ray, P., Reddy, S.S., Banerjee, T.: Various dimension reduction techniques for high dimensional data analysis: a review. Artif. Intell. Rev. **54**(5), 3473–3515 (2021). https://doi.org/10.1007/s10462-020-09928-0

18. Salam, M.A., Azar, A.T., Elgendy, M.S., Fouad, K.M.: The effect of different dimensionality reduction techniques on machine learning overfitting problem. Int. J. Adv. Comput. Sci. Appl. (IJACSA) **12**, 641–655 (2021). https://doi.org/10.14569/IJACSA.2021.0120480

19. Jovic, A., Brkic, K., Bogunovic, N.: In: 2015 38th International Convention on Information and Communication Technology, Electronics and Microelectronics (MIPRO), pp. 1200–1205

20. Massafra, R., et al.: Radiomic feature reduction approach to predict breast cancer by contrast-enhanced spectral mammography images. Diagnostics **11**(4), 684 (2021)

21. Muthukrishnan, R., Rohini, R.: In: 2016 IEEE International Conference on Advances in Computer Applications (ICACA), pp. 18–20

22. Shaikh, F., Franc, B., Mulero, F.: Clinical Nuclear Medicine. In: Ahmadzadehfar, H., Biersack, H.-J., Freeman, L.M., Zuckier, L.S. (eds.), pp. 193–207. Springer International Publishing (2020)

23. Alongi, P., et al.: Radiomics analysis of brain [18F]FDG PET/CT to predict alzheimer's disease in patients with amyloid PET positivity: a preliminary report on the application of SPM cortical segmentation, pyradiomics and machine-learning analysis. Diagnostics **12**, 933 (2022)

24. Ying, X.: An overview of overfitting and its solutions. J. Phys: Conf. Ser. **1168**, 022022 (2019). https://doi.org/10.1088/1742-6596/1168/2/022022

25. Poloni, K.M., Ferrari, R.J.: Automated detection, selection and classification of hippocampal landmark points for the diagnosis of Alzheimer's disease. Comput. Methods Programs Biomed. **214**, 106581 (2022)

26. Lian, C., Liu, M., Zhang, J., Shen, D.: Hierarchical fully convolutional network for joint atrophy localization and Alzheimer's disease diagnosis using structural MRI. IEEE Trans. Pattern Anal. Mach. Intell. **24**(4), 880–893 (2020)

27. Liu, M., et al.: A multi-model deep convolutional neural network for automatic hippocampus segmentation and classification in alzheimer's disease. Neuroimage **208**(1), 116459 (2020)

28. Liu, M., Zhang, J., Adeli, E., Shen, D.: Landmark-based deep multi-instance learning for brain disease diagnosis. Med. Image Anal. **43**, 157–168 (2018)

29. Zhang, J., Gao, Y., Gao, Y., Munsell, B.C., Shen, D.: Detecting anatomical landmarks for fast Alzheimer's disease diagnosis. IEEE Trans. Med. Imaging **35**(12), 2524–2533 (2016). https://doi.org/10.1109/TMI.2016.2582386

30. Korolev, S., Safiullin, A., Belyaev, M., Dodonova, Y.: Residual and plain convolutional neural networks for 3D brain MRI classification. In: Proceedings of the International Symposium on Biomedical Imaging (ISBI), pp. 835–838. IEEE, Melbourne, Australia (2017)

31. Zhang, J., Liu, M., An, L., Gao, Y., Shen, D.: Alzheimer's disease diagnosis using landmark-based features from longitudinal structural MR images. IEEE J. Biomed. Health Inform. **21**(5), 1607–1616 (2017)

32. Liu, X., Yang, J., Zhao, D., Huang, M., Zhang, J., Zaiane, O.: Nonlinearity-aware based dimensionality reduction and over-sampling for AD/MCI classification from MRI measures. Comput. Biol. Med. **91**, 21–37 (2017)

33. Ye, T., Zu, C., Jie, B., Shen, D., Zhang, D.: Discriminative multi-task feature selection for multi-modality classification of Alzheimer's disease. Brain Imaging Behav. **10**(3), 739–749 (2015). https://doi.org/10.1007/s11682-015-9437-x

34. Liu, S., Liu, S., Cai, W., Pujol, S., Kikinis, R., Feng, D.: Early diagnosis of Alzheimer's disease with deep learning. In: Proceedings of the IEEE 11th International Symposium on Biomedical Imaging - ISBI, pp. 1015–1018. IEEE, Beijing, China (2014)

35. Ponisio, M.R., Iranpour, P., Benzinger, T.L.S.: Amyloid imaging in dementia and neurodegenerative disease. In: Franceschi, A.M., Franceschi, D. (eds.) Hybrid PET/MR Neuroimaging: A Comprehensive Approach, pp. 99–110. Springer International Publishing, Cham (2022). https://doi.org/10.1007/978-3-030-82367-2_11

36. Syaifullah, A.H., et al.: Machine learning for diagnosis of AD and prediction of MCI progression from brain MRI using brain anatomical analysis using diffeomorphic deformation. Front. Neurol. **11**, 576029 (2021)
37. Basaia, S., et al.: Automated classification of Alzheimer's disease and mild cognitive impairment using a single MRI and deep neural networks. NeuroImage: Clinical **21**, 101645 (2019)
38. Gupta, Y., et al.: Prediction and classification of Alzheimer's disease based on combined features from apolipoprotein-E genotype, cerebrospinal fluid, MR, and FDG-PET imaging biomarkers. Front. Comput. Neurosci. **13**, 72 (2019)
39. Cai, Y., et al.: Initial levels of β-amyloid and tau deposition have distinct effects on longitudinal tau accumulation in Alzheimer's disease. Alzheimer's Research & Therapy **15**(1), 1–14 (2023)
40. NIH Homepage,. https://www.nia.nih.gov/health/what-happens-brain-alzheimers-disease. last accessed 2023/03/08
41. Abbasian Ardakani, A., Bureau, N.J., Ciaccio, E.J., Acharya, U.R.: Interpretation of radiomics features–a pictorial review. Comput. Methods Programs Biomed. **215**, 106609 (2022). https://doi.org/10.1016/j.cmpb.2021.106609
42. Frix, A.-N., et al.: Radiomics in lung diseases imaging: state-of-the-art for clinicians. J. Personalized Med. **11**, 602 (2021)
43. Yip, S.S.F., Aerts, H.J.W.L.: Applications and limitations of radiomics. Phys. Med. Biol. **61**, R150–R166 (2016). https://doi.org/10.1088/0031-9155/61/13/r150
44. Sanduleanu, S., et al.: Tracking tumor biology with radiomics: a systematic review utilizing a radiomics quality score. Radiother. Oncol. **127**(3), 349–360 (2018)
45. Van Griethuysen, J.J., et al.: Computational radiomics system to decode the radiographic phenotype. Can. Res. **77**(21), e104–e107 (2017)
46. Yüksel, C., Rasi, R., Güveniş, A.: A new method for diagnosing alzheimer's disease and monitoring its severity using FDG-PET. In: 2022 Medical Technologies Congress (TIPTEKNO), pp. 1–4 (2022)
47. Zwanenburg, A., et al.: The image biomarker standardization initiative: standardized quantitative radiomics for high-throughput image-based phenotyping. Radiology **295**(2), 328–338 (2020)
48. Bogowicz, M., et al.: Post-radiochemotherapy PET radiomics in head and neck cancer–the influence of radiomics implementation on the reproducibility of local control tumor models. Radiother. Oncol. **125**(3), 385–391 (2017)
49. Rasi, R., Guvenis, A.: RAB-PET (2023).https://doi.org/10.5281/zenodo.7859694
50. Zwanenburg, A. et al.: The image biomarker standardization initiative: standardized quantitative radiomics for high-throughput image-based phenotyping. Radiology **295**, 328–338 (2020). https://doi.org/10.1148/radiol.2020191145

Biomedical Signal Analysis

Deep Learning for Automatic Electroencephalographic Signals Classification

Nadia N. Sánchez-Pozo[1]([✉]) [iD], Samuel Lascano-Rivera[1] [iD],
Francisco J. Montalvo-Marquez[1] [iD], and Dalia Y. Ortiz-Reinoso[2] [iD]

[1] Universidad Politécnica Estatal del Carchi, Tulcán 040101, Ecuador
nadiasanchez239@gmail.com
[2] Universidad de Guayaquil, Guayaquil 090510, Ecuador

Abstract. Automated electroencephalographic (EEG) signals classification using deep learning algorithms is an emerging technique in neuroscience that has the potential to detect brain pathologies such as epilepsy efficiently. In this process, deep learning algorithms are trained with labeled EEG signal datasets. However, due to the highly complex nature of EEG signals and the large amount of irrelevant information they contain, feature extraction techniques must be applied to reduce their dimensionality and focus on relevant information. This paper presents a comparative study on feature extraction methods for the classification of EEG recordings. The results demonstrate that the proposed classification algorithms and characterisation techniques are effective and suitable, as the accuracy metrics reach a value of 99.27%. The results presented in this paper contribute to the further development of automatic EEG signal classification methods based on deep learning.

Keywords: Epilepsy diagnosis · Feature extraction · Classification · Convolutional Neural Networks (CNN) · Seizure detection

1 Introduction

Epilepsy is considered to be a common condition of chronic brain disorder, caused by a sudden abnormality of brain neurons and has affected almost 1% of the world's population [14]. Electroencephalography (EEG), in addition to being non-invasive, EEG is a technique for measuring the electrical activity in the brain that is used to detect and diagnose epilepsy [16].

The classification of EEG signals related to epilepsy is a complex problem that requires achieve high accuracy with a large amount of labeled data. There are different approaches and techniques that can be combined to achieve better performance [11]. However, visual analysis and EEG signal extraction are highly complex and time-consuming processes, and it is generally difficult to identify subtle variations in the brain. For this reason, waveform analysis of EEG signals is typically studied visually.

I. Rojas et al. (Eds.): IWBBIO 2023, LNBI 13919, pp. 259–271, 2023.
https://doi.org/10.1007/978-3-031-34953-9_20

Brain-computer interface (BCI) research has attracted considerable interest from a diverse group of researchers, including neuropsychologists, engineers, computer scientists, mathematicians and neuroscientists. The development of an effective BCI system requires a signal processing approach capable of extracting features and performing classifications. Recent research has focused on developing more efficient methods for signal processing and classification [12].

The process of extracting relevant information from EEG signals involves three key steps: preprocessing, feature extraction and classification. Identifying the most relevant features during the extraction stage is critical to achieving optimal performance in classifying EEG data [2]. Even if the classification method is highly efficient, classification performance will be compromised if the extracted feature is irrelevant to the problem at hand. Therefore, it is imperative to ensure that the extracted feature is well suited to the task at hand. Therefore, to achieve high performance classification, it is essential to identify and extract the appropriate features from the EEG signal.

In Iscan (2011), several feature extraction methods are described, including Fourier transform and wavelet transform (WT). However, WT-based analysis is more efficient than other methods because it takes into account the transient behaviour of the EEG signal. To enhance routine EEG applications and analysis, wavelet-based features have been integrated, including wavelet entropy, wavelet coefficients and statistical features such as mean, median and standard deviation of the wavelet transform (WT). [6,10,19]. In addition, Wang (2020) used machine learning classifiers and wavelet-based feature extraction to demonstrate the classification of EEG signals in cognitive tasks of a given workload from reference tasks [25].

Artificial Intelligence (AI) has proven to be a robust tool for classifying EEG signals in the context of epilepsy [23]. Using machine learning algorithms, models can be trained to identify specific patterns and features in EEG recordings. These models can be used to predict the presence of seizure activity in EEG signals and support the diagnosis of the disease [5]. Artificial neural networks are among the most commonly used machine learning algorithms to classify EEG recordings into different categories, such as the presence or absence of convulsive activity. In addition, techniques such as feature extraction and data preprocessing are essential to ensure the efficiency and accuracy of AI models [3].

This study compares the accuracy of automatic classification of EEG signals using different characterisation and deep learning methods. The results demonstrate the potential of deep learning algorithms for automatic classification of EEG recordings and their ability to outperform conventional methods. These findings not only advance research in the field of EEG recording analysis and automatic identification of EEG signals, but also have significant implications for the diagnosis and treatment of brain pathologies. Section 2 describes the methods and data used in the study, including EEG signal feature extraction and classification. Section 3 presents the experiments conducted, and details the results obtained. Finally, Sect. 4 describes the main conclusions of this study.

2   Methodology

2.1   Description of EEG Data

The data were retrieved from the UC Irvine (UCI) machine learning repository and stored in Google Drive for easy access and processing. The processing was carried out using the Google Colab tool, which allows programming in the Python language. This tool is particularly useful for research in the field of artificial intelligence, as it allows large amounts of data to be worked with and processed efficiently. In addition, Google Colab offers cloud computing resources to speed up data processing.

The dataset consists of 23.6 s of brain activity recordings from 500 subjects, with a total of 4097 data points. This dataset has been used in several studies to develop algorithms for detecting epilepsy-related pathology.

In this study we used a data set obtained by dividing 4097 data points into 23 fragments, resulting in 178 data points per second. Each data point corresponds to the EEG recording value at a particular time. This procedure resulted in a matrix with dimensions of 11500 × 179, where the last column (179) denotes the class label of the recording, described as 1, 2, 3, 4, 5.

1. Activity associated with seizures.
2. Electroencephalogram (EEG) from the tumour site.
3. Typical EEG patterns found in the healthy part of the brain.
4. EEG recorded with the patient's eyes closed.
5. EEG recorded with the patient's eyes open.

2.2   EEG Signal Preprocessing

EEG signal pre-processing is an important step in the analysis of EEG signals for epilepsy classification. Filtering, resampling, epoch and artefact removal are some of the steps that make up preprocessing.

– Filtering: EEG signals are typically filtered to remove unwanted frequencies, such as white noise, and to enhance frequencies of interest (e.g., alpha, beta, gamma, and theta bands) that are associated with seizures.
– Resampling: Resampling involves reducing the sampling frequency of EEG signals, which are often recorded at a high frequency (e.g. 500 Hz or 1000 Hz Hz), to simplify analysis and reduce computational burden. This process involves reducing the sampling rate to a lower level (e.g. 100 Hz).
– Epochs: EEG signals are usually divided into smaller segments, or epochs, corresponding to a specific window of time before, during and after a seizure episode. This allows EEG signals to be analysed on a very precise time scale.
– Artifact removal: Various types of artefacts, such as fibrillation, muscle movement, EEG sensor movement, electrical line interference, environmental factors, and others, can be incorporated into the EEG signal during recording [21]. Eliminating artifacts from the EEG signal is critical to extracting the

true characteristics of the signals. High-pass, low-pass, and band-pass filters are commonly used for this purpose. The choice of band-stop and notch filters depends on the frequency range of the artifacts to be removed [15].

- Normalization: The amplitude of EEG signals can vary between individuals and recordings, which can impact the analysis. To address this, it is common to normalize the signals to a fixed range, such as between -1 and 1, or to have zero mean and unit variance. This normalization process helps to standardize the signal amplitudes and enables meaningful comparisons across different recordings and individuals.

In this study, a normalisation process of EEG signals is performed using the MinMaxScaler method. This process involves scaling all features of the data within the range $[0, 1]$ or $[-1, 1]$ if there are negative values in the data. The scaling is performed to compress all internal values into a narrow interval $[0, 0.005]$, with the aim of eliminating offset levels. The Eq. (1) is used for the calculation.

$$Mm(std) = \frac{(Mm - Mm.min)}{(Mm.max - Mm.min)} \tag{1}$$

2.3   Feature Extraction

Feature extraction from EEG recordings is the process of identifying and selecting relevant attributes from EEG signals. The aim of this process is to simplify complex data and extract relevant information [7]. There are several feature extraction methods used to study EEG signals, including Fourier transform, principal component analysis, statistical feature extraction, discrete wavelet transform (DWT), and machine learning [20].

Statistical Features. Statistical features are a way to describe and summarize the properties of EEG signals. These features are used to describe the temporal and frequency characteristics of the EEG signal. In addition, when analyzing EEG signals for the diagnosis of epilepsy, statistical features have proven to be very effective. The following statistical characteristics are used, minimum value, maximum value, interquartile range, mean, energy, variance, log entropy energy, fractal dimension, second moment, third moment, fourth moment, kurtosis, skewness and median value, as suggested by [4].

Discrete Wavelet Transform. The wavelet transform is an efficient time-frequency tool for signals analysis [27]. The Discrete Wavelet Transform (DWT) is a method of decomposing the input signal H(t) into a set of functions called wavelets by scaling and modifying the parent wavelet function [28].

DWT is commonly used on electroencephalography (EEG) signals to identify patterns and features in brain activity. DWT is based on the idea of dividing a signal into different scales and decomposing it into components of different frequency bands, which allows detailed analysis of the signal on different time and frequency scales, facilitating the detection of patterns and abnormalities in brain activity. It can also be used to reduce signal size and to encode and compress data.

The Wavelet Transform is a signal processing technique that analyses signal characteristics in both the time and frequency domains. It does this by decomposing signals into multiple functions using a single mother function [1]. The mother function serves as the building block for the decomposition process Eq. (2).

$$DWT(s_n) = \sum_{k=0}^{n-1} h_k \cdot s_{n-k-1} \tag{2}$$

where s_n is the discrete signal and h_k are the wavelet coefficients of the mother wavelet db6.

To carry out this study, a four-level Discrete Wavelet Transform (DWT) using the db6 mother wavelet function was used to extract relevant features. The results of the four-level decomposition, including the detail coefficients CD1, CD2, CD3 and CD4, and the approximate coefficient CA4, are shown in Table 1. [4].

Table 1. DWT Decomposition level coefficients and frequency range

Level	Coefficient	Frequency Range (Hz)
1	CD1	64–128
2	CD2	32–64
3	CD3	16–32
4	CD4	8–16
4	CA4	0–8

2.4   Classification

EEG signal classification is the process of using machine learning techniques to identify patterns and features in EEG signals and assign them to one of the target classes. Deep learning is particularly useful for classifying large and complex amounts of data, including EEG signals [22].

There are several types of deep learning architectures that have been used to classify EEG signals, such as convolutional neural networks (CNNs) and recurrent neural networks (RNNs). These architectures can be trained on large amounts of labelled EEG recordings to learn to identify seizures and other EEG patterns associated with epilepsy [17].

This paper presents two models for classifying EEG signals. The choice of architecture is based on a thorough evaluation of performance, feature extraction capability and automated recognition.

Simple Model. This deep learning model consists of two main layers: input and output. The first is designed to receive input signals, while the second is used to make accurate predictions. The appropriate selection of the activation function is a critical factor in the design of deep learning models. For this study, different activation functions were used in each layer. The input layer uses a rectified linear unit (RELU) activation function, while the output layer uses a sigmoid function. Table 2 provides a detailed description of the neural network architecture. The combination of two different activation functions in each layer provides greater flexibility in processing the input signals and greater accuracy in the final predictions.

Table 2. Simple model architecture.

Layer	Activation Function	Shape Output
Dense	Relu	(N, 64)
Dense$_1$	Sigmoid	(N, 2)

The architecture of the simple two-levels model has been designed with the aim of conducting a comparative analysis with a hybrid model.

Hybrid Model. This deep learning model is composed of Conv1d convolutional layers, Max Pooling, Dropout, Flatten and Dense layers, using the Rectified Linear Unit (RELU) as the activation function. Figure 1 illustrates the architecture of the hybrid deep learning model for EEG signal classification.

The Conv1D layers extract relevant features from the input data, the Dense layer performs a linear transformation, and the MaxPooling layer reduces the dimensionality of the data to simplify the classification process [9]. Dropout is a regularization technique used in neural networks to prevent overfitting. Overfitting occurs when the neural network learns well from the data and thereby, generalizes poorly to new data. The dropout layer solves this problem by randomly disabling a percentage of neurons in each training iteration. The Flatten layer is an operation that flattens or converts multi-dimensional input data into a one-dimensional vector. This is done before the final layer. The dense layers are fully connected layers used to perform the final classification of the data. Each neuron in a dense layer receives a linear combination of the outputs of the neurons in the previous layer, followed by an activation function. The activation function mentioned in the architecture is a Rectified Linear Unit (RELU), which is used to introduce non-linearity into the model.

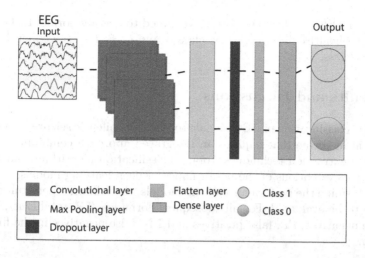

Fig. 1. Proposed hybrid model architecture for EEG signal classification.

Table 3 details the architecture of the hybrid model. The above architecture is a convolutional neural network structure that is used to extract features from the data, reduce dimensionality, prevent overfitting and perform the final classification of the data.

Table 3. Architecture of the proposed hybrid model.

Layer	Output Shape	Size Kernel	Step
Conv1d	(N, 178, 32)	(3,)	1
Conv1d$_1$	(N, 178, 64)	(3,)	1
Conv1d$_2$	(N, 178, 64)	(3,)	1
Max pooling1d	(N, 89, 64)	(3,)	3
Dropout	(N, 89, 64)	–	–
Flatten	(N, 5696)	–	–
Dense	(N, 64)	–	–
Dense$_1$	(N, 2)	–	–

In this paper, the data set consists of five categories, only one of which represents seizure activity. Therefore, several binary experiments were performed, with the primary objective being seizure activity in class 1 and normal activity in class 0. The experiments performed and their results are described below.

– Experiment 1 In this case, classes 2, 3, 4, and 5 were grouped into a single class **0**, which represented the Normal or non-epileptic class, while class **1** represented the class with epileptic seizures and served as the primary objective of classification.

– Experiment 2 For this experiment, we aimed to classify normal EEG activity from the healthy brain region as class **0** and class **1** as the epileptic seizure activity class.

3   Results and Discussions

Automatic classification of EEG signals to identify epileptic seizures is a relevant problem in medicine that requires an integrated approach combining appropriate feature extraction techniques, robust classification algorithms and rigorous quantitative evaluations of the performance of deep learning models.

To determine the performance of the models, the following performance metrics need to be evaluated. For all subsequent formulas, TP: total true positives, TN: true negatives, FP: false positives and FN: false negatives in the final classification.

– Accuracy: A measure of the accuracy of a classification model given by Eq. 3.

$$Accuracy = \frac{TN + TP}{TN + FP + FN + TP} * 100. \tag{3}$$

– Precision: This is an indicator of the quality of positive predictions. Equation 4.

$$Precision = \frac{TP}{FP + TP} * 100. \tag{4}$$

– Recall or Probability of classifying true positives, as shown in the Eq. 5:

$$Recall = \frac{TP}{TP + FN} * 100. \tag{5}$$

– F1, used to combine the measures of Precision and Recall into a single value, as shown in the Eq. 6:

$$F1 = \frac{2 \times Precision \times Recall}{Precision + Recall} * 100. \tag{6}$$

For the evaluation of the performance of the deep learning algorithm, a division of the dataset into two parts was carried out: one 70% was used for training and one 30 % for testing. The deep learning algorithm was trained using the training set until optimal performance was achieved.

3.1   Evaluation

In this stage, a comparative analysis is carried out to determine the most effective feature extraction technique for classifying EEG signals. Different techniques are examined, including statistical features and wavelet transform coefficients. Once the features have been extracted, the results of the different classification algorithms are evaluated. Accuracy, which refers to the ability of the model to correctly classify the signals, and training time, which measures the amount of time it takes to train a model, are compared between different algorithms. The purpose of this evaluation is to determine the optimal combination of feature extraction techniques and classification algorithms that gives the best results in terms of accuracy and training time. Figure 2 shows the loss curve corresponding to the algorithm with the best accuracy obtained in test 3 of the hybrid model. In this graph, the **y** axis represents the loss and the **x** axis corresponds to the number of epochs. It can be seen that the loss decreases continuously in both the training and validation data sets.

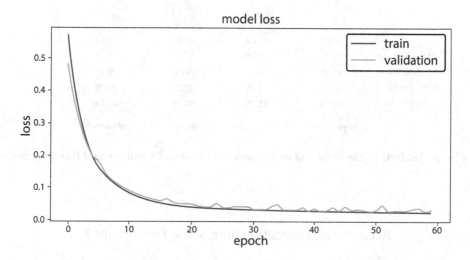

Fig. 2. Hybrid model loss curve test 4.

In this study, different classification models were evaluated and the results of the experiments are presented in Figs. 3 and 4, which show the accuracy and training time results of the models for Experiments 1 and 2, respectively. In addition, four tests were performed to train the models using different approaches for the normalized signal. The approaches for each test are as follows:

- Test 1: Normalized signal without performing any feature extraction.
- Test 2: Normalized signal using Discrete Wavelet Transform for feature extraction.
- Test 3: Normalized signal using statistical feature extraction.
- Test 4: Normalized signal using statistical feature extraction and Discrete Wavelet Transform.

In Experiment 1, the hybrid model achieved the highest percentage of accuracy, reaching 98.55% in Test 4. In Experiment 2, the results of the models were the same in Test 3, reaching an accuracy percentage of 99.27%. These results indicate that the hybrid model is a good option for classification tasks similar

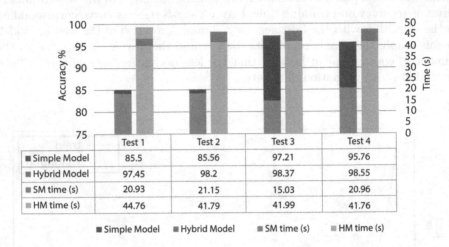

Fig. 3. Analysis of the results of experiment 1 in terms of accuracy and training time.

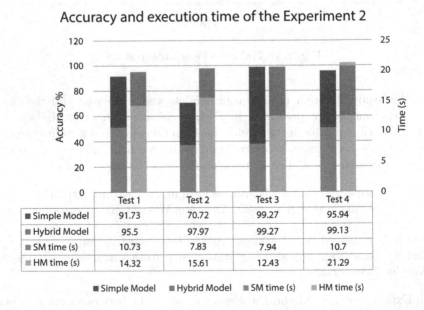

Fig. 4. Analysis of the results of experiment 2 in terms of accuracy and training time.

Table 4. Comparison of the results with other authors

Author	Year	Method	Accuracy	Reference
Tuncer et al.	2022	LSTM	98,08%	[24]
Jemal et al.	2021	CNN	90,9%	[13]
Zheng et al.	2020	Ensemble learning LSTMS-B	97,13%	[30]
Wu et al.	2019	LSTM	93,7%	[26]
Ghosh et al.	2018	Transfer Learning CNN	87,9%	[8]
Saeed et al.	2017	Multi-Objective Optimization	91,2%	[18]
Zhang et al.	2016	Deep Belief Networks	95,0%	[29]
Our method	2023	CNN	**99,27%**	

to those in Experiment 1, while each of the evaluated models is a good option for tasks like those in Experiment 2. It is important to note that the training times were similar in both experiments and no significant differences between the models were observed.

The performance of the results was compared to that of other authors who used different classifiers, The proposed models achieve an accuracy that is competitive with the accuracy of previous studies that employed computationally expensive and robust techniques (see Table 4).

From the analysis of the results obtained, it can be seen that the method proposed in this research, which uses a convolutional neural network (CNN) and statistical feature extraction, has the best performance in terms of accuracy, achieving a rate of **99.27%**. Compared to the other methods analyzed. These results indicate that the proposed technique, based on a convolutional neural network, is highly effective for the task at hand and represents a significant advance over the current state of the art in this area. In conclusion, a promising strategy for improving the accuracy of EEG signal classification is the use of a convolutional neural network and statistical feature extraction.

4   Conclusions and Future Work

In this paper, several experiments have been conducted in the area of recording segmentation and feature extraction. Although the Discrete Wavelet Transform (DWT) is an effective technique for obtaining features in EEG signal classification, this study shows that it is also possible to effectively classify these signals without the need for conventional characterization methods. By using deep learning techniques and statistical feature extraction, an accuracy of up to 99.27% was achieved.

Future work in the area of automatic classification of EEG signals using deep learning has great potential. One possibility is to explore other deep learning techniques, such as the use of convolutional neural networks (CNN) for EEG image classification, or the use of recurrent neural networks (RNN) to capture temporal patterns in the signal and improve classification accuracy. Other deep learning techniques, such as generative adversarial networks (GAN), can also be considered to generate synthetic data, which can be useful for training models with limited data sets.

References

1. Alturki, F.A., AlSharabi, K., Abdurraqeeb, A.M., Aljalal, M.: EEG signal analysis for diagnosing neurological disorders using discrete wavelet transform and intelligent techniques. Sensors **20**(9) (2020). https://doi.org/10.3390/s20092505.http://www.mdpi.com/1424-8220/20/9/2505
2. Amin, H.U., et al.: Feature extraction and classification for EEG signals using wavelet transform and machine learning techniques. Australas. Phys. Eng. Sci. Med. **38**(1), 139–149 (2015)
3. Asanza, V., Sánchez-Pozo, N.N., Lorente-Leyva, L.L., Peluffo-Ordóñez, D.H., Loayza, F.R., Peláez, E.: Classification of subjects with Parkinson's disease using finger tapping dataset. IFAC-PapersOnLine **54**(15), 376–381 (2021)
4. Bairagi, R.N., Maniruzzaman, M., Pervin, S., Sarker, A.: Epileptic seizure identification in EEG signals using DWT, ANN and sequential window algorithm. Soft Comput. Lett. **3**, 100026 (2021)
5. Burleigh, T.L., Griffiths, M.D., Sumich, A., Wang, G.Y., Kuss, D.J.: Gaming disorder and internet addiction: a systematic review of resting-state EEG studies. Addict. Behav. **107**, 106429 (2020)
6. Craik, A., He, Y., Contreras-Vidal, J.L.: Deep learning for electroencephalogram (EEG) classification tasks: a review. J. Neural Eng. **16**(3), 031001 (2019)
7. Fıçıcı, C., Telatar, Z., Eroğul, O.: Automated temporal lobe epilepsy and psychogenic nonepileptic seizure patient discrimination from multichannel EEG recordings using dwt based analysis. Biomed. Sig. Process. Control **77**, 103755 (2022)
8. Ghosh, S., Das, P., Nandi, S.: Transfer learning-based deep convolutional neural network for motor imagery EEG classification. J. Ambient Intell. Humanized Comput. **9**(5), 1669–1685 (2018). https://doi.org/10.1007/s12652-018-0858-z
9. Hamm, C.A., et al.: Deep learning for liver tumor diagnosis part i: development of a convolutional neural network classifier for multi-phasic MRI. Eur. Radiol. **29**, 3338–3347 (2019)
10. Hassouneh, A., Mutawa, A., Murugappan, M.: Development of a real-time emotion recognition system using facial expressions and EEG based on machine learning and deep neural network methods. Inf. Med. Unlocked **20**, 100372 (2020)
11. Iscan, Z., Dokur, Z., Demiralp, T.: Classification of electroencephalogram signals with combined time and frequency features. Expert Syst. Appl. **38**(8), 10499–10505 (2011)
12. Islam, M.K., Rastegarnia, A.: Recent advances in EEG (non-invasive) based BCI applications. Front. Comput. Neurosci. (2023)
13. Jemal, I., Mezghani, N., Abou-Abbas, L., Mitiche, A.: An interpretable deep learning classifier for epileptic seizure prediction using EEG data. IEEE Access **10**, 60141–60150 (2022)
14. Liu, Y.H., et al.: Epilepsy detection with artificial neural network based on as-fabricated neuromorphic chip platform. AIP Adv. **12**(3), 035106 (2022)

15. Mahjoub, C., Jeannès, R.L.B., Lajnef, T., Kachouri, A.: Epileptic seizure detection on EEG signals using machine learning techniques and advanced preprocessing methods. Biomed. Eng./Biomed. Tech. **65**(1), 33–50 (2020)
16. Mancha, V.R., Srinivasa, R.E., Ch, S.: Advanced convolutional neural network classification for automatic seizure epilepsy detection in EEG signal. IOP Conf. Ser.: Mater. Sci. Eng. **1074**(1), 012005 (2021)
17. Ouichka, O., Echtioui, A., Hamam, H.: Deep learning models for predicting epileptic seizures using iEEG signals. Electronics **11**(4), 605 (2022)
18. Saeed, H., Mohammadi, K.: A novel EEG feature extraction method using multi-objective optimization. Biomed. Sig. Process. Control **33**, 1–10 (2017). https://doi.org/10.1016/j.bspc.2016.10.005
19. Saeidi, M., et al.: Neural decoding of EEG signals with machine learning: a systematic review. Brain Sci. **11**(11), 1525 (2021)
20. Shoeibi, A., et al.: An overview of deep learning techniques for epileptic seizures detection and prediction based on neuroimaging modalities: methods, challenges, and future works. Comput. Biol. Med. 106053 (2022)
21. Shoka, A., Dessouky, M., El-Sherbeny, A., El-Sayed, A.: Literature review on EEG preprocessing, feature extraction, and classifications techniques. Menoufia J. Electron. Eng. Res **28**(1), 292–299 (2019)
22. Singh, K., Malhotra, J.: Smart neurocare approach for detection of epileptic seizures using deep learning based temporal analysis of EEG patterns. Multimed. Tools Appl. **81**(20), 29555–29586 (2022)
23. Tohidi, M., Madsen, J.K., Moradi, F.: Low-power high-input-impedance EEG signal acquisition SoC with fully integrated IA and signal-specific ADC for wearable applications. IEEE Trans. Biomed. Circ. Syst. **13**(6), 1437–1450 (2019)
24. Tuncer, E., Bolat, E.D.: Channel based epilepsy seizure type detection from electroencephalography (EEG) signals with machine learning techniques. Biocybernetics Biomed. Eng. **42**(2), 575–595 (2022)
25. Wang, F., et al.: Motor imagery classification using geodesic filtering common spatial pattern and filter-bank feature weighted support vector machine. Rev. Sci. Instrum. **91**(3), 034106 (2020)
26. Wu, J., Liu, H., Gao, X.: A semi-supervised deep clustering framework for EEG based motor imagery task. IEEE Trans. Neural Netw. Learn. Syst. **30**(12), 3663–3673 (2019). https://doi.org/10.1109/TNNLS.2019.2909198
27. Zarei, A., Asl, B.M.: Automatic seizure detection using orthogonal matching pursuit, discrete wavelet transform, and entropy based features of EEG signals. Comput. Biol. Med. **131**, 104250 (2021)
28. Zeng, W., Li, M., Yuan, C., Wang, Q., Liu, F., Wang, Y.: Identification of epileptic seizures in EEG signals using time-scale decomposition (ITD), discrete wavelet transform (DWT), phase space reconstruction (PSR) and neural networks. Artif. Intell. Rev. **53**(4), 3059–3088 (2020)
29. Zhang, X., Zhou, W., Li, Y., Li, L.: Combining deep belief network and support vector machine to classify motor imagery EEG signal. Neurocomputing **173**, 1500–1508 (2016). https://doi.org/10.1016/j.neucom.2015.09.080
30. Zheng, X., Chen, W., You, Y., Jiang, Y., Li, M., Zhang, T.: Ensemble deep learning for automated visual classification using EEG signals. Pattern Recognit. **102**, 107147 (2020)

Evaluation of Homogeneity of Effervescent Tablets Containing Quercetin and Calcium Using X-ray Microtomography and Hyperspectral Analysis

Michał Meisner[1,2] (iD), Piotr Duda[3] (iD), Beata Szulc-Musioł[4] (iD),
and Beata Sarecka-Hujar[2(✉)] (iD)

[1] Doctoral School of the Medical University of Silesia in Katowice, Faculty of Pharmaceutical Sciences in Sosnowiec, 41-200 Sosnowiec, Poland
[2] Department of Basic Biomedical Science, Faculty of Pharmaceutical Sciences in Sosnowiec, Medical University of Silesia in Katowice, 3 Kasztanowa Street, 41-200 Sosnowiec, Poland
bsarecka-hujar@sum.edu.pl
[3] Institute of Biomedical Engineering, Faculty of Science and Technology, University of Silesia, 39 Bedzinska Street, 41-200 Sosnowiec, Poland
[4] Department of Pharmaceutical Technology, Faculty of Pharmaceutical Sciences in Sosnowiec, Medical University of Silesia in Katowice, 3 Kasztanowa Street, 41-200 Sosnowiec, Poland

Abstract. Drug stability describes its ability to maintain physical, chemical, therapeutic, and microbiological properties during storage. The study aimed to assess whether there are any differences in the homogeneity of the effervescent tablets containing quercetin and calcium using hyperspectral imaging and X-ray microtomography. We analyzed unexpired, expired, as well as stressed tablets which were stored at 40 °C for 14 days. Both unexpired and expired tablets met the pharmacopoeial requirements. The homogeneity analysis showed significant differences between the three types of tablets. In addition, significantly higher reflectance values in the visible and near-infrared range were observed in unexpired tablets compared to expired and stressed tablets ($p = 0.001$ and $p < 0.001$, respectively). The X-ray microtomography showed that the densities of unexpired and stressed tablets were significantly higher compared to the density of expired tablets ($p < 0.001$). The changes in tablets' homogeneity may indicate possible physical changes in effervescent tablets during storage, especially in conditions deviating from those recommended by the manufacturer.

Keywords: effervescent tablets · stability · hyperspectral profiles · X-ray microtomography

1 Introduction

Effervescent tablets owe their popularity to the combination of the advantages of a solid and a liquid form of the active substance [1, 2]. They remain in a solid form during storage, which preserves the stability of the active substance and thus extends its shelf life. Prior

I. Rojas et al. (Eds.): IWBBIO 2023, LNBI 13919, pp. 272–282, 2023.
https://doi.org/10.1007/978-3-031-34953-9_21

to the patient's intake, the active substance is released into or suspended in solution, consequently accelerating the absorption and onset of action of active pharmaceutical ingredient API. The liquid form makes it easier to take the medicine for patients with swallowing difficulties and masks the unpleasant taste of the active substance, increasing the comfort of taking the medicine [3]. These facts give the impression that this drug form is more versatile compared with other oral dosage forms. However, some disadvantages may also be pointed out for effervescent tablets, including their hygroscopic nature and moisture-induced instability.

The active substances can be released from the effervescent tablets after the reaction of two typical components, an alkaline agent (most often sodium carbonate or sodium bicarbonate) and an acidic agent (citric acid, tartaric acid, malic acid, or maleic acid), which react with each other in the presence of water producing carbon dioxide (CO_2) [3, 4]. As a result of the gas release, the tablet disintegrates while the active substance is dispersed evenly.

New technologies popular in fields of science other than pharmacy and medicine turned out to be useful in learning about pharmaceutical products, especially in the context of their durability. These methods have in common among others, the speed of measurement, ease of sample preparation, and the preservation of unaltered samples. The stability of a drug formulation determines the length of time an active substance can be stored without compromising its therapeutic effect. It has a key role during the process of authorizing the finished preparation for use [5]. Imaging methods such as hyperspectral imaging or X-ray microtomography can be used to assess the stability of a solid drug formulation. The first technique enables the simultaneous acquisition of spectral and spatial information about the examined objects. The different interaction of different wavelengths with the sample allows to obtain a spectrum characteristic of the investigated structure and dependent on its composition [6]. In addition, the use of X-rays makes it possible to view the inside of the sample without destroying it. This method allows for capturing micro-cracks, air bubbles, or delaminations in the solid form of the drug [7, 8].

We aimed to analyze the difference in the homogeneity of the effervescent tablets using hyperspectral imaging and X-ray microtomography. Tablets containing calcium and quercetin were selected as a model preparation. Quercetin is one of the most common naturally occurring polyphenols. In plants, it is most often found in a glycosidic form [9]. This form is a common component of the human diet, with its greatest quantities found in apples and onions. According to various data, its daily intake is about 25 mg [10–12].

2 Materials and Methods

2.1 Tablets Selected for Analysis

The effervescent Alercal tablets were selected as a model preparation in the present analysis (Zdrovit; Natur Produkt Pharma sp. z o. o., Ostrów Mazowiecka, Poland). The tablets contain among others, 300 mg of calcium and 50 mg of quercetin, but also vitamin B12, folic acid, and selenium.

2.2 Evaluation of the Tablets

All tested tablets, i.e. unexpired (expiration date 04.2023; n = 20) and expired (expiration date 04.2021; n = 20) were carefully evaluated in terms of external appearance, weight, diameter, thickness, disintegration time, and hardness according to Polish Pharmacopeia [13]. They were accurately weighed using an analytical balance (Radwag, Poland). The moisture content was analyzed in the moisture analyzer WS-30 (Radwag, Radom, Poland) at 70 °C for 60 min. The hardness, diameter, and thickness of the tablets were determined in the MultiTest50 hardness tester (Pharmatron Dr. Schleuniger, Thun, Switzerland) for 5 tablets. To analyze *in vitro* disintegration time, 6 tablets were analyzed, one tablet at a time in a beaker that contained 200 ml of purified water at 15–25 °C.

To simulate the stressful conditions in which the tablets may be stored, 4 randomly selected unexpired tablets were placed at a temperature of 40 °C into the thermal chamber with forced air circulation (Memmert GmbH + Co. KG, Schwabach, Germany) for 14 days (stressed tablets).

2.3 Hyperspectral Imaging

To analyze the homogeneity of the expired and unexpired effervescent tablets, a hyperspectral analysis was performed using a Specim IQ hyperspectral camera (Spectral Imaging Ltd., Oulu, Finland). The images with a resolution of 512×512 pixels were obtained within the wavelength range from 400 nm to 1030 nm (i.e. across 204 spectral bands). The analyses of images, conversion of raw data into a matrix, as well as extraction of the selected features were performed using MATLAB Version 7.11.0.584 (R2010b) software.

To evaluate the homogeneity of the ingredients in the effervescent tablets, the hyperspectral profiles were compared to one another (unexpired vs expired, unexpired vs stressed, expired vs stressed), and the difference between the maximum and minimum reflectance of the hyperspectral profile was assessed. The greater the difference, the lower the homogeneity.

2.4 X-ray Microtomography

Two tablets of each analyzed type were scanned using an X-ray microtomography scanner (Phoenix vItomeIx, GE Sensing & Inspection Technologies GmbH, Wunstorf, Germany). The Phoenix DatosIx 2.0 software was used for image acquisition and image reconstruction.

The tablets were scanned at a voltage of 180 kV. A total of 2000 scans were recorded for each tablet, with a total scan time of 50 min and an amperage of 100 μA. To obtain the maximum image resolution, the appropriate distance between the sample and the matrix was established. Projections were acquired every 0.18°, with a total object rotation of 360°. The images with optimal contrast were registered with a resolution of 25 μm.

As the absorption of X-rays by an analyzed object is proportional to its density, we measured the samples' densities in a microtomographic image using the gray-scale level to compare unexpired, expired, and stressed tablets. To establish the grayscale level of the reference density, a calibration phantom (Micro-CT HA Phantom D32) was simultaneously scanned under the same conditions as the analyzed tablets. We assumed that "bright" pixels represented high-density areas, while "dark" pixels represented low-density areas. The average brightness value of pixels for the region of interest (ROI) was analyzed using the ImageJ software (ImageJ 1.53a; National Institutes of Health, USA). We determined the density vs mean intensity calibration curve and, consequently, we were able to determine the density of any area of the tablet slice.

2.5 Statistical Analysis

The obtained data were statistically analyzed using the Statistica 13.0 software (Stat-Soft; Statistica, Tulsa, OK, USA). Data were expressed as means (M) and standard deviations (SD). The normality of the distribution of quantitative data was evaluated by the Shapiro – Wilk W test. Mean values of quantitative data between the analyzed subgroups were carried out using the following tests: analysis of variance (ANOVA), when the distribution of data did not differ from the normal distribution, or the Kruskal-Wallis test when the distribution of quantitative data differed from the normal distribution. When the differences between the tablets were significant in ANOVA, Scheffe's post hoc test was used for pairwise comparisons. When the Kruskal-Wallis test showed a significant difference, the Mann-Whitney U test was used for post hoc comparisons. The results with $p \leq 0.05$ turned out to be statistically significant.

3 Results

3.1 Characteristics of the Tablets

All tablets were round in shape with a flat top and bottom surfaces. The tablets were yellow, and no discoloration, impurities, chips, or cracks were visible to the naked eye. All edges were preserved, with no visible damage.

Table 1. Parameters of the analyzed effervescent tablets.

Type of the effervescent tablets	Weight [g] M ± SD	Thickness [mm] M ± SD	Diameter [mm] M ± SD
Unexpired	4.227 ± 0.015	5.7 ± 0.05	25.3 ± 0.01
Expired	4.233 ± 0.016	5.7 ± 0.01	25.3 ± 0.01
p	0.307	0.551	0.259

M-mean; SD-standard deviation.

No differences in mean weight, thickness as well as diameter were observed between the unexpired and expired effervescent tablets (Table 1).

The mean values of the hardness factor, the force needed to crush the tablet, and the disintegration time are shown in Table 2.

The expired effervescent tablets with quercetin disintegrated in a significantly shorter time than unexpired tablets (44 s vs 47 s, $p = 0.013$). In turn, the mean force needed

Table 2. Comparison of the pharmacopoeial parameters, i.e. hardness factor, the force needed to crush the tablet, and disintegration time between unexpired and expired effervescent tablets

Type of the effervescent tablets	The force needed to crush the tablet [N] M ± SD	Hardness factor [N/m^2] M ± SD	Disintegration time [s] M ± SD
Unexpired	85.2 ± 3.15	5.93*10^5 ± 2.16*10^4	47.4 ± 1.68
Expired	106.0 ± 2.39	7.36*10^5 ± 1.57*10^4	44.5 ± 1.09
p	**0.002**	**0.002**	**0.013**

M-mean; SD-standard deviation. Significant differences are bolded.

to crush the tablet as well as the mean hardness factor were higher for expired than for unexpired tablets ($p < 0.001$ each). The hardness of the tablets did not correlate with the disintegration time both in unexpired and expired tablets.

The moisture content was the same in expired tablets as in unexpired tablets (0.2% each).

3.2 The Analysis of Tablets' Homogeneity Using Hyperspectral Imaging

The hyperspectral profiles of reflectance for unexpired, expired and stressed effervescent tablets are demonstrated in Fig. 1.

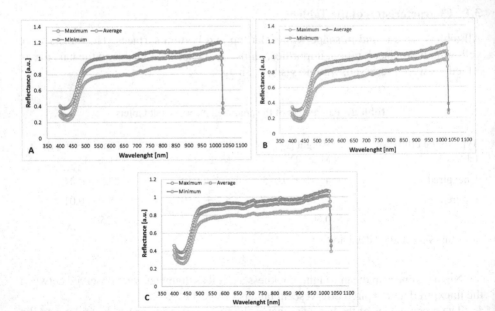

Fig. 1. Hyperspectral profiles of the tested tablets: A – unexpired; B – expired; C – stressed at 40 °C for 14 days.

In each image, the blue curve is the maximum reflectance, the orange/red curve is the average reflectance, and the grey curve corresponds to the minimum reflectance. The mean difference between the maximum and minimum value of reflectance corresponds to the homogeneity; the greater the difference, the lower the homogeneity.

We observed that the difference between the maximum and minimum reflectance varied significantly between the three analyzed types of tablets ($p < 0.001$). In general, the stressed tablets had the lowest mean difference between maximum and minimum reflectance compared to unexpired and expired tablets (0.145 vs 0.199 and 0.145 vs 0.198, respectively) The post hoc analysis demonstrated that significant differences were observed when the following pairs of effervescent tablets were compared: unexpired vs stressed ($p < 0.001$) and expired vs stressed ($p < 0.001$). Therefore, the homogeneity of ingredient distribution was similar between the unexpired and expired effervescent tablets containing quercetin but stressed tablets showed better homogeneity.

In addition, we analyzed reflectance obtained in the hyperspectral analysis within two spectral ranges, i.e. visible light (400–698 nm) and near-infrared (701–1030 nm). Figure 2 demonstrates the mean values of reflectance within the distinguished spectral ranges. As expected, the highest mean values of reflectance in both visible light and near-infrared were found for unexpired tablets. In turn, expired tablets showed the lowest mean values of reflectance within the visible light but within near-infrared, the stressed tablets showed the lowest value of reflectance.

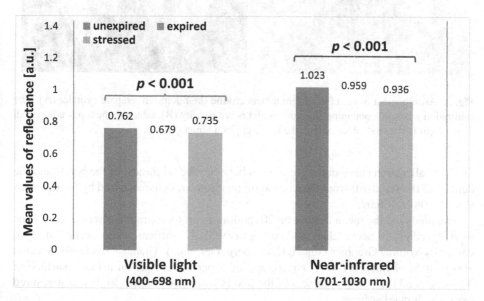

Fig. 2. Mean values of reflectance obtained in the hyperspectral imaging within the distinguished spectral ranges (i.e. visible light and near-infrared) between unexpired, expired and stressed tablets with quercetin.

In the post hoc analysis, the mean reflectance within the range of visible light differed significantly for the following pairs of tablets: unexpired vs stressed (p = 0.001), expired vs stressed (p = 0.036), and unexpired vs expired (p < 0.001). In turn, the mean difference between maximum and minimum reflectance differed between unexpired vs stressed (p < 0.001) and expired vs stressed (p < 0.001). Similar differences were demonstrated in the case of analysis in near-infrared ranges.

3.3 The Analysis of Inner Homogeneity Using X-ray Microtomography

The calibration phantom having areas with known reference density (W1 = 1.13 g/cm^3, W2 = 1.16 g/cm^3, W3 = 1.26 g/cm^3, W4 = 1.65 g/cm^3, and W5 = 1.90 g/cm^3) was scanned together with the analyzed effervescent tablets with quercetin. Figure 3 shows the arrangement of the effervescent tablets around the microtomographic cylinder as well as the calibration phantom.

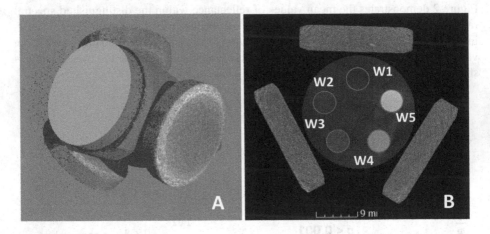

Fig. 3. Arrangement of the effervescent tablets around the microtomographic cylinder (**A**). The calibration phantom containing standards with known density (**B**): tablet on the top is unexpired, on the right is stressed tablet, and on the left is expired tablet.

The calibration curve of the correlation between the brightness of the pixels and the density of the standards from the calibration phantom was characterized by the equation y = 181.08x − 166.09.

We analyzed the mean density on 20 random microtomographic slices of each type of analyzed effervescent tablet made using myVGL 3.1 software (Viewer for Data Processed by Volume Graphics GmbH, Heidelberg, Germany). Then, we randomly selected 100 regions of interest (ROIs; an average of 5 per slice) with an average surface of 3.50 cm^2, and the mean brightness of the pixels within the marked ROIs was measured using the ImageJ software.

The mean brightness of the pixels as well as the mean densities of the tablets' inner structure are presented in Table 3.

Table 3. The mean density of the inner structure of the analyzed effervescent tablets.

Type of the Tablet	Brightness (a.u.), M ± SD	Density (g/cm^3), M ± SD
unexpired	93.748 ± 3.240	1.435 ± 0.018
expired	90.031 ± 2.981	1.414 ± 0.016
stressed	93.375 ± 4.659	1.433 ± 0.026
p	**< 0.001***	**< 0.001***

Significant differences are in bold. * In the post hoc analysis the unexpired and stressed tablet densities were significantly higher than the expired tablet density ($p < 0.001$ each).

The lowest mean density was found in the case of expired tablets (1.414 g/cm^3) while unexpired and stressed tablets showed comparable density (1.435 g/cm^3 and 1.433 g/cm^3, respectively). Thus, unexpired and stressed tablets had better homogeneity than expired tablets.

4 Discussion

Drug stability describes its ability to maintain physical, chemical, therapeutic, and microbiological properties during storage [14]. During the shelf life of a pharmaceutical preparation which is guaranteed by the manufacturer, the product remains stable and therefore maintains its quality and purity. However, it must be stored in accordance with the storage conditions stated on the label [15]. The Food and Drug Administration (FDA) requires stable data from pharmaceutical companies submitting applications for new drugs in order to establish shelf-life specifications [14]. The FDA verifies that the applicant's proposed expiry date is supported by appropriate studies conducted by the applicant.

In this study, the stability of effervescent tablets containing quercetin, vitamin B12, folic acid, and selenium was assessed using selected pharmacopoeial methods, and modern analytical methods, i.e. hyperspectral imaging and computed microtomography. Parameters of expired tablets stored at ambient temperature as well as tablets stressed in an elevated temperature for 14 days were compared with the parameters of unexpired tablets. It was shown that effervescent tablets, both unexpired and expired, met the pharmacopoeial requirements, and their weight, thickness, and diameter did not differ from one another. The expired tablets were characterized by a significantly shorter disintegration time and a higher value of the hardness coefficient compared to unexpired ones.

The basic element of pharmaceutical products is their homogeneity [16]. This parameter can be assessed using different novel techniques, including laser direct infrared reflectance imaging (LDIR), X-ray microtomography, or hyperspectral analysis [16–18]. Sacré et al. [16] evaluated the homogeneity of tablets made from a typical composition of two API and three excipients at different mixing time points using both Raman and LDIR. The authors obtained similar conclusions for both techniques, but only 7.5 min were needed to achieve the distribution map of acetylsalicylic acid with a step size of 100 μm for the LDIR method; in the case of Raman imaging, the total analysis lasted 4 h per tablet [16]. In our study, this parameter was tested using the hyperspectral camera in the wavelength range from 400 nm to 1030 nm. Surprisingly, the hyperspectral profiles showed that stressed tablets had better homogeneity than unexpired and expired tablets which may be related to the loss of moisture during storage at 40°. In turn, since the unexpired and expired tablets had the same moisture content (0.2% each) the homogeneity between these two types of tablets was spectroscopically comparable. The values of reflectance in the visible light (400–698 nm) and near-infrared (701–1030 nm) range were observed significantly higher ($p < 0.001$) for unexpired tablets compared to both expired and stressed tablets. This may indicate that a greater amount of light beam is transmitted into the expired and stressed tablets due to some physicochemical changes that occurred during the storage.

On the other hand, analysis of the tablet's inside in X-ray microtomography showed that all three analyzed tablets differed in average density. The densities of both the unexpired and stressed tablets were significantly higher than the density of the expired tablet. Thus, we may assume that expired tablets had lower homogeneity.

The rapid detection of physical changes which occurred in the medicines on the market is a new challenge for international agencies and inspectors of drug wholesalers/drug distributors. However, drug quality control is expensive and time-consuming. The solution to these inconveniences may be overwhelmed by modern spectral methods, including in the form of portable devices, for quick, screening assessment of various drug forms, also in the field.

Recently, Zambrzycki et al. [19] tested twelve technically diverse devices, from disposable single-use tests to portable spectrometers, and observed that all these devices showed high sensitivity (from 91.5 to 100.0%) in identifying medicines with no API or with the incorrect API. In the study by Ma et al. [20] distributional homogeneity index (DHI) was proposed to assess the distributional homogeneity of commercial tablets containing chlorpheniramine maleate. The authors visualized the distribution of the tablets from six brands with the use of near-infrared chemical imaging (NIR-CI). In turn, Zhu et al. [21] used X-ray microscopy to analyze both the particle size distribution of API agglomerates and an excipient and to detect and qualitatively evaluate porosity. The analysis of the images performed by the authors revealed regions of intra-particle porosity in the API particles. Another study also investigated the tablets' microstructure by X-ray microtomography, including pore size distribution, homogeneity, and porosity [17]. The authors observed however that X-ray microtomography can characterize tablet porosity only for the tablets with an ibuprofen concentration of 50 wt% while not for lower ibuprofen concentrations since the contrast between void phase and ibuprofen was too low.

The present study is one of the few studies that indicate the possibility of using new spectroscopic techniques and computed microtomography for the analysis of pharmaceutical preparations. Undoubtedly, the strength of the study is the fact that for the first time, hyperspectral analysis has been used to evaluate drugs during storage. However, the study has some limitations. The conducted research is preliminary. Further intensive studies on the application of new spectral methods in drug quality control are necessary. In the future, we plan further measurements for solid drug forms with a longer storage period at elevated temperatures and additional parameters that may be important for tablet stability, i.e. UV radiation and increased humidity using the methods mentioned above, as well as hemispherical directional reflection.

5 Conclusion

It has been shown that unexpired and expired effervescent tablets met the pharmacopoeial requirements. Unexpired and expired tablets differed in terms of homogeneity assessed with hyperspectral analysis. Expired tablets showed significantly lower reflectance in the visible and near-infrared range than unexpired tablets which may indicate greater transmission of the light beam into the inner of the expired tablets. In the X-ray microtomography analysis, the density of unexpired tablets was significantly higher than the density of expired tablets. The changes in tablets' homogeneity may indicate possible physical changes in the effervescent tablets during storage, especially in conditions deviating from those recommended by the manufacturer.

Funding. The research was funded by Medical University of Silesia in Katowice, Poland (project number PCN-1-058/K/2/O).

Conflicts of Interest. The authors declare no conflict of interest.

References

1. Herlina, Kuswardhani, N., Belgis, M., Tiara, A.: Characterization of physical and chemical properties of effervescent tablets temulawak (curcuma zanthorrhiza) in the various proportion of sodium bicarbonate and tartaric acid. In: E3S Web of Conferences, vol. 142, p. 03006 (2020)
2. Patel, S.G., Siddaiah, M.: Formulation and evaluation of effervescent tablets: a review. J. Drug Deliv. Ther. **8**(6), 296–303 (2018)
3. Aslani, A., Eatesam, P.: Design, formulation and physicochemical evaluation of acetaminophen effervescent tablets. J. Reports Pharm. Sci. **2**(2), 140–149 (2013)
4. Rosch, M., Lucas, K., Al-Gousous, J., Pöschl, U., Langguth, P.: Formulation and characterization of an effervescent hydrogen-generating tablet. Pharmaceuticals **14**(12), 1327 (2021)
5. Veronica, N., Liew, C.V., Heng, P.W.S.: Insights on the role of excipients and tablet matrix porosity on aspirin stability. Int. J. Pharm. **580**, 119218 (2020)
6. Elmasry, G., Kamruzzaman, M., Sun, D.W., Allen, P.: Principles and applications of hyperspectral imaging in quality evaluation of agro-food products: a review. Crit. Rev. Food Sci. Nutr. **52**(11), 999–1023 (2012)
7. da Silva, Í.B.: X-ray computed microtomography technique applied for cementitious materials: a review. Micron **107**, 1–8 (2018)
8. Landis, E.N., Keane, D.T.: X-ray microtomography. Mater. Charact. **61**(12), 1305–1316 (2010)
9. Andres, S., et al.: Safety aspects of the use of quercetin as a dietary supplement. Mol. Nutr. Food Res. **62**(1), 1–15 (2018)

10. Reyes-Farias, M., Carrasco-Pozo, C.: The anti-cancer effect of quercetin: molecular implications in cancer metabolism. Int. J. Mol. Sci. **20**(13), 1–19 (2019)
11. Ovaskainen, M.L., et al.: Dietary intake and major food sources of polyphenols in Finnish adults. J. Nutr. **138**(3), 562–566 (2008)
12. Pérez-Jiménez, J., et al.: Dietary intake of 337 polyphenols in French adults. Am. J. Clin. Nutr. **93**(6), 1220–1228 (2011)
13. Polish Pharmacopoeia, 12th ed. The President of the Office for Registration of Medicinal Products, Medical Devices and Biocidal Products, Warsaw (2020)
14. International Conference on Harmonisation of Technical Requirements for Registration of Pharmaceuticals for Human Use. Stability Testing of New Drug Substances and Products, Q1A(R2); ICH Harmonized Tripartite Guideline: Geneva, Switzerland (2003). https://www.ema.europa.eu/en/documents/scientific-guideline/ich-q-1-r2-stability-testing-new-drug-substances-products-step-5_en.pdf. (Assessed on 19.02.2023)
15. SUPAC-MR: Modified Release Solid Oral Dosage Forms; ScaleUp and Post-Approval Changes: Chemistry, Manufacturing and Controls. Vitro Dissolution Testing, and In Vivo Bioequivalence Documentation; Guidance for Industry, U.S. Department of Health and Human Services, Food and Drug Administration, Center for Drug Evaluation and Research (CDER), U.S. Government Printing Office: Washington, DC (1997). https://www.fda.gov/media/70956/download. (Assessed on 19.02.2023)
16. Sacré, P.Y., Alaoui Mansouri, M., De Bleye, C., Coïc, L., Hubert, P., Ziemons, E.: Evaluation of distributional homogeneity of pharmaceutical formulation using laser direct infrared imaging. Int. J. Pharm. **612**, 121373 (2022)
17. Schomberg, A.K., Diener, A., Wünsch, I., Finke, J.H., Kwade, A.: The use of X-ray microtomography to investigate the microstructure of pharmaceutical tablets: potentials and comparison to common physical methods. Int. J. Pharm. X **3**, 100090 (2021)
18. Sarecka-Hujar, B., Szulc-Musioł, B., Meisner, M., Duda, P.: The use of novel, rapid analytical tools in the assessment of the stability of tablets—a pilot analysis of expired and unexpired tablets containing nifuroxazide. Processes **10**(10), 1934 (2022)
19. Zambrzycki, S.C., et al.: Laboratory evaluation of twelve portable devices for medicine quality screening. PLoS Negl. Trop. Dis. **15**(9), e0009360 (2021)
20. Ma, L., et al.: Investigation of the distributional homogeneity on chlorpheniramine maleate tablets using NIR-CI. Spectrochim. Acta A Mol. Biomol. Spectrosc. **204**, 783–790 (2018)
21. Zhu, A., et al.: Investigation of quantitative x-ray microscopy for assessment of API and excipient microstructure evolution in solid dosage processing. AAPS PharmSciTech **23**(5), 117 (2022). https://doi.org/10.1208/s12249-022-02271-3

The Effect of Biofeedback on Learning the Wheelie Position on Manual Wheelchair

Antonio Pinti[1]($\boxtimes$), Atef Belghoul[1], Eric Watelain[2], Zaher El Hage[3], and Rawad El Hage[4]

[1] Laboratoire LARSH DeVisu, Université Polytechnique Hauts-de-France (UPHF), Valenciennes, France
antonio.pinti@uphf.fr
[2] Laboratoire IAPS, UR 201723207F, Université de Toulon, Toulon, France
[3] Department of Psychology Education and Physical Education, Faculty of Humanities, Notre Dame University, Louaize, Lebanon
[4] Department of Physical Education, Faculty of Arts and Sciences, University of Balamand, Kelhat El-Koura, Balamand, Lebanon

Abstract. The aim of this study was to investigate the impact of biofeedback (BFB) on manual wheelchair learning. The researchers conducted training sessions with two groups of participants, one using BFB and the other group without it (NBFB). The hypothesis was that BFB would reduce the learning time and help participants to achieve balance positions more quickly. The study enrolled 24 participants aged 24 ± 6 years old; they were divided into two groups of 12 subjects each (BFB and NBFB). The researchers also collected additional information about the participants, such as the sport they practiced, for future investigations. The data was collected using a non-contact electronic angular system placed directly on the wheelchair, measuring spatiotemporal parameters such as the angle between the wheelchair and the ground and the time at which this angle is reached. The results which are statistically significant ($p < 0.05$) were only obtained between early falling, learning time and number of trials. The study found that BFB did not seem to accelerate the learning time for the wheelie skill on manual wheelchair (BFB group). However, the BFB method could potentially reduce the number of trials using the manual wheelchair under (NBFB). In conclusion, the study showed that biofeedback may not necessarily accelerate the learning time for the wheelie skill on manual wheelchair but can help individuals to maintain balance positions with fewer trials. Further studies are required to confirm these results, as they only involved a small sample size. This study highlights the potential for using biofeedback as an effective tool for wheelchair training and could improve the quality of life of individuals with mobility impairments.

Keywords: Wheelie · Biofeedback · Motor Learning · Wheelchair · Manual Wheelchair

I. Rojas et al. (Eds.): IWBBIO 2023, LNBI 13919, pp. 283–291, 2023.
https://doi.org/10.1007/978-3-031-34953-9_22

1 Introduction

The wheelie position is used to overcome an obstacle in the environment which limits the mobility of the wheelchair user. To perform this, the user must intentionally lift the front wheels while the rear wheels remain in contact with the ground (Kirby et al., 2006). This skill is considered essential and must be carefully taught. In addition, the risk of falling while using wheelchairs is great when maintaining a balanced position as the back remains unstable, which increases the fear of falling. One hypothesis of the study is that the fear of falling increases the learning time. Feeling confident can encourage the person to take risks and try things out and therefore master the wheelie position.

Kirby et al. (2004) reported 14 daily tasks in which the ability to perform the wheelie was useful. In addition, Bullard and Miller (2001) showed that the wheelie balance position improved the performance of subjects in an obstacle course. A preliminary study conducted by Medodjou et al. (2018) investigated the effect of different visual cues on maintaining balance in a manual wheelchair. Maintaining balance in the wheelie position can be challenging. Hence, visual cues can improve maintaining postural balance in the wheelie position.

According to Kirby et al. (2004), most wheelchair users do not master this skill, and there are at least 3 explanations for this. Many wheelchair users are elderly and find that the two-wheeled (wheelie) balance position is too dangerous or too difficult to learn. Also, many clinicians are unable to perform the maneuver and may therefore lack the knowledge or confidence to teach it. In addition, there is insufficient data in the literature regarding the acquisition of two-wheeled skills. The problem investigated in this study therefore aims to improve the mobility options of wheelchair users, thereby increasing their autonomy in everyday life and reducing the learning time for this motor ability. In detail, our research work aims to determine how to help this population to increase its success rate in the two-wheeled position and to understand how learning can be impacted by BFB, in particular on difficulty and learning time.

The use of biofeedback began in the field of psychophysiology in the late 1960s. Suwillo (1986) defined psychophysiology as "a scientific study of the relationship between mental and behavioral activities and bodily events".

Therefore, the relevance of biofeedback interventions for sports preparation based on the psychophysiological aspect presented by Green and Walters (1970) is still not clear.

Biofeedback (BFB) can be introduced by the use of special electronic devices with electrodes and sensors to assess, monitor, and return psychophysiological information to a person. The main idea of BFB is to provide the individual with information about their body or mind reactions to various situations (Schwartz, 1979). Today, thanks to modern technology, it is possible to measure physiological indicators such as heart rate, muscle activity, brainwave activity, respiration, blood pressure and skin temperature for a better understanding of this type of studies.

The proprioceptive awareness necessary to modify habitual movements (Freedman 1968) without any form of external assistance is often absent in patients suffering from long-term movement disorders. This is where biofeedback is particularly useful, as it increases the patients' awareness of their movements using external stimuli. Schmidt (1975) stated that "motor memory is the product of learning". We have found that this learning process can be accelerated using sports technology and biofeedback. Biofeedback can be given in the form of kinematic and kinetic parameters, providing the subject with information on speed, acceleration, force and tension. Here we use biofeedback to put the participant in their balanced position and stimulate proprioceptive feedback.

2 Material and Methods

2.1 Subjects and Study Design

This study involved 24 healthy participants (16 men and 8 women) divided into two learning groups. Each group included 12 participants. All were informed of the experimental conditions and signed a consent in accordance with the Helsinki Declaration.

Participants were aged 24 ± 6 years old and had no prior experience with using a wheelchair. The inclusion criteria were: untrained participants who were willing to learn the wheelie.

The participants had a normal weight (BMI 18.5 to 30) and did not have any neuromuscular pathology or took medication that could affect their ability to maintain balance on two wheels.

The participants were free from all musculoskeletal disorders, hearing disorders, or back pain. In addition, information regarding practiced sports was collected by using a questionnaire. The work described has been carried out in accordance with the declaration of Helsinki (regarding human experimentation developed for the medical community by the World Medical Association).

2.2 Materials

A standard manual wheelchair 'Invacare Action 4 NG' is used by all participants in this study (Fig. 1). The wheelchair has the following features: seat width: 46 cm; seat depth: 41 cm; seat height: 40 cm; Seat height/Floor: 44 cm; height of the FRM from the ground: 45 cm; Rear wheel diameter: 59 cm; max user weight: 125 kg.

An electronic angular system: An angular measuring system (mini-computer, screen, sensor) without contact on Raspberry Pi 4 (Fig. 2) is connected to a telemetric sensor, adaptable to a wheelchair. The touch screen allows the operation and visualization of the system's operating status and the measured angle.

The angle was measured in real time and stored in memory. This system allowed us to export the recorded data intuitively on a USB key. The data was later processed in Excel (Pinti et al., 2022).

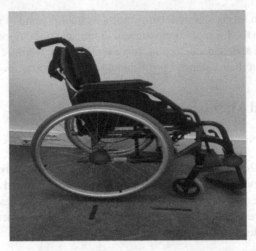

Fig. 1. Manual Wheelchair 'Invacare Action 4 NG'

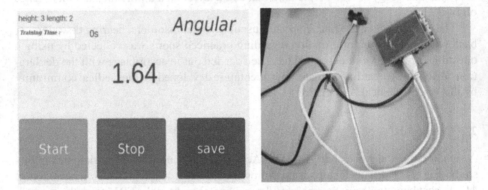

Fig. 2. Electronic Angular system 'Raspberry Pi 4'

A laptop was used to film the participants during learning (HP Notebook 650 G1,39.I5) in addition to a professional webcam (Full HD with a tripod support). The laptop was equipped by a video analysis software to observe and analyze learning (Kinovea version 0.9.52) (Fig. 3).

A mobile application: To schedule the learning sessions, a user friendly application was created for this study. It included several parameters to set: Warm-up time, number of training sessions, training duration, and rest duration.

Fig. 3. The laptop & professional webcam

2.3 Methods

The experiments took place at the University of Valenciennes "UPHF", France, during the COVID crisis where a strict health protocol was in place. Wearing a mask and respecting social distancing were mandatory during the entire protocol. Also, an alcohol-based gel was available for participants to clean their hands with. The participants filled out a consent form explaining the objective, protocol, and progression of the experiment. Once completed, a profile was created for the participant on an Excel file; for privacy reasons, the participant files were renamed. Before starting the acquisitions, the participant was invited to warm up for 2 min within a 3-m perimeter, by making back and forth trips. During the warm up, the participants had both wheels on the ground; they were free to talk and to try moving in all directions to familiarize themselves with the wheelchair. The average duration of a session was 30 min. If the participant was unable to perform 2 sets of 30 s in the wheelie position during the 1st session, one or several 30-min session(s) were performed between 1 and 4 days later depending on availability. According to Bonaparte (2001), the starting instructions were that participants must "roll slightly back and then quickly forward to lift your front wheels off the ground." This explanation was accompanied by instructions. The demonstration was performed by the examiner and the instructions were counted in the total training time to ensure a realistic representation of the total training time. Both groups followed the same training procedures and received the same instruction from the researcher. A learning session lasted about 30 min with 3 min learning periods followed by 30 s rests. During the rest period, the participants of the experimental group who trained with biofeedback were placed in their balance position while being told "You are in a balanced state" (Fig. 4). This way, they could feel and improve their balance position safely. During the learning, the researcher stood behind the wheelchair to ensure the safety of the subject by blocking the chair in case of tipping. It is important to note that the warm-up was not included in the training time. At the end of the learning session, the participant completed perceived exertion according to the scale of Borg.

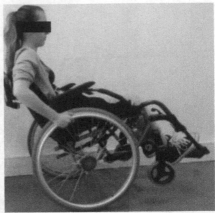

Fig. 4. The wheelie position

2.4 Statistical Analysis

The data was processed on excel, calculating the mean and standard deviation of the variables such as the number of trials per participant, the total learning time, the time when participants first fell and the angles. The T-student test was used for the comparisons between the 2 groups BFB and NBFB. Spearman correlation coefficients were calculated in order to explore relationships among the studied variables. A level of significance of $p < 0.05$ was set. Statistical analyses were performed using the SigmaStat 3.1 Program (Jandel Corp., San Rafael, CA).

3 Results

There were no significant differences between the two groups regarding the following parameters: number of falls, time to first fall, number of trials, total time and perceived exertion (Table 1).

Table 1. Mean values in the two groups (NBFB: Non-biofeedback group; BFB: Biofeedback group).

	NBFB group (n = 12)	BFB group (n = 12)	
	Mean ± SD	Mean ± SD	p-value
Number of falls	11.66 ± 13.43	14.25 ± 12.38	0.62
Time to first fall (min)	10.3 ± 12.8	6.5 ± 9.2	0.43
Number of trials	154.5 ± 113.4	107.5 ± 67.9	0.23
Total time (min)	35.6 ± 11.4	36.5 ± 13.9	0.86
Perceived Exertion (scale 0–20)	12.3 ± 2.3	12.0 ± 2.6	0.74

In the NBFB group, total time was positively correlated to the number of trials while perceived exertion was positively correlated to the number of trials and total time (Table 2).

Table 2. Correlation matrix among the different parameters in the NBFB group (*p < 0.05; **p < 0.01; ***p < 0.001).

NBFB group (n = 12)	Number of falls	Time to first fall (min)	Number of trials	Total time (min)	Perceived Exertion (scale 0–20)
Number of falls	1				
Time to first fall (min)	−0.22	1			
Number of trials	0.28	0.42	1		
Total time (min)	0.48	0.21	**0.77****	1	
Perceived Exertion (scale 0–20)	0.17	0.35	**0.85*****	**0.63***	1

In the BFB group, the number of falls was positively correlated to the number of trials while total time was positively correlated to the time of first fall and the number of trials (Table 3).

Table 3. Correlation matrix among the different parameters in the BFB group (*p < 0.05; **p < 0.01; ***p < 0.001).

BFB group (n = 12)	Number of falls	Time to first fall (min)	Number of trials	Total time (min)	Perceived Exertion (scale 0–20)
Number of falls	1				
Time to first fall (min)	−0.19	1			
Number of trials	**0.73****	0.44	1		
Total time (min)	0.43	**0.58***	**0.86*****	1	
Perceived Exertion (scale 0–20)	−0.34	−0.17	−0.32	−0.03	1

In the whole population, the number of trials was positively correlated to the number of falls and the time to first fall while total time was positively correlated to the number of falls, time to first fall and the number of trials (Table 4).

Table 4. Correlation matrix among the different parameters in the whole population (*p < 0.05; **p < 0.01; ***p < 0.001).

Whole population (n = 24)	Number of falls	Time to first fall (min)	Number of trials	Total time (min)	Perceived Exertion (scale 0–20)
Number of falls	1				
Time to first fall (min)	−0.17	1			
Number of trials	**0.43***	**0.46***	1		
Total time (min)	**0.43***	**0.46***	**0.80*****	1	
Perceived Exertion (scale 0–20)	−0.08	0.05	0.18	0.21	1

4 Discussion

Participants under BFB condition do not learn any faster than those using simple learning methods, with timing of 36 min and 58 s for the BFB group and 35 min and 20 s for the NBFB group. However, we were able to reveal a relationship between the first fall and the trials repetition for the BFB condition. This relationship was not found under the NBFB condition. Thus, participants who fall early complete their training more quickly than those who fear falling. Knowing that there is no risk of injury from falling increases confidence and allows participants to take more risks and make more attempts. In addition, participants who learned to use the BFB method needed fewer attempts to succeed, with 107 attempts compared to 147 for the other group.

During this study, we noticed that the participant profiles varied widely and did not allow us to correlate sports to wheelie performance. A larger sample would likely reveal a better result. The fall detection is also subject to debate, as the angle chosen to detect a fall is 45°. Most of the time, when participant exceed this angle, they do fall, but it is possible for atypical participants or those with particular behaviors to recover the seat without falling. We could not increase this angle since some participants were quickly caught. Moreover, the exceeded angles depend on participants' anthropometrics and morphology.

In order to improve this study, a larger sample size is needed. In addition, participants practicing the exact same sports could help to reduce the impact of a profile's effect.

5 Conclusion

This study revealed that using a biofeedback training method as opposed to a simple method is not more helpful for faster learning. However, using the biofeedback method can reduce the number of attempts to learn the wheelie. The study also shows that participants who fell early during training completed the training more quickly than those who were afraid of falling. For future studies, we suggest using a larger sample size and considering the sports which are practiced by the participants in order to draw a final conclusion.

Disclosure of Interest

None of the authors reported a conflict of interest related to the study.
The authors would like to thank all the participants in this study.

References

Kirby, R.L., Smith, C., et al.: The manual wheelchair wheelie: a review of our current understanding of an important motor skill. Disabil. Rehabil. Assist. Technol. **1**(1–2), 119–127 (2006)

Medodjou, A.-O., Watelain, E., Pinti, A., Faupin, A.: Effect of different visual cues on balance maintenance in manual wheelchair users on two wheels: a preliminary study. Congress of the Francophone Association in Adapted Physical Activities (AFAPA) Sci. Sports **33**, S26–S27 (2018)

Kirby, R.L., Dupuis, D.J., MacPhee, A.H., et al.: The wheelchair skills test (version 2.4): measurement properties. Arch. Phys. Med. Rehabil. **85**(5), 794–804 (2004)

Bullard, S., Miller, S.: Comparison of teaching methods for learning a wheelchair tilt and balance skill. Percept. Mot. Skills **93**(1), 131–138 (2001)

Schwartz, G.E.: Dysregulation and systems theory: a biobehavioral framework for biofeedback and behavioral medicine. In: Birbaumer, N., Kimmel, H.D. (eds.) Biofeedback and Self-Regulation, pp. 19–48. Lawrence Erlbaum Associates, Inc., Hillsdale, New Jersey (1979)

Pinti, A., Belghoul, A., Watelain, E., El Hage, R.: Measurement system of the sun-seat angle for learning manual wheelchair two-wheeling: preliminary study. In: 12th Handicap Conference, ISBN 978-2-9571218-2-3, June 8–10, Paris, France, pp. 180–183 (2022). (in French)

Freedman, S.J.: The Neuropsychology of Spatially Oriented Behavior, Chapter 3, pp. 37–55. Dorsey Press, Homewood, Illinois (1968)

Schmidt, R.A.: A schema theory of discrete motor skill learning. Psychol. Rev. **82**(4), 225–260 (1975)

Preliminary Study on the Identification of Diseases by Electrocardiography Sensors' Data

Rui João Pinto[1], Pedro Miguel Silva[1], Rui Pedro Duarte[1],
Francisco Alexandre Marinho[1], António Jorge Gouveia[1], Norberto Jorge Gonçalves[1],
Paulo Jorge Coelho[2,3], Eftim Zdravevski[4], Petre Lameski[4], Nuno M. Garcia[5,6],
and Ivan Miguel Pires[5,7(✉)]

[1] Escola de Ciências E Tecnologia, University of Trás-Os-Montes E Alto Douro, Quinta de Prados, 5001-801 Vila Real, Portugal
{al70648,al70649,al70650,al71518}@alunos.utad.pt, njg@utad.pt
[2] Polytechnic of Leiria, Leiria, Portugal
paulo.coelho@ipleiria.pt
[3] Institute for Systems Engineering and Computers at Coimbra (INESC Coimbra), Coimbra, Portugal
[4] Faculty of Computer Science and Engineering, University Ss Cyril and Methodius, 1000 Skopje, Macedonia
{eftim.zdravevski,petre.lameski}@finki.ukim.mk
[5] Instituto de Telecomunicações, 6201-001 Covilhã, Portugal
impires@it.ubi.pt
[6] Faculdade de Ciências, University of Lisboa, 1749-016 Lisbon, Portugal
nmgarcia@fc.ul.pt
[7] Polytechnic Institute of Santarém, Santarém, Portugal

Abstract. An electrocardiogram (ECG) is a simple test that checks the heart's rhythm and electrical activity and can be used by specialists to detect anomalies that could be linked to diseases. This paper intends to describe the results of several artificial intelligence methods created to automate identifying and classifying potential cardiovascular diseases through electrocardiogram signals. The ECG data utilized was collected from a total of 46 individuals (24 females, aged 26 to 90, and 22 males, aged 19 to 88) using a BITalino (r)evolution device and the OpenSignals (r)evolution software. Each ECG recording contains around 60 s, where, during 30 s, the individuals were in a standing position and seated down during the remaining 30 s. The best performance in identifying cardiovascular diseases with ECG data was achieved with the Naive Bays classifier, reporting an accuracy of 81.36%, a precision of 26.48%, a recall of 28.16%, and an F1-Score of 27.29%.

Keywords: Electrocardiography sensors · Cardiovascular Diseases · Disease Classification · Artificial Intelligence · Sensors

I. Rojas et al. (Eds.): IWBBIO 2023, LNBI 13919, pp. 292–304, 2023.
https://doi.org/10.1007/978-3-031-34953-9_23

1 Introduction

Cardiovascular diseases represent some of the leading causes of mortality worldwide [8, 38], and it is necessary to predict these health complications as early as possible to help reduce their mortality rate [5, 23]. For this purpose, we hold that the automatic analysis of electrocardiograms (ECG) should be used to help professionals diagnose cardiovascular diseases [14, 24, 26]. However, several kinds of diseases can be detected early using ECG data, where the relation of different parameters can be detected, such as diabetes and allergies [18, 33, 34, 36].

As presented in the literature, the ECG data can be used to detect a kind of diseases [2, 3, 32]. We verified the best-used methods in the literature to apply the different techniques for our data. However, there needs to be more code available in the literature, and it is an opportunity to promote the creation of a commodity diagnosis tool. Furthermore, some studies show that the authors reach high precision rates [16, 29, 37]. For this reason, we decided to conduct our study utilizing a previously acquired dataset [22], using some of the best-performing methods we could find in our research.

We employed eight different methods [1], including Nearest Neighbors, Linear SVM, RBF SVM, Decision Tree, Random Forest, Neural Net, AdaBoost, and Naive Bayes, to test their ability to classify ECG recordings into one of 8 different classes representing diseases, including Allergies, Hypertension, Cholesterol, Diabetes, Arrhythmia, Asthma, Unspecified heart problems, and Unspecified brain problems.

2 Methods

2.1 Study Design and Participants

The ECG recordings utilized during this study were collected from a total of 46 individuals belonging to the continental region of Portugal using a BITalino (r)evolution device [12] and the OpenSignals (r)evolution software [9]. Of these 46 subjects, 22 were male, aged 19 to 88 years old, with an average age of 58, and 24 were females, aged 26 to 90, with an average age of 59. The Ethics Committee from Universidade da Beira Interior approved the study with CE-UBI-Pj-2021-041. The dataset used in this research is publicly available at [22].

All the participants followed throughout this study disclosed at least one previously diagnosed health complication, such as allergies, hypertension, cholesterol, diabetes, arrhythmia, asthma, unspecified heart problems, and unspecified brain problems, with the most commonly available problems being, as seen in Table 1, hypertension, and diabetes. The process starts with the instrumentation of the individual with the ECG electrodes, and the data acquisition was initiated. The ECG recordings are around 60 s, during which the patient spends 30 s seated down and 30 s standing up. Detailed information about the data acquisition of the dataset this study was based on is available online [13]. For more details, it is recommended to the reader follow this publication.

Table 1. The number of times each disease was disclosed.

Diseases	Number of individuals
Allergies	6
Hypertension	19
Cholesterol	9
Diabetes	13
Arrhythmia	3
Asthma	4
Unspecified heart problems	6
Unspecified brain problems	3

2.2 Feature Extraction

For this study, we utilized the NeuroKit python package [40] to extract relevant features from the ECG recordings, such as P, Q, R, S, and T peaks and the onsets and offsets of P, T, and R waves. The remaining features used during this study were manually calculated based on the features that were automatically extracted:

- RR interval $\rightarrow PeakR_N - PeakR_{N-1}$
- PP interval $\rightarrow PeakP_N - PeakP_{N-1}$
- P duration $\rightarrow OffsetP - OnsetP$
- PR interval $\rightarrow OnsetR - OnsetP$
- PR segment $\rightarrow OnsetR - OffsetP$
- QRS duration $\rightarrow OffsetR - OnsetR$
- ST segment $\rightarrow OnsetT - OffsetR$
- ST-T segment $\rightarrow OffsetT - OffsetR$
- QT duration $\rightarrow OffsetT - OnsetR$
- TP interval $\rightarrow OnsetP - OffsetT$
- R amplitude $\rightarrow PeakR_N - PeakS_N$
- T amplitude $\rightarrow PeakT_N - PeakS_N$
- P amplitude $\rightarrow PeakP_N - PeakQ_N$

2.3 Data Classification

The data acquisition consisted of machine learning methods for classifying the mapped data on the dataset, where cross-validation techniques were used. Validation techniques are the most recommended in the literature [25]. A total of 8 different classification methods, such as Nearest Neighbors, Linear SVM, RBF SVM, Decision Tree, Random Forest, Neural Net, AdaBoost, and Naive Bayes, were applied to this study's dataset. Finally, the results were implemented using Python in a Jupyter Notebook [28].

Nearest Neighbors. K-Nearest Neighbors (K-NN) is a non-parametric algorithm that uses data with several classes to predict the classification of the new sample point [35].

This algorithm does not assume the studied data or use the training data points to make any generalization.

Linear SVM. Support Vector Machine (SVM) is a supervised learning algorithm that attempts to generate the best line or decision boundary to divide n-dimensional space into classes to get associated with the correct category when new data points are added [7]. The extreme points/vectors that help this algorithm generate the best decision boundary are called support vectors and give the name to the method [7].

RBF SVM. The Radial Basis Function (RBF) is the default kernel function used in various kernelized learning algorithms [27]. It is similar to the K-NN, so it overcomes the space complexity problem with the K-NN advantages [15]. However, the RBF SVM must store the support vectors during training [21].

Decision Tree. The decision tree algorithm is a rule-based supervised machine learning classifier that creates questions based on features of the dataset and, based on the answers, can assign new entries to their proper category [6]. It is a tree-based algorithm because every question it generates has a binary result, thus constantly dividing the database into halves [11].

Random Forest. The Random Forest is a Supervised Machine Learning Algorithm extensively used in classification and regression problems that construct decision trees on different samples at training time [4].

Neural Networks. The neural network consists of an input, a hidden, and an output layer [17, 20]. The last two are neurons acting as neurons and utilizing a nonlinear activation function. In addition, a multilayer perceptron uses backpropagation for training and can classify non-linearly separable data [30].

AdaBoost. The Adaboost is an ensemble learning method that combines the results of different classifier methods into a weighted sum to improve their efficiency and predictive power [19].

Naive Bayes. It is a probabilistic machine learning algorithm classifier based on the Bayes theorem [31, 39]. It is a simple algorithm without a complicated iterative parameter estimation, making it useful in medical science for diagnosing cardiac patients [10].

3 Results

The average values of the features that were measured during the feature extraction stage are presented in Table 2.

Table 3 compares the results of each classifier utilized during this study. The RBF SVM algorithm achieved the highest accuracy at 87.92%, Linear SVM achieved the highest precision and recall at 34.80%, and Naïve Bayes achieved far the highest F1-score at 27.29%.

Table 2. Average of features extracted.

Features	Average	Standard Deviation
RR Interval (ms)	797.68	120.42
PP Interval (ms)	797.33	119.97
P Duration (ms)	43.62	14.39
PR Interval (ms)	157.37	34.03
PR Segment Interval (ms)	114.76	37.33
QRS Duration (ms)	106.37	36.21
ST Segment Interval	171.07	52.59
ST-T Segment Interval (ms)	235.87	42.02
QT Duration (ms)	342.24	41.20
TP Interval (ms)	499.61	65.91
R Amplitude (mV)	268.96	142.22
T Amplitude (mV)	86.87	52.84
P Amplitude (mV)	60.71	41.20

Table 3. Performance comparison of the various classifiers.

Method	Accuracy (%)	Precision (%)	Recall (%)	F1-Score (%)
K-NN	80.00	20.00	20.00	20.00
Linear SVM	81.44	34.80	34.80	34.80
RBF SVM	87.92	18.72	18.72	18.72
Decision Tree	78.16	12.48	12.48	12.48
Random Forest	80.48	21.84	21.84	21.84
Neural Net	80.16	21.84	21.84	21.84
AdaBoost	80.56	21.84	21.84	21.84
Naive Bayes	81.36	26.48	28.16	27.29

The Nearest Neighbors algorithm applied to our dataset could correctly classify 2 out of the 6 cases of allergies and 4 out of the 19 cases of hypertension experienced by this study's participants. In addition, it achieved an accuracy of 80%, a precision of 20%, a recall of 20%, and an F1-score of 20%. More details of this method's performance are visible in the confusion matrix presented in Table 4.

Table 4. Confusion matrix of Nearest Neighbors classifier.

	allergies	hypertension	cholesterol	diabetes	arrhythmia	asthma	heart problems	brain problems
allergies	2	0	0	0	0	0	0	0
hypertension	1	4	0	2	1	0	0	0
cholesterol	2	4	0	0	0	0	0	0
diabetes	1	6	0	0	0	0	0	0
arrhythmia	0	1	0	0	0	0	0	0
asthma	1	1	0	0	0	0	0	0
heart problems	0	2	0	0	0	0	0	0
brain problems	0	2	0	0	0	0	0	0

As seen in Table 5, the Linear SVM classifier correctly identified 8 out of the 19 cases of hypertension present in our dataset, making it the method that identified the highest number of cases of any singular disease. Furthermore, this algorithm reached an accuracy of 81.44% and precision, recall, and F1-score of 34.8%.

Table 5. Confusion matrix of Linear SVM classifier.

	allergies	hypertension	cholesterol	diabetes	arrhythmia	asthma	heart problems	brain problems
allergies	0	1	0	0	0	0	0	0
hypertension	0	8	0	0	0	0	0	0
cholesterol	1	5	0	0	0	0	0	0
diabetes	0	7	0	0	0	0	0	0
arrhythmia	0	1	0	0	0	0	0	0
asthma	0	2	0	0	0	0	0	0
heart problems	0	3	0	0	0	0	0	0
brain problems	0	3	0	0	0	0	0	0

From the ECG recordings present in our dataset, the RBF SVM method was capable, as seen in Table 6, of accurately predicting 6 cases of hypertension. Furthermore, it achieved the highest accuracy of 87.92%, a precision of 18.72%, a recall of 18.72%, and an F1-score of 18.72%.

Table 6. Confusion matrix of RBF SVM classifier.

	allergies	hypertension	cholesterol	diabetes	arrhythmia	asthma	heart problems	brain problems
allergies	0	2	0	0	0	0	0	0
hypertension	0	6	0	2	0	0	0	0
cholesterol	0	6	0	0	0	0	0	0
diabetes	0	6	1	0	0	0	0	0
arrhythmia	0	1	0	0	0	0	0	0
asthma	1	1	0	0	0	0	0	0
heart problems	0	3	0	0	0	0	0	0
brain problems	0	3	0	0	0	0	0	0

The Decision Tree algorithm could identify the correct disease in 4 instances. Table 7 shows that the method correctly classified 3 of the recordings as belonging to someone with hypertension and one as belonging to someone with allergies. It achieved an accuracy of 78.16% and precision, recall, and F1-score of 12.48%.

Table 7. Confusion matrix of Decision Tree classifier.

	allergies	hypertension	cholesterol	diabetes	arrhythmia	asthma	heart problems	brain problems
allergies	1	0	0	0	0	0	1	0
hypertension	1	3	0	3	1	0	0	0
cholesterol	2	4	0	0	0	0	0	0
diabetes	1	4	2	0	0	0	0	0
arrhythmia	0	0	0	1	0	0	0	0
asthma	2	0	0	0	0	0	0	0
heart problems	0	1	0	2	0	0	0	0
brain problems	0	3	0	0	0	0	0	0

The Random Forest classifier correctly predicted cases of three different types of diseases. Table 8 shows that the method identified 2 cases of allergies, 4 cases of hypertension, and 1 case belonging to the class of unspecified heart problems. This method reached an accuracy of 80.48%, a precision of 21.84%, a recall of 21.84%, and an F1-score of 21.84%.

Table 8. Confusion matrix of Random Forest classifier.

	allergies	hypertension	cholesterol	diabetes	arrhythmia	asthma	heart problems	brain problems
allergies	2	0	0	0	0	0	0	0
hypertension	0	4	0	3	1	0	0	0
cholesterol	1	5	0	0	0	0	0	0
diabetes	3	3	1	0	0	0	0	0
arrhythmia	0	1	0	0	0	0	0	0
asthma	2	0	0	0	0	0	0	0
heart problems	0	2	0	0	0	0	1	0
brain problems	0	2	0	1	0	0	0	0

As seen in Table 9, the Neural Network classifier correctly identified 2 in 6 cases of allergies, 4 in 19 cases of hypertension, and 1 in 13 cases of diabetes in our dataset. Furthermore, this algorithm achieved an accuracy of 80.16% and precision, recall, and F1-score of 21.84%.

Table 9. Confusion matrix of Neural Net classifier.

	allergies	hypertension	cholesterol	diabetes	arrhythmia	asthma	heart problems	brain problems
allergies	2	0	0	0	0	0	0	0
hypertension	1	4	0	3	0	0	0	0
cholesterol	2	3	0	0	0	0	1	0
diabetes	1	4	1	1	0	0	0	0
arrhythmia	0	1	0	0	0	0	0	0
asthma	1	1	0	0	0	0	0	0
heart problems	0	2	0	1	0	0	0	0
brain problems	0	3	0	0	0	0	0	0

The AdaBoost algorithm, as seen in Table 10, was capable of correctly classifying 5 in 19 cases of hypertension and 2 in 9 cases of cholesterol experienced by the participants. In addition, it achieved an accuracy of 80.56% and precision, recall, and F1-score of 21.84%.

Table 10. Confusion matrix of AdaBoost classifier.

	allergies	hypertension	cholesterol	diabetes	arrhythmia	asthma	heart problems	brain problems
allergies	0	1	1	0	0	0	0	0
hypertension	0	5	2	0	0	0	1	0
cholesterol	0	4	2	0	0	0	0	0
diabetes	0	5	2	0	0	0	0	0
arrhythmia	0	1	0	0	0	0	0	0
asthma	0	0	2	0	0	0	0	0
heart problems	0	3	0	0	0	0	0	0
brain problems	0	3	0	0	0	0	0	0

The Naive Bayes classifier could correctly identify cases belonging to 4 of the 8 diseases considered in this study. As seen in Table 11, the algorithm accurately identified 2 cases of allergies, 5 cases of hypertension, 1 case of diabetes, and 1 case belonging to the class of unspecified heart problems. It was the best-performing method applied during our study, correctly identifying the highest number of instances of disease present in our dataset.

Table 11. Confusion matrix of Naive Bayes classifier.

	allergies	hypertension	cholesterol	diabetes	arrhythmia	asthma	heart problems	brain problems
allergies	2	0	0	0	0	0	0	0
hypertension	1	5	0	2	0	0	0	0
cholesterol	1	4	0	1	0	0	0	0
diabetes	2	3	1	1	0	0	0	0
arrhythmia	0	0	1	0	0	0	0	0
asthma	1	1	0	0	0	0	0	0
heart problems	0	1	1	0	0	0	1	0
brain problems	0	3	0	0	0	0	0	0

4   Discussion and Conclusions

We employed 8 classifiers to a dataset of ECG recordings to test their capacity to classify them into one of 8 classes representing different diseases.

Applying these methods to our dataset resulted in confusion matrixes, which we could utilize to determine the accuracy, precision, recall, and F1-score of each of these 8 methods. For example, the RBF SVM method achieved the highest accuracy at 87.92%, and the best precision and recall were conducted by the Linear SVM method, both at 34.80%, and the best F1-score was 27.29% and was reached by the Naive Bayes method.

Considering these results, we concluded that the best-performing method was Naive Bayes, correctly classifying 9 ECG signals from 4 out of 8 classes.

Data from ECG sensors can provide vital details about heart health and be used to identify several disorders. The following are some advantages of utilizing ECG data to identify diseases:

- ECG readings can identify anomalies in the heart's electrical activity even before a disease's symptoms manifest. Healthcare professionals may intervene and offer the proper therapy before the condition worsens with early detection;
- ECG examinations are painless and non-invasive. Electrodes must be applied to the patient's skin, and the test may be completed quickly and efficiently. It makes ECG testing a reliable and secure method of diseases screening;
- ECG data can give comprehensive information on the heart's rhythm, pace, and other crucial aspects. Arrhythmias, heart attacks, and heart failure are just a few of the disorders that may be identified using this information;
- Screening for the cardiac disease may be done at a relatively low cost using ECG testing. It is an efficient tool for finding people who might need additional testing or therapy;
- ECG information can be utilized as a part of a proactive strategy for healthcare. For example, healthcare professionals can collaborate with patients to build plans for avoiding disease and enhancing general health by identifying people at risk for heart disease or other disorders.

Individuals and medical professionals can benefit significantly from using ECG data to identify disorders. In addition, patients can have improved health outcomes and a higher quality of life by recognizing possible health concerns early and offering appropriate treatment.

The results of our tests achieved in this study did not meet our initial expectations, with the best-performing method only being able to classify a small portion of the ECG recordings correctly. However, this lackluster outcome is a result of the small database we were able to utilize for this study. In the future, we hope to repeat the experiment with a larger sample size.

Acknowledgments. This work is funded by FCT/MEC through national funds and co-funded by FEDER – PT2020 partnership agreement under the project **UIDB/50008/2020**.

This work is also funded by FCT/MEC through national funds and co-funded by FEDER – PT2020 partnership agreement under the project **UIDB/00308/2020**.

This article is based upon work from COST Action CA19101 - Determinants of Physical Activities in Settings (DE-PASS), supported by COST (European Cooperation in Science and Technology). More information on www.cost.eu.

References

1. Abdulhussein, A.A., Hassen, O.A., Gupta, C., Virmani, D., Nair, A., Rani, P.: Health monitoring catalogue based on human activity classification using machine learning. Int. J. Electr. Comput. Eng. (2088–8708) **12**, 3970 (2022)
2. Ahsan, M.M., Siddique, Z.: Machine learning-based heart disease diagnosis: a systematic literature review. Artif. Intell. Med., 102289 (2022)
3. Alarsan, F.I., Younes, M.: Analysis and classification of heart diseases using heartbeat features and machine learning algorithms. J. Big Data **6**(1), 1–15 (2019). https://doi.org/10.1186/s40 537-019-0244-x
4. Alazzam, H., Alsmady, A., Shorman, A.A.: Supervised detection of IoT botnet attacks. In: Proceedings of the Second International Conference on Data Science, E-Learning and Information Systems, pp. 1–6 (2019)
5. Ali, F., et al.: A smart healthcare monitoring system for heart disease prediction based on ensemble deep learning and feature fusion. Inf. Fusion **63**, 208–222 (2020)
6. Almuhaideb, S., Menai, M.E.B.: Impact of preprocessing on medical data classification. Front. Comp. Sci. **10**(6), 1082–1102 (2016). https://doi.org/10.1007/s11704-016-5203-5
7. Amarappa, S., Sathyanarayana, S.V.: Data classification using support vector machine (SVM), a simplified approach. Int. J. Electron. Comput. Sci. Eng. **3**, 435–445 (2014)
8. Balakumar, P., Maung-U, K., Jagadeesh, G.: Prevalence and prevention of cardiovascular disease and diabetes mellitus. Pharmacol. Res. **113**, 600–609 (2016)
9. Batista, D., Plácido da Silva, H., Fred, A., Moreira, C., Reis, M., Ferreira, H.A.: Benchmarking of the BITalino biomedical toolkit against an established gold standard. Healthc. Technol. Lett. **6**, 32–36 (2019)
10. Celin, S., Vasanth, K.: ECG signal classification using various machine learning techniques. J. Med. Syst. **42**, 1–11 (2018)
11. Chio, C., Freeman, D.: Machine Learning and Security: Protecting Systems with Data and Algorithms. O'Reilly Media, Inc. (2018)
12. Da Silva, H.P., Guerreiro, J., Lourenço, A., Fred, A.L., Martins, R.: BITalino: a novel hardware framework for physiological computing. In: International Conference on Physiological Computing Systems (PhyCS), pp. 246–253 (2014)
13. Duarte, R.P., et al.: Extraction of notable points from ECG data: a description of a dataset related to 30-s seated and 30-s stand up. Data Brief **46**, 108874 (2023). https://doi.org/10.1016/j.dib.2022.108874
14. Escobar, L.J.V., Salinas, S.A.: e-Health prototype system for cardiac telemonitoring. In: 2016 38th Annual International Conference of the IEEE Engineering in Medicine and Biology Society (EMBC), pp. 4399–4402. IEEE, Orlando, FL, USA (2016)
15. García, V., Mollineda, R.A., Sánchez, J.S.: On the k-NN performance in a challenging scenario of imbalance and overlapping. Pattern Anal. Appl. **11**, 269–280 (2008)
16. Gardes, J., Maldivi, C., Boisset, D., Aubourg, T., Vuillerme, N., Demongeot, J.: Maxwell®: an unsupervised learning approach for 5P medicine. Stud. Health Technol. Inf. **264**, 1464–1465 (2019). https://doi.org/10.3233/SHTI190486
17. Gautam, M.K., Giri, V.K.: A neural network approach and wavelet analysis for ECG classification. In: 2016 IEEE International Conference on Engineering and Technology (ICETECH), pp. 1136–1141. IEEE, Coimbatore, India (2016)

18. Gupta, S.: Evaluation of ECG abnormalities in patients with asymptomatic type 2 diabetes mellitus. JCDR **11**, OC39 (2017). https://doi.org/10.7860/JCDR/2017/24882.9740
19. Hastie, T., Rosset, S., Zhu, J., Zou, H.: Multi-class AdaBoost. Statistics and Its Interface **2**, 349–360 (2009). https://doi.org/10.4310/SII.2009.v2.n3.a8
20. Haykin, S.: Neural Networks: A Comprehensive Foundation, 1st edn. Prentice Hall PTR, USA (1994)
21. Hsu, C.-W., Lin, C.-J.: A comparison of methods for multiclass support vector machines. IEEE Trans. Neural Netw. **13**, 415–425 (2002)
22. Pires, I.M., Garcia, N.M., Pires, I., Pinto, R., Silva, P.: ECG data related to 30-s seated and 30-s standing for 5P-Medicine project. Mendeley Data (2022). https://data.mendeley.com/dat asets/z4bbj9rcwd/1
23. Jindal, H., Agrawal, S., Khera, R., Jain, R., Nagrath, P.: Heart disease prediction using machine learning algorithms. In: IOP Conference Series: Materials Science and Engineering, p. 012072. IOP Publishing (2021)
24. Kakria, P., Tripathi, N.K., Kitipawang, P.: A real-time health monitoring system for remote cardiac patients using smartphone and wearable sensors. Int. J. Telemed. Appl. **2015**, 1–11 (2015). https://doi.org/10.1155/2015/373474
25. Kalkstein, N., Kinar, Y., Na'aman, M., Neumark, N., Akiva, P.: Using machine learning to detect problems in ECG data collection. In: 2011 Computing in Cardiology, pp. 437–440. IEEE (2011)
26. Kannathal, N., Acharya, U.R., Ng, E.Y.K., Krishnan, S.M., Min, L.C., Laxminarayan, S.: Cardiac health diagnosis using data fusion of cardiovascular and haemodynamic signals. Comput. Methods Programs Biomed. **82**, 87–96 (2006). https://doi.org/10.1016/j.cmpb.2006.01.009
27. Maji, S., Berg, A.C., Malik, J.: Classification using intersection kernel support vector machines is efficient. In: 2008 IEEE Conference on Computer Vision and Pattern Recognition, pp. 1–8. IEEE (2008)
28. Pires, I.: Jupyter Notebooks ECG Data (2022)
29. Pires, I.M., et al.: Mobile 5P-medicine approach for cardiovascular patients. Sensors **21**, 6986 (2021). https://doi.org/10.3390/s21216986
30. Pires, I.M., Garcia, N.M., Flórez-Revuelta, F.: Multi-sensor data fusion techniques for the identification of activities of daily living using mobile devices. In: Proceedings of the ECMLPKDD (2015)
31. Prescott, G.J., Garthwaite, P.H.: A simple Bayesian analysis of misclassified binary data with a validation substudy. Biometrics **58**, 454–458 (2002)
32. Ramaraj, E.: A novel deep learning based gated recurrent unit with extreme learning machine for electrocardiogram (ECG) signal recognition. Biomed. Signal Process. Control **68**, 102779 (2021)
33. Rivas, R.G., Domínguez, J.J.G., Marnane, W.P., Twomey, N., Temko, A.: Real-time allergy detection. In: 2013 IEEE 8th International Symposium on Intelligent Signal Processing, pp. 21–26. IEEE (2013)
34. Swapna, G., Soman, K.P., Vinayakumar, R.: Diabetes detection using ECG signals: an overview. In: Dash, S., Acharya, B.R., Mittal, M., Abraham, A., Kelemen, A. (eds.) Deep Learning Techniques for Biomedical and Health Informatics. SBD, vol. 68, pp. 299–327. Springer, Cham (2020). https://doi.org/10.1007/978-3-030-33966-1_14
35. Tran, T.M., Le, X.-M.T., Nguyen, H.T., Huynh, V.-N.: A novel non-parametric method for time series classification based on k-nearest neighbors and dynamic time warping barycenter averaging. Eng. Appl. Artif. Intell. **78**, 173–185 (2019)

36. Twomey, N., Temko, A., Hourihane, J.O., Marnane, W.P.: Allergy detection with statistical modelling of HRV-based non-reaction baseline features. In: Proceedings of the 4th International Symposium on Applied Sciences in Biomedical and Communication Technologies, pp. 1–5 (2011)
37. Villasana, M.V., Sá, J., Pires, I.M., Albuquerque, C.: The New Era of Technology Applied to Cardiovascular Patients: State-of-the-Art and Questionnaire Applied for a System Proposal, pp. 267–278. Springer International Publishing, Cham (2021)
38. Vogel, B., et al.: The Lancet women and cardiovascular disease commission: reducing the global burden by 2030. The Lancet **397**, 2385–2438 (2021)
39. Webb, G.I., Boughton, J.R., Wang, Z.: Not so naive bayes: aggregating one-dependence estimators. Mach. Learn. **58**, 5–24 (2005). https://doi.org/10.1007/s10994-005-4258-6
40. Neurophysiological Data Analysis with NeuroKit2 — NeuroKit2 0.2.1 documentation. https://neuropsychology.github.io/NeuroKit/. Accessed 10 Jul 2022

Computational Proteomics

Exploring Machine Learning Algorithms and Protein Language Models Strategies to Develop Enzyme Classification Systems

Diego Fernández[1], Álvaro Olivera-Nappa[2,3], Roberto Uribe-Paredes[1], and David Medina-Ortiz[1,2(✉)]

[1] Departamento de Ingeniería en Computación, Universidad de Magallanes, Av. Pdte. Manuel Bulnes, 01855 Punta Arenas, Chile
david.medina@umag.cl
[2] Departamento de Ingeniería Química, Biotecnología y Materiales, Universidad de Chile, Beauche 851, Santiago, Chile
[3] Centre for Biotechnology and Bioengineering, Universidad de Chile, Beauchef 851, Santiago, Chile

Abstract. Discovering functionalities for unknown enzymes has been one of the most common bioinformatics tasks. Functional annotation methods based on phylogenetic properties have been the gold standard in every genome annotation process. However, these methods only succeed if the minimum requirements for expressing similarity or homology are met. Alternatively, machine learning and deep learning methods have proven helpful in this problem, developing functional classification systems in various bioinformatics tasks. Nevertheless, there needs to be a clear strategy for elaborating predictive models and how amino acid sequences should be represented. In this work, we address the problem of functional classification of enzyme sequences (EC number) via machine learning methods, exploring various alternatives for training predictive models and numerical representation methods. The results show that the best performances are achieved by applying representations based on pre-trained models. However, there needs to be a clear strategy to train models. Therefore, when exploring several alternatives, it is observed that the methods based on CNN architectures proposed in this work present a more outstanding facility for learning and pattern extraction in complex systems, achieving performances above 97% and with error rates lower than 0.05 of binary cross entropy. Finally, we discuss the strategies explored and analyze future work to develop integrated methods for functional classification and the discovery of new enzymes to support current bioinformatics tools.

Keywords: Machine learning algorithms · protein language models · EC number classifications · convolutional neural networks · enzyme discovery

© The Author(s), under exclusive license to Springer Nature Switzerland AG 2023
I. Rojas et al. (Eds.): IWBBIO 2023, LNBI 13919, pp. 307–319, 2023.
https://doi.org/10.1007/978-3-031-34953-9_24

1   Introduction

Enzymes play a fundamental role in organisms, being the universal catalysts of biological reactions [2]. These proteins have particular characteristics, such as having an active site, which is the site where reactions occur, and allosteric sites that can serve as sites of inhibition or loss of function [8]. Enzymes have acquired significant relevance in industries such as fishing, mining, agriculture, and textile, being the focus of study for biotechnology and improving their activities or properties through protein engineering [15]. Due to numerous functions and processes performed by enzymes, they have been grouped into categories called enzyme commission numbers, which allow describing general functionalities of enzymes to group them into families with similar functions, which are divided into more specific properties depending on the level of the EC number considered, with enzymatic functions such as hydrolases, ligases, and transferases at a first level, and at a last level indicating the specificity related to the activity or substrate processed [13].

Discovering enzyme functions in newly reported or identified organisms has been one of the significant challenges for bioinformatics. The gold standard of functional annotation employs phylogeny to relate annotated sequences in the literature with high relatedness to unknown sequences and annotation target, being supported by tools such as AGTK [18], Anotatornia [22], and MyMiner [27], among an extensive list of available systems [21], which are supported by biological databases such as UniProt [7] or Protein Data Bank (PDB) [4]. However, there is a gap between functional knowledge and reported sequences, so many of them are annotated as unknown or with unknown functions, which limits the annotation processes. Another common problem in useful identification is the expert judgment to work with the required identity percentages and the skills to select the most likely options [28]. Finally, in many cases, protein sequences do not necessarily meet the portion of identity. Yet, they do have the same function, and more tools or methodologies must be needed to overcome these problems [11].

Following the problems detected using phylogeny-based functional annotation systems, machine learning-based methods have emerged to elaborate functional classification strategies for proteins or enzymes [3,10], being specific tools for multiple types of functions or activities such as the identification of DNA-binding proteins [23], recognition of biological activities in peptide sequences [24], and functional classification models for EC numbers [19]. In this context, several tools have been built for EC number prediction. Among them is DEEPre [16], which uses one hot encoding information to encode sequences, combined with a position-specific scoring matrix, solvent accessibility information, secondary structure representation, and functional domain information to develop its CNN and RNN-based predictor system. DeepEC [26] uses CNN architectures for pattern extraction and learning model development, combining embedding methods with learning layers. EFICAz^2 [1] has employed enzymatic classification strategies emulating the functionally discriminating residues methods (FDRs) with support vector machine algorithms for the elaboration of classification systems and so on, as well as various computational tools for predicting EC numbers reported in the literature [30].

One of the critical steps during the development of predictive models using machine learning algorithms consists of the numerical representations for protein sequences [20]. Initial methods are related with the characterization of the protein sequences, employed similar strategies to proposed in DEEPre [16]. Alternatively, methods such as amino acid encodings based on physicochemical properties have been proposed [20]. Besides, the use of Fourier transforms for spatial transformations has been explored, allowing elaborate representations that emulate the structural behaviors of proteins [5,20,29]. Recently, representation learning methods based on Natural Language Processing (NLP) techniques have been studied, generating different pre-trained models and computational tools such as bio-embedding [9], TAPE [25], or ECNet [17]. However, there is yet to be a clear alternative for employing these methods. Their use in predictive models of protein interaction has only considered the concatenation of information, not being explored in methods of transformations and combinations through techniques such as Principal Component Analysis or similar.

In this work, we sought to design and implement functional classification models by evaluating Enzyme commission numbers (EC) using machine learning methods. First, the sequences of enzymes with function reported in the KEGG database were collected. Then, using statistical methods, the dataset was processed and cleaned. A total of 3.8 million sequences were used to work on the development of the predictive models. Once the datasets were processed, 22 numerical representation methods were used to encode the sequences and generate vectors interpretable by the machine learning algorithms. Amino acid encoding strategies via physicochemical properties, transformations to frequency space using Fourier Transforms, and pre-trained models based on learning representations using Natural Language Processing methods available in the bio-embedding tool were explored. Classical machine learning algorithms such as KNN or Random Forest were applied to train the predictive models, and eight CNN architectures were designed to evaluate deep learning methods. As a result, it was obtained that regardless of the training method, strategies based on numerical representations with pre-trained models achieved a better performance for classifying the EC number for the predictive model training strategies. It was observed that the architectures designed in this work based on CNN-1D and the models trained with KNN or Random Forest achieved performances above 94% accuracy and overfitting rates lower than 0.2. The choice of algorithm strategy was based on achieving better performance, lower overfitting rate, and higher scalability when extracting complex patterns in higher datasets, with CNN-1D architectures being selected. The learning power and generalization capacity in epoch times were evaluated, showing that an architecture of three pattern extraction blocks and two learning blocks working with the numerical representations based on the Esm1B pre-trained achieves a performance of 99% precision, an overfitting rate of 0.06 and a binary cross entropy value of 0.02, being these results better than the methods reported in the literature, with the need of not requiring different characterization methods and additional information. Finally, we discuss the usability of the explorative methodologies and how they can be used to predict new activities or discover sequences with unknown functions, fostering support for bioinformatics tools for functional annotations.

2 Methods

2.1 Numerical Representation Strategies

The numerical representation of the protein sequences is necessary to apply machine learning algorithms. This work explored three strategies: I) amino acid encoding methods via physicochemical properties. Eight amino acid encodings were used for coding the protein sequences. The encoders were proposed in [20] and represent semantic clusters of physicochemical properties obtained from the AAIndex database [14]. Zero-padding strategies were applied to guarantee the creation of codified sequences with the same length. ii) Applying frequency space representation methods via Fast Fourier Transform. The FFT was applied for all results generated using amino acid encoding using as suggested in [5,20,29]. In this case, the coded vector was expanded via zero-padding methods to guarantee the requirements of the FFT applications, and iii) employing pre-trained models developed using natural language processing strategies. Six pre-trained models were evaluated using the bio-embedding tool. The package was installed to apply the pre-trained models, and the pre-trained were downloaded. Then a reduction strategy available on the bio-embedding tool [9] was used to create vectors in a 1D space.

2.2 Collecting and Processing Datasets

The datasets were collected from the KEGG database [12] using the download API enabled on its platform and Biopython [6] scripts for processing. Optimization of the download process was developed by applying cloud computing methods by creating virtual machines in Azure for distributed extraction and processing of the collected records. Then, processing methods were used to remove sequences with non-canonical residues in their records. Statistical tests were developed to apply sequence filters by the number of residues to facilitate sequence coding. Finally, the data sequences are represented numerically by applying the three methods mentioned in the previous section and generating operational coded data sets for training predictive models.

2.3 Developing Functional Classification Models

A standard predictive model development protocol was utilized for training the enzyme functional classification models. First, the dataset is divided into training and validation datasets, applying a 70:30 ratio. Then the training dataset is used to fit the algorithm with the hyperparameters, and the classification model is trained. The training stage considers a k-fold cross-validation process with a value of $k = 10$, estimating the training performance as a weighting of each cross-validation stage to prevent over-fitting. With the trained model, the validation dataset is applied, predictions are obtained, and the performances of the generated models are estimated. The performance metrics correspond to precision, recall, accuracy, and f1-score. Finally, overfitting evaluation rates are developed by comparing the training and validation stage performances using the formula $over\ fitting\ rate = 1 - (training \times testing)$. This work explored classical supervised learning algorithms such as K-NN, Support vector machine, decision tree, Random Forest, AdaBoost, and Gradient Tree Boost. In addition, deep learning methods based on CNN architectures were explored. Table 1 summarizes the CNN architectures implemented in this work. In the case of classic supervised learning algorithms, they are not explained due to the use only of default hyperparameters defined in the scikit-learn library.

Table 1. CNN architectures designed for exploring predictive models in protein-protein network tasks.

Architecture	Pattern extraction configuration	Learning configuration	Optimization	Loss Function
A	A block of three Conv1D layers with MaxPooling activated by ReLU function	Two dense layers activated by tanh and linear functions	Adam	MSE
B	A block of three Conv1D layers with MaxPooling and dropout, activated by ReLU function	Two dense layers activated by tanh and linear functions	Adam	MSE
C	Pattern extraction blocks of two Conv1D and one MaxPooling, activated with ReLU function	One dense layer activated by linear function	Adam	MSE
D	Pattern extraction blocks of two Conv1D, one MaxPooling, and one droput, activated with ReLU function	One dense layer activated by linear function	Adam	MSE

2.4   Implementation Strategies

All scripts, methods, and functionalities implemented in this work were performed by the Python programming language version 3.9.16, helping with classical data science and machine learning packages like Pandas, Numpy, scikit-learn, and scipy. The bio-embedding tool was employed to apply the pre-trained models for the numerical representations. The TensorFlow package was used to implement the predictive models. A Conda environment was created to facilitate replicating all experiments processed in this work. Finally, all source code is available for non-commercial use in the GitHub repository https://github.com/ProteinEngineering-PESB2/ECNumberClassModels licenced by GNU General Public Licence v3.0.

3   Results and Discussions

3.1   Collecting and Processing Datasets

To facilitate the coding of amino acid sequences via physicochemical properties or Fourier transforms, statistical tests were applied to filter sequences by length. A total of 7.7 million amino acid sequences of enzymes were collected from the KEGG database, and sequences containing non-canonical residues were removed to avoid errors at the coding stage, leaving 7.6 million records. All enzymes smaller than 1024 were released, generating a processed dataset of 3.8 million records.

3.2   Protein Language Models Achieved Better Performances Than Amino Acid Encoding Strategies

One of the crucial steps in developing classification models corresponds to the numerical processing of protein sequences to be interpreted by machine learning algorithms. This work explored three types of methodologies associated with amino acid coding methods using physicochemical properties, transformations in the signal space applying Fourier Transforms, and pre-trained models based on Natural Language Processing techniques, generating 22 datasets prepared for the training process. All datasets were subjected to the same training strategy described in the methodology, exploring different algorithms and hyperparameters, together with eight deep learning architectures based on CNN-1D or CNN-2D. A total of 348 classification models were generated, and their performance was evaluated by applying classical metrics such as precision, recall, or f1-score. Figure 1 shows the performance distributions for the classification model development process, compared by type of numerical representation strategy, and visualizes the performances in both the training and validation stages. Models generated by applying numerical representations based on pre-trained models achieved better performance than models generated with coding and FFT methods without considering the type of algorithm or training strategy developed, demonstrating their usefulness for generating predictive models concerning bioinformatics tasks.

Comparing the trained models with the PLM methods showed that the best classification models are achieved by applying the pre-trained models Esm1B, bepler, and plus rnn. However, the performances of the models generated with Esm1B and plus rnn show a higher tendency to present better performance results compared to the rest of the pre-trained models used in this work independent of the strategy used for training the models (See Fig. 2).

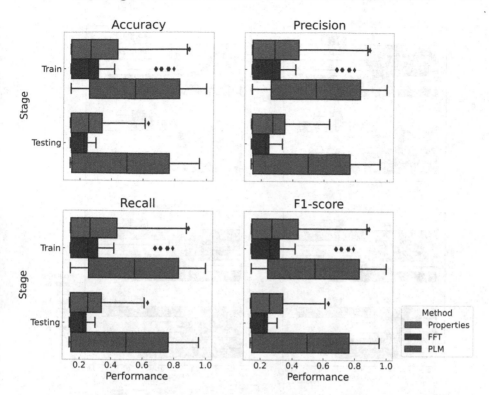

Fig. 1. Performance distribution of classification models trained with pre-trained modeling strategies (PLM), physicochemical properties (Properties), and signal space representations (FFT). Regardless of the training strategy, PLM models present a distribution with higher performance measures compared to the values of models generated by amino acid encodings or signal space representations. Despite outliers in classification models developed with FFT or properties, they do not exceed quartile 3 of the performance distribution of PLM-generated models, achieving a higher generality by the pre-trained model strategies and a higher transfer of learning.

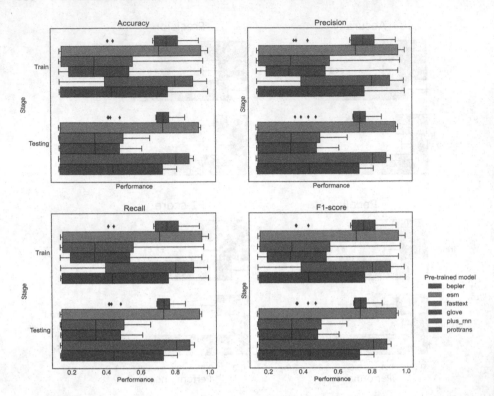

Fig. 2. Comparison of performance distributions for the different pre-trained models explored in this work. When using the pre-trained models bepler, Esm1B, and plus rnn, better performances are achieved than the rest of the strategies. Bepler presents a more bounded and homogeneous distribution in its performances, which implies that it is not affected by the type of algorithm. When applying methods such as Esm1B and plus rnn, better performances are achieved, and the highest values reached are with this type of numerical representation. However, by applying any of these three pre-trained models, the performance in trend will be higher than 70% of accuracy independent of the predictive model training strategy.

3.3 Models Based on CNN Architectures Show Better Performance. But, They Tend to Overfit

Classification model training strategies were based on exploring classical machine learning algorithms with default hyperparameters and using the CNN architectures described in Table 1. In addition, modifications were made to the CNN-1D architectures to work with matrix representations and employ CNN-2D layers. These matrix transformations were created from the initial numerical representation using reshape methods to generate *ntimesm* matrices. Figure 3 summarizes the overfitting distributions of the performances achieved by the classification models for each training strategy explored in this work.

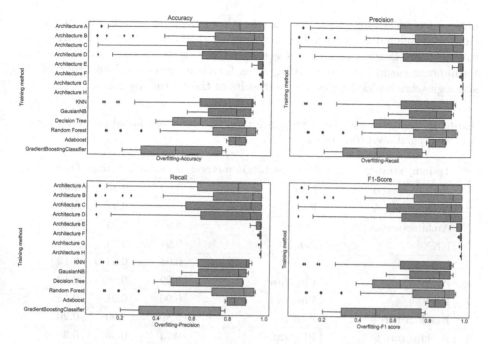

Fig. 3. Distribution of performance obtained by the classification models for the training strategies explored in this work. All the architectures explored in this work present a higher tendency to be overfitted compared to classical ML methods. However, the outliers of the overfitting distributions of the CNN models show results close to 1, implying that there are no significant differences between validation and training and that the performances are high in both stages. Methods such as the Gradient tree boost and decision tree achieve distributions with lower overfitting but lower performance than the rest of the strategies. Algorithms such as KNN and Random Forest show better results than the rest of the classical models, with no transparent system for training classification models.

In general, the overfitting distributions show a clear tendency to overfit by the training strategies based on CNN architectures. However, there are outliers in their distributions that show high performance and low overfitting rate, observed in CNN-1D architectures. Models generated via classical machine learning algorithms perform less than models trained with CNN architectures. However, they tend to exhibit a lower overfitting rate. Models based on Gradient Tree Boost and Decision tree present a more homogeneous distribution. But with fewer good performances. The performance distributions generated by the KNN and Random Forest algorithms show a lower tendency to overfit and, in some cases, achieve comparable performances with CNN methods. However, combining this algorithm with PLM representations achieves accuracy above 90%.

In contrast to the results observed in comparing the performance distributions for the numerical representation strategies, the model training strategies must show a clear trend regarding which strategy or strategies are the best alternatives for developing functional classification models. However, several aspects can be considered when deciding on a specific strategy. Table 2 summarizes the top 10 best classification models generated in this work, considering the performance measures and overfitting rates for their selection. The order of the strategies was decided based on the results of the overfitting rates.

Table 2. Top 10 classification models for enzyme functions based on performances (precision) and overfitting rates.

#	Training strategy	Representation method	Training	Testing	Overfitting
1	Architecture B	Esm1B	0.99	0.95	0.06
2	Architecture A	Esm1B	0.99	0.92	0.08
3	Architecture C	Esm1B	0.98	0.93	0.08
4	KNN	Esm1B	0.94	0.95	0.1
5	Random Forest	Esm1B	0.94	0.94	0.12
6	Architecture A	Plus rnn	0.99	0.88	0.13
7	KNN	Plus rnn	0.90	0.91	0.18
8	Random Forest	Plus rnn	0.89	0.90	0.20
9	Architecture C	Plus rnn	0.94	0.86	0.20
10	Gradient Tree Boosting	Esm1B	0.89	0.90	0.20

In the pre-trained models, using Esm1B is the best alternative to the other strategies explored in this work. In the training methods, architectures B, A, and C showed the highest performance measures and the lowest overfitting rates. However, it can be observed that the difference between training and validation is more considerable compared to the KNN or Random Forest methods, which show lower performance but do not show significant differences compared to the CNN methods. Despite this, the performance of the KNN and Random Forest models is higher than 94% precision, demonstrating a high predictive power and generalization capability. When considering performance alone as a decision criterion, CNN methods are better than classical ML methods. However, suppose one wishes to view an explainability of the predictions or an interpretability of how the model works. In that case, KNN methods are a better alternative since they are based on distance comparison in metric spaces to classify new objects. This work selects alternatives based on CNN architectures as the best strategy because of their better performance and lower overfitting rate. In addition, the versatility of deep learning-based methods favors the extraction of complex patterns when making predictions. The fact of discarding the KNN and Random Forest methods is that although the predictions can be explained, an explanation in terms of sequences or critical amino acids for the classification of one type of enzyme over another will not be achieved, without making extensive analyses of the representations, since the pre-trained models are developed under deep learning methods.

Finally, the two architectures with better performances and lower overfitting rates are compared regarding epochs needed to achieve learning, minimize the loss function, and stabilize good performance. Figure 4 shows the training history results for architectures A and B with the Esm1B representation. Architecture A offers a higher stabilization over a smaller number of epochs or decreases the loss function and optimizes the performance of the models reaching a minimum maximum of 0.02 for the loss function over the 100 training epochs. In contrast, the training results for architecture B show that the optimal performance is reached at a higher number of epochs and that the loss function value is higher than that achieved by architecture A.

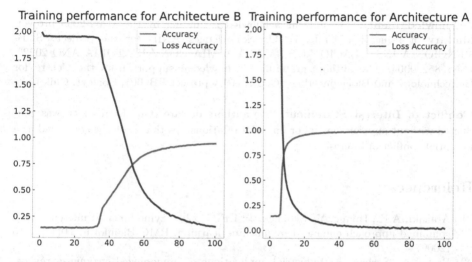

Fig. 4. Performance history for training CNN-based models using 100 epochs. The performance history during training is compared for the three selected architectures based on the best performance. The performances obtained by Architecture A (right panel) show a better generalization and lower error when training the models, compared to architecture B (left panel). Architecture A shows that fewer epochs are required to stabilize the model performances and achieve learning. At the same time, architecture B needs a minimum of 60 epochs to achieve stable performance.

4   Conclusions

This work explored strategies for developing functional classification models using amino acid sequences, numerical representation methods, and machine learning or deep learning techniques. Of the three types of process explored for numerically representing enzyme sequences, representation learning methods performed better than amino acid coding strategies and FFT transformations to frequency space, independent of the predictive model training strategy. The classification models were trained by applying different machine-learning algorithms and eight architectures designed for this work. All methods were compared based on overfitting rates, and it was observed that architectures A and

B, together with the KNN and Random Forest algorithms, presented better performances, exceeding 94% accuracy. However, architectures A and B performed better, exceeding 97% accuracy. Then, the learning capacity in the number of epochs is compared, demonstrating that architecture A is superior. Finally, a functional classification model for enzymes based on learning representations via pre-trained model Esm1B in combination with architecture A is proposed in this work, achieving a performance of 99% accuracy and an overfitting rate of 0.06. With these advances, it is expected to continue using this predictive model development guide to delve into the remaining subclassifications or levels of the EC number to develop an integrated functional classification system for enzymes.

Acknowledgments. The authors acknowledge funding by the MAG-2095 project, Ministry of Education, Chile. DMO acknowledges ANID for the project "SUBVENCIÓN A INSTALACIÓN EN LA ACADEMIA CONVOCATORIA AÑO 2022", Folio 85220004. The authors gratefully acknowledge support from the Centre for Biotechnology and Bioengineering - CeBiB (PIA project FB0001, Conicyt, Chile).

Conflict of Interest Statement. The authors declare that the research was conducted without any commercial or financial relationships that could be construed as a potential conflict of interest.

References

1. Arakaki, A.K., Huang, Y., Skolnick, J.: EFICAz2: enzyme function inference by a combined approach enhanced by machine learning. BMC Bioinform. **10**(1), 1–15 (2009)
2. Basso, A., Serban, S.: Industrial applications of immobilized enzymes-a review. Mol. Catal. **479**, 110607 (2019)
3. Bonetta, R., Valentino, G.: Machine learning techniques for protein function prediction. Proteins: Struct. Function Bioinform. **88**(3), 397–413 (2020)
4. Burley, S.K., Berman, H.M., Kleywegt, G.J., Markley, J.L., Nakamura, H., Velankar, S.: Protein data bank (PDB): the single global macromolecular structure archive. In: Protein Crystallography: Methods and Protocols, pp. 627–641 (2017)
5. Cadet, F., et al.: A machine learning approach for reliable prediction of amino acid interactions and its application in the directed evolution of enantioselective enzymes. Sci. Rep. **8**(1), 16757 (2018)
6. Cock, P.J., et al.: Biopython: freely available python tools for computational molecular biology and bioinformatics. Bioinformatics **25**(11), 1422–1423 (2009)
7. UniProt Consortium: Uniprot: a worldwide hub of protein knowledge. Nucleic Acids Res. **47**(D1), D506–D515 (2019)
8. Copeland, R.A.: Enzymes: A Practical Introduction to Structure, Mechanism, and Data Analysis. Wiley, Hoboken (2023)
9. Dallago, C., et al.: Learned embeddings from deep learning to visualize and predict protein sets. Curr. Protoc. **1**(5), e113 (2021)
10. Gao, W., Mahajan, S.P., Sulam, J., Gray, J.J.: Deep learning in protein structural modeling and design. Patterns **1**(9), 100142 (2020)

11. Greener, J.G., Kandathil, S.M., Moffat, L., Jones, D.T.: A guide to machine learning for biologists. Nat. Rev. Mol. Cell Biol. **23**(1), 40–55 (2022)
12. Kanehisa, M., Furumichi, M., Tanabe, M., Sato, Y., Morishima, K.: KEGG: new perspectives on genomes, pathways, diseases and drugs. Nucleic Acids Res. **45**(D1), D353–D361 (2017)
13. Kanehisa, M., Sato, Y., Kawashima, M.: KEGG mapping tools for uncovering hidden features in biological data. Protein Sci. **31**(1), 47–53 (2022)
14. Kawashima, S., Pokarowski, P., Pokarowska, M., Kolinski, A., Katayama, T., Kanehisa, M.: Aaindex: amino acid index database, progress report 2008. Nucleic Acids Res. **36**(Suppl. 1), D202–D205 (2007)
15. Kuo, C.H., Huang, C.Y., Shieh, C.J., Dong, C.D.: Enzymes and biocatalysis (2022)
16. Li, Y., et al.: DEEPre: sequence-based enzyme EC number prediction by deep learning. Bioinformatics **34**(5), 760–769 (2018)
17. Luo, Y., et al.: ECNet is an evolutionary context-integrated deep learning framework for protein engineering. Nat. Commun. **12**(1), 1–14 (2021)
18. Maeda, K., Strassel, S.M.: Annotation tools for large-scale corpus development: using AGTK at the linguistic data consortium. In: LREC (2004)
19. Mazurenko, S., Prokop, Z., Damborsky, J.: Machine learning in enzyme engineering. ACS Catal. **10**(2), 1210–1223 (2019)
20. Medina-Ortiz, D., et al.: Generalized property-based encoders and digital signal processing facilitate predictive tasks in protein engineering. Front. Mol. Biosci. **9** (2022)
21. Neves, M., Ševa, J.: An extensive review of tools for manual annotation of documents. Brief. Bioinform. **22**(1), 146–163 (2021)
22. Przepiórkowski, A.: XML text interchange format in the national corpus of polish. In: The Proceedings of Practical Applications in Language and Computers PALC 2009 (2009)
23. Qu, K., Wei, L., Zou, Q.: A review of DNA-binding proteins prediction methods. Curr. Bioinform. **14**(3), 246–254 (2019)
24. Quiroz, C., et al.: Peptipedia: a user-friendly web application and a comprehensive database for peptide research supported by machine learning approach. Database **2021** (2021)
25. Rao, R., et al.: Evaluating protein transfer learning with tape. In: Advances in Neural Information Processing Systems, vol. 32 (2019)
26. Ryu, J.Y., Kim, H.U., Lee, S.Y.: Deep learning enables high-quality and high-throughput prediction of enzyme commission numbers. Proc. Natl. Acad. Sci. **116**(28), 13996–14001 (2019)
27. Salgado, D., et al.: MyMiner: a web application for computer-assisted biocuration and text annotation. Bioinformatics **28**(17), 2285–2287 (2012)
28. Sapoval, N., et al.: Current progress and open challenges for applying deep learning across the biosciences. Nat. Commun. **13**(1), 1728 (2022)
29. Siedhoff, N.E., Illig, A.M., Schwaneberg, U., Davari, M.D.: PyPEF-an integrated framework for data-driven protein engineering. J. Chem. Inf. Model. **61**(7), 3463–3476 (2021)
30. Tao, Z., Dong, B., Teng, Z., Zhao, Y.: The classification of enzymes by deep learning. IEEE Access **8**, 89802–89811 (2020)

A System Biology and Bioinformatics Approach to Determine the Molecular Signature, Core Ontologies, Functional Pathways, Drug Compounds in Between Stress and Type 2 Diabetes

Md. Abul Basar[1], Md. Rakibul Hasan[2(✉)], Bikash Kumar Paul[1], Khairul Alam Shadhin[2], and Md. Sarwar Mollah[3]

[1] Department of Information and Communication Technology (ICT), Mawlana Bhashani Science and Technology University (MBSTU), Santosh, Tangail 1902, Bangladesh

[2] Department of Software Engineering (SWE), Daffodil International University (DIU), Ashulia, Dhaka, Bangladesh
hasan35-148@diu.edu.bd

[3] Department of Computing and Information System (CIS), Daffodil International University (DIU), Ashulia, Dhaka, Bangladesh

Abstract. Bioinformatics is the application of computer science and information technology to the field of biology and medicine. It involves the analysis of large amounts of biological data, such as DNA sequences, protein structures, and gene expression patterns. Bioinformatics is used to develop new methods for understanding and analyzing biological data, as well as to develop new tools and technologies for biological research. Bioinformatics is used in a variety of fields, including genomics, proteomics, and drug discovery. In this study, focus on two severe diseases which affect millions of people globally such as stress and type 2 diabetes. Stress can have a significant impact on people with type 2 diabetes. Stress can cause blood sugar levels to rise, making it difficult to manage diabetes. The purpose of this research is to use various bioinformatics methods to discover potential therapeutic drugs and functional pathways between stress and type 2 diabetes. The microarray datasets GSE183648 and GSE20966 are used for the analysis of stress and type 2 diabetes samples respectively. After the datasets have been preprocessed and filtered through the use of the R programming language, identified the common DEGs. The depiction of common DEGs is shown by venn diagram. Next, the most active genes are identified through topological properties, and PPIs are built from the similar differential expressed genes (DEGs). These five genes NTRK2, SOCS3, NEDD9, MAP3K8, and SIRPA are the most important hub genes within the interaction network of protein-protein. According to the common DEGs, GO terms molecular function (MF), KEGG and WikiPathways are shown in this study. Gene-miRNA interaction, TF-gene regulatory network, module analysis, GO terms (Biological Process, Cellular Component), Pathways (Reactome, BioCarta, BioPlanet) are all things that

I. Rojas et al. (Eds.): IWBBIO 2023, LNBI 13919, pp. 320–331, 2023.
https://doi.org/10.1007/978-3-031-34953-9_25

could be done with this research work in the future. In last, a therapeutic drug compounds are recommended on the basis of common DEGs.

Keywords: Stress · Microarray · System Biology · Transcriptomic · Differentially expressed genes · Protein-protein interactions · Gene Ontology · TF-gene interaction · Hub gene · drug compounds · Type 2 Diabetes

1 Introduction

In today's world, stress has become one of the most common as well as dangerous diseases [1]. Unchecked and uncontrolled stress has been linked to a wide range of negative health outcomes that put individuals' mental, physical, and social health at risk. The body adapts to any kind of demand is to engage its natural defense system, which is known as stress. It's really essential that both negative and positive situations to contribute to the issue [2]. When things get much more challenging socially, it can be tough for some people, and this can cause stress. According to the World Health Organization, an estimated 3.4 billion people worldwide are affected by stress [3]. In the United States, approximately 8 in 10 adults report feeling stressed at least occasionally. Additionally, 1 in 5 adults report feeling extreme levels of stress [4]. Stress is a known risk factor for type 2 diabetes [5]. When you experience stress, your body releases hormones such as adrenaline and cortisol, which can cause physical reactions such as increased heart rate, sweating, and muscle tension. These hormones can also affect your blood sugar levels, making it more difficult for your body to regulate them. This can lead to an increased risk of developing type 2 diabetes. Additionally, stress can lead to unhealthy behaviors such as overeating, which can also increase your risk of developing type 2 diabetes [6]. Type 2 diabetes is a chronic condition in which the body does not produce enough insulin or does not use insulin effectively. This can lead to high blood sugar levels, which can cause a variety of health problems [7]. Common symptoms of type 2 diabetes include increased thirst, frequent urination, fatigue, blurred vision, and slow healing of cuts and bruises. Treatment for type 2 diabetes typically includes lifestyle changes such as eating a healthy diet and exercising regularly, as well as medications to help control blood sugar levels [8]. According to the World Health Organization, approximately 422 million people worldwide were living with diabetes in 2014. Of those, 90–95% had type 2 diabetes. In the United States, approximately 30.3 million people, or 9.4% of the population, had diabetes in 2015. Of those, approximately 23.1 million had type 2 diabetes [9]. High throughput methods have been drastically enhanced by the study of microarray data along with the data provided by that expression dataset [10]. Following the screening of genes taken from datasets GSE183648 and GSE20966, a total of 47 differentially expressed genes (DEGs) that were shared by both datasets were identified. For this study, the protein-protein interaction (PPI) is the most significant thing, so we will proceed by analyzing that. Through a degree-topological analysis, the PPIs network found hub genes and ranked the 10 most important ones. Various bioinformatics investigations have been carried out to merge DEGs with shared features and to use DEGs to pinpoint specific drug compounds. Analysis of frequently occurring DEGs may also help identify Gene Ontology (GO) terms and other functional pathways. Therapeutic studies including the recommended drugs for stress and type 2 diabetes should yield positive results. Microarray data contain biological information that may be gleaned via computer investigation, which is useful for the work of biological researchers. The overarching goal of this study is to use gene-based research to uncover the biological connection between stress and type 2 diabetes, and to then use that information to identify promising biomarkers. Differentially expressed genes must be discovered before the genes underlying stress and type 2 diabetes may be identified. The relationship between the DEGs is then visualized using a PPIs network. The next step is to analyze the biological system's pathways using the KEGG database and then draw conclusions about the system's overall functionality. We propose therapeutic compounds targeting the common DEGs in stress and type 2 diabetes once hub genes have been identified as shared among both conditions. The procedure that was followed for this research can be seen in Fig. 1.

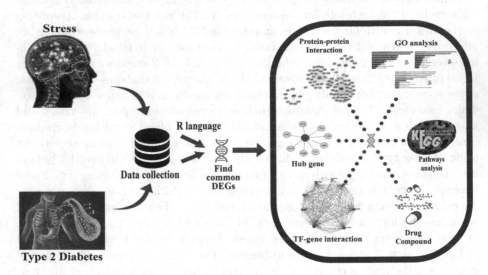

Fig. 1. Analysis of data and research procedure shows the flow chart. We combed through the samples (normal and infected) in the GSE183648 and GSE20966 databases to find what we required. Loss of p53 prevents replicating strain Damage to dna in ATRX-deficient neuroblastoma, as shown by the microarray dataset GSE183648. Conversely, GSE20966 is a microarray dataset that was obtained by collecting beta-cell enriched tissue from persons with type 2 diabetes and then analyzing their gene expression levels by Laser Capture Microdissection. We found the most prevalent DEGs by using R to analyze the two data sets. PPIs network TF-gene interaction, Hub genes, GO keywords, KEGG and WikiPathways, TF-gene interactions, and therapeutic drug compounds can all be found by using the common DEGs.

2   Proposed Mehtodology

2.1   Data Collection

In this study, we selected two diseases one is stress and other type 2 diabetes. We selected two datasets form NCBI are GSE183648 and GSE20966. Gene Expression Omnibus (GEO) database was used to gather for the GSE183648 and GSE20966 datasets, respectively [11]. Microarray and high-throughput sequencing datasets are both included in the GEO database, which is hosted on the platform that the NCBI oversees [12]. Microarray dataset GSE183648 demonstrates that p53 deficiency prevents DNA damage inflicted by replication stress in ATRX-deficient neuroblastoma cancer. The dataset is analyzed using GPL21185, an agilent-072363 SurePrint G3 Human GE v3 8x60K Microarray 039494 platform. GSE183648 has a total of 12 samples. Four of them are normal cells, while the remaining eight are stress cells. GSE183648 is comprised of a total of twelve samples, four of which are representative of normal cells and the remaining eight representing stress cells. Dataset GSE183648 was imparted by J. Akter et al. [13]. Besides, GSE20966 is a microarray dataset that was collected beta-cell enriched tissue from people who had type 2 diabete and their gene expression profiles were analyzed using Laser Capture Microdissection. Affymetrix Human X3P Array GPL1352 [U133 X3P] platform was utilized to evaluate this dataset. Additionally, there were 10 samples taken from people who did not have diabetes and 10 samples taken from diabetics that were used in the microarray dataset. The GSE20966 dataset was contributed by Marselli et al., who gathered pancreatic beta cells via LCM [14].

2.2   Data Filtering and Finding of DEGs, and Identification of Concordant DEGs Between Stress and Type 2 Diabetes

Transcriptomic datasets GSE183648 for Stress and GSE20966 for Type 2 Diabetes is employed for this study. This study begins by retrieving DEGs from the datasets GSE183648 and GSE20966. With the assistance of a R programming language, identify the differential expressed genes (DEGs). Preprocessing procedures were carried out on both sets of data with an adjusted p-value $<$ 0.05 and a $|logFC| > 1$. The Benjamini-Hochberg approach was applied to the GSE183648 and GSE20966 datasets with the purpose of reducing the false discovery rate [15]. Identify the DEGs from both datasets and also find the common genes between them that scenario is depicted in a Venn diagram.

2.3   Design of Protein-Protein Interaction Network and Analysis

The study of protein interactions, which is considered as the first step in drug discovery and systems biology, yields significant information about the functions of proteins [16]. PPIs networks [17] are being thoroughly investigated to find out how numerous complex biological processes there are [18]. Stress and Type 2 Diabetes DEGs that were found to be common were put into the String [19] database with the help of the NetworkAnalyst tool [20]. Cytoscape (https://cytoscape.org/) was used to perform additional modifications to the network. Cytoscape software is a good way to combine how proteins interact with each other and how genes interact with each other.

2.4 Analysis of Hub Genes

The protein-protein interaction network shows the relationship between nodes and edges, it also illustrates the connections between the nodes. A hub gene is a node in a network that has the most connections to other nodes. Identifying the functional genes for this work is accomplished by the use of the degree topological method. Cytoscape [21] has been used in recent studies to do an analysis of the networks of PPIs. It is feasible to discover the genes that function as hubs for the related protein interaction network by applying the Cytoscape bioinformatics software plugin called as cytoHubba [22]. That plugin may be found at (http://apps.cytoscape.org/apps/cytohubba).

2.5 Enrichment Analysis: Gene Ontology and Pathways

Enrichment analysis of gene is a computational and statistical technique for determining if a group of genes is significantly enriched in one biological context compared to another [23]. For the annotation of gene products, it is referred to as the Gene Ontology (GO) word, which is categorized into three subheadings: biological process, molecular function, and cellular component [24]. When it comes to learning about metabolic pathways, the KEGG pathway is quite popular due to the many benefits it offers in comparison to more conventional gene annotation approaches. Alongside the use of the KEGG pathway, the WikiPathways are also utilized. A web-based tool called Enrichr (https://amp.pharm.mssm.edu/Enrichr/) supplied all the pathways and GO keywords for the shared genes discovered in the initial step. Enrichr is a web service that performs enrichment analysis on gene sets for genes that have been subjected to genome-wide analysis [25].

2.6 TF-Gene Interactions

TF-gene interactions with similar DEGs analyze TF's impact on functional pathways and gene expression [26]. After determining which genes are similar throughout organisms, the NetworkAnalyst (https://www.networkanalyst.ca/) platform is used to find the TFs that interact with those genes. NetworkAnalyst is a powerful web tool that permits users to do gene expression for a wide variety of species, as well as meta-analysis. The NetworkAnalyst software utilizes the ENCODE [27] data to build the TF-gene interaction network.

2.7 Finding of Candidate Drug

Finding new drug molecules is a crucial aspect of ongoing study. A number of therapeutic molecules can be estimated using the DSigDB database and the

Enrichr platform, both of which are based on common differential expressed genes. There are a total of 22527 gene sets, 19531 genes, and 17389 distinct substances in DSigDB database [28]. DSigDB is mainly responsible for the prediction of drugs through the use of gene expression-based datasets, and each group of genes is taken into consideration whenever a drug is investigated.

3   Results

3.1   Findings of Differential Expressed Genes and Common DEGs Among Stress and Type 2 Diabetes

A comprehensive analysis of the GSE183648 dataset revealed 1547 genes to be differentially expressed. Total of 1547 genes, 654 of them are up-regulated, while 893 are down-regulated. Another dataset GSE20966 was analyzed, and it was found that 538 genes exhibit differential expression. Of these genes, 180 are up-regulated, and 358 are down-regulated. In the analysis of those two datasets, we found two common up-regulated genes and 23 common down-regulated genes. There were 2038 DEGs found with opposite behaviors among two diseases. Among the DEGs 47 were identified as common differential expressed genes between two datasets, and this filtered data is being used to perform a more comprehensive analysis of the ongoing research. The common DEGs among two datasets depicts by venn diagram in Fig. 2.

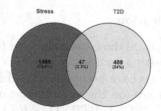

Fig. 2. Common DEGs from the stress and T2D dataset depicted by that figure. There are a total of 1547 DEGs in the stress dataset, while there are 538 DEGs in the type 2 diabetes dataset. Of these, 47 were found to be highly significant. Only 2.3% of the 2085 DEGs were between the two data sets.

3.2   Analysis of Protein-Protein Interaction Network

As an input, the common differential expressed genes (DEGs) were provided to NetworkAnalyst. This study focused mostly on the protein-protein interaction network, analyzing hub genes and specific modules in relation to a network. NetworkAnalyst, a web-based tool, is used in connection with the STRING database to produce SIF files for the network diagram. The protein-protein interaction network and top 5 five hub gene connection are shown in Fig. 3.

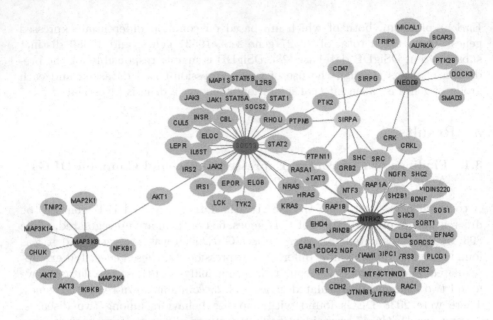

Fig. 3. Protein-protein interactions (PPIs) network created by the 47 common DEGs. Orange/red color nodes indicates hub genes, and yellow nodes represents the connections between the hub genes and other. The PPIs network has a total of 97 nodes, 106 edges, and 6 seeds. (Color figure online)

3.3   Hub Gene Identification

The nodes in a network that are the most connected to other nodes are referred to as hub nodes. After doing an analysis of the PPIs network with cytohubba, the top ten most active genes have been determined for the purposes including further study. These ten genes are the most important ones: NTRK2, SOCS3, NEDD9, MAP3K8, SIRPA, BCL3, PTPN11, AKT1, HRAS, and SHC1. The degree value of NTRK2 is greater than that of any other DEG in the network. Cytoscape's PPI network is examined with that program's Network Analyzer in order to determine the topological characteristics of the network. Table 1 contains an overview of the topological analysis performed on the top five genes.

Table 1. The topological properties of the five most important hub genes, as examined through cytoscape.

Hub gene	Degree	Stress	Closeness Centrality	Betweenness Centrality
NTRK2	43	23438.0	58.95476	5683.97922
SOCS3	32	22570.0	53.38333	5898.04545
NEDD9	10	2652.0	36.4381	1333.92078
MAP3K8	10	8076.0	32.61667	2448.0
SIRPA	7	2142.0	33.77143	480.05455

3.4   According to the Common DEGs Analyze GO and Pathways

Following the identification of shared DEGs between stress and type 2 diabetes, several databases (KEGG, WikiPathways) were used to discover GO keywords and cell pathways. To further understand the relationships between these three GO terms, we take a look at their molecular components in Fig. 4(A). According to the results presented in the molecular function section, a transition metal ion binding factor is magnificently implicated in the widespread DEGs. KEGG and WikiPathways are also analyzed by the common DEGs between two diseases and illustrates in Fig. 4(B).

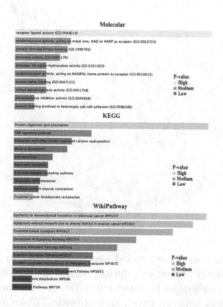

Fig. 4. (A) Identification of Molecular Component-Related GO Terms Based on a Weighted Average of Their Individual Scores. If an ontology has a high enrichment score, then a large proportion of relevant genes are represented in that ontology. (B) Locating KEGG and WikiPathways pathway analysis outcomes. The cumulative score was used to determine which pathway terms had yielded significant results.

3.5   TF-Gene Interactions

TF-gene interactions mainly show the connections between TF-gene and common genes between those two diseases. The TF-gene is determined by the common DEGs using NetworkAnalyst online tool. In Fig. 5, the similar DEGs and the TF regulators that interact with them are displayed. There are 104 nodes, 321 edges, and 42 seeds in this network by using JASPAR database. The network of interactions between TFs and common genes is shown in Fig. 5.

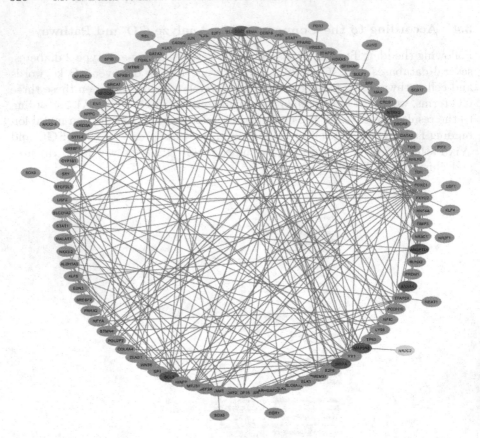

Fig. 5. Commonly differentially expressed genes connected to a network of TF-gene interactions. Genes that are shared by multiple organisms are represented by the green node, whereas the other yellow nodes represent different TF genes. There are 104 nodes in the network, 321 edges between them, and 42 seeds. (Color figure online)

3.6   Identification of Drug Compounds

The Enrichr online program is used to extract the drug compounds from the DSigDB database to provide the desired results. The following drugs were suggested based on a combination of the p-value and the modified p-value. The table below describes potential therapeutic compounds for stress and type 2 diabetes that combine common DEGs. Table 2 is utilized to evaluate which drug compounds have been the most successful by employing the DEGs that are most commonly encountered.

Table 2. Drug compounds estimates based on the degree's most heavily linked to stress and type 2 diabetes.

Drugs	p-value	Adjusted p-value	Genes
STOCK1N-35696 MCF7 UP	5.69E-06	0.003823864	ALDH1A3, GDF15, CYP1B1, NEAT1; ANGPTL4
Caspan CTD 00000180	1.46E-05	0.003823864	NHLH2, CADM2, ANXA4, MTUS1, NEDD9, CYP1B1, ARHGAP29, TFCP2L1, SULF1, NEAT1
trichostatin A CTD 00000660	2.01E-05	0.003823864	FOXC1, PCDH10, GDF15, ANXA4, COL23A1
puromycin MCF7 UP	2.09E-05	0.003823864	GDF15,BCL3,NEDD9,MAP3K8
GW-8510 MCF7 UP	2.32E-05	0.003823864	NHLH2, ALDH1A3, GDF15, CYP1B1

4   Discussion

Stress can have a significant impact on people with type 2 diabetes. Stress can cause blood sugar levels to rise, making it difficult to manage diabetes. Stress can also lead to unhealthy behaviors, such as overeating, which can further increase blood sugar levels. Additionally, stress can lead to depression and anxiety, which can make it difficult to manage diabetes. To help manage stress, people with type 2 diabetes should practice relaxation techniques, such as deep breathing, yoga, and meditation. In a component of the ongoing study, bioinformatics analysis was performed to combine shared DEGs between stress and type 2 diabetes, and protein-protein interaction network analysis was conducted once the common genes were identified. The drug compounds are proposed for the therapeutic of stress and type 2 diabetes based on the differential expressed genes (DEGs). Analysis of differential expressed genes (DEGs) between GSE183648 and GSE20966 datasets, identified 47 common DEGs among these two diseases. Owing of its necessity to the ongoing study, the PPI network is the next target of this evaluation. Hub genes as well as the 10 most prominent hub genes were found using the PPIs network. Using a degree topological strategy, the PPIs network singled out 10 major hub genes. Therapeutic drug compounds included in the top ten include those for STOCK1N-35696 MCF7 UP, Caspan CTD 00000180, trichostatin A CTD 00000660, puromycin MCF7 UP, and GW-8510 MCF7 UP. Several bioinformatics analyses have been carried out to merge similar DEGs and to use them to pinpoint specific drug compounds. Stress and type 2 diabetes therapeutic are promising areas for the suggested drugs.

5   Conclusion

There has never been a study of stress and type 2 diabetes in the field of transcriptomic analysis. In genetic innovation, the bioinformatics field has made substantial developments. Using the technique of system biology and the functional annotation method, we were able to identify the primary pathways and biomolecules that are responsible for stress and type 2 diabetes. A diverse range of bioinformatics analysis methods are utilized in the process of gene filtering, which is an indispensable step in the field of systems biology. After that, the genes are examined with one another to find a regular healing drug for stress and type 2 diabetes. According to similar differentially expressed genes

(DEGs), a protein-protein interaction network was designed and also found the top ten responsible genes (NTRK2, SOCS3, NEDD9, MAP3K8, SIRPA, BCL3, PTPN11, AKT1, HRAS, and SHC1) depending on their degree. The components of mass-produced successful remedial drugs can be identified when two diseases contain the same DEGs because the hub DEGs are highly harmful. The most essential five biomarkers were prognostic, which helped make a good therapeutic drug molecule for stress and type 2 diabetes. Through our research work, we identified novel molecular biomarkers and provided guidelines for the effective diagnosis and treatment of stress and type 2 diabetes at an early stage.

Acknowledgements. This paper is neither published nor being considered for publication anywhere else at this time. Each and every person who helped with this study is greatly appreciated by the authors.

Funding. This work is not financially supported.

References

1. Rahal, A., et al.: Oxidative stress, prooxidants, and antioxidants: the interplay. BioMed Res. Int. (2014)
2. Selye, H.: Stress without distress. In: Serban, G. (ed.) Psychopathology of Human Adaptation, pp. 137–146. Springer, Boston (1976). https://doi.org/10.1007/978-1-4684-2238-2_9
3. Basar, M.A., Hosen, M.F., Paul, B.K., Hasan, M.R., Shamim, S.M., Bhuyian, T.: Identification of drug and protein-protein interaction network among stress and depression: a bioinformatics approach. Inform. Med. Unlocked 101174 (2023)
4. Theodore, W.H., et al.: Epilepsy in North America: a report prepared under the auspices of the global campaign against epilepsy, the International Bureau for Epilepsy, the International League Against Epilepsy, and the World Health Organization. Epilepsia **47**(10), 1700–1722 (2006)
5. Cosgrove, M.P., Sargeant, L.A., Caleyachetty, R., Griffin, S.J.: Work-related stress and Type 2 diabetes: systematic review and meta-analysis. Occup. Med. **62**(3), 167–173 (2012)
6. Hosen, M.F., Basar, M.A., Paul, B.K., Hasan, M.R., Uddin, M.S.: A bioinformatics approach to identify candidate biomarkers and common pathways between bipolar disorder and stroke. In: 2022 12th International Conference on Electrical and Computer Engineering (ICECE), pp. 429–432. IEEE (2022)
7. Joint National Committee on Prevention, Evaluation, Treatment of High Blood Pressure and National High Blood Pressure Education Program: Report of the Joint National Committee on Prevention, Detection, Evaluation, and Treatment of High Blood Pressure, vol. 6. Public Health Service, National Institutes of Health, National Heart, Lung, and Blood Institute (1997)
8. Goyal, S., Morita, P., Lewis, G.F., Yu, C., Seto, E., Cafazzo, J.A.: The systematic design of a behavioural mobile health application for the self-management of type 2 diabetes. Can. J. Diabetes **40**(1), 95–104 (2016)
9. Gesinde, B.: An Avatar video intervention on type 2 diabetes for women of color using brief motivational interviewing: predictors of self-efficacy post-video for performing the American Association of Diabetes Educator's Seven Self-care Behaviors. Doctoral dissertation, Teachers College, Columbia University (2019)

10. Hasan, M.R., Paul, B.K., Ahmed, K., Bhuyian, T.: Design protein-protein inter-action network and protein-drug interaction network for common cancer diseases: a bioinformatics approach. Inform. Med. Unlocked **18**, 100311 (2020)
11. Clough, E., Barrett, T.: The gene expression omnibus database. In: Statistical Genomics, pp. 93–110. Humana Press, New York (2016)
12. Edgar, R., Domrachev, M., Lash, A.E.: Gene expression omnibus: NCBI gene expression and hybridization array data repository. Nucleic Acids Res. **30**(1), 207–210 (2002)
13. Akter, J., et al.: Loss of P53 suppresses replication stress-induced DNA damage in ATRX-deficient neuroblastoma. Oncogenesis **10**(11), 73 (2021)
14. Marselli, L., et al.: Gene expression profiles of Beta-cell enriched tissue obtained by laser capture microdissection from subjects with type 2 diabetes. PLoS ONE **5**(7), e11499 (2010)
15. Benjamini, Y., Hochberg, Y.: Controlling the false discovery rate: a practical and powerful approach to multiple testing. J. Roy. Stat. Soc.: Ser. B (Methodol.) **57**(1), 289–300 (1995)
16. Shadhin, K.A., et al.: Analysis of topological properties and drug discovery for bipolar disorder and associated diseases: a bioinformatics approach. Cell Mol. Biol. (Noisy-le-grand) **66**(7), 152–160 (2020)
17. Šikić, M., Tomić, S., Vlahoviček, K.: Prediction of protein-protein interaction sites in sequences and 3D structures by random forests. PLoS Comput. Biol. **5**(1), e1000278 (2009)
18. Pagel, P., et al.: The MIPS mammalian protein-protein interaction database. Bioin-formatics **21**(6), 832–834 (2005)
19. Mering, C.V., Huynen, M., Jaeggi, D., Schmidt, S., Bork, P., Snel, B.: STRING: a database of predicted functional associations between proteins. Nucleic Acids Res. **31**(1), 258–261 (2003)
20. Xia, J., Gill, E.E., Hancock, R.E.: NetworkAnalyst for statistical, visual and network-based meta-analysis of gene expression data. Nat. Protoc. **10**(6), 823–844 (2015)
21. Shannon, P., et al.: Cytoscape: a software environment for integrated models of biomolecular interaction networks. Genome Res. **13**(11), 2498–2504 (2003)
22. Chin, C.H., Chen, S.H., Wu, H.H., Ho, C.W., Ko, M.T., Lin, C.Y.: cytoHubba: identifying hub objects and sub-networks from complex interactome. BMC Syst. Biol. **8**(4), 1–7 (2014)
23. Subramanian, A., Kuehn, H., Gould, J., et al.: GSEA-P: a desktop application for gene set enrichment analysis. Bioinformatics **23**(23), 3251–3 (2007)
24. Delfs, R., Doms, A., Kozlenkov, A., Schroeder, M.: GoPubMed: ontology-based literature search applied to Gene Ontology and PubMed. In: German Conference on Bioinformatics 2004, GCB 2004. Gesellschaft fur Informatik eV (2004)
25. Kuleshov, M.V., et al.: Enrichr: a comprehensive gene set enrichment analysis web server 2016 update. Nucleic Acids Res. **44**(W1), W90–W97 (2016)
26. Ye, Z., et al.: Bioinformatic identification of candidate biomarkers and related transcription factors in nasopharyngeal carcinoma. World J. Surg. Oncol. **17**(1), 1–10 (2019)
27. Davis, C.A., et al.: The encyclopedia of DNA elements (ENCODE): data portal update. Nucleic Acids Res. **46**(D1), D794–D801 (2018)
28. Yoo, M., Shin, J., Kim, J., et al.: DSigDB: drug signatures database for gene set analysis. Bioinformatics **31**(18), 3069–71 (2015)

Recent Advances in Discovery of New Tyrosine Kinase Inhibitors Using Computational Methods

Vesna Rastija[1]([⊠]) [iD] and Maja Molnar[2] [iD]

[1] Faculty of Agrobiotechnical Sciences Osijek, University of Osijek, V. Preloga 1, 31000 Osijek, Croatia
vrastija@fazos.hr

[2] Faculty of Food Technology Osijek, University of Osijek, F. Kuhača 18, 31000 Osijek, Croatia

Abstract. Tyrosine kinases are enzymes that phosphorylate tyrosine residues in specific substrates, and their activities are involved in the pathophysiology of cancer. The inhibitors of tyrosine kinases block their oncogenic activation in cancer cells, therefore presenting a target for the development of new anticancer drugs. The computational methods in drug discovery and development minimize the time and cost needed in drug designing process. We have reviewed the recent advance in the quantitative structure-activity relationship (QSAR) study and molecular docking related to the new antitumor agents, such as amidine derivatives of 3,4-ethylenedioxythiophene, quinoline-arylamidine hybrids, 7-chloro-4-aminoquinoline-benzimidazole hybrids, rhodanine derivatives, and flavonoids isolated from the leaves of *Cupressus sempervirens*. The QSAR studies revealed important physicochemical and structural requirements for the antitumor activity and generated models for the prediction of antitumor activity of future potent molecules. Molecular docking allows rapid screening of a large number of compounds to determinate of potential binders of the target protein or enzyme, which is related to the anticancer activity and possible mechanism of action.

Keywords: Cheminformatic · Tyrosine kinase · Antiproliferative activity

1 Introduction

Tyrosine-protein kinases catalyse chemical reactions that transfer phosphate group from adenosine triphosphate (ATP) to a tyrosine residue in a protein. There are two major type of tyrosine kinase: receptor tyrosine kinase (RTK) and cytoplasmic/non-receptor tyrosine kinases (C/NRTK). C/NRTK, such as Src, Abl, Fak and Janus kinase, act as regulatory proteins, and playing key roles in cell differentiation, motility, proliferation, and survival. Src family of C/NRTK consists of nine members. Three of them (Src, Yes and Fyn) are inherent in all, while rest (Lck, Fgr, Blk, Lyn, Yrk, and Hck) are limited in specific types of tissues. Recent advances have implicated the role of tyrosine kinases in neoplastic development and progression [1, 2]. Thus, Hck is confined into cells of the myeloid and B-lymphocyte lineages, and their enhanced activity is associated with several types of leukemia and many solid malignancies, such as breast and colon cancer [3].

© The Author(s), under exclusive license to Springer Nature Switzerland AG 2023
I. Rojas et al. (Eds.): IWBBIO 2023, LNBI 13919, pp. 332–337, 2023.
https://doi.org/10.1007/978-3-031-34953-9_26

Selective tyrosine kinase inhibitors can block their oncogenic activation in cancer cells. There are four different kind of tyrosine kinase inhibitors: competitive inhibitors that bind at ATP binding pocket (type I inhibitors); inhibitors that bind to the ATP binding pocket in the Asp-Phe-Gly (DFG)-flipped conformation (type II inhibitors); not-competitive inhibitors that bind opposite (type III inhibitors) or any remote site from the ATP binding pocket (type IV inhibitors) [4, 5]. Thus, imatinib, (brand name Gleevec, Novartis), used for chronic myelogenous leukemia treatment, is II type of inhibitors, binds tightly to the Abl kinase domain. Series of inhibitors, pyridinyl triazine derivatives (DSA compounds), which are based on the central chemical scaffold of imatinib, recognized inactive kinase conformations and equally inhibit c-Src and Abl variants [6].

The development of new inhibitors of tyrosine kinase and the study of their mode of action is a necessary part of the discovery of a new cancer therapy. However, drug discovery and development is a very complex, time-consuming, and expensive process. Rational drug design methods, which comprise computational techniques, minimize that time and costs. In this article, we reviewed recent advances in the study of tyrosine kinase inhibitors performed by computational techniques, such as quantitative structure-activity relationship (QSAR), and molecular docking.

2 Structure of SRC-Protein Kinase Domain

Src-protein kinase domain consists of small amino-terminal (N) lobe made by residues 267–337 and large carboxyl-terminal (C) lobe made by residues 341–520, which are connected by a "hing" region. The smaller N-lobe has antiparallel β-sheet structures, and contains the glycine (G)-rich loop that forms part of the nucleotide-phosphate binding site. N-lobe is involved in anchoring and orienting ATP. The larger C-lobe is predominantly α-helical and is responsible for binding the protein substrate. Flexible activation (A) loop is placed between the N- and C-lobe nearby Tyr416. Phosphorylation of Tyr-416 in the activation loop positively regulates catalytic activity, whereas phosphorylation of Tyr-527 in the carboxy-terminal region repressed the activity. The ATP binding pocket of Src lies deep in cleft between the two lobes (Fig. 1). Cleft exists in open and close form. Open cleft allows free access of ATP to the catalytic site and release of ADP, while the closed form brings residues into the catalytically active state. Processes that block the interconversion of the open and closed form may cause the inhibition of enzyme [2]. A "gatekeeper" is a single residue, threonine, located in the hinge region that separates the ATP binding pocket from an adjacent hydrophobic pocket. This residue controls access of the inhibitor to the active site [7].

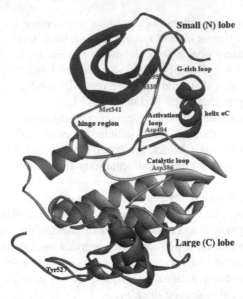

Fig. 1. Structure of Src-protein kinase domain.

3 The Role of Computational Methods in Discovery of New Inhibitors of Src-Protein Kinase

The computational methods have become an essential tool in drug discovery since they enable the prediction of physicochemical, pharmacokinetic, and toxicological properties of the leading compound of the future drug. *In silico* techniques like quantitative structure-activity relationships (QSAR), pharmacophore, docking and virtual screenings are playing crucial roles for the design of "better" molecules that may later be synthesized and biological assayed. QSAR techniques provides insight on relationships between chemical structure and biological activity and presents an alternative pathway for design and development of new molecules with improved activity. Using this relationship, the QSAR model is used to predict the activity of new leading compounds as active components of the future drug [8].

Molecular docking is a molecular modeling technique that is used to describe the interactions between receptor (enzyme, protein) and ligand (molecule). This technique allows rapid screening of a large number of compounds to determinate of potential binders through modeling/simulation and visualization techniques. This technique allows rapid screening of a large number of compounds to determine the potential of the inhibitor of the target protein or enzyme, which is related to the biological activity. Since the tyrosine kinase has role in the control of cellular growth and thus participates in human neoplastic diseases, tyrosine kinase inhibitors are targets for design of new mode of cancer therapy [1].

Among the newly synthesized amidine derivatives of 3,4-ethylenedioxythiophene (3,4-EDOT) the most efficient all six cell lines (AGS, gastric adenocarcinoma; MIA-PaCa2, pancreatic carcinoma; CaCo2, colon adenocarcinoma; HEp2, larynx carcinoma;

HeLa, cervix adenocarcinoma; and NCIH358, bronchioalveolar carcinoma) was cyano-amidine derivate. According to the results of QSAR analysis new potentially active compound were proposed. Antitumor activity of proposed compound ($logIC_{50} = 0.71$) against the pancreatic carcinoma cells has been predicted by means of the predictive model: $logIC_{50} = -2.86 + 5.55\ BIC5 - 3.30\ MATS7e$, where $BIC5$ presents bond information content of 5-order of symmetry, and $MATS7e$ is Moran autocorrelation descriptor weighted by atomic Sanderson electronegativities [9]. Quinoline-arylamidine hybrids showed higher efficacy against a human chronic myelogenous leukemia cell line (K562) and lymphoblastoid cell line (Raji) than carcinoma cells [10]. The QSAR study of the same quinoline-arylamidine hybrids generated predictive QSAR model $logGI_{50} = 5.69 - 6.94\ BIC1 - 0.11\ RDF110m - 0.2\ DIPX$, where descriptor $BIC1$ is bond information content index; $RDF110m$ is radial distribution function) descriptor weighted by atomic mass, and $DIPX$ is dipole moment along x axis. Obtained model revealed that enhanced 3D molecular distribution of mass calculated at radius 11 Å from the center of molecule, a higher number of terminal electronegative atoms, extension of the molecules' central linker between quinoline and arylamidine, higher ratio of single bonds and total number of atoms, and symmetric charge distribution are important physicochemical and structural requirements for the enhanced antitumor activity. Molecular docking study was applied to ensure the anticancer activity affinity to the binding site of the tyrosine-protein kinase (c-SRC) (PDB ID: 3G6H). Compounds were ranked by total energy of predicted pose in the binding site. Compounds that obtained the high energy-based scoring functions had also the highest antitumor activities. It was confirmed that the most active compound binds on the pocket between the small and large lobes of c-SRC, mostly throughout the hydrogen bonds and van der Waals interactions, and implied on the importance of the terminal nitrile group and amino groups from the 2-aminoethanol linker for enhanced binding affinity toward tyrosine kinase [11].

The results of the *in vitro* studies of the novel 7-chloro-4-aminoquinoline-benzimidazole hybrids on the antiproliferative activity toward one normal and seven cancer cell lines suggest that the amidine substituents on the benzimidazole ring decrease the activity of the novel compounds against normal and tumour cells [12]. Molecular docking performed on c-Src (PDB: 3G6H) has shown that compound, which demon-strated a significant effect on the growth of the HuT78 cell line, has also shown the highest binding energy, forming hydrogen bonds with Glu310 and Asp404 in the active site and other residues with van der Waals interactions.

Four polyphenols isolated from the leaves of *Cupressus sempervirens* showed signif-icant cytotoxicity against human hepatocellular liver carcinoma HepG2 cells, MCF-7, HC116, and A549. The Frontier Molecular Orbital analysis gives an insight into the reactivity of the molecule. The flavonoid molecule with the highest occupied molecular orbital (HOMO) energy and the lowest unoccupied molecular orbital (LUMO) energy, has stronger electron-donating abilities and showed the highest inhibition of cancer cell growth. Docking simulation results revealed the intractability of all structures with the enzyme, but especially with compound apigenin 4'-geranyl-8-glucopyranosyl-7-O-α-glucopyranoside, which showed the lowest total energy overtaking the original ligand pyridinyl triazine from the crystal structure (DSA1, PDB ID: G6H) and drug reference Doxorubicin [13].

A series of rhodanine derivatives synthesized in the Knoevenagel condensation of rhodanine and different aldehydes using deep eutectic solvent (DES), choline chloride:urea (1:2). Benzylidene derivative, with two hydroxyl groups in 3,4-position on the phenyl ring, has shown the highest 1,1-diphenyl-2-picrylhydrazyl radical (DPPH) scavenging activity (71.2%). Interaction of 5-arylidenerhodanines and haematopoietic cell tyrosine kinase (Hck) (PDB ID: 2HCK) *in silico* has evaluated by molecular docking based on binding mode of quercetin as inhibitor. The study confirmed that inhibitor, benzylidene derivatives of the 5-arylidenerhodanines, forms the key interactions, the four H bond with Met341, Ser345 (one with the main chain, and the other one with the side chain), and with Asp348 [14]. Cytotoxic effect on cell proliferation (CaCo-2, HeLa, MDCK-1, Hut-78, K562) in vitro was evaluated by a series of rhodanine derivatives synthesized via Knoevenagel condensation in DESs [15]. The best inhibiting activity toward all cell lines have showed derivatives that possessing only one group in the C2 of the phenyl ring. Molecular docking was performed into the binding site of the target enzyme tyrosine-protein kinase c-Src (PDB ID: 3G6H). The ligand with highest binding energy was located in the hydrophobic pocket between the N-lobe and C-lobe. The key interactions with binding site residues were achieved through oxygen atoms from phenoxy and rhodanine groups and rhodanine sulphur atoms. The findings of the molecular docking study performed on the c-Src are in accordance with the results of the QSAR study. The best QSAR model obtained for the cytotoxic activity against Human T cell lymphoma (Hut-78) is: $\log IC_{50} = 2.51 - 3.46\,MATS2e - 1.05\,MATs7e - 0.15\,RDF060p$; where descriptors $MATS2e$ and $MATs7e$, are Moran two-dimensional autocorrelation descriptors weighted by Sanderson electronegativity, and $RDF060p$ is radial basis functions (RDF) descriptor centred on the interatomic distance 6 Å and weighted by atomic polarizabilities. QSAR study on the cytotoxic activity against Human T cell lymphoma revealed the importance of the presence of atoms with higher polarizability in the outer region of molecules.

4 Conclusion

Tyrosine-protein kinases are a favourable target for the screening of the compounds as future active components of anticancer drugs. Cheminformatics techniques, such as QSAR and molecular docking can aid in rational drug design and provide identify the important physicochemical properties and structural attributes for the anticancer activity of analysed compounds, as well as, explained their binding mode on the tyrosine-protein kinases. QSAR study provides information about the structural characteristics related to the enhanced cytotoxic activity and reveals the mechanisms of action within a series of chemicals. Also, QSAR models could predict the anticancer activity of the future molecules. Molecular docking studies can provide insights into the potential of the synthesized compounds as ligands for c-Src and possible mechanism of action. Computational methods in drug design rationalize the time and cost of experiments in comparison to traditional drug discovery methods.

References

1. Paul, M.K., Mukhopadhyay, A.K.: Tyrosine kinase – role and significance in cancer. Int. J. Med. Sci. **1**(2), 101–115 (2004)
2. Roskoski, R.: Src protein–tyrosine kinase structure and regulation. Biochem. Biophys. Res. Commun. **324**, 1155–1164 (2004)
3. Poh, A.R., O'Donoghue, R.J.J., Ernst, M.: Hematopoietic cell kinase (HCK) as a therapeutic target in immune and cancer cells. Oncotarget **6**(18), 15752–15771 (2015)
4. Vijayan, R.S.K., et al.: Conformational analysis of the DFG-out kinase motif and biochemical profiling of structurally validated type II inhibitors. J. Med. Chem. **58**, 466–479 (2015)
5. Gavrin, L.K., Saiahm, E.: Approaches to discover non-ATP site kinase inhibitors. MedChem-Comm **4**(1), 41–51 (2012)
6. Seeliger, M.A., et al.: Equally potent inhibition of c-Src and Abl by compounds that recognize inactive kinase conformations. Cancer Res. **69**, 2384–2392 (2009)
7. Azam, M., Seeliger, M.A., Gray, N.S., Kuriyan, J.K., Daley, G.Q.: Activation of tyrosine kinases by mutation of the gatekeeper threonine. Nat. Struct. Mol. Biol. **15**(10), 1109–1118 (2008)
8. Winkler, D.A.: The role of quantitative structure–activity relationship (QSAR) in biomolecular discovery. Brief. Bioinform. **3**, 73–86 (2002)
9. Jukić, M., et al.: Antitumor activity of 3,4-ethylenedioxythiophene derivatives and quantitative structure–activity relationship analysis. J. Mol. Struct. **1133**, 66–73 (2017)
10. Krstulović, L., et al.: New quinoline-arylamidine hybrids: synthesis, DNA/RNA binding and antitumor activity. Eur. J. Med. Chem. **137**, 196–210 (2017)
11. Rastija, V., et al.: Investigation of the structural and physicochemical requirements of quinoline-arylamidine hybrids for the growth inhibition of K562 and Raji leukemia cells. Turkish J. Chem. **43**(1), 251–265 (2019)
12. Krstulović, L., et al.: Novel 7-chloro-4-aminoquinoline-benzimidazole hybrids as inhibitors of cancer cells growth: synthesis, antiproliferative activity, in silico ADME predictions, and docking. Molecules **28**(2), 540 (2023)
13. Elsharkawy, E.R., Almalki, F., Ben Hadda, T., Rastija, R., Lafridi, H., Zgou, H.: DFT calculations and POM analyses of cytotoxicity of some flavonoids from aerial parts of Cupressus sempervirens: docking and identification of pharmacophore sites. Bioorg. Chem. **100**, 103850 (2020)
14. Molnar, M., et al.: Environmentally friendly approach to knoevenagel condensation of rhodanine in choline chloride: urea deep eutectic solvent and QSAR studies on their antioxidant activity. Molecules **23**, 1897 (2018)
15. Molnar, M., Lončarić, M., Opačak-Bernardi, T., Glavaš-Obrovac, L., Rastija, V.: Rhodanine derivatives as anticancer agents – QSAR and molecular docking studies. Anti-Cancer Agents Med. Chem. **23**, 839–846 (2023)

The Coherent Multi-representation Problem with Applications in Structural Biology

Antonio Mucherino[✉]

IRISA, University of Rennes, Rennes, France
antonio.mucherino@irisa.fr

Abstract. We introduce the Coherent Multi-representation Problem (CMP), whose solutions allow us to observe simultaneously different geometrical representations for the vertices of a given simple graph. The idea of graph multi-representation extends the common concept of graph embedding, where every vertex can be embedded in a domain that is unique for each of them. In the CMP, the same vertex can instead be represented in multiple ways, and the main aim is to find a general multi-representation where all the involved variables are "coherent" with one another. We prove that the CMP extends a geometrical problem known in the literature as the distance geometry problem, and we show a preliminary computational experiment on a protein-like instance, which is performed with a new Java implementation specifically conceived for graph multi-representations.

1 Introduction

In several applications, given a certain number of constraints involving a given set of objects, the main goal is to identify suitable geometrical representations for such objects. This work takes as a starting point the works in the context of Distance Geometry (DG), where embeddings of a given graph G in a Euclidean space need to be defined in such a way to satisfy a certain number of distance constraints, where the distance metric is generally the Euclidean norm [10].

Among the DG applications, the traditional ones (i.e. the ones that mostly appear in the scientific literature) are the applications in structural biology and sensor networks. Experimental techniques such as Nuclear Magnetic Resonance (NMR) can in fact provide estimates on the proximity between atom pairs in a given molecule, so that looking for molecular conformations satisfying these distance constraints is a problem falling in the context of DG [3,15]. In sensor networks, the radio signals between two mobile antennas can also provide proximity information, so that localizing the sensors from the estimated distances is basically this very same problem, where the only difference can be found in the quality of the distances that come to play in the definition of the constraints [2,13]. Other emerging DG applications include acoustic networks [4] and their use in robotics [5], as well as the recent adaptive maps [8].

The motivation for this work comes from the observation that distances are not the only type of information that is given for the definition of the geometric constraints. In structural biology, not only distances are available, but also a certain number of torsion angles formed by quadruplets of atoms in the molecule [11], named *dihedral angles*. Under certain assumptions, these torsion angles may be converted to distances [7], and the obtained distances may simply be used together with the original distance information. However, conversion procedures cannot guarantee that the performed conversions are lossless. For this reason, we found the need to define a more general problem, where one can deal with several types of information at the same time.

Even though the focus of this work is mostly on the application in structural biology, it is important to point out that the necessity to consider different types of information for the definition of the geometric constraints is also valid for other applications. In sensor networks, in spite of the fact that several works only focus on distances (see for example [9]), there is another type of information that can be measured: the *angle-of-arrival* of the radio signal on the antennas [13].

The main DG problem, known in the literature under the acronym DGP, asks whether it is possible to find an embedding $x : V \rightarrow \mathbb{R}^K$ for a simple weighted undirected graph $G = (V,E,d)$ such that the distances between two embedded vertices u and $v \in V$ correspond to the edge weights given by the weight function d. It is generally supposed that G is connected, otherwise the DGP could simply be divided in as many sub-problems as the number of disconnected components in the graph. In some special situations, the properties of G allow us to discretize the search space for potential embeddings x of the graph [14], and this is possible through the introduction of ad-hoc vertex orders on G [6,18]. In this work, we will suppose that a vertex order on G is available, and that it is given through the orientation of the edges of the graph. In other words, we will suppose that our graphs are directed, and hence denote their arcs with the symbol $(u,v) \in E$.

The Coherent Multi-representation Problem (CMP) is therefore introduced in this work to provide new theoretical basis for the solution of problems arising in the applications mentioned above where different types of information can be employed at the same time. Differently from the DGP, the CMP does not make any net distinction between known (e.g. the distances) and unknown information (e.g. the embedding), but it actually represents the full piece of information in one unique multi-representation system. Thus, a solution to the CMP is a set of values for the internal variables for the several employed representations that turns out to be "coherent" (see Sect. 2).

This work comes with the initial commit of a new public GitHub repository containing some Java classes implementing the multi-representation system (see Sect. 3). Throughout the paper, we make reference to types of representations for the vertices that are typical in applications in structural biology, and a very preliminary computational experiment on a protein-like helical model is commented in Sect. 4. Finally, some future research directions are given in Sect. 5.

We point out that the idea of employing multi-representations is not completely new in the scientific literature. One important example is given by the works in education and learning [1], where the idea to have complementary representations for the same object, and to simultaneously exploit the advantages that each of them can give, is already explored. In computer science, an example of multi-representation can be found in [20], where some geographic maps at different levels of resolution are managed in one unique database. Some map representations, however, may not be compatible, and they may need to be reused independently from each other. In this work, there is one important aspect that comes to play: we attempt to have the several representations for a single object to be compatible to one another at all times. To this purpose, we introduce the idea of "coherence" for all involved vertex representations. A similar idea was previously exploited in the context of education in [19].

2 Coherent Multi-representation Problem

Let $G = (V, E)$ be a simple directed graph, where the orientation of its arcs gives us the information about the ordering of the vertices in V. We exploit in this work the fact that each vertex of G can be embedded in the traditional Euclidean space by employing different types of coordinate systems. The most common naturally is the Cartesian coordinate system, but others are also possible. One alternative coordinate system that we will use in the following for our case study in dimension 3 is given by the well-known spherical coordinates, where the location of the vertex $v \in V$ is relative to another vertex u, and it is defined through the distance and the relative orientation of v w.r.t. u in the given Cartesian system. Another representation that we use is through torsion angles [12]. In order to switch from one coordinate system to another, coordinate transformations can be applied.

We use the symbol Y for the definition domain of a given coordinate system. Notice that the dimension of Y may differ from the dimension of the original Euclidean space $\mathbb{R}^K$. An example where these two dimensions are different is given for example by the representation through torsion angles, where the Euclidean space has dimension 3 and the torsion domain has only one dimension. We refer to a coordinate transformation, capable to convert the coordinates from one system to another, as a triplet (P, Y, f), where P is the set of parameters, Y is the coordinate domain, and f is the mapping:

$$f : (p, y) \in P \times Y \longrightarrow x \in \mathbb{R}^K,$$

that performs the transformation from the variables $y \in Y$ to the standard Cartesian coordinates, denoted by x as in the Introduction.

We remark that the trivial transformation (P, Y, f), where $P = \emptyset$, $Y = \mathbb{R}^K$ and f is the identity function, maps Cartesian coordinates into the same Cartesian coordinates. The spherical coordinates require a reference vertex u for a given vertex $v \in V$, so that the Cartesian coordinates of u need, as a consequence, to be included in P. For the definition of torsion angles, the parameter set P contains much more information, which we omit to comment here for lack of space; the interested reader can find all the necessary details in [12]. In the following, we will refer to the triplets (P, Y, f) as transformations, as well as "representations" for a given vertex v.

The definition domains Y can encapsulate geometric constraints, so that the corresponding y variables are only able to cover restricted regions. For example, simple constraints in $\mathbb{R}^K$ can define box-shaped regions. In dimension 3, simple constraints on the distance involved in the definition of the spherical coordinates allow us to control the relative distance between the two involved vertices. Also, simple constraints on the possible values of torsion angles allow us to define arc-shaped regions of the original

Euclidean space. For values for y that are not in the delimited region Y, we can consider the existence of a projection operator that projects those values on the allowed coordinate space Y.

One peculiarity in this work consists in associating vertex representations to the arcs of the graph G, and not its vertices. Let $\hat{E}$ be the set obtained as the union of E with the set $\{(v,v) : v \in V\}$; we will use the subscripts $e = (u,v) \in \hat{E}$ to make reference to vertex representations (P_e, Y_e, f_e) that are related to a specific arc of the graph. The predecessor u of v in the arc e is supposed to play the role of reference in the representation of the vertex v. Notice that, for the torsion angle case, there are actually three reference vertices, but it makes sense to associate the torsion angle to the arc (u,v) where u is the farthest reference in the vertex order induced by the orientation of the arcs. Finally, representations which make use of no reference vertices, as it is the case for Cartesian coordinates, can be associated to the added arcs of the type $(v,v) \in \hat{E}$.

For a given arc $e \in \hat{E}$, we can define multiple representations for this same arc; we will use superscripts to distinguish among the various employed representations. The number of representations for the each arc e can vary: there can be only one, or several, or even no one. However, this last situation is generally to be avoided, for some information encoded by the graph would not be exploited in this case. Naturally, all functions f_e involved in the vertex representations need to have a common codomain, which is predetermined and set to $\mathbb{R}^K$ in this work.

Definition 2.1. *Given a vertex $v \in V$ and one of the arcs $e = (u,v) \in \hat{E}$, a "vertex multi-representation" in dimension $K > 0$ and w.r.t the arc e is the set*

$$\Xi_e = \{(P_e^1, Y_e^1, f_e^1), \dots, (P_e^r, Y_e^r, f_e^r)\}$$

containing r different vertex representations related to the arc e. The "expected" Cartesian coordinates for Ξ_e are given by the function:

$$\xi_e : (y_e^1, \dots, y_e^r) \in Y_e^1 \times \dots Y_e^r \longrightarrow \frac{1}{r} \sum_{i=1}^{r} f_e^i(p_e^i, y_e^i) \in \mathbb{R}^K, \tag{1}$$

where $p_e^i \in P_e^i$, for each i. We say that the set of internal coordinates $y_e = (y_e^1, \dots, y_e^r)$ is "coherent" if

$$\forall i \in \{1, \dots, r\}, \quad \xi_e(y_e) = f_e^i(p_e^i, y_e^i).$$

Vertex multi-representations allow us to simultaneously associate several types of representations to the same vertex v of the graph. For example, the use of spherical coordinates can help controling the distance between v and its reference u, while v can at the same time also be represented through a torsion angle. Recall that the domain Y associated to the transformations makes it possible to restrict the region of feasibility for the variables. In some special cases, the domain Y may allow some variables to take only value, thus indicating that the value for this variable is actually known (or imposed).

Definition 2.1 can be trivially extended to all the arcs in the edge set $\hat{E}$, so that a multi-representation for all the vertices $v \in V$, and for all arcs $e = (u,v) \in \hat{E}$, can be defined. When we extend to the entire graph G, we can notice that there are actually two levels of multiplicity for the representations. Firstly, for a given arc $(u,v) \in \hat{E}$, several different representations can be defined, as commented a few lines above. Moreover, the second level of multiplicity comes from the fact that at least one representation is expected to be defined for every arc in $\hat{E}$, and hence there will be at least as many representations for v as there are arcs $(u,v) \in \hat{E}$ having v as a destination vertex.

When all arcs in $\hat{E}$ are involved, we say that the set of vertex multi-representations $\Xi_G = \cup_{e \in \hat{E}} \Xi_e$ forms a "graph multi-representation". The function in Eq. (1) can also be extended to the entire graph, and we will refer to it with the symbol ξ_G. The input of ξ_G is a vector $\bar{y}$ combining all the variables y_e^i, for each arc $e \in \hat{E}$ and each representation (indexed by i). We say therefore that $\bar{y}$ is *coherent* if all its components are coherent (see Definition 2.1).

The only possibility for the function ξ_G to be equivalent to a standard graph embedding is when the graph is trivial. This situation is naturally of no interest, and we will therefore suppose that our graphs G are non-trivial. The graph multi-representation evidently allows for a richer representation of the graph, which explicitly exploits the connectivity information encoded by the graph, which is instead not taken into consideration by standard graph embeddings.

Definition 2.2. *Given a simple directed graph $G = (V,E)$ and a graph multi-representation Ξ_G in dimension $K > 0$, the Coherent Multi-representation Problem (CMP) asks whether there exists a vector $\bar{y}$, composed by the internal variables for the coordinates of the vertex multi-representations in Ξ_G, that is coherent.*

The following result relates the DGP (see Introduction) with the new CMP. An immediate consequence of this result is that the CMP is NP-hard.

Proposition 2.3. *The DGP is a special case of the CMP.*

Proof. Let $G^{dgp} = (V,E,d)$ be a simple weighted undirected graph representing a generic instance of the DGP. Let $G = (V,E)$ be the same graph G^{dgp} without the weight function d, but with directed edges for encoding a suitable vertex order (see Introduction). If no information about vertex orders on V is available, any order can be used.

We will proceed by constructing a graph multi-representation Ξ_G. For every $v \in V$, we assign a representation by Cartesian coordinates to each arc of the type $(v,v) \in \hat{E}$. For every $(u,v) \in E$, moreover, we assign a representation by spherical coordinates, where the domain Y encodes the bounds on the distance values given by the weights of the graph G^{dgp}. Notice that more than one distance may be known for a given vertex v, and hence the number of introduced representations by spherical coordinates is likely to vary vertex per vertex.

By definition, a solution to the so-constructed CMP is a vector $\bar{y}$ containing the internal coordinates for all these vertex representations (the introduced Cartesian and spherical coordinates). The vector $\bar{y}$ is supposed to be coherent, in the sense that, for every $e \in \hat{E}$, all functions f_e^i are supposed to give the same result. As a consequence, the expected Cartesian coordinates for the generic vertex v must satisfy all distance constraints encoded by the spherical coordinates, and they form therefore a solution to the original DGP. □

3   An Object-Oriented Implementation

A quick search on the Internet can reveal the existence of several freely distributed implementations for storing graph structures, in several low and high level programming languages. For this work, we opt for a completely new implementation which does not only allow us to store the graph structure, but also the several vertex representations that we can associate to its arcs. Moreover, our implementation exploits an internal notification system for keeping all vertex representations updated. Naturally, it is not always possible to keep all vertex representations in a coherent state. We have implemented specific methods for this verification.

The notification system acts at two levels: it makes sure that all representations for the same vertex are synchronized (as far as this is possible), and it notifies other vertex representations, that use the modified vertex as a reference, of the performed change. These other representations will have to update their internal state in order to consider the new information provided by their references.

The language we have chosen is Java for its object-oriented paradigm and for its relative simplicity; specific syntax related to more recent versions of Java have been avoided, so that translations to other languages supporting classes, encapsulation and inheritance will potentially be easy to perform. The Java codes are available on a public GitHub repository[1]. The current implementation focuses on coordinate systems in a three-dimensional Euclidean space only.

The main class in our Java code is the `Coordinates` class. As its name suggests, this class is supposed to hold information, and to perform actions on, the numerical values employed for the various vertex representations. The peculiarity of this class is that it contains a certain number of private sub-classes, each defining a particular representation and some basic methods for their manipulation. We point out that here we relax the encapsulation principle for the access to the attributes of the sub-classes. We found in fact that a stricter encapsulation would have led to a more complex and less efficient code without really adding any level of access security to the data. Notice however that the encapsulation principle is respected at the level of the `Coordinates` class. This class, moreover, contains a certain number of private methods that are not supposed to be invoked from the outside, which ensure the realization of the notification system mentioned above.

Every instance of `Coordinates` can hold several representation types, and for every type, it can hold several instances for the given type. The only exception is for the standard Cartesian coordinates, because two instances of the Cartesian coordinates that are coherent in the same `Coordinates` object can only be identical. This is however not true for the spherical coordinates, and for the torsion angles.

Suppose for example that an instance of `Coordinates`, call it C1, contains three inner representations: one of Cartesian type, one of spherical type, and another of torsion type. Suppose we wish to position C1 in specific Cartesian coordinates. Our class provides a method to this purpose, but it does not only limit itself to change the values of the Cartesian coordinates, it also attempts to adapt all other representations to the new imposed Cartesian coordinates. In practice, the variables used for the spherical and

---

[1] https://github.com/mucherino/DistanceGeometry, commit be4e33b, folder `javaCMP`.

torsion types are updated for keeping them compatible with the new Cartesian coordinates. However, as mentioned above, this is not always possible, because the definition domain of the internal variables may be constrained (through the set Y introduced in Sect. 2, see the BoundedDouble class). When the update fails, then we say that C1 is not in a coherent state.

Now suppose for simplicity that the current state of C1 is coherent. Suppose that the reference Coordinates instance for C1, in both spherical and torsion types, is C0. C0 admits only one representation, whose type is Cartesian. When the Cartesian coordinates of C0 are modified, our notification system sends a "signal" to C1, which is now supposed to update some of its internal variables. First of all, we can recompute the Cartesian coordinates of C1, by using the information given by the spherical type. However, C0 takes also part in the definition of the torsion angle, which may now be incompatible with the spherical type. All possible updates, for all types in C1, are attempted, but, again, the procedure may fail in leaving C1 in a coherent state. The little experiment commented in the next section was instead conceived in such a way to maintain the coherent state for its representations.

The Coordinates class contains methods for verifying whether an instance is in a coherent state or not, and it can provide, in both situations, its "expected" Cartesian coordinates via a devoted Java method implementing the formula in Definition 2.1. The interested reader is invited to look directly at the code for additional technical details of this implementation.

4   A Simple Experiment

We present in this section a very preliminary experiment performed with our new Java classes. The details of the experiment can be found on the GitHub repository in the class named Experiments (please make sure to refer to the commit indicated in the previous section). Our experiment makes use of the torsion type, which is particularly useful in the construction of the typical helical structures in protein conformations. The reader can find other smaller experiments, involving the spherical type as well, in the set of automatic tests implemented in the main method of the Coordinates class.

In Experiments, we initially construct our helical conformation by simply instantiating a multi-representation for the underlying graph where standard Cartesian coordinates and torsion angles co-exist. We do not give values to the Cartesian coordinates, we basically only provide the torsion angles, and this action automatically gives us the model depicted at the bottom of Fig. 1. The interesting point about the experiment is that the user is only supposed to initialize and set up the values of the torsion angles, whereas the calculations that allow the construction of the model are automatically performed by our Java code in an attempt to keep the various representations coherent with one another.

The second model, at the top of Fig. 1, was obtained with one line of code in the Experiments class, consisting in changing the value for one torsion angle. In this case, not only the change in one representation type started the updating system within the same Coordinates instance (the one whose torsion angle was changed), but it also notified, in a chain, all *subsequent* Coordinates instances, which updated their inner variables as well.

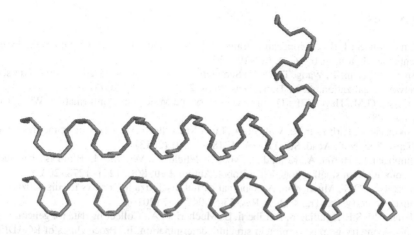

Fig. 1. Two helix-like confirmations obtained with our new Java classes.

This automatic synchronization of the various representations can be particularly useful when implementing solution methods that mainly act on specific coordinate systems, because it allows for simultaneously controling other vertex representations, by immediately revealing, for example, that some geometric constraints are consequently violated.

5   Conclusions

This work introduced the CMP, a new decision problem that extends another very common and widely studied problem, named the DGP. We showed that the DGP is included in the CMP, so that the latter inherits the NP-hardness of the former. We have briefly presented a new Java implementation allowing for vertex multi-representations that are at the core of the CMP.

Future works will be mainly devoted to the development of solvers for the CMP, by taking as a starting point some previous works that we have performed in the context of the DGP (see for example [16]). Moreover, it is also our intention to include other vertex representations in our implementations, in order to tackle a larger variety of applications. The realization of this future work is likely to be particularly delicate, for it may reveal some limitations of our multi-representation approach. In fact, some types of information may not be capable to uniquely reconstruct the Cartesian coordinates of the vertices, or at least not with the same precision as the spherical and torsion types can do. Finally, we also plan to study, in a near future, the use of the CMP in the context of dynamical problems [17].

Acknowledgments. We wish to thank the three reviewers for their fruitful comments. This work is partially supported by ANR French funding agency (MULTIBIOSTRUCT project ANR-19-CE45-0019).

References

1. Ainsworth, S.: DeFT: a conceptual framework for considering learning with multiple representations. Learn. Instr. **16**(3), 183–198 (2006)
2. Biswas, P., Lian, T., Wang, T., Ye, Y.: Semidefinite programming based algorithms for sensor network localization. ACM Trans. Sens. Netw. **2**(2), 188–220 (2006)
3. Crippen, G.M., Havel, T.F.: Distance Geometry and Molecular Conformation. Wiley, Hoboken (1988)
4. Dokmanić, I., Parhizkar, R., Walther, A., Lu, Y.M., Vetterli, M.: Acoustic echoes reveal room shape. Proc. Natl. Acad. Sci. **110**(30), 12186–12191 (2013)
5. Dümbgen, F., Hoffet, A., Kolundzija, M., Scholefield, A., Vetterli, M.: Blind as a bat: audible echolocation on small robots. IEEE Robot. Autom. Lett. **8**(3), 1271–1278 (2023)
6. Gonçalves, D.S., Mucherino, A.: Optimal partial discretization orders for discretizable distance geometry. Int. Trans. Oper. Res. **23**(5), 947–967 (2016)
7. Hengeveld, S.B., Malliavin, T., Liberti, L., Mucherino, A.: Collecting data for generating distance geometry graphs for protein structure determination. In: Proceedings of ROADEF23, Rennes, France, 2 p. (2023)
8. Hengeveld, S.B., Plastria, F., Mucherino, A., Pelta, D.A.: A linear program for points of interest relocation in adaptive maps. In: Geometric Science of Information (GSI 2023). LNCS (2023, to appear)
9. Krislock, N., Wolkowicz, H.: Explicit sensor network localization using semidefinite representations and facial reductions. SIAM J. Optim. **20**, 2679–2708 (2010)
10. Liberti, L., Lavor, C., Maculan, N., Mucherino, A.: Euclidean distance geometry and applications. SIAM Rev. **56**(1), 3–69 (2014)
11. Malliavin, T.E., Mucherino, A., Lavor, C., Liberti, L.: Systematic exploration of protein conformational space using a distance geometry approach. J. Chem. Inf. Model. **59**(10), 4486–4503 (2019)
12. Malliavin, T.E., Mucherino, A., Nilges, M.: Distance geometry in structural biology: new perspectives. In: [15], pp. 329–350. Springer, New York (2013). https://doi.org/10.1007/978-1-4614-5128-0_16
13. Mao, G., Fidan, B., Anderson, B.D.: Wireless sensor network localization techniques. Comput. Netw. **51**(10), 2529–2553 (2007)
14. Mucherino, A., Lavor, C., Liberti, L.: The discretizable distance geometry problem. Optim. Lett. **6**(8), 1671–1686 (2012)
15. Mucherino, A., Lavor, C., Liberti, L., Maculan, N. (eds.): Distance Geometry: Theory, Methods and Applications. Springer, New York (2013). https://doi.org/10.1007/978-1-4614-5128-0
16. Mucherino, A., Lin, J.-H., Gonçalves, D.S.: A coarse-grained representation for discretizable distance geometry with interval data. In: Rojas, I., Valenzuela, O., Rojas, F., Ortuño, F. (eds.) IWBBIO 2019. LNCS, vol. 11465, pp. 3–13. Springer, Cham (2019). https://doi.org/10.1007/978-3-030-17938-0_1
17. Mucherino, A., Omer, J., Hoyet, L., Robuffo Giordano, P., Multon, F.: An application-based characterization of dynamical distance geometry problems. Optim. Lett. **14**(2), 493–507 (2020)
18. Omer, J., Mucherino, A.: The referenced vertex ordering problem: theory, applications and solution methods. Open J. Math. Optim. **2**, 1–29 (2021). Article No. 6
19. Seufert, T.: Supporting coherence formation in learning from multiple representations. Learn. Instr. **13**(2), 227–237 (2003)
20. Zhou, S., Jones, C.B.: A multi-representation spatial data model. In: Hadzilacos, T., Manolopoulos, Y., Roddick, J., Theodoridis, Y. (eds.) SSTD 2003. LNCS, vol. 2750, pp. 394–411. Springer, Heidelberg (2003). https://doi.org/10.1007/978-3-540-45072-6_23

Computational Study of Conformational Changes in Intrinsically Disordered Regions During Protein-Protein Complex Formation

Madhabendra Mohon Kar, Prachi Bhargava, and Amita Barik$^{(\boxtimes)}$ ⓘ

Department of Biotechnology, National Institute of Technology, Durgapur 713209, India
amita.barik@bt.nitdgp.ac.in

Abstract. Intrinsically Disordered Regions (IDRs) even though they cannot form a defined three-dimensional structure play a pivotal role in modulating cellular processes and signalling pathways. In the present study, we analyse the conformational changes in IDRs upon complex formation using a non-redundant dataset of binary, X-ray solved 356 protein-protein (P-P) complexes and their corresponding unbound forms. IDRs are prevalent in both unbound and complex proteins and after comparing them in both groups they were categorised into three classes: (a) Disordered-Ordered (D-O), where IDRs present in first group were observed to be ordered in the second group (b) Disordered-Partial Ordered (D-PO), where IDRs present in the first group were found to be partially ordered in the second group and (c) Disordered-Disordered (D-D), where IDRs present in one group remained disordered in the other group. The study of secondary structures of residues in the D-O category reveals that majority of IDRs upon complexation form coils followed by helices and strands. Though majority of residues of IDRs in the D-O class are located at the surface of P-P complexes, we observe a significant number of residues form the interface suggesting that they contribute to the stability of the complexes. Amino acids of IDRs under the D-O category are also involved in polar interactions making hydrogen bonds with other residues as well as water. There are some structured and partially structured regions in the unbound proteins which upon complexation become completely disordered. These findings provide fundamental insights into the underlying principles of molecular recognition by disordered regions in P-P complexes.

Keywords: Intrinsically Disordered Proteins (IDPs) · Intrinsically Disordered Regions (IDRs) · protein-protein complexes · interface · hydrogen bonds

1 Introduction

Intrinsically Disordered Regions (IDRs) are polypeptide segments that can perform their function, not only despite being disordered but because of being disordered [1–5]. IDPs and IDRs exist as a dynamic structural ensemble, and thus disordered proteins do not lie at a single opposite position from ordered proteins, but instead ordered and disordered proteins lie at different points on a continuum [1, 2, 6]. Studies have reported that IDPs and proteins containing IDRs are abundant in nature [7] and their occurrence increases with organismal complexity [2].

I. Rojas et al. (Eds.): IWBBIO 2023, LNBI 13919, pp. 347–363, 2023.
https://doi.org/10.1007/978-3-031-34953-9_28

The popularity and relevance of the structure-function paradigm initially led to observations like that of missing electron density in X-ray crystallography structures of proteins, the in-vivo increased sensitivity of some proteins to proteolysis and abnormalities during the purification process being relegated as mere experimental errors [6, 8]. Eventually, the disorder-function paradigm was formulated to account for such observations [8].

An IDP or IDR can have an amount of heterogeneity comprising different segments like ordered segments (foldons), partially folded segments with transient residual structures (semi-foldons), segments that can fold upon interaction with specific binding partners (inducible foldons), and segments whose function is dependent on being intrinsically disordered (nonfoldons) [9, 10]. The heterogeneity of IDRs influences the various modes of binding of a protein with its specific partner [10, 11]. On binding, the IDRs may adopt a well-defined tertiary structure, retain some degree of disorder, or remain completely disordered, forming fuzzy complexes [11, 12]. The fuzzy complexes may be formed due to the protein regions forming different ordered conformation with different partners, known as polymorphism, or the disordered regions may adopt a transient secondary structural element, under certain environmental conditions, known as conditional folding or the intermolecular contacts between the binding partners are occasionally sampled, known as dynamic binding [13, 14]. In some cases, after binding to its specific partner an ordered region in a protein becomes disordered, and thus becomes functional [9].

The conformational flexibility of IDRs allows them to interact promiscuously with different protein targets on different occasions and thus, proteins enriched in IDRs serve as hubs in protein-protein interaction networks [7, 15]. The IDPs and IDRs can bind to their specific targets with high specificity but moderate avidity, which allows them to transiently associate with the partners and give them the ability to dissociate spontaneously and rapidly, conferring kinetic advantage in cell signalling pathways [7, 15]. They thus, play a critical role in the cell signalling pathways and in crucial cellular processes, like the regulation of transcription, translation, and the cell cycle. The abundance and the overall characteristics of the IDPs and IDRs in the cell are under tight regulation, as mutations or any alterations in their abundance are associated with diseases [7]. Several studies have also shown that IDRs could play a vital role in drug design application, as structural disorders could guide ligand selection during drug development [16–18]. Thus, studying the structural and functional role of IDRs in the protein-protein complexes is very important.

Over the years, computational and experimental studies on protein disorders have been conducted with the aim of understanding various aspects related to IDRs and IDPs, like sequence, abundance, structure, function, evolution, regulation, etc. [8]. There are a limited number of experimental studies on disordered-to-ordered and ordered-to-disordered transitions that have been made to understand the underlying molecular principles [19–22]. It has been reported that disorder is encoded in the amino acid sequence of IDRs, and thus has a biased amino acid composition, disfavoring the order promoting amino acids [23]. However, such studies have been carried out with limited datasets that might not be representative of all types of IDRs [24]. Several databases such as DisProt [25], IDEAL [26, 27], and MobiDB [28] have been developed for curating the disordered regions in proteins. Also, many predictors such as IUPred [29, 30], ANCHOR [31], DISOPRED [32], DEPICTER [33] have been developed to predict IDRs in protein sequences.

In the present study, we have carried out a computational analysis of the IDRs using the structural information available in the Protein Data Bank (PDB) [34]. The number of experimentally derived three-dimensional (3D) structures of proteins is rapidly increasing in the PDB and majority of them (about 85%) are derived from X-Ray crystallography experiments. Because of the lack of stability, IDRs are often missing in the electron density maps of X-ray crystallography studies and thus their information is not available in the 3D structures. Although, this might be a result of crystal packing irregularities, a long stretch of missing residues is likely to be a disorder segment. We have curated high resolution X-ray solved 3D structures of protein-protein (P-P) complexes and their unbound or free forms from the PDB and have identified the IDRs in these proteins. A number of structural, and physicochemical parameters were calculated to understand how the IDRs present in the unbound forms of proteins undergo a disordered-to-ordered transition upon complex formation. This analysis may provide more insights into the underlying principles of molecular recognition by disordered proteins. Moreover, the knowledge of how ordered and disordered proteins work together will provide a broader picture of the functions of proteins.

2 Methods

2.1 Dataset Curation

Three-dimensional (3D) atomic coordinates of binary protein-protein complexes solved through X-ray crystallography were curated from the PDB [34]. 3D structures having a 3 Å or better resolution with protein subunits having at least 30 amino acids in length, with global asymmetry were collected for the present study. When the protein sequence in two complexes had more than 35% sequence identity, the one with the better resolution was retained. To remove the redundant data, CD-HIT software [35, 36] was used to cluster the sequences with 90% identity cut-off, and the structure with the best resolution was retained among the different clusters to create a non-redundant dataset.

For obtaining the unbound protein subunits corresponding to the bound protein complexes, pBLAST [74] was used, against the non-redundant PDB database, with sequence identity $\geq 90\%$, query coverage $\geq 95\%$, and an E-value $\leq 10^{-3}$. The entry having the highest sequence similarity and maximum alignment length with the bound structure was selected. The structures having resolution of ≤ 3 Å were retained.

2.2 IDRs in the Bound and Complex Proteins

The SEQATOMs [37] entries for all protein complexes and unbound subunits were compared using in-house programs, and the missing residues along with their position and length were obtained for all the protein subunits. All the internal missing regions, having a length of 2 residues or more were considered to be IDRs. For the terminal regions, a missing region was considered to be IDRs if it had a minimum length of 5 residues. This assumption was made as the structures used in this analysis were of high resolution, and previous works support such an assumption [28, 38–43].

2.3 Classification of IDRs

The IDRs present in unbound forms were compared with those present in the complexes and vice versa and were classified into 3 categories: (a) Disordered-Ordered (D-O), where IDRs present in the former were observed to be ordered in the later ($\geq$80% ordered residues) (b) Disordered-Partial Ordered (D-PO), where IDRs present in the former were found to be partially ordered in the later (21–79% ordered residues) and (c) Disordered-Disordered (D-D) ($\leq$20% ordered residues), where IDRs found in former remained disordered in the later. When there was more than one chain for a subunit, the chain that had the best alignment with the unbound protein was considered. There are certain cases where the amino acid sequences of the IDRs did not match after comparing the bound and unbound proteins and such cases were kept under the miscellaneous category.

2.4 Interface Region and Hydrogen Bond Calculation

For the P-P complexes, the interface residues were obtained using the solvent accessible surface area (SASA) approach. The software NACCESS [45] which implements the Lee and Richards algorithm [46] was used to calculate the SASA of the complex and its subunits and thus the interface area and residues were calculated. HBPLUS program [47] was used to calculate the number of hydrogen bonds (H-bonds) present in P-P complexes. A H-bond can be classified as water-mediated or direct H-bond, based on its formation between the atom of an amino acid and water or between the atoms of two different amino acids respectively. For all the IDRs of the unbound proteins, that were classified as D-O, the total number of water-mediated and direct H-bonds was calculated.

2.5 Secondary Structural Elements

The software PSIPRED [48, 49] was used to predict the secondary structural element of each residue present in the protein sequence. The secondary structural elements were classified as Helix (H), Coil (C) and Strands (E). For all the IDRs of the unbound proteins, that were classified as D-O, the secondary structure of the residues was mapped and classified into H, C and E category.

2.6 Study of Conformational Changes

The software DALI [50] was used for comparing the bound and unbound proteins. For the D-O category, where disordered regions in one protein undergo a transition from disorder-to-order was analysed and the structural variation was mapped. The quantification of superimposition was done using the root mean square deviation (RMSD) and the Z-score. The superimposed structures were further visualized using the software PyMOL [44]. MD simulation of one P-P complex (PDB id:5V6U) and its unbound protein subunit1 (PDB id:1I4O) was run for 100 ns using GROMACS software [51, 52], utilizing the force field a99SBdisp [53, 54]. Based on the simulation, graphs for RMSD, RMSF (Root Mean Square fluctuation), and Radius of Gyration (Rg) were constructed. The SWISS-MODEL [55, 56] server was used for modelling the proteins, to fill up the internal missing residues.

3   Results

3.1   The Dataset of Protein-Protein Complexes and Their Unbound Forms

The non-redundant dataset of protein-protein (P-P) complexes curated from the PDB consists of 899 structures. Of these 899 complexes, for 356 complexes, we obtained the unbound or free forms of the constituent proteins. Each protein complex in our dataset consists of two subunits: subunit1 (s1) and subunit2 (s2). Here we have represented the s1 and s2 of complex as cpx_s1 and cpx_s2 respectively while for the unbound forms, they are denoted as ub_s1 and ub_s2 respectively. The unbound forms of s1 and s2 (ub_s1 and ub_s2) were obtained for 267 and 155 P-P complexes respectively after performing BLASTp (refer Materials and Methods section). In 66 P-P complexes, unbound forms of both s1 and s2 were obtained.

3.2   Intrinsically Disordered Regions in Protein-Protein Complexes and Their Unbound Forms

Missing residues or the IDRs were identified in both the subunits of the P-P complexes. In case of 184 cpx_s1, 562 IDRs and in 82 cpx_s2, 187 IDRs were identified. IDRs were located in the unbound forms as well. A total of 683 IDRs were found in 199 ub_s1 while 407 IDRs were located in 82 ub_s2. The length of the IDRs varied from 2 to >1000 residues in both complex and unbound forms. Figure 1 shows the variation of length of IDRs present in the unbound (Fig. 1A) and complexes (Fig. 1B).

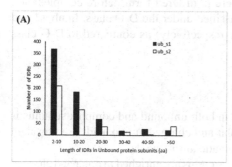 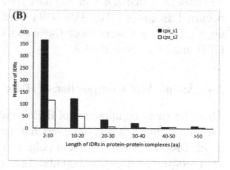

Fig. 1.   Variation of length of IDRs in (A) unbound subunits and (B) complex subunits.

Majority of the IDRs (>50%) identified in our dataset are of length 2–10 amino acids (aa) long. Less than 10% of IDRs present in unbound and complexes were found to be >50 aa in length. In crystal structure of *Drosophila melanogaster* CENP-C cumin domain (PDB id: 6XWU), 90% of the protein was identified as disordered [57]. Here the length of the protein chain is 1411 aa and the reported IDR has a length of 1273 aa. The average length of IDRs in ub_s1 and ub_s2 was found to be 17 and 21 aa respectively while in cpx_s1 and cpx_s2, the average length of IDRs was observed to be 20 and 13 aa respectively.

3.3 Classification of IDRs in Unbound and Complex Proteins

IDRs in both unbound and complex proteins were classified into 3 main classes as mentioned in Materials and Methods section. Table 1 lists the frequency of IDRs in each category (D-O, D-PO, D-D and miscellaneous) in both the subunits of the unbound and complex proteins. While in all the cases, the IDRs when compared between unbound and complex forms remain disordered ($\geq 30\%$), the number of IDRs in D-O category was found to be more prevalent than the D-PO class in both unbound and complex proteins.

Table 1. Frequency present in IDRs of unbound and complex proteins in different classes

Category of IDRs	IDRs in unbound proteins (%)		IDRs in complex proteins (%)	
	Subunit 1 (ub_s1)	Subunit 2 (ub_s2)	Subunit 1 (cpx_s1)	Subunit 2 (cpx_s2)
D-O	20	25	33	34
D-PO	12	16	12	17
D-D	30	44	42	30
Miscellaneous	39	16	14	18

A significant number of IDRs in complexes were found to be structured in the corresponding unbound forms. In cpx_s2, out of 187 IDRs, 64 were classified as D-O (34%) which depicts that IDRs in complex form were in ordered form before complexation occurred. In cpx_s1 also, 33% IDRs were classified under the D-O class. In ub_s1 and ub_s2, D-D cases were more (30% and 44% respectively) as compared to D-O class (20% and 25% respectively).

3.4 Amino Acid Composition of IDRs

The amino acid composition of IDRs located in both unbound and complex subunits is shown in Table 2. It was observed that the polar and charged amino acids are prevalent in all IDRs followed by the non-polar and aromatic amino acids.

IDRs present in both unbound and complex proteins are enriched in polar and charged amino acids like Ser (12%), Asp (11%), Glu (10%), Lys (9%) while depleted in Cys (1–2%). Gly (9–13%) is abundant in IDRs of both unbound and complex proteins. Aromatic amino acids such as Trp, Tyr and Phe contribute less (2–3%) in the IDRs.

3.5 Distribution of IDRs After Complex Formation

The distribution of the residues in IDRs located in unbound proteins which get ordered after complex formation was analysed. The average interface area of a given protein-protein complex is 3706.8 ± 3701.4 Å^2. The IDRs of unbound proteins classified under D-O category were mapped with the interface and surface residues of the complexes. It was observed that majority of residues of such IDRs of unbound proteins form the

Table 2. Amino acid composition of IDRs of unbound and complex proteins in different classes

Category of IDRs	AA Composition of IDRs in unbound proteins (%)		AA Composition of IDRs in complex proteins (%)	
	Subunit 1 (ub_s1)	Subunit 2 (ub_s2)	Subunit 1 (cpx_s1)	Subunit 2 (cpx_s2)
D-O				
Polar & charged[a]	61	56	59	56
Non polar[b]	33	40	37	40
Aromatic[c]	6	4	5	5
D-PO				
Polar & charged[a]	57	65	62	56
Non polar[b]	40	32	34	43
Aromatic[c]	3	4	5	1
D-D				
Polar & charged[a]	59	61	61	58
Non polar[b]	37	37	36	39
Aromatic[c]	4	2	3	3

[a]Polar & charged amino acids: Lys, Arg, His, Glu, Asp, Ser, Thr, Cys, Asn, Gln
[b]Non polar amino acids: Gly, Ala, Pro, Val, Leu, Ile, Met
[c]Aromatic amino acids: Phe, Trp, Tyr

surface of the bound form after complex formation. While 76% of residues in such IDRs of ub_s1 contributed to the surface region, 24% formed the interacting surface in the complex. A similar trend was observed in ub_s2 where 74% of IDRs residues formed the surface while only 26% of IDRs were located at the interface region after complex formation. Figure 2 shows an example of distribution of IDRs upon complex formation. Figure 2(A) illustrates the distribution of an IDR of length 5 aa of unbound protein (PDB id: 4AU8) in the CDK5-p25(nck5a) complex (PDB id: 1H4L) where it forms the interface region of the complex. In Fig. 2(B), the IDRs that form the surface part of P-P complex after complexation is shown using the crystal structure of interleukin-2 with its alpha receptor (PDB id: 1Z92). The IDR of the unbound subunit (PDB id: 1M47) of the protein has a length of 4 aa.

3.6 Secondary Structural Elements

The secondary structures formed in the P-P complexes were identified and the IDRs of unbound proteins which get ordered after complex formation (D-O category) were analysed. It was observed that IDRs upon complex formation mostly form turns and coils. In both the subunits of unbound proteins, the IDRs were analysed and it was observed that 81% IDRs in subunit1 and 78% IDRs in subunit2 formed turns and coils upon binding to another protein. There were some cases where the IDRs form helices and

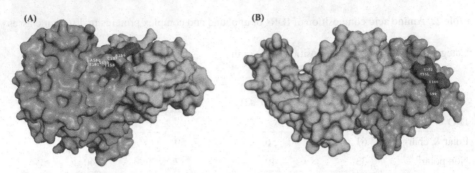

Fig. 2. Distribution of IDRs upon complex formation. (A) The IDR that was present in unbound subunit1 [PDB id: 4AU8] with residue number 156 to 161 has become ordered after complexation and is present at the interface of CDK5-p25(nck5a) complex (PDB id: 1H4L). (B) the IDR that was present in unbound subunit1 [PDB id: 1M47] with residue number 99 to 102 has become ordered after complexation and is present at the surface of interleukin-2 with its alpha receptor (PDB id: 1Z92). The images were generated using PyMOL where the subunits of P-P complex are represented in surface form and coloured in green and cyan. The IDRs are shown in magenta (color figure online).

also beta strands. IDRs of ub_s1 and ub_s2 contribute 12% and 13% in forming helices respectively while 7% in ub_s1 and 9% in ub_s2 resulted in strands post complexation. Figure 3 illustrates an example of D-O class where the IDRs present in one protein forms secondary elements after binding to its partner protein. In Fig. 3(A), IDRs in unbound are forming a helix in the complex while in Fig. 3(B) IDRs tend to form a strand. However, such cases are very few in our dataset and mostly IDRs tend to form coil or turns after complexation.

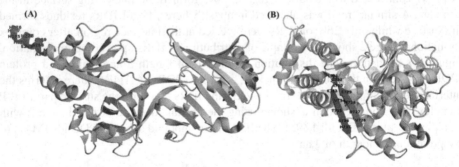

Fig. 3. IDRs upon complex formation form secondary elements. (A) In the P-P complex of *E. coli* LolA and periplasmic domain of LolC (PDB id: 6F3Z), the IDR of unbound protein (PDB id: 5NAA), gets ordered in the complex and has formed a helix (residues 266 to 277) (B) IDR of unbound protein (PDB id: 6J1Y), gets ordered and formed a strand (residues 726 to 737) in the complex WWP1 HECT and UbV P1.1 (PDB id: 5HPS). The images were generated using PyMOL where the subunits of P-P complex are represented in cartoon form and coloured in green and cyan. The IDRs are shown in magenta (color figure online).

3.7 IDRs in Polar Interactions

Residues present in IDRs under D-O class of unbound dataset are involved in polar interactions in the corresponding complexes. Of 136 IDRs present in ub_s1, except for 8 IDRs, rest 128 IDRs form a total of 1645 H-bonds. Of these 70% are direct H-bonds while 30% are water-mediated H-bonds. In IDRs of ub_s2, 43 IDRs form 440 H-bonds. A similar trend is observed here as well where the frequency of direct H-bonds is more (83%) as compared to the water-mediated H-bonds (17%). 3 IDRs of ub_s2 were found to make no H-bonds.

3.8 Structural Superimposition of Unbound and Complex Proteins

Structural alignment between unbound and bound proteins of two cases is discussed below. Figure 4 shows the superimposition of an unbound and complex protein. In Fig. 4(A) catalytic domain of human deubiquitinase DUBA in complex with ubiquitin aldehyde (PDB id: 3TMP) is superimposed with its unbound protein subunit1 (PDB id: 3TMO) and a Z-score of 16 and RMSD of 1.7 was obtained from the DALI server. It was observed that the IDR in the unbound protein of length 13 (residues 262–278) formed a structured strand along with a coil region. Figure 4(B) shows an enlarged picture of the disordered-to-ordered transition of the residues 262 to 278.

Figure 4(C) shows the superimposed structure of human caspase-7 soaked with allosteric inhibitor 2-[(2-acetylphenyl) sulfanyl] benzoic acid (PDB id:5V6U) and its corresponding unbound protein subunit1 XIAP/Caspace-7 (PDB id:1I4O). The IDR in unbound of 12 residues (274 to 285) formed a structured helix along with a coil region. Figure 4(D) shows an enlarged picture of the disordered-to-ordered transition of the residues 274 to 285.

The MD simulation results for unbound (PDB id: 5V6U) and bound protein (PDB id: 1I4O) is shown in Fig. 5. The mean RMSD for the unbound protein was found to be 0.532 ± 0.115 nm while for the complex it was 0.203 ± 0.031 nm. The variation of RMSF for Cα atoms of the residues of the protein throughout the course of the simulation was also noted and the mean RMSF for the unbound protein was found to be 0.140 ± 0.169 nm while for complex it was 0.089 ± 0.082 nm. The residues which composed the IDRs, like 187–211, 226–230, and 274–285 showed a higher fluctuation in the RMSF graph. The variation of Rg for Cα atoms of the residues of the protein showed a little difference in the unbound and bound protein. While in case of unbound protein, a mean Rg of 1.8 ± 0.029 nm was observed, for the complex it was 1.807 ± 0.009 nm.

4   Discussion

The present study provides a comprehensive analysis of different physicochemical features of the IDRs present in protein-protein complexes and their free or unbound forms. Here 356 non-redundant X-ray structures of binary protein-protein complexes are analysed and the findings are correlated with the results of the previous studies available in the literature. We have restricted our study to high resolution structures to rule out the concept that missing residues in the protein structures can be due to low electron

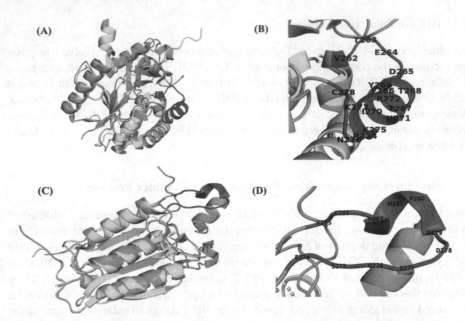

Fig. 4. Structural superimposition of complex and the unbound protein. (A) Catalytic domain of human deubiquitinase DUBA in complex with ubiquitin aldehyde (PDB id: 3TMP) is super-imposed with its unbound protein subunit1 (PDB id: 3TMO) (B) Shows an enlarged picture of the disordered-to-ordered transition of the residues 262 to 278 (C) Human caspase-7 soaked with allosteric inhibitor 2-[(2-acetylphenyl) sulfanyl] benzoic acid (PDB id:5V6U) superimposed with its corresponding unbound protein subunit1 XIAP/Caspace-7 (PDB id:1I4O) (D) Shows an enlarged picture of the disordered-to-ordered transition of the residues 274 to 285. The images were generated using PyMOL where the structures are shown in cartoon form and the bound form of the protein is shown in green; the unbound form is shown in yellow, and the residues undergoing D-O transition are shown in magenta (color figure online).

density. Moreover, IDRs in both subunits of the complexes as well as in the unbound forms were analysed and categorised into 3 major classes: D-O, D-PO, and D-D. There were IDRs which could not be mapped between the complex and unbound forms and were kept under miscellaneous category.

It was observed that the disordered regions in our dataset are mostly 2 to 20 aa long. Though long IDRs of length >30 aa were observed in both subunits of unbound and complexes, their frequency is less (<15%). For example, in our unbound protein dataset, in the crystal structure of UVRB protein of *Thermus thermophilus* HB8 (PDB id: 1D2M) a long stretch of IDRs is reported. Here, in the C-terminal, 82 residues (584–665) are disordered due to its large motion, which may be restricted by the interaction with UvrC [58]. Such cases are also found in P-P complexes of our dataset. The crystal structure of the ferredoxin protease FusC in complex with its substrate plant ferredoxin (PDB id: 6B03) has an internal IDR and disorder in ferredoxin residues 38–85 is reported in the same [59].

The analysis of the amino acid composition in different classes of IDRs reveals a biased amino acid composition, consistent with the previous findings [60, 61]. The IDRs

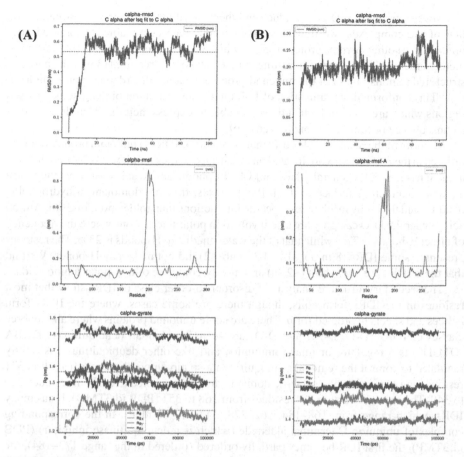

Fig. 5. Molecular dynamics simulation results for unbound and bound protein. Variation of RMSD, RMSF and Rg is shown for the (A) unbound protein: human caspase-7 soaked with allosteric inhibitor 2-[(2-acetylphenyl) sulfanyl] benzoic acid (PDB id: 5V6U) and for the (B) the XIAP/Caspace-7 complex (PDB id: 1I4O).

are enriched in polar (Ser, Thr, Cys, Asn, Gln) and charged (Lys, Arg, His, Glu, Asp) aa (~60%). The frequency of hydrophobic amino acids (Ala, Pro, Val, Leu, Ile, Met) is lower but cannot be neglected (30–40%). It is observed that Ser is abundantly found (12%) within the IDRs, which could be due to its low hydrophobicity, higher flexibility in backbone conformations, and the ability to interact easily with the solvent [62]. The contribution of Gly (9–13%) is also significant in both unbound and complex proteins since it lacks side chains and thus its flexibility entropically distavours formation of ordered structures [63]. IDRs in our dataset are depleted in aromatic amino acids (Tyr, Trp, Phe) and Cys, as aromatic amino acids and Cys are major structure promoting-residues [64]. The abundance of polar and charged amino acids and paucity of hydrophobic amino acids results in a weak hydrophobic environment which is the key factor of polypeptides folding into stable tertiary structures [65].

After mapping the IDRs's residues and their coordinates of unbound proteins with that of the complexes, we find most IDRs in our dataset remained disordered after binding to another partner protein (25%), while some became partially (9%) or completely ordered (16%). It was interesting to note that some structured (33%) and partially structured regions (14%) in the unbound proteins become disordered upon complexation. This conformational transition of IDRs indicates that upon binding some ordered regions which are dormantly disordered are able to express their disorderedness and as a consequence become functionally active [9].

Majority of the residues (74–76%) that were part of IDRs in unbound proteins and are ordered in the P-P complex are located on the surface of the complex, 24–26% are located at the interface. IDRs generally lack bulky hydrophobic amino acids but when they form a part of interacting surface in the P-P complexes, they are dominant in hydrophobic residues and they may result in less specific interactions than folded proteins [66]. Amino acids under the D-O category are also involved in polar interactions where the frequency of direct H-bonds is 77% while that of the water-mediated H-bonds is 23%. The residues present in some IDRs (8 in ub_s1 and 3 in ub_s2) did not make any H-bonds. We find that these IDRs are very short (~2 aa) and mostly comprise of non-polar amino acids.

The secondary structure analysis of disorder-to-order class of IDRs show that most residues in the IDRs form coils, though there are some cases, where the IDRs form helices and sheets in ordered form. There are some unbound proteins where all the three classes of IDRs (D-O, D-PO and D-D) are seen. The human deubiquitinase DUBA (OTUD5) is a regulator of innate immunity, and like other deubiquitinase, is tightly regulated to control the removal of ubiquitin in a spatiotemporal manner [67]. It is 571 residue long consisting of a catalytic domain which is predicted to be from residues 220 to 340. The crystal structure has residues from 168 to 351 (PDB id:3TMO). It has many IDRs, which range from 168–184, 262–278, 343–351 residues. In the corresponding complex of ubiquitin C-terminal aldehyde (which is a deubiquitinase inhibitor) (PDB id:3TMP), the first IDR becomes partially ordered (ordered in the range 173–184), the second region gets completely ordered, and the third region remains disordered. It is interesting to note that a major part of the predicted catalytic domain gets ordered in the complex. It has been reported that phosphorylation at Ser177 activates DUBA, otherwise, it remains inactive [67]. The region around pSer177 lies in a partially disordered region in a complex and forms an alpha helix [67].

Similarly, in Caspase-7, we observe both D-PO and D-O conditions. Caspase-7 is an executor caspase, which belongs to a class of enzymes that are responsible for controlling the final steps of inflammation and apoptosis, and thus their activity is tightly regulated [68]. When the IDRs present in chain A of the human caspase-7 soaked with allosteric inhibitor 2-[(2-acetylphenyl) sulfanyl] benzoic acid (PDB id:5V6U), was compared with the regions present in the corresponding complex, the XIAP/Caspase-7 complex (PDB id: 1I4O), it was observed that the first IDR region become partially ordered (ordered in the range 187–196) and the second (226–230) and third (274–285) IDRs form order structures. X chromosome linked IAP (XIAP) belongs to a type of inhibitor of apoptosis proteins (IAPs), that effectively suppress apoptosis [69]. XIAP consists of the BIR2 domain and its linker region. As reported, only the linker region is present in the crystal structure and the BIR2 region remains completely disordered [69]. The linker region interacts with the substrate groove in such a way that it blocks the active site, while the BIR2 domain acts as a regulatory domain, strengthening the linker interaction for caspase [69]. It is seen from the Fig. 4(C), that the third IDR undergoes D-O transition and forms a structured strand along with coils. It was also found that this region was in the interface and is thus interacting with the linker of XIAP.

The MD simulation results on Caspase-7 protein confirmed our findings. The complex protein was found to have lower RMSD than that of the unbound form, indicating that the complex form of the protein has a more stable conformation. The RMSD indicates the average displacement of the atoms at an instant of the simulation relative to a reference structure, which is the crystallographic structure [70, 71]. It measures the difference between the backbones of a protein from its initial structural conformation to its final position [70, 71]. A relatively large conformational change is indicative of less stability while a small conformational change indicates more overall stability. The overall RMSF of the protein complex was also found to be less than that of the unbound protein indicating that post complex formation, the conformational flexibility of the protein reduces. Moreover, the IDRs present in the unbound subunit which got partially or completely ordered in the protein complex, have less RMSF value than that of the unbound form. The RMSF measures the displacement of a particular atom, or group of atoms, relative to the reference structure, averaged over the number of atoms. It indicates the extent of variation of position of a specific residue around the average structure to further describe the extent of conformational flexibility of the protein [70, 71]. Although the mean Rg for both the complex and the unbound form is almost similar, the standard deviation of the unbound form is relatively higher than that of the complex, indicating less overall variation in the compactness of the protein in complex form than in the unbound form. Rg can be defined as the distribution of atoms of a protein around its axis and can reveal if there is any conformational change and thus describing the overall compactness of the protein [72, 73].

5 Conclusion

Proteins are very heterogenous molecules and interact with other molecules to perform large variety of biological functions. IDRs in the proteins undergo a disordered-to-ordered transition upon binding to a suitable partner. However, little is known about the structural mechanism of folding of such regions. A structure-based study to analyse the conformational changes occurring in disordered regions upon complex formation will provide a structural perspective of molecular mechanism of IDRs and IDPs. This will further shed light on protein folding mechanisms. The information gained from the analysis of structural and physico-chemical features of disordered and ordered regions can also be incorporated in disorder-based predictors to improve their accuracies. Furthermore, with increasing flexibility of the proteins, there is an escalation in complexity of developing efficient docking tools. We believe our findings related to conformational changes in disordered regions will be useful in designing novel flexible docking algorithms. Furthermore, it has been reported that numerous IDPs are associated with human diseases, and the outcomes of this study will help in developing novel strategies for drug discovery based on IDPs. The results of our study will thus envisage the protein functioning and will help illuminating the 'dark proteome'.

Acknowlegements. A.B. acknowledges the support from SERB, DST, India for SRG. M.M.K. is a recipient of GATE research fellowship from MHRD, India. P.B. is a recipient of junior research fellowship from NIT Durgapur, India.

References

1. Babu, M.M.: The contribution of intrinsically disordered regions to protein function, cellular complexity, and human disease. Biochem. Soc. Trans. **44**, 1185 (2016)
2. Van Der Lee, R., et al.: Classification of intrinsically disordered regions and proteins. Chem. Rev. **114**, 6589–6631 (2014)
3. Oldfield, C.J., Uversky, V.N., Dunker, A.K., Kurgan, L.: Introduction to intrinsically disordered proteins and regions. In: Intrinsically Disordered Proteins: Dynamics, Binding, and Function, pp. 1–34. Elsevier (2019)
4. Ferrie, J.J., Karr, J.P., Tjian, R., Darzacq, X.: "Structure"-function relationships in eukaryotic transcription factors: the role of intrinsically disordered regions in gene regulation. Mol. Cell **82**, 3970–3984 (2022)
5. Misiura, M.M., Kolomeisky, A.B.: Role of intrinsically disordered regions in acceleration of protein-protein association. J. Phys. Chem. B **124**, 20–27 (2020)
6. DeForte, S., Uversky, V.N.: Order, disorder, and everything in between (2016)
7. Wright, P.E., Dyson, H.J.: Intrinsically disordered proteins in cellular signalling and regulation (2015)
8. Trivedi, R., Nagarajaram, H.A.: Intrinsically disordered proteins: an overview (2022)
9. Uversky, V.N.: Functional roles of transiently and intrinsically disordered regions within proteins. FEBS J. **282**(7), 1182–1189 (2015). https://doi.org/10.1111/febs.13202
10. Bondos, S.E., Dunker, A.K., Uversky, V.N.: Intrinsically disordered proteins play diverse roles in cell signaling (2022)
11. Fuxreiter, M.: Classifying the binding modes of disordered proteins. Int. J. Mol. Sci. **21**, 1–9 (2020)
12. Olsen, J.G., Teilum, K., Kragelund, B.B.: Behaviour of intrinsically disordered proteins in protein–protein complexes with an emphasis on fuzziness. Cell. Mol. Life Sci. **74**(17), 3175–3183 (2017). https://doi.org/10.1007/s00018-017-2560-7
13. Morris, O.M., Torpey, J.H., Isaacson, R.L.: Intrinsically disordered proteins: modes of binding with emphasis on disordered domains. Open Biol. **11**(10), 210–222 (2021). https://doi.org/10.1098/rsob.210222
14. Tompa, P., Fuxreiter, M.: Fuzzy complexes: polymorphism and structural disorder in protein–protein interactions. Trends Biochem. Sci. **33**, 2–8 (2008)
15. Uversky, V.N., Oldfield, C.J., Dunker, A.K.: Intrinsically disordered proteins in human diseases: introducing the D2 concept. Annu. Rev. Biophys. **37**(1), 215–246 (2008). https://doi.org/10.1146/annurev.biophys.37.032807.125924
16. Blundell, T.L., Gupta, M.N., Hasnain, S.E.: Intrinsic disorder in proteins: relevance to protein assemblies, drug design and host-pathogen interactions (2020)
17. Ruan, H., Sun, Q., Zhang, W., Liu, Y., Lai, L.: Targeting intrinsically disordered proteins at the edge of chaos. Drug Discov. Today **24**(1), 217–227 (2019). https://doi.org/10.1016/j.drudis.2018.09.017
18. Metallo, S.J.: Intrinsically disordered proteins are potential drug targets. Curr. Opin. Chem. Biol. **14**, 481 (2010)
19. Sridhar, A., Orozco, M., Collepardo-Guevara, R.: Protein disorder-to-order transition enhances the nucleosome-binding affinity of H1. Nucleic Acids Res. **48**, 5318–5331 (2020)

20. Moritsugu, K., Terada, T., Kidera, A.: Disorder-to-order transition of an intrinsically disordered region of sortase revealed by multiscale enhanced sampling. J. Am. Chem. Soc. **134**, 7094–7101 (2012)
21. Ahmad, J., et al.: Disorder-to-order transition in PE–PPE proteins of Mycobacterium tuberculosis augments the pro-pathogen immune response. FEBS Open Biol. **10**, 70–85 (2020)
22. Nishi, H., Fong, J.H., Chang, C., Teichmann, S.A., Panchenko, A.R.: Regulation of protein-protein binding by coupling between phosphorylation and intrinsic disorder: analysis of human protein complexes. Mol. Biosyst. **9**, 1620–1626 (2013)
23. Uversky, V.N.: Intrinsically disordered proteins and their "mysterious" (meta)physics (2019)
24. Seoane, B., Carbone, A.: The complexity of protein interactions unravelled from structural disorder. PLoS Comput. Biol. **17**, e1008546 (2021)
25. Quaglia, F., et al.: DisProt in 2022: improved quality and accessibility of protein intrinsic disorder annotation. Nucleic Acids Res. **50**, D480–D487 (2022)
26. Fukuchi, S., et al.: IDEAL: Intrinsically disordered proteins with extensive annotations and literature. Nucleic Acids Res. **40** (2012)
27. Fukuchi, S., et al.: IDEAL in 2014 illustrates interaction networks composed of intrinsically disordered proteins and their binding partners. Nucleic Acids Res. **42** (2014)
28. Piovesan, D., et al.: MobiDB: intrinsically disordered proteins in 2021. Nucleic Acids Res. **49**, D361–D367 (2021)
29. Erdos, G., Mátyás, P., Dosztányi, D.: IUPred3: prediction of protein disorder enhanced with unambiguous experimental annotation and visualization of evolutionary conservation. Nucleic Acids Res. **49**, W297–W303 (2021)
30. Mészáros, B., Erdös, G., Dosztányi, Z.: IUPred2A: context-dependent prediction of protein disorder as a function of redox state and protein binding. Nucleic Acids Res. **46**, W329–W337 (2018)
31. Dosztányi, Z., Mészáros, B., Simon, I.: ANCHOR: web server for predicting protein binding regions in disordered proteins. Bioinformatics **25**, 2745 (2009)
32. Jones, D.T., Cozzetto, D.: DISOPRED3: precise disordered region predictions with annotated protein-binding activity. Bioinformatics **31**, 857–863 (2015)
33. Barik, A., Katuwawala, A., Hanson, J., Paliwal, K., Zhou, Y., Kurgan, L.: DEPICTER: intrinsic disorder and disorder function prediction server. J. Mol. Biol. **432**, 3379–3387 (2020)
34. Berman, H.M., et al.: The protein data bank. Nucleic Acids Res. **28**, 235–242 (2000)
35. Fu, L., Niu, B., Zhu, Z., Wu, S., Li, W.: CD-HIT: accelerated for clustering the next-generation sequencing data. Bioinformatics **28**, 3150–3152 (2012)
36. Li, W., Godzik, A.: Cd-hit: a fast program for clustering and comparing large sets of protein or nucleotide sequences. Bioinformatics **22**, 1658–1659 (2006)
37. Brandt, B.W., Heringa, J., Leunissen, J.A.M.: SEQATOMS: a web tool for identifying missing regions in PDB in sequence context. Nucleic Acids Res. **36**, W255–W259 (2008)
38. Monzon, A.M., et al.: Experimentally determined long intrinsically disordered protein regions are now abundant in the Protein Data Bank. Int. J. Mol. Sci. **21**, 1–13 (2020)
39. Oldfield, C.J., et al.: Utilization of protein intrinsic disorder knowledge in structural proteomics. Biochim. Biophys. Acta **1834**, 487 (2013)
40. Gall, T.L., Romero, P.R., Cortese, M.S., Uversky, V.N., Dunker, A.K.: Intrinsic disorder in the Protein Data Bank. J. Biomol. Struct. Dyn. **24**, 325–341 (2007)
41. Zhang, Y., Stec, B., Godzik, A.: Between order and disorder in protein structures – analysis of "dual personality" fragments in proteins. Structure **15**, 1141 (2007)
42. Baruah, A., Rani, P., Biswas, P.: Conformational entropy of intrinsically disordered proteins from amino acid triads. Sci. Rep. **5** (2015)
43. Ferron, F., Longhi, S., Canard, B., Karlin, D.: A practical overview of protein disorder prediction methods. Proteins Struct. Funct. Bioinform. **65**, 1–14 (2006)

44. Schrödinger LLC: The PyMOL Molecular Graphics System, Version 2.5 (2015)
45. Hubbard, S.J., Thornton, J.M.: 'NACCESS', Computer Program. Department of Biochemistry and Molecular Biology, University College, London (1993). www.bioinf.manchester.ac.uk/naccess/
46. Lee, B., Richards, F.M.: The interpretation of protein structures: estimation of static accessibility. J. Mol. Biol. **55**(3), 379 (1971). https://doi.org/10.1016/0022-2836(71)90324-X
47. McDonald, I.K., Thornton, J.M.: Satisfying hydrogen bonding potential in proteins. J. Mol. Biol. **238**, 777–793 (1994)
48. Jones, D.T.: Protein secondary structure prediction based on position-specific scoring matrices. J. Mol. Biol. **292**, 195–202 (1999)
49. Buchan, D.W.A., Jones, D.T.: The PSIPRED protein analysis workbench: 20 years on. Nucleic Acids Res. **47**, W402–W407 (2019)
50. Holm, L.: Dali server: structural unification of protein families. Nucleic Acids Res. **50**, W210–W215 (2022)
51. Abraham, M., et al.: GROMACS 2023.1 Manual (2023)
52. Abraham, M.J., et al.: GROMACS: high performance molecular simulations through multi-level parallelism from laptops to supercomputers. SoftwareX **1–2**, 19–25 (2015)
53. Robustelli, P., Piana, S., Shaw, D.E.: Developing a molecular dynamics force field for both folded and disordered protein states. Proc. Natl. Acad. Sci. U.S.A. **115**, E4758–E4766 (2018)
54. Shrestha, U.R., Smith, J.C., Petridis, L.: Full structural ensembles of intrinsically disordered proteins from unbiased molecular dynamics simulations. Commun. Biol. **4**(1), 243 (2021). https://doi.org/10.1038/s42003-021-01759-1
55. Bienert, S., et al.: The SWISS-MODEL repository – new features and functionality. Nucleic Acids Res. **45**, D313–D319 (2017)
56. Waterhouse, A., et al.: SWISS-MODEL: homology modelling of protein structures and complexes. Nucleic Acids Res. **46**, W296–W303 (2018)
57. Medina-Pritchard, B., et al.: Structural basis for centromere maintenance by Drosophila CENP-A chaperone CAL1. EMBO J. **39** (2020)
58. Nakagawa, N., Sugahara, M., Masui, R., Kato, R., Fukuyama, K., Kuramitsu, S.: Crystal structure of Thermus thermophilus HB8 UvrB protein, a key enzyme of nucleotide excision repair. J. Biochem. **126**, 986–990 (1999)
59. Grinter, R., et al.: FusC, a member of the M16 protease family acquired by bacteria for iron piracy against plants. PLOS Biol. **16**(8), e2006026 (2018). https://doi.org/10.1371/journal.pbio.2006026
60. Zhao, B., Kurgan, L.: Compositional bias of intrinsically disordered proteins and regions and their predictions. Biomolecules **12**(7), 888 (2022)
61. Campen, A., Williams, R.M., Brown, C.J., Meng, J., Uversky, V.N., Dunker, A.K.: TOP-IDP-Scale: a new amino acid scale measuring propensity for intrinsic disorder. Protein Pept Lett. **15**, 956 (2008)
62. Uversky, V.N.: The intrinsic disorder alphabet: III. Dual personality of serine. Intrins. Disord. Proteins **3**(1), e1027032 (2015). https://doi.org/10.1080/21690707.2015.1027032
63. Cheng, S., Cetinkaya, M., Gräter, F.: How sequence determines elasticity of disordered proteins. Biophys. J. **99**, 3863 (2010)
64. Structural and functional analysis of "non-smelly" proteins|Enhanced Reader. Accessed 29 Apr 2023
65. Ahmed, S.S., et al.: Characterization of intrinsically disordered regions in proteins informed by human genetic diversity. PLOS Comput. Biol. **18**(3), e1009911 (2022). https://doi.org/10.1371/journal.pcbi.1009911
66. Vacic, V., et al.: Characterization of molecular recognition features, MoRFs, and their binding partners. J. Proteome Res. **6**, 2351–2366 (2007)

67. Huang, O.W., et al.: Phosphorylation-dependent activity of the deubiquitinase DUBA. Nat. Struct. Mol. Biol. **19**(2), 171–175 (2012)
68. Vance, N.R., Gakhar, L., Spies, M.A.: Allosteric tuning of caspase-7: a fragment-based drug discovery approach. Angew. Chem. Int. Ed. Engl. **56**, 14443 (2017)
69. Abhari, B.A., Davoodi, J.: A mechanistic insight into SMAC peptide interference with XIAP-Bir2 inhibition of executioner caspases. J. Mol. Biol. **381**, 645–654 (2008)
70. Aier, I., Varadwaj, P.K., Raj, U.: Structural insights into conformational stability of both wild-type and mutant EZH2 receptor. Sci. Rep. **6**(1), 1–10 (2016)
71. Martínez, L.: Automatic identification of mobile and rigid substructures in molecular dynamics simulations and fractional structural fluctuation analysis. PLoS One **10**(3), e0119264 (2015). https://doi.org/10.1371/journal.pone.0119264
72. Sneha, P., Priya Doss, C.G.: Molecular dynamics: new frontier in personalized medicine. Adv. Protein Chem. Struct. Biol. **102**, 181–224 (2016)
73. Funari, R., Bhalla, N., Gentile, L.: Measuring the radius of gyration and intrinsic flexibility of viral proteins in buffer solution using small-angle X-ray scattering. ACS Meas. Sci. Au **2**, 547–552 (2022)
74. Altschul, S.F., Gish, W., Miller, W., Myers, E.W., Lipman, D.J.: Basic local alignment search tool. J. Mol. Biol. **215**(3), 403–410 (1990). https://doi.org/10.1016/S0022-2836(05)80360-2

Computational Support for Clinical Decisions

Predicting and Detecting Coronary Heart Disease in Patients Using Machine Learning Method

Michał Woś[1]([⊠]) [iD], Bartłomiej Drop[1] [iD], and Bartłomiej Kiczek[2] [iD]

[1] Department of Medical Informatics and Statistics with e-learning Lab, Medical University of Lublin, Lublin, Poland
michal.wos@umlub.pl
[2] Institute of Physics, Maria Curie-Skłodowska University, Lublin, Poland

Abstract. Machine learning creates new opportunities for medicine and public health, especially in the field of helping medical examiners. The prospect of generating hints for the diagnosis of a particular disease for physicians is widely considered in this field. The work presents only the possibility of predicting coronary heart disease (CHD), which is classified as a civilization disease, that threatens an increasing number of people in the world. The dataset consisting of more than 11,000 patient records was used for the study. In total each record of the dataset contained 16 features (which were analysed by the model) such as BMI, weight, total cholesterol, years of smoking. Several families of regressors were browsed in search of the best fit. Starting from different variants of the linear model, with various regularisation factors, to support vector machines, ensemble learning (RandomForest) and tree-based solutions with gradient boosting. It turned out that extreme gradient boosted trees (XGBoost) proved to provide the best results. After the model was trained a series of backward searches, such as a visual analysis of the tree models, model estimated weights and an additional LIME-based (Local Interpretable Model-agnostic Explanations) explanation were performed.

The obtained results look promising in terms of a deployment of this model, but also similar models aimed in different diseases, in software supporting public health professionals. Such machine learning solution can serve as an as an assistant for a physician, which can be a part of a bigger medical data management system.

Keywords: Machine learning · Ischemic heart disease · heart disease ML methods · illness prediction

1 Introduction

Coronary heart disease (CHD) is a condition that occurs when the blood vessels that supply the heart with oxygen and nutrients become narrowed or blocked by a buildup of plaque. CVDs refer to a range of conditions affecting the heart and blood vessels, including coronary heart disease, cerebrovascular disease, peripheral arterial disease,

I. Rojas et al. (Eds.): IWBBIO 2023, LNBI 13919, pp. 367–377, 2023.
https://doi.org/10.1007/978-3-031-34953-9_29

rheumatic heart disease, congenital heart disease, and deep vein thrombosis and pulmonary embolism. These conditions can cause acute events such as heart attacks and strokes, which are often caused by blockages in the blood vessels due to fatty deposits. In some cases, strokes can also be caused by bleeding from a blood vessel in the brain or blood clots. Plaque is made up of cholesterol, fat, calcium, and other substances found in the blood. As the plaque buildup increases, it can restrict blood flow to the heart muscle, causing chest pain (angina), shortness of breath, and other symptoms. In some cases, a blood clot can be formed on the surface of the plaque and completely block the blood flow to the heart muscle and cause a heart attack. Risk factors for CHD include high blood pressure, high cholesterol, smoking, diabetes, obesity, lack of physical activity, and a family history of heart disease. Treatment for CHD may include lifestyle changes (such as eating a healthy diet, exercising regularly, and quitting smoking), medications (such as cholesterol-lowering drugs and blood thinners), and in some cases, surgery (such as angioplasty or bypass surgery) to restore blood flow to the heart.

Coronary heart disease (CHD) is a common condition worldwide. According to the World Health Organization (WHO), CHD is the leading cause of death globally, accounting for over 7 million deaths in 2016. According to the latest available data from the European Heart Network, in 2017, coronary heart disease (CHD) responsible for approximately 1.8 million deaths in Europe [12]. This corresponds to 20% of all deaths in this region. The prevalence of CHD varies by country and region within Europe. In general, countries in Western Europe have higher rates of CHD than those in Eastern Europe. For example, in 2017, the age-standardized death rate from CHD in the European Union (EU) was 136 per 100,000 population, but this rate varied from 51 per 100,000 in Bulgaria to 214 per 100,000 in Ireland. It's important to note that these figures are from 2017 and may have changed in the years since then. However, they provide a general idea of the prevalence of CHD in Europe and how it varies by country (Fig. 1).

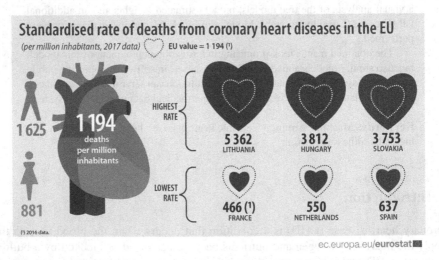

Fig. 1. Rate of deaths from coronary heart diseases in the EU.

The COVID-19 pandemic has resulted in a significant number of severe illnesses and deaths in the United States. As of March 2021, the country has recorded approximately 545,000 deaths, which translates to about 166 cases per 100,000 individuals. Higher death rates were observed in counties with metropolitan areas, with a high percentage of non-Hispanic Black population, Hispanic population, or people living in poverty. The death rate was approximately 185 per 100,000 in counties with metropolitan areas, 200 per 100,000 among non-Hispanic Black population, 219 per 100,000 among Hispanic population, and 211 per 100,000 among people living in poverty.

Because of the high COVID-19 mortality rates, life expectancy in the United States for the year 2020 has been estimated to decline with disproportionate impacts on populations with high COVID-19 mortality rates. Provisional US life expectancy estimates for January to June 2020 indicate that between 2019 and the first half of 2020, life expectancy decreased from 74.7 to 72.0 years for NH Black individuals, from 81.8 to 79.9 years for Hispanic individuals, and from 78.8 to 78.0 years for NH White individuals [18].

Studies have shown a significant association between myocardial injury and fatal outcomes of COVID-19. In a study involving 187 COVID-19 positive patients, 52% had myocardial injury, which was detected by higher levels of troponin T (TnT). Elevated TnT levels were also observed in cases of mortality from COVID-19. Therefore, it is important to focus on protecting the cardiovascular system when treating COVID-19 patients, as severe acute respiratory syndrome coronavirus 2 (SARS-CoV-2) infection can cause acute myocardial injury and chronic damage to the cardiovascular system. According to the National Health Commission of China (NHC), 35% of COVID-19 patients had hypertension and 17% had a history of coronary heart disease (CHD), suggesting that cardiovascular diseases can exacerbate pneumonia and other symptoms in SARS-CoV-2 infected patients [19].

The prevalence of CHD varies by age, gender, and other factors. Men are more likely to develop CHD than women, and the risk increases with age. Other factors that in-crease the risk of CHD include high blood pressure, high cholesterol, diabetes, obesity, smoking, and a family history of heart disease [13].

Clinical decision support systems can help patients make better assessments than medical examiners [1, 2]. In a world dominated by cloud and fog computing, it is almost impossible to extract healthcare requirements and healthcare needs without complete, comprehensive, and linked health data [3]. The security of this data is also an enormous challenge [4, 5]. It turned out that clinical decision support systems could be promising tools that help patients make more conscious decisions about their health. Patient feedback can be helpful and very valuable in the process of pattern classification that determines patient's health status and disease severity. An expert system can use this information to determine whether a patient has a disease or not. Then, doctors can make medical decisions based on accurate features or measurements.

In the modern era, there has been a significant increase in the use of information technology and the Internet of Things (IoT) in medical diagnosis, disease prevention and patient satisfaction [5, 6]. It has been observed that most affluent countries have a lower prevalence of ischemic heart disease than underdeveloped countries. As a result of the complexity and unpredictability of these sectors, intelligent architectures such as fuzzy system and artificial neural network, as well as genetic algorithms, have been developed to address these challenges.

The IHME studies also show that the proportion of patients is increasing year by year in the European population. Due to the increasing number of patients, additional problems such as: staff shortages and increasing queues to specialist doctors can now be observed. The solution to these issues is found in domain-specific information systems that use machine learning to detect or predict the possibility of contracting the disease.

Several methodologies were used during the analysis of the various aspects associated with the application to complete the study, including the use of data from the UCI machine learning library to achieve high accuracy grading. Automatic detection of cardiac disorders using an artificial immune recognition system (AIRS) with a fuzzy resource allocation mechanism and k-NN (nearest neighbor) weighted pre-classification algorithms are used for most clinical diagnoses of coronary artery disease [8, 9]. RA company has achieved 50.00% accuracy using the IB1-4 algorithm with the ToolDiag tool. Using InductH, WEKA, RA has an assessment accuracy of 58.5%, while using RBF, ToolDiag, RA has an assessment accuracy of 60.00% [10]. ToolDiag used the MLP + BP algorithm, which had a success rate of up to 65.00%. WEKA and K*, T2, 1R, IB1c and RA had classification accuracy of 68.10%, 71.40%, 74.00% and 76.70%, respectively.

Such system automatically detects heart disease using an artificial immune recognition system (AIRS) with a fuzzy resource allocation mechanism and k-nan (nearest neighbor) weighted preprocessing classification algorithms [7]. This system is widely used in the clinical diagnosis of coronary artery disease. Various methodologies have been used to achieve high accuracy evaluation, including the use of data from the UCI machine learning library. In addition, a fuzzy expert system for heart disease was proposed in 2007, in which a fuzzy expert system was used to assess the risk of coronary heart disease (CHD) in patients [11]. The machine predicted the hazard ratio and proposed one of three outcomes: staying in usual method, nutrition and pharmaceutical treatment. In addition, 79.00% of the results are matched with an expert. Another approach to predicting heart disease is to use data mining techniques. Data mining methods can estimate patient survival and adjust operations accordingly. Fuzzy classification and data mining techniques are used to accurately diagnose heart disease, and unstructured data have been discovered as huge datasets in the history of medicine. Large datasets have been used to predict people with heart disease in a variety of circumstances.

The K-nearest-neighbor approach has also been used in predicting cardiac patients, offering a screening approach for cardiovascular disease based on real clinical evidence. A total of 450 articles were prepared and used, each included study criteria. There are 36 disease types in the HKH dataset, of which 29 are cardiovascular diseases and seven are other diseases.

2   Methods

The use of machine learning is beneficial from two perspectives. First, it can establish a reliable software assistant, which will be helpful to medical practitioners in day-to-day diagnosis. In addition, it could be used in various places where self-diagnosis could take place. In this approach, the patient could share his or her data with the software and give specific instructions, such as recommending a medical appointment. From the other perspective, understanding how ML algorithms learn the data and predict the decision, can be beneficial in general medical methodology.

The starting point of quantitative research is the dataset which consists of 11.000 cases. Each of them was examined by a group of physicians and probabilities of morbidity were estimated. These were aggregated into a single probability by averaging. The dataset consisted of sixteen features. First eight features are typically collected in the interview, these are patient's sex, age, weight, height, BMI, the fact of smoking, years of smoking and the assessed daily number of cigarettes. The remaining eight features came from the examination, and these are total cholesterol, low density lipoprotein (LDL), high density lipoprotein (HDL), three measurements of glucose conducted day by day. Finally, we have systolic and diastolic pressures.

In the pre-processing stage the smoking feature was rejected from the dataset as it was recognised as redundant, with non-zero smoking years proving that patient was a smoker. Naturally, there is a possibility of a currently non-smoking patient with smoking history, however such cases were not considered in this dataset. The next step included the train-test split, with 80–20 ratio and the standardization of continuous numeric variables. This process left us with prepared data, ready for modelling.

Various models were tested for the considered dataset, starting with linear models up to tree-based solutions. At this point, a question may be asked, why neural networks or more complex models were not used? In fact, the point in this research was to find a golden mean between model score and easiness of explanation. At the end of the day, a black box model that perfectly predicts the diagnosis would not bring much educational value. The model which satisfactory met these conditions was the XGBoost regressor.

XGBoost (eXtreme Gradient Boosting) is a powerful library for gradient boosting decision trees, which is a machine learning technique used for solving both regression and classification problems [14]. The method belongs to the of ensemble learning techniques, with other notable methods [15, 16] such as random forest [17, 18].

XGBoost uses gradient boosting to subsequently add new trees to the set, in order to correct the error of previously corrected trees. The algorithm uses gradient of the loss function, and other metrics, to fit another tree to the ensemble.

In addition, XGBoost provides built-in regularization techniques to prevent overfitting, such as L1 and L2 regularization, and can handle missing data by automatically learning the best imputation strategy. XGBoost also includes a feature importance calculation method, which helps users identify the most relevant features for their models. The last feature is extremely important for us, as it allows for a global understanding of factors for the coronary heart disease morbidity probability.

The model was fitted by using grid search with 5-fold cross validation on the training set. The training grid swiped through different values of number of trees in the model, maximum depth of the tree, shrinkage rate (or learning rate), data subsampling rate and column subsampling rate.

The most optimal values for our model proved to be:

- Number of trees: 100,
- Maximum depth: 5,

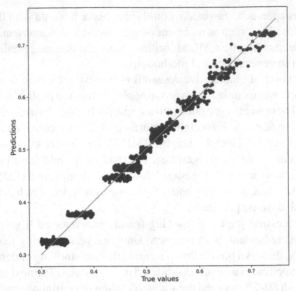

Fig. 2. Predictions versus true values of the test set targets

- Shrinkage rate: 0.1,
- Data subsampling rate: 1,
- Column subsampling rate: 1.

In Fig. 2. On this figure predicted targets compared with actual targets on the test set can be seen. It can be noticed that there are certain levels of predicted values, which are the result of the tree-based background of the model – the output can never be fully continuous. Nevertheless, the overall Pearson score R2 = 0.98 on the test set.

To test the results more precisely we introduce thresholding to the probabilities and inspect them as a classification problem. We use standard 0.5 value to distinguish between a sick and healthy patient.

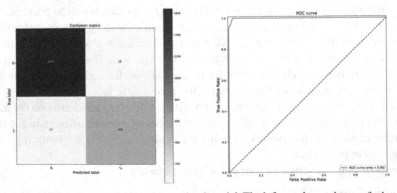

Fig. 3. Classification characteristics of the trained model. The left panel contains confusion matrix and the right panel the ROC curve.

Figure 3 contains two panels. On the left panel can be seen the confusion matrix, where the classification results are more than satisfactory. On 2255 samples in the test set only 45 were misclassified. In the right panel a ROC curve of the model is drawn. The area under curve (AUC) equals 0.99. Classification metrics for the test set are following:

- Accuracy: 0.98,
- Precision: 0.98,
- Recall: 0.97,
- F1-score: 0.97,

which summarize the model as extremely accurate.

To get a good understanding of how the trained model predicts the probability of coronary heart disease, an analysis of the model was carried out We approached the problem in two ways, namely globally and locally.

The global approach is made possible by the easy interpretability of the XGBoost model, as it is possible to estimate the effect of individual characteristics on the increase in probability determined by the model. The specific values are presented in Table 1. We can see that the greatest influence on the high probability of CHD is high LDL levels. The advanced age of the patient is situated in the second place. Systolic blood pressure, a high value of which contributes significantly to the likelihood of CHD is almost equally important as the age.

Much weaker effect, by about 3.5 times, is found for male sex. The remaining characteristics have a marginal effect on CHD incidence – their contributions to the final outcome are 3 orders of magnitude weaker than the most important factors.

As mentioned earlier, the model consists of a set of decision trees, 100 to be exact, one of which can be seen in Fig. 4. This tree was showed for illustrative purposes only, so that the reader can see how the model generates a contribution to the overall outcome. Of course, some of the trees may behave differently, and this particular tree should not be completely generalized. Nevertheless, it is possible to see similar weights for specific features, as is the case in Table 1. It is worth noting at this point that the numbers displayed in the tree diagram are normalized feature values, not their original values.

On the other hand the model on a local basis, using the LIME (Local Interpretable Model-Agnostic Explanations) method was analyzed. This method consists of generating local explanations for individual predictions by fitting an interpretable model to a dataset sampled around the prediction of interest. This local interpretation model is usually a very simple model, such as a decision tree or linear regression, which can be very easy to interpret.

For the LIME analysis, two contrasted subsets were drawn, i.e. 100 points with low maturity probability and 100 points with high maturity probability. We perform a LIME analysis on each of these subsets and then sum the impact scores of the individual characteristics.

The results of the analysis are presented in Fig. 5. The left panel shows the results for the subset with high maturity probability. We can see that, in this case, the same three characteristics play the largest role, but in a different order than was the case in the global analysis. LDL comes first, systolic blood pressure second and patient age third.

Table 1. Global contributions of features in the XGBoost model.

Feature name	Normalised importance score
Sex	0.0859
Age	0.2875
Weight	0.0006
Height	0.0006
BMI	0.0006
Smoking years	0.0006
Daily number of cigarettes	0.0007
Total cholesterol	0.0007
LDL	0.3659
HDL	0.0006
Glucose 1	0.0006
Glucose 2	0.0007
Glucose 3	0.0007
Systolic pressure	0.2627
Diastolic pressure	0.0010

Things are similar in the second subset, the results of which can be seen in the right-hand panel. The order of influence of the characteristics on the low probability of CHD is the same.

While the three most important characteristics are the same in both analyses conducted, their order is different. It is undisputed that LDL level is the most important risk factor. However, the global analysis puts age in second place, and the local analysis puts systolic blood pressure in second place. This is interesting because the difference in weights shown by the local analysis is quite large, whereas in the global view they are relatively similar.

3 Summary

In this paper we have studied a dataset containing records of coronary heart disease patients via machine learning methods. The target value – the estimated probability of CHD morbidity – was modeled by 16 predictors, such as body mass, cholesterol levels, age or blood pressure. Our analysis and numerical experiments showed that a tree based XGBoost model provided the best fit to the data. The model was fitted, validated, and analyzed for further understanding of the logic it gathered from the data. Surprisingly, the ML model requires only three features to accurately predict the CHD morbidity probability. These are the level of LDL, systolic blood pressure and patient's age. The remaining features, including smoking and glucose levels, had minor effect on the final result.

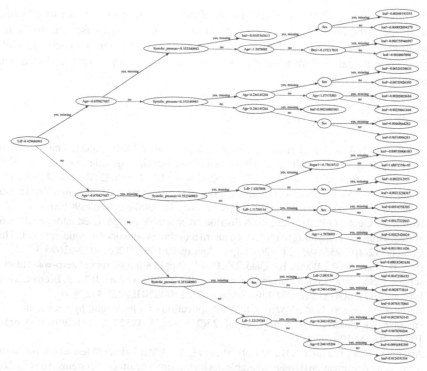

Fig. 4. An example decision tree of the total XGBoost ensemble.

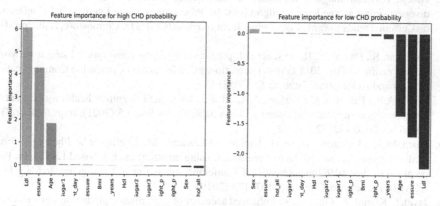

Fig. 5. Feature importance from LIME analysis for high (left panel) and low (right panel) probabilities of CHD.

Determining whether a person is at risk of developing coronary heart disease (CHD) can be done by assessing three key characteristics: age, cholesterol level (especially LDL level), and BMI level. Diastolic pressure level can also be considered as an additional factor. By utilizing well-designed machine learning models, faster diagnosis suggestions can be made, leading to the implementation of effective preventive programs. Furthermore, the described model can also be applied to analyze other diseases within the field of cardiology, such as atherosclerosis and myocarditis. Implementing such a model in hospital systems can significantly improve medical data analysis and lead to quicker diagnosis.

References

1. Revathi, A., Kaladevi, R., Ramana, K., Jhaveri, R.H., Kumar, M.R., Sankara Prasanna Kumar, M.: Early detection of cognitive decline using machine learning algorithm and cognitive ability test. Secur. Commun. Netw. 13 (2022). https://doi.org/10.1155/2022/4190023.4190023
2. Sagar, R., Jhaveri, R., Borrego, C.: Applications in security and evasions in machine learning: a survey. Electronics 9(1), 97 (2020). https://doi.org/10.3390/electronics9010097
3. Kumar, P., Gupta, G.P., Tripathi, R.: A distributed ensemble design based intrusion detection system using fog computing to protect the internet of things networks. J. Ambient Intell. Hum. Comput. 12(10), 9555–9572 (2020). https://doi.org/10.1007/s12652-020-02696-3
4. Han, B., Jhaveri, R., Wang, H., Qiao, D., Du, J.: Application of robust zero-watermarking scheme based on federated learning for securing the healthcare data. IEEE J. Biomed. Health Inform. 27(2), 804–813 (2023). https://doi.org/10.1109/JBHI.2021.3123936
5. Barini, G.O., Ngoo, L.M., Mwangi, R.W.: Application of a fuzzy unit hypercube in cardiovascular risk classification. Soft. Comput. 23(23), 12521–12527 (2019). https://doi.org/10.1007/s00500-019-03802-0
6. Arikumar, K.S., Prathiba, S.B., Alazab, M., et al.: FL-PMI: federated learning-based person movement identification through wearable devices in smart healthcare systems. Sensors 22(4), 1377 (2022). https://doi.org/10.3390/s22041377
7. Dinesh, K.G., Arumugaraj, K., Santhosh, K.D., Mareeswari, V.: Prediction of cardiovascular disease using machine learning algorithms. In: Proceedings of the International Conference on Current Trends towards Converging Technologies (ICCTCT), Coimbatore, India, pp. 1–7 (2018)
8. Ambekar, S., Phalnikar, R.: Disease risk prediction by using convolutional neural network. In: Proceedings of the 2018 Fourth International Conference on Computing Communication Control and Automation, Pune, India (2018)
9. Javed, A.R., Fahad, L.G., Farhan, A.A., et al.: Automated cognitive health assessment in smart homes using machine learning. Sustainable Cities Soc. 65 (2021). https://doi.org/10.1016/j.scs.2020.102572.102572
10. Islam, M.S., Muhamed Umran, H., Umran, S.M., Karim, M.: Intelligent healthcare platform: cardiovascular disease risk factors prediction using attention module based LSTM. In: Proceedings of the 2019 2nd International Conference on Artificial Intelligence and Big Data (ICAIBD), Chengdu, China, pp. 167–175 (2019)
11. Javid, I., Khalaf, A., Ghazali, R.: Enhanced accuracy of heart disease prediction using machine learning and recurrent neural networks ensemble majority voting method. Int. J. Adv. Comput. Sci. Appl. 11 (2020). https://doi.org/10.14569/ijacsa.2020.0110369
12. Deaths due to coronary heart diseases in the EU. https://ec.europa.eu/eurostat/web/products-eurostat-news/-/edn-20200928-1. Accessed 01 Apr 2023
13. Price, S., Katz, J., Kaufmann, C.C., Huber, K.: The year in cardiovascular medicine 2021: acute cardiovascular care and ischaemic heart disease. Eur. Heart J. 43(8), 800–806 (2022). https://doi.org/10.1093/eurheartj/ehab908
14. Chen, T., Guestrin, C.: XGBoost: a scalable tree boosting system. https://arxiv.org/abs/1603.02754

15. Breiman, L.: Bagging predictors. Mach. Learn. **24**(2), 123–140 (1996)
16. Greenwell, B., Boehmke, B., Cunningham, J.: GBM Developers. Gbm: Generalized Boosted Regression Models (2019)
17. Ho, T.K.: Random decision forests. In: Proceedings of 3rd International Conference on Document Analysis and Recognition, 1, pp. 278–82. IEEE (1995)
18. Venables, W.N., Ripley, B.D.: Modern Applied Statistics with S. Springer, New York, NY (2002)
19. Centers for Medicare & Medicaid Services. Decision memo for supervised exercise therapy (SET) for symptomatic peripheral artery disease (PAD) (CAG-00449N) (2017). https://www.cms.gov/medicare-coverage-database/details/nca-decision-memo.aspx? NCAId=287. Accessed 01 July 2021
20. Zheng, Y.Y., Ma, Y.T., Zhang, J.Y., Xie, X.: COVID-19 and the cardiovascular system. Nat. Rev. Cardiol. **17**, 259–260 (2020)

Systematic Comparison of Advanced Network Analysis and Visualization of Lipidomics Data

Jana Schwarzerová[1,2](✉) ⓘ, Dominika Olešová[3,4](✉) ⓘ, Aleš Kvasnička[5] ⓘ,
David Friedecký[5] ⓘ, Margaret Varga[6] ⓘ, Valentine Provazník[1] ⓘ,
and Wolfram Weckwerth[2,7] ⓘ

[1] Department of Biomedical Engineering, Faculty of Electrical Engineering and
Communication, Brno University of Technology, Technicka 12, 616 00 Brno, Czech Republic
Jana.Schwarzerova@vut.cz
[2] Molecular Systems Biology (MOSYS), University of Vienna, Vienna, Austria
[3] Institute of Experimental Endocrinology, Biomedical Research Center, Slovak Academy of
Sciences, Dúbravská Cesta 9, 845 10 Bratislava, Slovak Republic
dominika.olesova@savba.sk
[4] Institute of Neuroimmunology, Slovak Academy of Sciences, Dúbravská Cesta 9, 845 10
Bratislava, Slovak Republic
[5] Department of Clinical Biochemistry, Palacký University Olomouc and University Hospital
Olomouc, Olomouc, Czech Republic
[6] Department of Biology, University of Oxford, Oxfordshire, GB, UK
[7] Vienna Metabolomics Center (VIME), University of Vienna, Vienna, Austria

Abstract. Comprehensive analysis of lipids is becoming a forefront of clinical
data analysis. Due to significant technical advancements, lipidomics is emerging
in clinical diagnostics for improvement and earlier detection of a broad range of
diseases. However, in order to understand the biological complexities and interrelationships between the molecules, it is important to have a correct representation
of the data and visualizations that enable good interpretability of the lipidomic
data. Therefore, the present study systematically compares different visualization
methods for lipidomic data, based on different computational relations between
the selected lipids and supplemented with known biological information. Networks were reconstructed, and an analysis was performed to objectively compare
the visualizations.

Keywords: Comprehensive Analysis · Networks Analysis · Lipids · Network
Visualization

1 Introduction

Lipidomics is a new emerging subdiscipline of metabolomics, mainly due to ever improving analytical techniques based on mass spectrometry [1]. The main goal of this omics
field is the full characterization, identification, and quantification of lipid species and
their biological functions [2]. Markus R. Wenk [3] defines lipidomics as system-level
analysis and characterization of lipids and their interacting moieties. This definition

© The Author(s), under exclusive license to Springer Nature Switzerland AG 2023
I. Rojas et al. (Eds.): IWBBIO 2023, LNBI 13919, pp. 378–389, 2023.
https://doi.org/10.1007/978-3-031-34953-9_30

implies that lipids – small-molecule metabolites – have important functions in physio-logical processes. The well-coordinated lipid metabolism presents a base for molecular stability [4, 5], signalling pathways [6], and membrane structure [7]. Moreover, lipids have been shown to be involved, and contribute to the pathogenesis of various diseases [8, 9].

Currently, an important application of lipidomics is in pharmaceutical industry. Drug research targeting lipid pathways include cholesterol-lowering substances, cyclooxyge-nase inhibitors and many others. Furthermore, research also focuses on specific regula-tors of multiple lipidic targets such as phosphatidylinositol 3-kinases, sphingosine, and ceramide kinases for therapeutic interventions in diseases ranging from inflammation and cancer to metabolic disorders [4]. An important milestone in this field is FTY720 [10] approved for the treatment of multiple sclerosis [4].

Lipids represent an enormous number of chemically distinct molecular species aris-ing from the various combinations of fatty acyls on backbone structures. In recent time, lipid research has been significantly improved by novel analytical chemistry technolo-gies, and bioinformatic tools for lipidomic data analysis [2]. Experimental approaches in lipidomics usually include chromatography, mass spectrometry (MS) and nuclear mag-netic resonance (NMR). Various MS-based lipidomic workflows already exist and are widely applied to measure the metabolic networks of certain lipid classes. Nowadays, among the commonly known and used chromatographic techniques are gas chromatog-raphy (GC), high-performance liquid chromatography (HPLC), and supercritical fluid chromatography (SFC) [2].

However, due to the relatively recent emergence of this omics field, lipidomic research lacks specialized tools for tailored analysis of this specific kind of data. Present study focuses on developing new approaches for lipidomics data analysis, focusing on more targeted approaches for selected compounds of lipid metabolism, such as ceramides (Cer) and sphingomyelins (SM) and others.

This work is based on classical approaches such as calculation of the Pearson corre-lation coefficient (cc), or Mutual Information (MI) to obtain undirected information for network visualization [11–15], and novel approaches based on prediction algo-rithms that are applied for visualization of directed networks to make their interpretabil-ity easier. Prediction methods are commonly used for the biomarker screenings using metabolomics [16–18]. Even using ordinary linear regression principles, we can model quality predictions based on metabolite interactions [19].

Recently, there has been an increase in the use of random forest (RF) in compu-tational biology, due to its nonparametric, interpretable, efficient and high prediction accuracy properties [20]. Therefore, the present study focuses mainly on RF prediction methodology which offers many unique advantages in dealing with small sample size, high-dimensional feature space, and complex data structures [20].

Currently, lipidomics is a promising area of bio-medical research. Thus, the combina-tion of lipidomics analysis with prediction methods and appropriate choice of visualiza-tion techniques is a very important foundation package for improving clinical diagnostics in disease research.

2   Materials and Methods

2.1   Dataset

In this study we used previously published lipidomic dataset from a study by Andreyev AY et al. [21] which is available in LION/web (http://www.lipidontology.com/) [22]. This study includes 229 individual isobaric species, with positively identified 163 glycerophospholipids, 48 sphingolipids, 13 sterols, and 5 prenols. In the present study, we focused on sphingolipid metabolic pathways.

Data were collected from immortalized mouse macrophage-like RAW264.7 cells from ATCC and PBS. The fetal calf serum with low endotoxin content was used [21]. KLA (Kdo2-Lipid A) and lipid standards were from Avanti Polar Lipids, and all other reagents and kits were from Sigma-Aldrich. More detailed information is available in [21].

Study [21] used LC–MS (Liquid Chromatography-Mass Spectrometry) technique with the utilization of synthetic odd-carbon phospholipid standards.

2.2   Methods

The main contribution of our methodology shows the advantages and disadvantages of different representations of lipid data using undirected and directed analytical approaches. The whole concept of our methodology is summarized in Fig. 1.

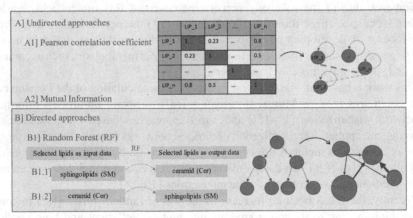

Fig. 1. Schematic visualization of our methodology; A] represents an overview of undirected approaches for reconstruction network visualization in lipids using Pearson correlation coefficient and Mutual Information, B] represents directed methods applying random forest methodology that provides feature importance and prediction accuracy information

The section A] in Fig. 1 describes undirected approaches divided into two different methods which are often associated with the search for reciprocal relationships. Pearson correlation coefficient (cc) is widely spread in lipidomic analysis [23, 24]. Moreover, we also used Mutual Information (MI) [25]. Section B] presents an advanced approach based on a machine learning algorithm, namely Random Forest (RF) [20].

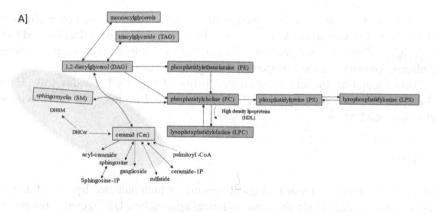

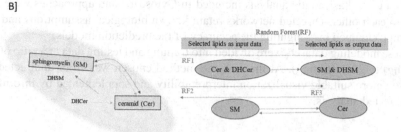

Fig. 2. Lipids pathway created using literature and KEGG databases, yellow is the highlighted area of interest for prediction methods; RF is Random Forest prediction method which takes into account biology predisposition of causality (color figure online)

The biological application of this prediction method, depending on the selection of input and output data, is based on the available literature and searching KEGG databases [26], see Fig. 2. Figure 2A] shows the information about lipids gathered from the literature and available databases. The area of interest for present study is highlighted in yellow and the direction of the predictions methods is indicated by the green arrow.

2.3 Undirected Approaches

Firstly, the cc and MI matrix were obtained for sphingomyelin (SM), ceramide (Cer), dihydrosphingomyelin (DHSM), dihydroceramide (DHCer). The cc matrix was calculated by R/ Hmisc package [27] as correlation matrix with significance levels (p-value = 0.05). The MI matrix was calculated by R/ infotheo package [28]. Both computed matrices were subsequently reconstructed into networks using the Cytoscape tool [29] and the network analysis for the undirected network was performed in Cytoscape.

2.4 Directed Approaches

Next calculation method was RF. By accurately determining the input and output data for prediction methods, we can verify, confirm or refute the causality of lipid components.

In total, three different tasks were modelled. The first model focused on prediction of SM from Cer. The second model predicts Cer from SM and last model was based on Cer and DHCer information for prediction SM and DHSM information. Thus, to preserve the already known biological properties shown in Fig. 2.

These prediction models were created using numpy [30], pandas [30, 31] and scikit-learn [32] in python version 3.10.2. The achieved results were visualised using Cytoscape, see Sect. 3.

3   Results

In total, we reconstructed four biological networks which included lipids and their biochemical relationships using different analytical approaches. Undirected networks were analyzed using basic static analysis included in Cytoscape and approaches were compared to each other. Directed networks retain known biological assumptions and have been supplemented by weighting the accuracy of the prediction models.

These modeling datasets were divided into training and testing data at a ratio of 80% (training) to 20% (testing). At the end, the method chosen was from undirected part based on more suitable visualization interpretability and supplemented by information from directed analysis.

3.1   Undirected Analytical Methods for Network Visualization

The cc network was reconstructed based on the correlation matrix calculated using Pearson correlation coefficient. In total, there were four lipid's classes: Cer, DHCer, SM, and DHSM. Each class encompasses 8 different lipids, which are represented as nodes in the network. The results of the correlation analysis are shown in the Fig. 3. Subsequently, a basic network analysis was performed using Cytoscape. A similar procedure was performed for MI analytic approach, see Fig. 4.

In Fig. 3A, we can see two strong correlation areas which reflected chemical and biological properties of lipids. The first area is connected to SM. The second area includes Cer and DHCer. From the comparison of the correlation between SH and DHSM area and Cer and DHCer, we can conclude that Cer and DHCer affect each other much more. This could indicate a possible feedback loop within these lipids which is much more apparent from this correlation than for DHSM and SM components.

Undirected analysis based on MI may reveal which classes should not be considered as good biomarkers for further study. This means those that have low MI from others and thus, there is a high probability that within biological systems these lipids can be influenced by something else, such as SM C24 or Cer C26, see Fig. 4A.

Reconstructed networks were analyzed using Analyse Network tool in Cytoscape. Both these networks include 32 nodes and 1024 edges. Clustering coefficient were 1.00. These parameters present a complete connection of all nodes. Therefore, the transparency of the edges was set depending on the size of the calculated value, both the correlation coefficient and the MI, see Fig. 5.

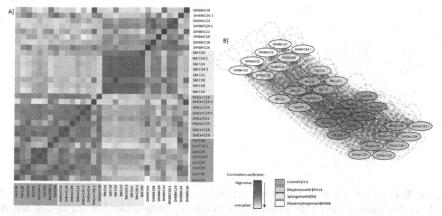

Fig. 3. Results from correlation analysis; A] shows heatmap (where the darkest red colors represent the largest values of the correlation coefficient; conversely, light red colors indicate the smallest values of the correlation coefficient.) which is used as core for B] correlation network in which edge transparency presents the value of the correlation coefficient (color figure online)

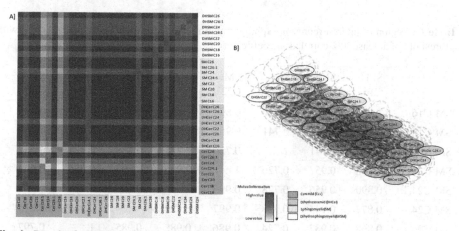

Fig. 4. Results from MI analysis; A] shows heatmap (where the darkest red colors represent the largest values of the MI; conversely, light red colors indicate the smallest values of the correlation coefficient); It is used as core for B] MI network in which edge transparency presents the MI-value (color figure online)

Figure 5 shows an important relationship between SM class which is much more separated from DHSM than Cer and DHCer among themselves in the correlation network. Conversely, by using MI, we can also notice a certain relationship between more distant objects, such as DHSM and DHCer.

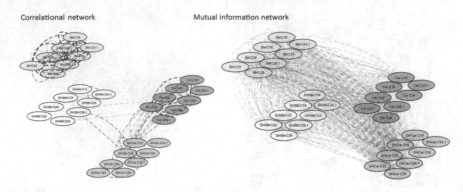

Correlational network   Mutual information network

Fig. 5. Reconstructed networks from undirected information using correlation analysis and MI analysis. The transparency of the edge is set as weight reporting values from calculated analyses and self-loops were remove

In Table 1 the strong correlations among SM. The maximum is 0.997 between SM C22 and SM C24.

Table 1. Region of interest represents sphingomyelin obtained from network visualization with a threshold of at least 70% correlation coefficient

	SM C16	SM C18	SM C20	SM C22	SM C24.1	SM C24	SM C26.1	SM C26
SM C16	1	0.957	0.670	0.898	0.890	0.877	0.868	0.623
SM C18	0.957	1	0.741	0.957	0.956	0.944	0.937	0.739
SM C20	0.670	0.741	1	0.726	0.728	0.728	0.724	0.538
SM C22	0.898	0.957	0.726	1	0.995	0.997	0.986	0.809
SM C24.1	0.890	0.956	0.728	0.995	1	0.990	0.988	0.805
SM C24	0.877	0.944	0.728	0.997	0.990	1	0.985	0.8147
SM C26.1	0.868	0.937	0.724	0.986	0.988	0.985	1	0.797
SM C26	0.623	0.739	0.539	0.809	0.805	0.814	0.797	1

3.2   Directed Analytical Methods for Network Visualization

The evaluation of performance, which is calculated by correlation coefficient, represents accuracy of regression prediction methods. In total, we had 32 variables (8 lipids components in 4 different lipid's class) and 42 observations representing: three replicates for Nuclei Control, three replicates for Nuclei Kdo2-Lipid A (KLA), three replicates of Mitochondria Control, Mitochondria KLA, endoplasmic reticulum (ER) Control, and ER KLA, Plasmalemma Control, Plasmalemma KLA, Dense Microsomes Control, Dense

Microsomes KLA, Cytosol Control, Cytosol KLA, Whole cell homogenate Control and Whole cell homogenate KLA.

Reconstruction visualization were based on two obtained information from RF prediction. Weight edges represents accuracy of regression prediction and size of nodes presents feature importance, see Table 2, calculated during RF prediction for each observation, i.e. lipids components. The directed visualization based on RF prediction modelling is shown in Fig. 6.

Table 2. Feature importance calculated by RF for each lipid

	C16	C18	C20	C22	C24.1	C24	C26.1	C26
Cer	0.04248	0.10056	0.24927	0.07027	0.07314	0.09654	0.11772	0. 25002
SM	0.19397	0.11930	0.18371	0.04686	0.09071	0.05365	0.11094	0.20086

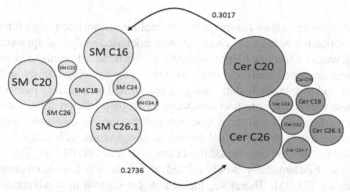

Fig. 6. Visualization reconstructed by RF prediction; edges represents accuracy of regression prediction and size nodes represents feature importance for modelled predictions

At the end of this study we applied interconnection of undirected and directed tested approaches, see Fig. 7. We created three different RF models. Firstly, we calculated model from Cer to predict SM, secondly, SM to Cer and at the end we modelled RF prediction from Cer and DHCer (orange square) to SM and DHSM (yellow square).

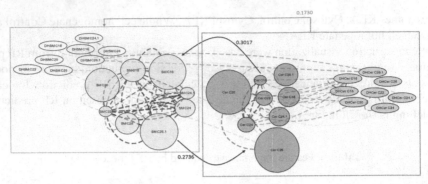

Fig. 7. Interconnection of undirected and directed approaches for visualization of lipids relationships. Directed edges represent prediction based on RF; undirected edges represent correlation analysis

4 Discussion

The present study presents a comparison and final interconnection of different types of data visualization methods using advanced, network based analytical approach. The aim of this work was to determine lipid relationships from different points of view, that were obtained from advanced statistical analysis. Due to the growing quality and reliability of lipidomic and metabolomic data, research has shifted to revealing interrelationships between these two fields, or to find new ways based on predictive algorithms to screen for new biomarkers. This innovative research will be directed by correct understanding of relationships among the individual molecular components such as lipids.

Our study has focused on selected lipid classes such as Cer, DHCer, SM, and DHSM. These lipids are biochemically closely related, and previously have been connected with human diseases [33–35]. Therefore, the search for various new advanced analytical techniques which could reveal, confirm or refuse relationships, correlations or causalities among lipids is desirable.

Firstly, we searched the available literature and database [26, 36], and we selected lipid pathways connected to our area of interest. Based on these biological assumptions, we were able to create a methodology that includes both lipid correlation analysis and predictive modelling. The first part of analysis uses correlation and MI networks for each selected lipid components and reveals undirected relationships. Even though MI does not require linear dependence – which cannot be guaranteed for lipids, and the calculation of MI depends on entropy, for lipid analysis it seems more advantageous to use classical correlation analysis as analysis for subsequent reconstructions of networks as could be seen in Fig. 5 because denser nets are harder to interpret. Therefore, the interconnection approach which is shown in Fig. 7, includes correlation analysis with RF prediction modelling.

RF prediction was used according to biological assumptions. The correlation coefficient representing prediction accuracy is 0.3017 in one direction (Cer → SM) and 0.2736 in the opposite direction (SM → Cer). Since the numbers are around 30% for both predictions, we can state that the biological assumption that enables balance based on homeostasis here is confirmed by this prediction model, see Figs. 6 and 7. As part of the visualization, RF modelling also brings additional value regarding feature importance. The most important node according to feature importance was Cer C26 which is known as a biomarker for the diagnosis of Farber Disease [37]. Thanks to these biologically confirmed assumptions in our visualizations, we can assume the right direction for other network visualization based on advanced analytic approaches. Innovated visualization can improve and complete detailed information about the location of biochemical lipids reactions which is crucial for understanding their roles in function and dysfunction concept.

5 Conclusion

Lipidomic is currently one of the most developing scientific fields, mainly due to its wide range of clinical and biomedical applications. Predictive modeling will come to the fore in this research area in the near future, and it is necessary to start thinking about its correct application. In this study, we provide a first insight at different types of non-conventional visualization techniques for lipid pathways, which rely on biological data and assumptions based on available databases. The study presents two different approaches for reconstruction network which is divided into undirected methods such as correlation analysis and MI, and directed method represents RF prediction. Moreover, the final visualization is created using interconnection of tested methods. Due to an innovated approach for visualization we can reveal new avenues for understanding the overall complex of lipidomics.

Acknowledgements. Computational resources were supplied by the Ministry of Education, Youth and Sports of the Czech Republic under the Projects CESNET (Project No. LM2015042) and CERIT-Scientific Cloud (Project No. LM2015085) provided within the program Projects of Large Research, Development and Innovations Infrastructures.

References

1. Blanksby, S.J., Mitchell, T.W.: Advances in mass spectrometry for lipidomics. Annu. Rev. Anal. Chem. **3**, 433–465 (2010)
2. Sethi, S., Brietzke, E.: Recent advances in lipidomics: analytical and clinical perspectives. Prostagland. Other Lipid Mediat. **128**, 8–16 (2017)
3. Wenk, M.: The emerging field of lipidomics. Nat. Rev. Drug Discov. **4**, 594–610 (2005). https://doi.org/10.1038/nrd1776
4. Wenk, M.R.: Lipidomics: new tools and applications. Cell **143**(6), 888–895 (2010)
5. Jordan, S.D., Könner, A.C., Brüning, J.C.: Sensing the fuels: glucose and lipid signaling in the CNS controlling energy homeostasis. Cell. Mol. Life Sci. **67**, 3255–3273 (2010)
6. Hou, Q., Ufer, G., Bartels, D.: Lipid signalling in plant responses to abiotic stress. Plant Cell Environ. **39**(5), 1029–1048 (2016)
7. Quinn, P.J., Joo, F., Vigh, L.: The role of unsaturated lipids in membrane structure and stability. Prog. Biophys. Mol. Biol. **53**(2), 71–103 (1989)
8. Leray, C.: Lipids. CRC Press (2014). https://doi.org/10.1201/b17656

9. Schmitt, F., Hussain, G., Dupuis, L., Loeffler, J.P., Henriques, A.: A plural role for lipids in motor neuron diseases: energy, signaling and structure. Front. Cell. Neurosci. **8**, 25 (2014)

10. Brinkmann, V., Billich, A., Baumruker, T., Heining, P., Schmouder, R., Francis, G., Aradhye, S., Burtin, P.: Fingolimod (FTY720): discovery and development of an oral drug to treat multiple sclerosis. Nat. Rev. Drug Discov. **9**(11), 883–897 (2010)

11. Steuer, R., Morgenthal, K., Weckwerth, W., Selbig, J.: A gentle guide to the analysis of metabolomic data. Methods Mol. Biol. **358**, 105–126 (2007)

12. Weckwerth, W.: Metabolomics in systems biology. Annu. Rev. Plant Biol. **54**, 669–689 (2003)

13. Weckwerth, W., Loureiro, M.E., Wenzel, K., Fiehn, O.: Differential metabolic networks unravel the effects of silent plant phenotypes. Proc. Natl. Acad. Sci. U.S.A. **101**, 7809–7814 (2004)

14. Morgenthal, K., Wienkoop, S., Scholz, M., Selbig, J., Weckwerth, W.: Correlative GC-TOF-MS-based metabolite profiling and LC-MS-based protein profiling reveal time-related systemic regulation of metabolite–protein networks and improve pattern recognition for multiple biomarker selection. Metabolomics **1**, 109–121 (2005)

15. Muller-Linow, M., Weckwerth, W., Hutt, M.T.: Consistency analysis of metabolic correlation networks. BMC Syst. Biol. **1**, 44 (2007)

16. Taylor, J.M., Ankerst, D.P., Andridge, R.R.: Validation of biomarker-based risk prediction models. Clin. Cancer Res. **14**(19), 5977–5983 (2008)

17. Sidak, D., Schwarzerová, J., Weckwerth, W., Waldherr, S.: Interpretable machine learning methods for predictions in systems biology from omics data. Front. Mol. Biosci. **9**, 926623 (2022). https://doi.org/10.3389/fmolb.2022.926623

18. Bachmann, G., Sun, X., Jaeger, W., Kautzky-Willer, A., Weckwerth, W.: Combined metabolomic analysis of plasma and urine reveals AHBA, tryptophan and serotonin metabolism as potential risk factors in gestational diabetes mellitus (GDM). Front. Mol. Biosci. **4**, 84 (2017)

19. Schwarzerova, J., Pierides, I., Sedlar, K., Weckwerth, W.: Linear predictive modeling for immune metabolites related to other metabolites. In: Rojas, I., Valenzuela, O., Rojas, F., Herrera, L.J., Ortuño, F. (eds.) Bioinformatics and Biomedical Engineering: 9th International Work-Conference, IWBBIO 2022, Maspalomas, Gran Canaria, Spain, Proceedings, Part I, pp. 16–27. Springer International Publishing, Cham (2022). https://doi.org/10.1007/978-3-031-07704-3_2

20. Qi, Y.: Random forest for bioinformatics. In: Ensemble Machine Learning: Methods and Applications, pp. 307–323. Springer, US, Boston, MA (2012)

21. Andreyev, A.Y., et al.: Subcellular organelle lipidomics in TLR-4-activated macrophages 1 [S]. J. Lipid Res. **51**(9), 2785–2797 (2010)

22. Molenaar, M.R., Jeucken, A., Wassenaar, T.A., van de Lest, C.H., Brouwers, J.F., Helms, J.B.: LION/web: A web-based ontology enrichment tool for lipidomic data analysis. GigaScience **8**(6), giz061 (2019). https://doi.org/10.1093/gigascience/giz061

23. Yetukuri, L., Katajamaa, M., Medina-Gomez, G., Seppänen-Laakso, T., Vidal-Puig, A., Orešič, M.: Bioinformatics strategies for lipidomics analysis: characterization of obesity related hepatic steatosis. BMC Syst. Biol. **1**(1), 1–15 (2007)

24. Yu, J., et al.: Lipidomics and transcriptomics analyses of altered lipid species and pathways in oxaliplatin-treated colorectal cancer cells. J. Pharm. Biomed. Anal. **200**, 114077 (2021)

25. Mahony, S., Auron, P.E., Benos, P.V.: Inferring protein–DNA dependencies using motif alignments and mutual information. Bioinformatics **23**(13), i297–i304 (2007)

26. Kanehisa, M.: The KEGG database. In: 'In Silico' Simulation of Biological Processes: Novartis Foundation Symposium, vol. 247, pp. 91–103. John Wiley & Sons, Ltd., Chichester, UK

27. Harrell, F.E., Jr., Harrell, M.F.E., Jr.: Package 'hmisc'. CRAN2018, pp. 235–236 (2019)

28. Meyer, P.E., Meyer, M.P.E.: Package 'infotheo'. R Packag. version, 1 (2009)
29. Shannon, P., et al.: Cytoscape: a software environment for integrated models of biomolecular interaction networks. Genome Res. **13**(11), 2498–2504 (2003)
30. McKinney, W.: Python for Data Analysis: Data Wrangling with Pandas, NumPy, and IPython. O'Reilly Media, Inc. (2012)
31. McKinney, W.: pandas: a foundational Python library for data analysis and statistics. Python High Perform. Sci. Comput. **14**(9), 1–9 (2011)
32. Pedregosa, F., et al.: Scikit-learn: machine learning in Python. J. Mach. Learn. Res. **12**, 2825–2830 (2011)
33. Bouwstra, J.A., Dubbelaar, F.E.R., Gooris, G.S., Weerheim, A.M., Ponec, M.: The role of ceramide composition in the lipid organisation of the skin barrier. Biochim. Biophys. Acta: Biomembranes **1419**(2), 127–136 (1999). https://doi.org/10.1016/S0005-2736(99)00057-7
34. Silva, L.C., de Almeida, R.F., Castro, B.M., Fedorov, A., Prieto, M.: Ceramide-domain formation and collapse in lipid rafts: membrane reorganization by an apoptotic lipid. Biophys. J. **92**(2), 502–516 (2007)
35. Sankaram, M.B., Thompson, T.E.: Interaction of cholesterol with various glycerophospholipids and sphingomyelin. Biochemistry **29**(47), 10670–10675 (1990)
36. Fahy, E., Sud, M., Cotter, D., Subramaniam, S.: LIPID MAPS online tools for lipid research. Nucleic Acids Res. **35**(Suppl. 2), W606–W612 (2007). https://doi.org/10.1093/nar/gkm324
37. Cozma, C., et al.: C26-ceramide as highly sensitive biomarker for the diagnosis of Farber disease. Sci. Rep. **7**(1), 1–13 (2017)

Comparison of Image Processing and Classification Methods for a Better Diet Decision-Making

Maryam Abbasi[1,2], Filipe Cardoso[3,5], and Pedro Martins[4(✉)]

[1] Polytechnic Institute of Coimbra, Institute of Applied Research, Coimbra, Portugal
`maryam@dei.uc.pt`
[2] Centre for Informatics and Systems of the University of Coimbra,
Department of Informatics Engineering, University of Coimbra, Coimbra, Portugal
[3] INESC Coimbra, Coimbra, Portugal
`filipe.cardoso@esg.ipsantarem.pt`
[4] CISeD - Research Centre in Digital Services, Polytechnic of Viseu, Viseu, Portugal
`pedromom@estgv.ipv.pt`
[5] Polytechnic Institute of Santarem, Santarem, Portugal

Abstract. This paper aims to explore the use of different deep learning techniques, specifically convolutional neural networks (CNNs), for dietary assessment through image food recognition and compare their performance to the human visual system (HVS). Currently, there are three main techniques for using CNNs in this task: training a network from scratch; using an off-the-shelf pre-trained network; and performing unsupervised pre-training with supervised adjustments. In this study, the authors evaluate the performance of three CNN models with varying numbers of parameters (5,000 to 160 million) based on dataset size and spatial image context.

The authors also consider human knowledge and classification to compare the performance of the CNNs to the HVS. They find that while the CNNs make errors across different food classes, the HVS tends to make semantic errors with specific food classes. As a result, the HVS shows more consistency in its answers. Overall, the findings suggest that the HVS is more accurate when the dataset is diverse, while the CNN performs better when the dataset is focused on a particular niche.

In conclusion, this study provides empirical evidence that machine learning can be more efficient than the HVS in certain tasks but also highlights the strengths and limitations of both approaches. The authors suggest that combining CNNs with other classification techniques, such as bag-of-words, may be a promising approach for improving the accuracy of dietary assessment through image food recognition.

Keywords: Compare · Image processing · Machine learning · Image Classification · GoogLeNet · Inception-v3 · ResNet · Bag-of-words

I. Rojas et al. (Eds.): IWBBIO 2023, LNBI 13919, pp. 390–403, 2023.
https://doi.org/10.1007/978-3-031-34953-9_31

1 Introduction

Object recognition involves extracting features from images that are then used to identify objects. These features can be global, such as colour histograms or circular shapes, or local, such as pixel colour or SIFT features. The selection of features plays a crucial role in the accuracy of the classifier. Several approaches have been proposed, including the combination of colour and texture features as suggested in [11], or the integration of contour, motion, texture, and colour features as proposed in [14]. Some studies have also focused on using different colour spaces, such as RGB, HSV, and LAB, or texture properties to recognize objects, as discussed in [13]. Other research has utilized Bag of Features (BoF) algorithms, as seen in [4], or the Pittsburgh Fast-Food Image Database (PFID) with a baseline algorithm using a bag of SIFT features, as outlined in [7], to identify food items.

The key advantage of using deep learning methods, is their ability to handle a wide range of image categories and features and to automatically adjust their processing of images, as noted in [16]. Few studies have applied deep learning to the problem of food recognition using neural networks. There are several machine learning and CNN libraries available, including CudaConvNet, Torch, Theano, and Caffe. In this work, we used MatConvNet, a MATLAB tool that implements CNNs specifically designed for computer vision tasks. CNNs have revolutionized computer vision by replacing traditional image processing techniques such as SIFT, Jseg, and DBscan. However, training CNNs to perform well requires learning through adjusting coefficients using backpropagation on large amounts of data, often millions of images. MatConvNet is optimized for this task and includes optimizations that enable automatic parallelism and support the use of GPUs (using CUDA DevKit and a compatible NVIDIA GPU) to accelerate specific mathematical computations.

Mobile devices with sufficient computational capabilities are crucial for computer vision applications, including the assessment of dietary habits for the promotion of healthy lifestyles. In the process of food recognition, the first step is to identify the plate containing the food and ignore the surrounding environment. The next step is to isolate the individual food items on the plate, taking into account their individual components (e.g., rice and beans, the yolk and white of an egg), and classify each item. With this information, it is possible to track the caloric intake of consumed food and make recommendations for improved lifestyle choices.

Convolutional Neural Networks (CNNs) have enabled image classification and object recognition with high accuracy, and the combination of CNNs with modern mobile computing capabilities has made it possible to replace manual identification of food with automatic classification simply by pointing the camera of a smartphone at a plate of food. However, it is important to determine whether CNNs can match or surpass human performance in this task and whether they are more efficient than traditional methods. Therefore, it is necessary to study the capabilities of CNNs in the context of food recognition.

This study compares the performance of a traditional approach, the bag-of-words model, with state-of-the-art deep learning approaches using CNN architectures (GoogLeNet, Inception-v3, and Resnet101) in the task of food recognition. The human visual system is also tested as a baseline for comparison. The study involves training on and identifying new dishes, specifically European foods, and includes a survey to assess the food recognition abilities of a selected group of humans. The paper describes the bag-of-words pipeline and CNN architectures and presents the results of the training and testing.

This study uses real-world food image datasets (UEC-256 and Food-101) to evaluate the performance of three CNNs (GoogLeNet, Inception-v3, and ResNet-101) in food detection and classification. Using the UECFood-256 dataset, the authors train and evaluate the networks with different CNN architectures and configuration parameters. They compare the accuracy of the CNNs to that of human vision-based classification using a study group of approximately 100 university students. The results show that the CNNs achieve an accuracy of 70.68% when trained on the full food dataset, compared to 80.6% for the human visual system. However, when the CNNs are trained on a smaller set of 16 food classes, their accuracy increases to 89.89%. Additionally, increasing the number of training epochs from 6 to 20 further improves the accuracy of the CNNs to 93.86%.

2   Related Work

In general, object recognition involves extracting features from images that are then fed to a classifier to identify the objects in the image. These features can be global, such as color histograms or circular shapes, or local, such as pixel color or SIFT features. The selection of features plays a crucial role in the accuracy of the classifier. There have been a variety of approaches proposed, including combining color and texture features as in [11], or integrating image features such as contour, motion, texture, and color as in [15]. Other studies have focused on using different color spaces, such as RGB, HSV, and LAB, or texture properties, as discussed in [13], for object recognition.

Other research in the field of food recognition has focused on the use of Bag of Features (BoF) algorithms, as seen in [2], and the Pittsburgh Fast-Food Image Database (PFID) with a baseline algorithm using a bag of SIFT features, as outlined in [6]. The key advantage of using deep learning methods, as demonstrated in [1], is their ability to handle a wide range of image categories and features and to automatically adjust their processing of images. However, few studies have applied deep learning to the problem of food recognition using neural networks. In [8], features are combined using Fisher Vectors (FVs) from 1000 food-related categories retrieved from ImageNet. Other architectures derived from Inception are explored in [3], where the modules are modified by introducing 1×1 convolutional layers to reduce the input dimension for the next layer. The VGG-16 network is used with a multitask loss in [17], where the problem of food ingredient recognition is also addressed, and a conditional random field is applied to optimize the probability distribution of ingredient identification.

In a study published in 2022, Chen et al. [5] compared three different variations of the ResNet architecture: Inception-ResNet-v1, Inception-ResNet-v2, and Inception-v4. The authors found that the inclusion of residual connections in Inception-ResNet-v1 and Inception-ResNet-v2 led to faster training times, while still maintaining good recognition performance. In contrast, Inception-v4, which did not have residual connections, had similar recognition performance but took longer to train.

Other researchers, such as Salim et al. [11], have used a combination of machine learning and statistical techniques to improve the accuracy of food classification systems. They considered the spatial relationships between ingredient labels in order to increase the accuracy of their systems.

Sharma et al. [12] proposed an automatic food image recognition system for tracking people's eating habits. They used the Multiple Kernel Learning method to integrate various types of image features, including color, texture, and SIFT, in order to achieve a classification rate of 61.34% for 50 different types of food. Mao et al. [9] used a similar approach, combining bag-of-words with various image features such as SIFT, spatial pyramid, histograms, and Gabor texture. Their experiments showed an accuracy of 56% when classifying ten different types of food.

In [8], the authors propose a method for improving food recognition accuracy using a deep convolutional neural network (DCNN) combined with hand-crafted image features (Fisher Vectors with HoG and Color patches). They evaluated their method on the UEC-FOOD100 dataset and achieved top-1 accuracy of 72.26% and top-5 accuracy of 92.00%, which outperforms the previous best reported classification accuracy of 59.6% on this dataset.

In [10], the authors investigated the effectiveness of using a DCNN for food recognition. They explored different combinations of pre-training with large-scale ImageNet data, fine-tuning, and activation features extracted from a pre-trained DCNN. They found that pre-training the network on a dataset of 1000 food-related categories and fine-tuning on the UEC-FOOD100 dataset resulted in the highest accuracy, achieving 78.77%. When using the UECFOOD256 dataset, the accuracy was 67.57%.

3 Experimental Setup

The goal of this work was to evaluate the performance of three convolutional neural networks (CNNs), GoogleNet, Inception v3, and ResNet 101, on the task of food recognition using the UEC FOOD 256 data-set. The data-set consists of 32,000 images, each with a bounding box indicating the location of the food item, and belongs to 256 food categories, most of which are popular in Japan. The data-set was split into three parts: 60% for training, 20% for testing, and 20% for validation. The study was conducted using an Intel i5 3.4 GHz processor, 16 GB RAM, an Asus NVIDIA GeForce GTX 1070 8 GB graphics card, and MatLab R2018a on a Windows 10 64-bit system. The same parameters were used for all three CNNs, as detailed in Table 1.

Table 1. Testing configuration for all CNNs

Training Cycle	
Epoch	6
Iterations	15264
Iterations per epoch	2544
Validation	
Frequency	3 iterations

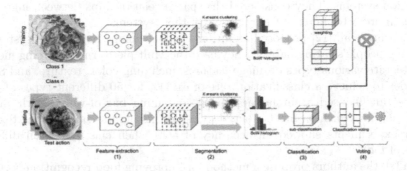

Fig. 1. Bag-of-words architecture sketch

It appears that this text is describing a study that was conducted to compare the performance of convolutional neural networks (CNNs) with the human visual system (HVS) in recognizing and classifying different types of Japanese food dishes. The study included a small group of 20 subjects, 10 of whom were female and 10 of whom were male, all between the ages of 25 and 40 and of European nationality. Since the subjects were not familiar with Japanese food dishes, the study was divided into two stages: a training stage and an image classification stage. In the training stage, the subjects were shown a set of 98 slides featuring images of different food categories, and were asked to identify the image that did not belong to the category. In the image classification stage, the subjects were shown a set of images from the trained classes and asked to select the correct label for each image. The HVS dataset included 686 food photos, with 43 photos in each class.

3.1   Bag-of-Words Architecture

Bag-of-words is a machine learning technique used for image recognition and classification. It represents images as a histogram of the visual features that occur in the image, and uses this representation to create a classifier. The process involves extracting features from a training dataset to represent the image, creating a visual vocabulary (a "bag" of features), and using this vocabulary to classify the image.

Figure 1 depicts the bag-of-words architectural steps to extract features and classify an image. In (1) is represented the feature extraction, for both training and testing pipelines. These features are described in the bag-of-words approach, by vectors of features. In this step, oriented to the proposed research, were extracted features such as texture (GLCM binary patterns), color histograms, geometry features of regions (6 layers the start of step (2)), and also SURF. Using k-means algorithm, images features from the training dataset are extracted and clustered, and this allows obtaining k feature vectors. The k feature vectors represent the centroid of each class feature from the training dataset (2) visual words are grouped and separated by similar characteristics and defined as a vocabulary histogram. In the classification step (3), a trained artificial neural network combines all weighted vectors and saliencies that are analyzed based on a ReLU function. Feature visualization is the same as the forward pass of the deep convolutional network, which only leaves the input's positive components. After step (3), the already trained neural network feeds the testing images in step (4) and label pictures. Once the BOW trained, the method identifies and removes features from the training images provided every new picture for classification and generates the histogram for the picture occurring in codewords. Afterwards, the qualified classifier is used to label the picture as one of the groups (step (4)).

3.2   Deep Learning Architecture(s)

A group of networks is chosen for testing, including GoogleNet, Inception-v3, and Resnet101, with the aim of assessing the accuracy of CNNs. Prior to Resnet, neural networks had difficulty dealing with gradient issues that were neglected and later deleted during the back-propagation learning process, affecting the number of layers. As a result, Resnet architecture was developed to enhance how deep learning networks train using stochastic gradient descent. The sub-architectural blocks known as residual modules feed both the layer below and layers located two or three hops away. Deep learning networks are better equipped to reduce vanishing gradient issues and provide more accurate classification when they continue to use the remaining modules rather than deleting them.

Resnet is made up of layers, and the output from the previews is added at each layer. Resnet uses the same full 3×3 convolutional layer design as VGG. Two 3×3 convolutional layers with the equal amount of output channels make up the residual block. A batch normalization layer and a ReLU activation function are placed after each convolutional layer. The input is then added following these two convolution operations, just before the last ReLU activation function. In order to be joined together, this type of design necessitates that the output of the two convolutional layers match the geometry of the input.

GoogleNet first introduces the Inception architecture concept, and later the CNN inception improves it. Inceptionv3 CNN introduces a special inception module is adopted to improve model performance, which is a multi-level feature extraction that computes 1×1, 3×3 and 5×5 convolutions, all in the same network module. In the inception module, convolutional layers with different filter sizes are computed in parallel. Resulting features are concatenated before passed to the next layer. The increase in features significantly increases the learning power of the model. Also, pooling operations are essential for the Inception convolutional network; hence, a parallel pooling path is added to reduce the amount of data and the computation time (grid size reduction). Softmax activation outputs the probabilities of each class. The architecture for the inceptionv3 model is described in Fig. 2. Note that the outputs of the inception filters are all stacked and used as input to the next layer.

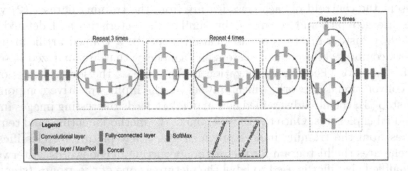

Fig. 2. Simplified inception architecture, ConvNets model

In the experimental setup, three CNNs (Resnet, Inception, and Googlenet) were trained to classify images from the UECFOOD256 food dataset into either 256 or 16 randomly selected categories. The number of layers for the architectures were 101, 48, and 22, respectively, and training was performed for 300 epochs to ensure convergence. The CNNs were pre-trained on a selected set of images and the training rate was initially set to 0.05, with validation performed every four interactions. Results showed that accuracy converged in every run. To classify new food images, the CNNs automatically extract features from the image and assign learned weights to each feature. The food is then classified based on these features and probabilities are generated indicating the likelihood that the food belongs to one of the previously learned categories. An independent dataset is used to evaluate the accuracy of the classification.

3.3   Baseline Survey

A survey in Asian foods was mounted and delivered to humans without previews knowledge in these food classes. A test group of 20 subjects was accepted (the ones not knowing at least 80% of the dishes) all in the age rank between 25 and 40 years old, all with European nationality. As control questions, two items in the survey were universal foods, such as Pizza.

Training is necessary before testing humans. The challenge then is to train the subject to be able to pick the correct name of food dishes presented to him, and the accuracy is measured as the percentage of the right choices. For this reason, the survey was divided into two stages.

A training stage was given to the human subjects, consisting of showing a set of total 98 screens for the 16 food categories (a number that was deemed sufficiently small to allow humans to keep the attention and simultaneously not too small to be too easy). Each screen shows seven images for the chosen food category and just one image from other food categories, in random order. With that, the test subject indicates which one does not belong to the food category, learning each food category's name and critical characteristics.

In the second stage, image classification, a set of images, based on 32 screens, is shown in random order, from the 16 trained food dish classes, each with 16 class label options, where only one is correct. The test subjects need to select the right label for the image.

4   Comparison Study

Using the UEC Food 256 full data-set, the following CNNs were trained from the scratch to classify different types of foods: GoogleNet; Inception v3; ResNet 101.

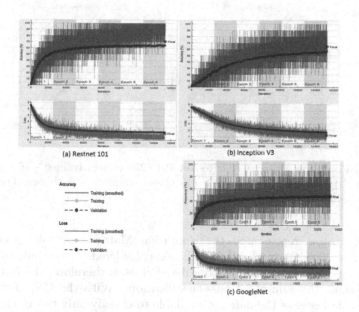

Fig. 3. CNN training and loss comparison. (a) ResNet CNN. (b) InceptionV3 CNN. (c) GoogleNet CNN

Figure 3 shows the training progress for the three used CNNs. From these charts it is possible to observe how each network evolves, epoch after epoch, to achieve its best, given the same configurations for all.

Table 2. Results comparison, validation accuracy, validation loss, and training time

	Results GoogleNet	Results Inception v3	ResNet 101
Validation accuracy	47.73%	65.43%	70.68%
Validation loss	2%	1.4%	1.2%
Training time	6105 min 12 s	5290 min 42 s	7678 min 17 s

Table 2 shows a summary of the comparison of the obtained accuracy and training time results.

When comparing all charts, Fig. 3, ResNet 101, and GoogleNet converge faster (in fewer epochs) to the final result, while Inception V3 "climbs" to the final result/accuracy more gradually over the training time (more epochs are necessary). On the other hand, Inception v3, shows more potential to improve the accuracy performance, if extended training is performed. Nevertheless, with the same test conditions, ResNet 101 obtained better accuracy results.

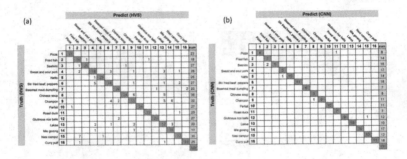

Fig. 4. (a) Confusion Matrix CNN ResNet 101 (256 classes trained - 16 classes classification); (b) Confusion Matrix HVS (16 classes trained - 16 classes classification)

In Fig. 4(a) is represented the Confusion Matrix (confusion matrix) for the CNN, resNet 101, with all 256 classes considered. Note that, most miss-classifications fell outside the 16 considered classes, therefore, the last column, sum represents the sum of all miss-classifications. With the CNN ResNet 101, with all food classes of the data-set available to classify only the 16 classes used in the HVS survey inquiry, this CNN reached an accuracy of 67.7%.

Figure 4(b) shows the Confusion Matrix for the survey inquiry performed to our test group. Note that, as happens with humans, during the test/inquiry, the CNN also has a wide knowledge of other foods (256 types of food), which as in humans, can influence classification. HVS was able to reach an accuracy of 80.6%, 13.4% more than the CNN.

Table 3. CNN vs. HVS results

	HVS	CNN ResNet 101
Validation accuracy	80.6%	67.2%
Training time (avg)	17 min 56 s	7678 min 17 s

Table 3, resumes the Confusion Matrix in Fig. 4(a) (16 food classes classification, with the knowledge of all 256 food classes) and Fig. 4(b) (16 food classes classification of 16 food classes training, with a lifetime of food knowledge). From obtained results, when considering all trained food classes for both CNN and human test subjects, HVS performed better than the CNN. However, we must consider that both make mistakes when:

- The training data-set does not cover comprehensively the classification images.
- When the image to classify is too different from the learned ones.
- Other dishes similarities, influence the classification (i.e., over-fitting).

In the previous experiments, the CNNs were trained with 256 food categories. In the next experiment, the three CNNs were trained again from the scratch. This time only considering the training and classification images used during the HVS inquiry training. Results are shown in the following subsection.

4.1   Additional Results, CNN (16 Classes) vs. HVS

In this subsection, additional tests were performed with the three CNNs, using only 16 food classes (and the same 6 epochs). With these results it is possible to evaluate which CNN is more efficient, with a similar training as the HVS, and how the results compare with HVS in this case. For this purpose, the same images used during HVS inquiries, for training and classification, were used to train the three CNNs from scratch. The CNN training data-set had the same images as the HVS training set, total 686 food photos, each class with a total of 43 photos.

Table 4. Results comparison, 16 food classes, 6 epoch: validation accuracy, validation loss, and training time

	Results GoogleNet	Results Inception v3	ResNet 101
Validation accuracy	85.92%	85.92%	89.89%
Validation loss	0.45%	0.5%	0.3%
Training time	15 min 18 s	73 min 27 s	94 min 26 s

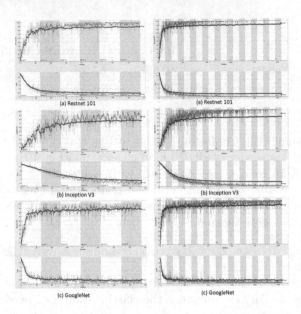

Fig. 5. CNN, 16 class training, 16 classes classification, 6 epochs (left) and 20 epochs (right). (a) ResNet. (b) InceptionV3. (c) GoogleNet.

Figure 5 on the left, shows the CNNs training progress (accuracy and loss). In Table 4, results are summarized. Both GoogleNet and InceptionV3 reached the same accuracy (85.92%), and ResNet101 was the most efficient (89.89%). However, ResNet101 was the CNN that took more time to train, and GoogleNet the fastest. For this case (Fig. 5 on the left), an analysis of the CNN confusion matrix, with 16 class training, and 16 classes classification, shows that:

- HVS errors are constant and visually/syntactically related.
- CNNs errors are much more disperse and (for us humans) without visual relation.

4.2   More (20) Epochs Impact on CNN Accuracy

Next, we increased the number of training epochs from 6 to 20. The data-set was the same as in the previous and the HVS experiments (16 classes).

Figure 5 on the right, shows the training progress for the three CNNs using 20 epochs. From these images, it is possible to conclude that after approximately 10 epochs the CNNs do not significantly improve their accuracy. Table 5 compares accuracy, loss and training times of the three CNNs. When using 20 epochs, instead of 6, CNNs can improve significantly their accuracy, almost stabilizing in a maximum around the 90% accuracy. From the comparison of the confusion matrix, of the CNNs with 20 epochs, with the HVS. Similar to what happens with 6 epochs, CNN confusion matrix's errors are dispersed and not directly related, as happens with the HVS.

Table 5. Results comparison, 16 food classes, 20 epochs: validation accuracy, validation loss, and training time

	Results GoogleNet	Results Inception v3	ResNet 101
Validation accuracy	89.17%	90.25%	93.86%
Validation loss	0.30%	0.30%	0.25%
Training time	49 min 31 s	237 min 22 s	288 min 28 s

4.3   Human, Resnet, Inception and Bag-of-Words

The training time for Resnet101 was 8018 min for 256 food categories and nearly 226 min for 16 classes. Figure 6 compares the accuracy of several methods: Human, Resnet101, GoogleNet, Inception v3, and two approaches using Bag-of-words (500 and 1000 words vocabulary), all with two approaches, 16 food categories and 256 food categories (except for the human survey). The results for bag-of-words show that increasing the number of codewords does not significantly improve accuracy, and the bag-of-words method performs worse than CNNs and humans. The results also show that for 16 food categories, CNNs perform better than humans. However, a more realistic test using 256 food categories shows that the performance of CNNs decreases significantly. Among the CNNs, Resnet had the highest accuracy, with 95% for 16 food categories and 75% for 256 food categories.

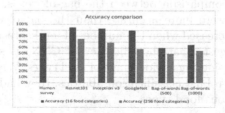

Fig. 6. Accuracy comparison.

The Fig. 6 compares the class precision values of Humans, Resnet, Inceptionv3, GoogleNet, and Bag-of-words in the classification of various food classes. Humans were tested on 16 food classes, while the neural networks were tested on 256 categories. The results in Fig. 7 suggest that humans and neural networks have distinct behaviors when classifying different types of food, with humans performing better in certain categories.

A global analysis of the results from both Fig. 6 and 7 reveals that CNNs are more accurate than bag-of-words, but as the number of categories increases from 16 to 256, the accuracy decreases. This indicates that humans may have an advantage over CNNs in terms of identifying different types of food.

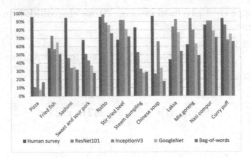

Fig. 7. Classification accuracy's for different sample food

5 Conclusions

The results of the comparison between the CNN and HVS on 266 food classes showed that the HVS performed better than the CNN on a limited number of classes (16), with an accuracy of 80.6% versus 67.2% for the CNN. However, when the CNN was trained on only 16 classes, its accuracy significantly improved, reaching 89.89%. Further tests revealed that increasing the number of epochs to 20 resulted in even higher accuracy for the CNN, at 93.86%. A notable difference between the CNN and HVS in terms of classification is that the CNN made errors with a variety of food classes, while the HVS consistently made semantic errors with the same food classes, suggesting that the HVS showed more consistency in its answers. The comparison between the bag-of-words, CNNs, and humans showed that CNNs were more accurate than the bag-of-words. However, the accuracy of CNNs decreased when the number of food categories increased, from 16 to 256. This suggests that the efficiency of CNNs is surpassed by human intelligence when considering a wide range of food types. While both humans and CNNs demonstrated robust classification abilities, they had distinct approaches when classifying various food types.

Acknowledgements. "This work is funded by National Funds through the FCT - Foundation for Science and Technology, I.P., within the scope of the project Ref. UIDB/05583/2020. Furthermore, we would like to thank the Research centre in Digital Services (CISeD) and the Instituto Politécnico de Viseu for their support."

Maryam Abbasi thanks the National funding by FCT - Foundation for Science and Technology, P.I., through the institutional scientific employment program-contract (CEECINST/00077/2021).

References

1. Abdel-Jaber, H., Devassy, D., Al Salam, A., Hidaytallah, L., El-Amir, M.: A review of deep learning algorithms and their applications in healthcare. Algorithms **15**(2), 71 (2022)
2. Agarwal, R., Shekhawat, N.S.: Enhanced bag of features using AlexNet and henry gas solubility optimization for soil image classification. In: Saraswat, M., Roy, S., Chowdhury, C., Gandomi, A.H. (eds.) Proceedings of International Conference on Data Science and Applications. LNNS, vol. 287, pp. 493–503. Springer, Singapore (2022). https://doi.org/10.1007/978-981-16-5348-3_39
3. Al-Talib, G.A., Saeed, Y.Y.: Comparative studying for extracting food contents using machine learning algorithms. In: AIP Conference Proceedings, vol. 2386, pp. 050008. AIP Publishing LLC (2022)
4. Anthimopoulos, M.M., Gianola, L., Scarnato, L., Diem, P., Mougiakakou, S.G.: A food recognition system for diabetic patients based on an optimized bag-of-features model. IEEE J. Biomed. Health Inform. **18**(4), 1261–1271 (2014)
5. Chen, F., Wei, J., Xue, B., Zhang, M.: Feature fusion and kernel selective in inception-v4 network. Appl. Soft Comput. **119**, 108582 (2022)
6. Chen, M., Dhingra, K., Wu, W., Yang, L., Sukthankar, R., Yang, J.: PFID: Pittsburgh fast-food image dataset. In: 2009 16th IEEE International Conference on Image Processing (ICIP), pp. 289–292. IEEE (2009)
7. Farooq, M., Sazonov, E.: Feature extraction using deep learning for food type recognition. In: Rojas, I., Ortuño, F. (eds.) IWBBIO 2017. LNCS, vol. 10208, pp. 464–472. Springer, Cham (2017). https://doi.org/10.1007/978-3-319-56148-6_41
8. Khan, R., Kumar, S., Dhingra, N., Bhati, N.: The use of different image recognition techniques in food safety: a study. J. Food Qual. **2021**, 1–10 (2021)
9. Mao, R., He, J., Shao, Z., Yarlagadda, S.K., Zhu, F.: Visual aware hierarchy based food recognition. In: Del Bimbo, A., et al. (eds.) ICPR 2021. LNCS, vol. 12665, pp. 571–598. Springer, Cham (2021). https://doi.org/10.1007/978-3-030-68821-9_47
10. Ohri, K., Kumar, M.: Review on self-supervised image recognition using deep neural networks. Knowl.-Based Syst. **224**, 107090 (2021)
11. Salim, N.O., Zeebaree, S.R., Sadeeq, M.A., Radie, A., Shukur, H.M., Rashid, Z.N.: Study for food recognition system using deep learning. In: Journal of Physics: Conference Series, vol. 1963, p. 012014. IOP Publishing (2021)
12. Sharma, P., Sharma, A., et al.: Hybrid approach for food recognition using various filters. Int. J. Adv. Comput. Technol. **11**(1), 1–5 (2022)
13. Tahir, G.A., Loo, C.K.: A comprehensive survey of image-based food recognition and volume estimation methods for dietary assessment. In: Healthcare, vol. 9, p. 1676. Multidisciplinary Digital Publishing Institute (2021)
14. Wang, R., Chen, S., Ji, C., Fan, J., Li, Y.: Boundary-aware context neural network for medical image segmentation. Med. Image Anal. **78**, 102395 (2022)
15. Wang, W., Min, W., Li, T., Dong, X., Li, H., Jiang, S.: A review on vision-based analysis for automatic dietary assessment. Trends Food Sci. Technol. **122**, 223–237 (2022)
16. Xiong, J., Yu, D., Liu, S., Shu, L., Wang, X., Liu, Z.: A review of plant phenotypic image recognition technology based on deep learning. Electronics **10**(1), 81 (2021)
17. Zhu, Z., Dai, Y.: Food ingredients identification from dish images by deep learning. J. Comput. Commun. **9**(4), 85–101 (2021)

A Platform for the Study of Drug Interactions and Adverse Effects Prediction

Diogo Mendes[1] and Rui Camacho[1,2]([✉]) [iD]

[1] Faculdade de Engenharia da Universidade do Porto, Rua Dr Roberto Frias, s/n,
4200-465 Porto, Portugal
up201605360@edu.fe.up.pt
[2] INESC-TEC, Rua Dr Roberto Frias, s/n, Porto, Portugal
rcamacho@fe.up.pt

Abstract. This article reports on the development of a Web platform for the study of Adverse Drug Events (ADEs). The platform is able to import ADE episodes from official Web sites, like OpenFDA, analyse the chemistry of the drugs involved, together with patient data, and produce a potential explanation based on the drugs interactions. Each study uses chemical knowledge to enrich the information on the molecules involved in the episodes. Data Mining is then used to construct models that can help in the explanation of the ADE occurrence and to predict future events. This paper reports on the Web portal developed and the Data Mining experiments conducted to evaluate the quality, and potential explanations of the forecasted adverse reactions, using real reports of drug administration and the subsequent adverse events. The results showed that it was possible to predict the outcomes of ADEs based on the structure of the molecules of the drugs involved and the data collected from real reports of drug administration up to an accuracy of 79%, while also predicting, with high accuracy, the severity of events where the outcome is the death of the patient (with a precision of 98.9%). The platform provides a less expensive and more accurate way of predicting adverse drug reactions compared to traditional methods. This study highlights the importance of understanding drug interactions at a molecular level and the usefulness of utilising Data Mining techniques in predicting ADEs.

Keywords: Adverse Drug Event (ADE) · Data Mining · Machine Learning · Drug Interactions

1 Introduction

Adverse Drug Events (ADEs) can cause significant harm to patients and lead to increased healthcare costs. In recent years, the number of reported ADEs has increased, highlighting the need for a platform that can effectively handle and

With thank FEUP, DEI and M.EIC for the support on the work reported.

analyse ADE data. The aim of this study was to develop a platform that would provide the user with functionalities to parse, filter and enrich the ADE data with extra chemical characterisation, and to use Data Mining in combination with Machine Learning to predict and provide a potential explanation for the outcomes of adverse drug reactions.

The prediction of ADEs is a complex task that requires a thorough understanding of the molecular interactions between drugs and the biological systems they affect. Traditional methods of predicting ADEs rely on observational studies and clinical trials, which can be expensive and time-consuming. Data Mining offers a cost-effective and coherent alternative for predicting ADEs, as it can be used to analyse large amounts of data, leading to the identification of patterns and relationships that might otherwise be overlooked.

In this study, we used Data Mining techniques to enhance the ADE data and predict the outcomes of adverse drug reactions. The experiments were conducted using real reports of drug administration and the subsequent adverse events, allowing us to evaluate the quality of the forecasted adverse reactions. Results showed that there is great potential in predicting outcomes of ADRs based on the molecular characteristics of the drugs involved, as well as other relevant data collected from the real world reports.

All in all, this study highlights the importance of understanding drug interactions at a molecular level and the usefulness of utilising Data Mining mechanisms in predicting ADRs. The platform developed in this study provides a cost-free and accurate way of predicting ADRs, without having to resort to the more time-consuming and expensive traditional methods. The findings of this study have the potential to inform and improve the design of future systems for predicting ADRs, and could ultimately lead to improved patient safety and reduced healthcare costs.

2   Related Work

The prediction of adverse drug events (ADEs) is an important area of research in pharmacology, as it can help identify potential safety issues with drugs before they are approved for use. Computational methods are becoming increasingly popular for predicting ADEs, as they are more efficient and cost-effective than traditional methods. In this context, the article "Improving the Prediction of Adverse Drug Events Using Feature Fusion-Based Predictive Network Models" [1] presents a new approach for predicting drug-ADE associations using feature fusion-based predictive network models (FFPNMs) with three different machine learning (ML) methods.

The article presents the results of a study in which the authors used the Jaccard and Adamic–Adar indices to build FFPNMs with logistic regression (LR), random forest (RF), and support vector machine (SVM) ML models. The FFPNMs were built using a bipartite network consisting of 152 drugs and 633 ADEs, which were obtained from the FDA Adverse Event Reporting System (FAERS) 2010 dataset. The performance of the FFPNMs was evaluated using the Area Under the Receiver Operating Characteristic Curve (AUROC) value. The FFPNM with RF achieved the best predictive result with an AUROC value of 0.913, followed by the FFPNM with LR and SVM.

The authors concluded that FFPNMs with ML methods, especially RF, have superior prediction performance and robustness using only the topology features of the drug–ADE network. The results of this study can help identify potential safety issues with drugs before they are approved for use, and can aid in the development of more efficient and cost-effective computational methods for predicting ADEs.

Data mining has many applications in the healthcare industry, as showcased by the article "Healthcare data mining: Predicting hospital length of stay of dengue patients" [2], that proposes a data mining approach to build a model that can predict inpatient length of stay of dengue patients at the time of admission, which can be used for effective decision-making that can lead to better clinical and resource management in hospitals.

The authors applied the C4.5 algorithm, a decision tree classifier, to a dataset of dengue patients, and obtained an accuracy of 71.57% along with a ROC curve value of 0.761. Furthermore, a prototype of a prediction system using the resulting model was also developed, providing clinicians with a practical tool that predicts the length of stay of dengue patients.

Overall, this study contributes to the growing body of research on data mining and predictive analytics in healthcare. It demonstrates the potential of data mining techniques in solving healthcare related problems, with results potentially being helpful for healthcare providers to better manage patient flow and allocate resources more efficiently.

Finally, a research on the application of data mining data in pharmacovigilance [3], reaffirms that, in pharmacovigilance, the detection of adverse drug events (ADEs) is critical to ensuring patient safety. While current systems exist for this purpose, there is a need for more efficient methods capable of detecting potential ADEs. Knowledge discovery in databases (KDD), a technique that involves the selection of data variables and databases, data preprocessing, data mining, and data interpretation, has been proposed as a potential solution to this problem.

Data mining techniques such as cluster analysis, link analysis, deviation detection, and disproportionality assessment can all be used to determine the presence of ADE signals and assess their strength.

This study points out that given the importance of ADEs and the development of massive data storage systems along with powerful computer systems, the use of data mining techniques in knowledge discovery in medical databases is likely to become increasingly important in pharmacovigilance, as they have the potential to detect signals earlier than current methods.

In summary, the reviewed literature suggests that various computational approaches can be used to predict and identify potential ADEs, including network-based predictive models and data mining techniques. These methods have shown potential to improve the efficiency and accuracy of pharmacovigilance and lead to the detection of ADEs. Data mining techniques have also shown promise in other areas of prediction beyond pharmacovigilance, indicating their broad applicability to other fields. As such, this paper aims to continue exploring the potential of computational methods, in specific data mining, in the field of pharmacovigilance, and to identify how they can be used in conjunction with other methods for ADE detection and prevention.

3   Platform Description

Research on ADEs involve the collection and analysis of large volumes of data, which can be a challenging task. To aid in this process, a user-friendly web-based platform, Tamingo, has been developed. The platform is designed to simplify data collection and analysis for both researchers and participants, and it can be a useful tool in facilitating the study of ADEs.

Tamingo was built with 3 main technologies: Python, Django, and SQLite.

Python

Python is a popular language for scientific computing and data analysis due to its high-level programming and features, which allow for simple code writing and testing. The language benefits from a large and engaged community that has produced a variety of frameworks and libraries for data analysis and scientific computing, such as NumPy, SciPy, pandas, Matplotlib, seaborn, and scikit-learn, among others, which have been used in many data science projects, including those related to ADEs. Python's simple and readable syntax make it a convenient language for data collection and analysis, as its user-friendly syntax facilitates these processes [4].

Django

Django is an open-source web framework that is built in Python, which provides a high-level infrastructure to develop web applications. Following the Model-View-Controller (MVC) pattern, Django offers an easy-to-use approach for creating web-based applications. One of the main advantages of using Django is its focus on security, providing built-in features that help secure the application and data, which is particularly important in handling sensitive health information. Additionally, Django's built-in admin interface enables researchers to efficiently manage data, reducing the time and resources required for management tasks [5].

SQLite

SQLite is a type of Relational DataBase Management System (RDBMS) that is lightweight and operates using a file-based system. It is typically used as the primary database for web applications that are small to medium-sized. The advantage of using SQLite is that it does not need a separate server and is uncomplicated to set up and maintain. As a result, it is a desirable option for projects that are not expected to accommodate high traffic or are smaller in scale. One of the notable features of SQLite is its capability to handle moderate-sized datasets and concurrent requests effectively, making it a reliable and sturdy choice [6].

Layout and Design

The Tamingo platform is a user-friendly and intuitive system, designed for researchers, healthcare professionals, and others interested in ADE studies. The system is built using Django, a powerful web framework that enables the creation of complex and customizable web applications. The homepage of the Tamingo platform has a clean and modern design, presenting information in an easy-to-read format and featuring a slogan that sums up the platform's objective: "Empowering Drug Development with ADE Knowledge: Shaping the Future of Medicine through Research."

One of the platform's standout features is its sidebar, which is accessible to users at all times, allowing them to switch between tasks and navigate the system quickly. As explained in detail in the following section, the sidebar also provides users with access to various functionalities such as data collection, data visualisation, and data analysis, with options such as *Home*, *OpenFDA*, *Filter*, *SMILES*[1]$\rightarrow$*SDF*, *Descriptors*, *Predict*, and *Admin*.

The Tamingo platform's layout and design have been carefully considered to ensure a seamless and efficient user experience, optimised for desktop use. Overall, Tamingo provides a user-friendly solution for conducting ADE studies and making substantiated claims about drugs characteristics, simplifying data collection and analysis for anyone interested in the field.

Capabilities

The platform provides a range of tools for data cleaning, transformation, collection, visualisation, and analysis, making it suitable for users from different backgrounds. With a user-friendly interface and intuitive navigation, the platform allows easy access to all its capabilities. Furthermore, we present a use case diagram (see Fig. 1) that illustrates the primary use cases and interactions between the user and the system, including the actors and relationships involved.

---

[1] Simplified Molecular-Input Line-Entry System.

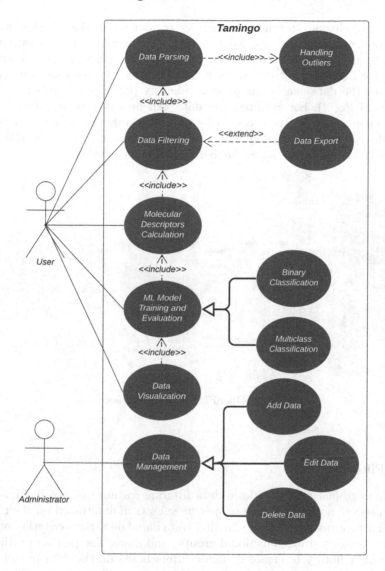

Fig. 1. Tamingo Use Case Diagram

4   User Controlled Data Manipulation

OpenFDA ADE Data Collection

The OpenFDA API provided us with valuable data on ADEs, including patient demographics, the severity of the event and the drugs involved in the reported episode. However, it is important to note that the data is based on voluntarily reported adverse events and may not represent the entire population of adverse events. Therefore, caution should be exercised in making causal inferences. Data parsing and cleaning techniques, such as regular expressions, string manipulation, and filtering, were applied to extract key elements of interest, such as patient gender, weight, severity of the event, patient reaction, reaction outcome, and active ingredient of the drugs involved. Once the relevant elements are extracted, the data can be analysed to identify patterns and trends in ADEs (as seen in Fig. 2). For example, the data can be grouped by category or by functional groups to arrive at conclusions about whether these influence the risk for ADEs. Overall, by gaining insights into patterns and trends in ADEs, it is possible to identify areas where further research is needed [7].

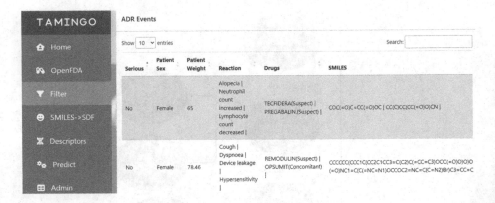

Fig. 2. OpenFDA Data Overview

Data Filtering

The Tamingo platform provides a data filtering feature (as seen in Fig. 3) that allows users to access and analyse specific subsets of data based on their preferences. This feature enables users to filter data based on various criteria, including specific reactions, drugs, functional groups, and more. The platform utilises the Django-filter library to create dynamic filters based on the data model, which generates a filtered queryset based on the user-specified criteria. The filtered data is then converted into a pandas dataframe that can be exported as a .csv file for further analysis. This user-friendly and flexible data filtering functionality allows users to focus on specific subsets of data relevant to their research or experimentation. The platform's filtering capabilities are available in eight different ways, including by severity, gender, weight, type of reaction, drugs, functional groups, categories, or outcome, and multiple filters can be applied at once [8].

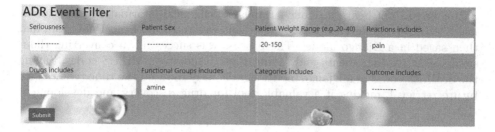

Fig. 3. Tamingo's Filtering Feature

Machine Learning

Our platform uses machine learning algorithms to predict the severity or outcome of an adverse drug event based on a dataset of collected reactions. For predicting the severity of a reaction, binary classification models like Logistic Regression, Decision Tree, and Random Forest classifiers are used, while for predicting the outcome of a reaction, multiclass classification models such as Random Forest, Ridge Regression, and Label Propagation are utilised. The scikit-learn library was used, which provided accurate results based on patient characteristics, functional groups, drug categories, and molecular descriptors. The platform presents the results in a table format, allowing users to compare and select the best model for their needs [9].

Data Management

This web application includes a data management feature that enables users to manually modify, delete, or add new data to the platform's database models. This feature is conveniently accessible through the app's sidebar and is designed to be user-friendly, allowing users to easily edit data without extensive technical knowledge. The data management feature is particularly valuable for ensuring data accuracy and completeness, keeping the data current and relevant to the user's needs, and allowing for customisation based on personal preference [10].

5   Automated Data Operations

SMILES Retrieval

The conversion of active ingredients from pharmaceutical drugs into their respective SMILES codes is crucial in the drug discovery process. SMILES codes are standardised formats used for representing the chemical structure of molecules and are utilised in various applications, including computer-aided drug design and virtual screening of chemical libraries. In this process, active ingredients from OpenFDA[2] were converted into their respective SMILES codes using Python

---

[2] https://open.fda.gov/.

API calls to PubChem[3], a public database of chemical information. The canonical SMILES code for each active ingredient was retrieved from PubChem, and then saved in a Django model. The process was successful in converting thousands of active ingredients and saving their respective SMILES codes. Saving the SMILES codes in a Django model significantly reduced the time required for future retrieval, making it a valuable tool for drug discovery and development [11].

Categorisation of Drugs

In this section, we present a method for retrieving the category of a drug based on its active ingredient using Python and the PUG VIEW API of PubChem.

To retrieve the category of a drug, we first use the "compound/name" endpoint of the PUG VIEW API to search for the compound ID by name. Once the PubChem Compound Identifier (CID) is obtained, we use the "compound/cid" endpoint with the Mesh Pharmacological Classification header to retrieve the category of the drug.

We demonstrate the utility of this method [12] by providing examples of drug categories in Table 1:

Table 1. Drug Categories Examples

Drug	Category
Atenolol	Anti-Arrhythmia Agents
Duloxetine	Analgesics
Clozaril	Antipsychotic Agents
Taxotere	Tubulin Modulators

The method presented in this section can be extended to retrieve other information about the compound, such as its chemical structure or toxicity data, by using other endpoints of the PubChem API. Additionally, the batch search functionality provided by the API can be used to retrieve information about multiple compounds at once.

Overall, our method provides a useful tool for researchers and healthcare professionals to retrieve important information about drugs and their categories, facilitating drug discovery and development.

Functional Groups Matching

Tamingo is a tool that utilises SMARTS codes and the Python RDKit library[4] to identify functional groups present in drugs. Functional groups are specific

---

[3] https://pubchem.ncbi.nlm.nih.gov/.
[4] https://www.rdkit.org/.

atoms or groups of atoms within a molecule that are responsible for its chemical properties and reactivity [13].

To identify functional groups, Tamingo collects a file of the most relevant SMARTS codes from Daylight[5] and creates a molecule object from the SMILES code of a given drug using the Chem.MolFromSmiles() function. The molecule object is then compared to query molecule objects created from the stored SMARTS codes using the Chem.MolFromSmarts() function, and the GetSubstructMatches() function is used to identify any instances of the functional groups represented by the SMARTS codes in the molecule. This process can be repeated for each functional group in the database, providing a comprehensive analysis of the functional groups present in the given drug.

By providing insights into the chemical structure of a drug and the role of functional groups in its properties and reactivity, Tamingo can aid in the development of safer and more effective pharmaceuticals.

Molecular Descriptors Calculation

Molecular descriptors are numerical values that describe the structural and physicochemical properties of a molecule, and are important in predicting the biological activity, toxicity, and other properties of a compound [14].

Obtaining molecular descriptors from SMILES is a crucial step in the analysis of chemical compounds. Mordred[6], a python library, is a powerful tool that can calculate more than 2000 molecular descriptors, including constitutional, topological, geometrical, and electronic descriptors.

Using Mordred, Tamingo is able of calculating thousands of descriptors, that can be used in various scenarios, such as drug discovery and toxicity prediction, as well as in machine learning models, as demonstrated further on this article. Mordred's compatibility with the Open Babel and RDKit libraries further enhances its utility in SMILES manipulation.

SMILES to SDF

Chemical compounds are often represented in distinct formats, including SMILES and/or SDF format. SDF format is commonly used in cheminformatics and drug discovery as it allows for the storage of not only the chemical structure of a compound but also additional information such as properties and experimental data [15]. Converting SMILES to SDF format is a crucial task in chem-informatics as it facilitates the integration of various data sources and enables the manipulation and analysis of data using various software programs.

When converting SMILES to SDF format, several considerations must be taken into account. The chosen method should correctly convert the SMILES representation to the SDF format while preserving the structural information and additional data. Stereo-chemistry, the three-dimensional arrangement of

---

[5] https://www.daylight.com/.
[6] https://github.com/mordred-descriptor/mordred.

atoms in a molecule, plays an essential role in the biological activity of a compound. Therefore, the chosen method should correctly handle the stereo-chemistry information present in the SMILES representation and convey it correctly in the SDF format.

In this study, we used PandasTools and its connection with RDKit to convert SMILES to SDF format. We used a CSV file that contained the substance name and respective SMILES code and loaded it into a Pandas dataframe. The AddMoleculeColumnToFrame function available in PandasTools was used to add a new column to the dataframe containing molecular information, such as SMILES strings or molecular descriptors. The function takes the dataframe to which the new column will be added and the name of the column in the dataframe that contains the SMILES strings of the molecules as arguments. The resulting dataframe was immediately modified by applying the WriteSDF function, which writes the contents of a Pandas dataframe to a file in the SDF format.

6   Case Study - Dataset Creation and Results Analysis

Severe ADEs are those that result in death, a life-threatening condition, hospital-isation, disability, congenital anomaly, or other severe condition. The prediction of severe ADEs is a critical task in pharmacovigilance, as it can help identify drugs that pose a high risk to patients and inform drug regulatory agencies about the safety of medications.

In this chapter, we present a case study on the prediction of the severity of ADEs using machine learning techniques. We collected OpenFDA ADE data from 2019 to 2021, and created two datasets: one containing around 21000 events where two drugs interact, and the other with almost 50000 events where two or more drugs were present in the reaction. A set of patient and drug's attributes was used, including the patient's gender and weight, drug categories and functional groups present, and molecular descriptors of the drugs in the form of the average and standard deviation of each descriptor. In the case of the first set the absolute difference of the values of each descriptor was also computed. Both datasets contained a 50–50 split of positive and negative target values, and were used to train and evaluate several machine learning models, using the scikit-learn tool and a variety of its capacities.

Data preprocessing techniques were applied to prepare the data for analysis, such as data scaling practices, and also feature selection techniques, like variance threshold, which was responsible for eliminating all attributes where the deviation is zero.

To optimise the performance of each machine learning model, we used ran-domised search cross-validation to find the best combination of hyperparameters. This technique randomly samples hyperparameters from a given distribution and evaluates the model performance using cross-validation.

For each model, we defined a search space of hyperparameters and their respective distributions, including the number of estimators for the random forest model, the regularisation parameter for the logistic regression model, or the learning rate for the neural network model. We then used 5-fold randomised search cross-validation to evaluate the model performance for each combination of hyperparameters and select the best one based on the average accuracy score.

We then trained several machine learning models, using different combinations of features and hyperparameters, while applying 5-fold cross-validation. The performance of each model was evaluated using various metrics, such as accuracy, global precision, precision when the outcome of the reaction is fatal, and the area under the receiver operating characteristic curve (AUC-ROC). The Table 2 shows the average of the obtained results on the first dataset involving reactions with 2 drugs, with the respective standard deviation values.

Table 2. 2 Drugs ADE Severity Prediction Results

Algorithms	Accuracy	Precision (P)	P when Fatal	AUC-ROC
Naive-Bayes	63.8 ± 0.01%	63.9 ± 0.01%	65.4 ± 0.01%	68.7 ± 0.01%
Logistic Reg.	76.3 ± 0.01%	77.6 ± 0.02%	71.7 ± 0.01%	83.2 ± 0.01%
K-NN	75.1 ± 0.01%	75.3 ± 0.01%	**98.2 ± 0.01%**	82.7 ± 0.01%
SVM	76.4 ± 0.02%	78.1 ± 0.02%	70.8 ± 0.02%	82.0 ± 0.02%
Decision Tree	70.0 ± 0.01%	72.5 ± 0.02%	74.5 ± 0.01%	72.3 ± 0.01%
Bagging Class.	76.2 ± 0.01%	**78.6 ± 0.02%**	69.4 ± 0.01%	82.3 ± 0.01%
AdaBoost Class.	75.6 ± 0.01%	75.8 ± 0.01%	**88.1 ± 0.01%**	81.8 ± 0.01%
Random Forest	75.8 ± 0.01%	76.4 ± 0.01%	81.5 ± 0.01%	83.1 ± 0.01%
MLP Class.	**76.8 ± 0.01%**	**78.6 ± 0.02%**	74.8 ± 0.02%	**83.2 ± 0.01%**
Voting Class.	**76.9 ± 0.01%**	**78.7 ± 0.01%**	77.3 ± 0.01%	**83.6 ± 0.01%**

When it comes to the second dataset containing ADEs where 2 or more drugs were involved the results are presented in Table 3.

Overall, the results demonstrate that machine learning can be an effective tool for predicting the severity of ADEs. The best-performing algorithm was the Voting Classifier, which achieved the highest Accuracy, Precision and AUC-ROC scores when trained with both datasets. The Voting Classifier combines the predictions of several different models, which may contribute to its high scores.

Other algorithms that performed well include the MLP Classifier, which achieved high precision and accuracy scores, and the K-NN algorithm, which achieved an astoundingly high precision score close to **100%**, when the outcome of the reaction was fatal, meaning that practically every single time a reaction leads to death, this model predicts that the adverse event will be severe! These results suggest that these algorithms are a great starting point and may be particularly useful in identifying severe ADEs.

It is worth noting that some algorithms, such as the Naive-Bayes and Decision Tree models, had lower accuracy and precision scores compared to the other

Table 3. 2 or more Drugs ADE Severity Prediction Results

Algorithms	Accuracy	Precision (P)	P when Fatal	AUC-ROC
Naive-Bayes	61.2 ± 0.01%	61.7 ± 0.01%	57.9 ± 0.01%	64.8 ± 0.01%
Logistic Reg.	77.1 ± 0.01%	78.9 ± 0.01%	64.5 ± 0.01%	83.9 ± 0.01%
K-NN	74.6 ± 0.01%	75.3 ± 0.01%	**98.9 ± 0.01%**	81.8 ± 0.01%
SVM	77.5 ± 0.01%	79.6 ± 0.01%	64.4 ± 0.01%	83.7 ± 0.01%
Decision Tree	67.8 ± 0.01%	69.8 ± 0.02%	71.8 ± 0.02%	69.7 ± 0.01%
Bagging Class.	77.2 ± 0.01%	**80.2 ± 0.01%**	63.2 ± 0.01%	83.7 ± 0.01%
AdaBoost Class.	76.0 ± 0.01%	76.3 ± 0.01%	**88.6 ± 0.01%**	82.8 ± 0.01%
Random Forest	76.3 ± 0.01%	76.8 ± 0.02%	84.1 ± 0.01%	83.4 ± 0.01%
MLP Class.	**77.6 ± 0.01%**	79.3 ± 0.01%	79.0 ± 0.01%	**84.7 ± 0.01%**
Voting Class.	**79.0 ± 0.01%**	**80.6 ± 0.01%**	81.3 ± 0.01%	**85.5 ± 0.01%**

models. This may be due to the relatively simplistic nature of these models, which may struggle to capture the complex relationships between patient and drug attributes that contribute to the severity of ADEs.

In summary, these results suggest that machine learning can be a valuable tool for predicting the severity of ADEs, and that a combination of different algorithms may be the most effective approach. However, further research and an intricate understanding of the algorithms is needed to validate these findings and determine the most effective way to integrate machine learning into pharmacovigilance practices.

Table 4. Random Forest Top 10 Features with the Highest Importance Score

Feature	Score	Description
Weight	0.04277	Patient's weight in kilograms
Gender	0.00444	Patient's gender (male or female)
GATS2v	0.00327	Topological distance between atom types
VSA_EState7	0.00264	Distribution of the atomic charges in a molecule
ATSC2p	0.00249	Topological and electronic properties of a molecule
GATS1m	0.00223	Atomic distribution of a molecule
AATSC2m	0.00222	Shape and size of a molecule
n9HRing	0.00220	Presence and number of 9-membered heterocycles
Si	0.00215	Electrophilicity of a molecule
AATSC0pe	0.00214	Polarizability and electronic properties of atoms

7   Conclusions

Based on the results obtained from this study, it can be concluded that Tamingo, along with the use of Data Mining techniques, was successful in predicting the outcomes of adverse drug events. The platform provides the user with the ability to parse, filter, and extend the ADE data with new chemical concepts, and the results showed that it was possible to predict the outcomes of ADEs with an accuracy of up to 79%. Furthermore, the platform was able to predict with a precision of 98.9% that an adverse reaction is severe when it in fact results in the death of the patient.

Using the capabilities of the Random Forest Classifier, to assign an importance score to each feature used in the model, which indicates how much each feature contributes to the overall performance of the model, we decided to collect the top 10 features (see Table 4) that according to this model contributed the most to the prediction of ADEs, in the hopes of shedding more light into which factors are more likely to trigger an unwanted reaction.

Patient's weight and gender were found to have the highest importance scores, which indicates that personalised dosing based on these factors, might improve drug safety. Additionally, molecular descriptors related to topological distance, electronic properties, or shape and size of the molecules also seem to be important predictors of ADEs. These findings suggest that a drug's molecular properties are key determinants of its safety profile and should be carefully considered during drug development and screening.

The results of this study are significant as they demonstrate the importance of understanding drug interactions at a molecular level and the usefulness of utilising Data Mining techniques in predicting ADEs. Given the increasing number of adverse events reported each year, which can have severe consequences for patient health and safety, the success of the Tamingo platform developed in this study can pave the way to a less expensive and more accurate way of predicting adverse drug events when compared to the methods of today.

8   Future Work

The Tamingo platform is a valuable tool in health informatics, but it has yet to reach its full potential. There are several areas for improvement that could enhance the platform's capabilities and provide even greater value to users.

Exploring additional molecular descriptors or attributes could improve prediction accuracy. Furthermore, applying advanced machine learning techniques like convolutional neural networks or recurrent neural networks could enhance the platform's performance.

Another area for future work is to refine the feature selection techniques used in the platform. Utilising more advanced techniques like genetic algorithms or particle swarm optimisation could help identify the most relevant attributes and molecular descriptors to make better predictions.

In summary, there are several areas for improvement that could take the Tamingo platform to the next level and offer even more insights to users!

References

1. Li, J., Ji, X., Hua, L.: Improving the prediction of adverse drug events using feature fusion-based predictive network models. IEEE Access **8**, 48812–48821 (2020). https://doi.org/10.1109/ACCESS.2020.2979452
2. Govindaraju, R., Salamah, S.Y.: Healthcare data mining: predicting hospital length of stay of dengue patients. HAYATI J. Biosci. **25**, 178–187 (2018). https://doi.org/10.4308/hjb.25.4.178
3. Wilson, A., Thabane, L., Holbrook, A.: Application of data mining data in pharmacovigilance. Br. J. Clin. Pharmacol. **57**, 127–34 (2004). https://doi.org/10.1046/j.1365-2125.2003.01968.x
4. Kumar, R.: Future for scientific computing using Python. Int. J. Eng. Technol. Manag. **2**(1), 30–41 (2015)
5. Li, H., Shen, S.: Construction of college students' physical health data sharing system based on Django framework. J. Sens. **2021**, 1–7 (2021). https://doi.org/10.1155/2021/3859351
6. Owens, M.: The definitive guide to SQLite (2010). https://doi.org/10.1007/978-1-4302-0172-4
7. Meng, L., Tang, X.-W., Ji, H.-H., Song, L., Niu, X.-D., Jia, T.: Mining and evaluation of statin-associated adverse events signals: data mining of the public version of the OpenFDA adverse event reporting system. Chin. J. New Drugs **28**, 244–248 (2019)
8. Haslwanter, T.: Data filtering. In: Haslwanter, T. (ed.) Hands-on Signal Analysis with Python, pp. 71–104. Springer, Cham (2021). https://doi.org/10.1007/978-3-030-57903-6_5
9. Ali, A., Amin, M.: Hands-on Machine Learning with Scikit-Learn. Amazon Kindle Direct Publishing (2019)
10. Rubio, D.: Django admin Management. In: Beginning Django (2017). https://doi.org/10.1007/978-1-4842-2787-9_11
11. Ratnawati, D.E., Marjono, Anam, S.: Prediction of active compounds from smiles codes using backpropagation algorithm. In: Coverage of Basic Sciences Toward the World's Sustainability Challanges. https://doi.org/10.1063/1.5062773
12. Gururaj, H.L., Flammini, F., Kumari, H.A.C., Puneeth, G.R., Kumar, B.R.S.: Classification of drugs based on mechanism of action using machine learning techniques. Discov. Artif. Intell. **1**(1), 1–14 (2021). https://doi.org/10.1007/s44163-021-00012-2
13. Ertl, P.: An algorithm to identify functional groups in organic molecules. J. Cheminform. (2017). https://doi.org/10.1186/s13321-017-0225-z
14. Winter, R., Montanari, F., Noé, F., Clevert, D.-A.: Learning continuous and data-driven molecular descriptors by translating equivalent chemical representations. Chem. Sci. **10**, 1692–1701 (2019). https://doi.org/10.1039/C8SC04175J
15. Sanches, P., et al.: Fitting structure-data files (.SDF) libraries to progenesis QI identification searches. J. Braz. Chem. Soc. (2023). https://doi.org/10.21577/0103-5053.20230016

A Machine Learning Approach to Predict MRI Brain Abnormalities in Preterm Infants Using Clinical Data

Arantxa Ortega-Leon[1]([✉]) [ID], Roa'a Khaled[2] [ID],
María Inmaculada Rodríguez-García[1] [ID], Daniel Urda[3] [ID],
and Ignacio J. Turias[1] [ID]

[1] Department of Computer Science, University of Cadiz, Algeciras, Spain
`arantxa.ortega@uca.es`
[2] Department of Computer Science, University of Cadiz, Puerto Real, Spain
[3] Grupo de Inteligencia Computacional Aplicada (GICAP), Departamento de Digitalización, Escuela Politécnica Superior, Universidad de Burgos, Av. Cantabria s/n, 09006 Burgos, Spain

Abstract. Preterm infants are prone to several neurodevelopmental impairments (NDI). Early and accurate diagnosis could cooperate in the treatment of their clinical manifestations. Clinical data from a cohort of preterm infants at the neonatal intensive care unit (NICU) of the Hospital Puerta del Mar, Cadiz, Spain, was used in this work to perform a classification task to predict abnormal magnetic resonance imaging (MRI) findings using machine learning models. The results in this analysis indicate that the best model able to predict abnormal MRI findings was the K-nearest Neighbor (KNN), with a recall of 0.80. This study represents an initial step towards developing a practical and reliable tool for predicting abnormal MRI findings in preterm infants using readily available clinical data.

Keywords: Machine learning · preterm infants · neuroimaging · MRI

1 Introduction

Preterm birth is defined as a birth that occurs before 37 weeks of gestation; its global prevalence rate is about 11% [1]. During their stay at the neonatal intensive care unit (NICU), preterm infants can experience neonatal illness and may need intensified support that might lead to affected brain development and consequently develop short and long-term adverse neurodevelopmental outcomes that include motor, cognitive, language and behavioral impairments [2,3].

Neurodevelopmental outcomes in preterm infants have been associated with different causes including antenatal, perinatal, and postnatal factors [4]. An early and accurate diagnosis of neurodevelopmental outcomes is key to treat neurodevelopmental outcomes considering that during early development the brain has greater plasticity potential [5].

I. Rojas et al. (Eds.): IWBBIO 2023, LNBI 13919, pp. 419–430, 2023.
https://doi.org/10.1007/978-3-031-34953-9_33

Clinical assessments of preterm infants during their stay at the NICU can include clinical examination, cranial ultrasound (CUS), electroencephalography (EEG), magnetic resonance imaging (MRI), etc. CUS is one of the most conventional neuroimaging techniques, this method aims to diagnose intracranial lesions such as intraventricular hemorrhage (IVH), cystic periventricular leukomalacia (PVL), ventriculomegaly, and white matter injury [6,7], using CUS has the advantages of being non-invasive, low cost and available at the bedside [7].

MRI is a neuroimaging technique that captures a detailed image of the brain, which in most cases is able to detect white matter injury, better delineation of deep structures, and cortical injury [4]. MRI has some advantages over CUS, for example, its higher sensitivity to diagnose white matter injury. However, MRI has some disadvantages, such as being expensive equipment, time-consuming, and the need for transporting and sedating the preterm infant in some cases [8]. These reasons prevent making MRI a routine test for preterm infants at NICUs, particularly in low and middle-income countries [9,10]. Therefore, a triage on performing MRI in preterm infants can have a positive impact on the prioritization of patients and the reduction of radiology waiting times.

By taking advantage of non-conventional techniques such as Artificial Intelligence (AI), it is now possible to analyze large datasets and seek for potential alternatives to assist clinical decision-making. Machine learning (ML) is a type of AI, that branches into supervised learning, unsupervised learning, and reinforcement learning. A common type of ML used in medical facilities and health-related studies is supervised learning, which branches into regression, classification, and ranking. Supervised learning algorithms learn from labeled samples, where the model learns from these patterns and makes predictions. Moreover, ML has been shown to have a significant impact in medicine and healthcare [11]. The use of AI has become an important tool to improve diagnosis, prediction of risk morbidity [12–14] and mortality [15–17] of preterm infants in different diseases and its implementation at hospitals has been growing across time. However, different approaches and scopes still need to be considered and evaluated. One of the greatest challenges of implementing AI in the medical context is the underrepresented conditions to be studied. As a consequence, it is widely common to encounter imbalanced classes while dealing with health data. In this sense, approaches such as synthetic patient data generation have been currently explored [18].

This current work aims to perform a classification task to predict abnormal MRI using clinical data and to determine the best models that work on this dataset. Insights from this work can help clinicians in the future in terms of triage and decision-making on performing MRI in preterm infants in low and middle-income countries. The main contributions of this work can be summarized as the following:

- Implementing well-known supervised ML models aiming at accurately predicting abnormal MRI using clinical data.
- Analyzing the effect on the predictors' performance of SMOTE-NC, which is a well-known technique to deal with data imbalance issues.

The rest of the paper is organized as follows: Sect. 2 describes the study design, data collection, and feature pre-processing. Then, Sect. 3 explains the methods selected in this research and the evaluation strategies that have been considered, including the generation of synthetic data. Section 4 describes and discusses the results that were obtained with the current methods, and discusses the potential future applications of this work. Finally, Sect. 5 gives an overview of the conclusions of this study and the aimed future work.

2   Dataset

2.1   Study Design and Participants

Data from preterm infants admitted to the NICU at Hospital Puerta del Mar, Cadiz, Spain was collected from May 2018 to January 2021. The clinical database used in this study includes prenatal, perinatal, and comorbidities records from preterm infants admitted at the NICU. Research and Ethics Committee approval and informed consent of participants were obtained for data acquisition in this research.

The clinical variables from this database,[1] include prenatal variables, which refer to the mother's health record before and during pregnancy. These include chorioamnionitis, gestational diabetes, preeclampsia, cesarean delivery, IV fertilization, etc. Perinatal variables are variables related to conditions of the preterm infant during the delivery. These include gestational age, sex, Apgar 1 and 5, Clinical Risk Index for Babies (CRIB), intubation, head circumference, small for gestational age, cardiac massage, cardiopulmonary resuscitation, etc. Comorbidities are the acquired medical conditions during the patient's stay at the NICU. These include invasive candidiasis, bronchopulmonary dysplasia, days of oxygen therapy, mechanical ventilation, etc.

Preterm infants from this cohort had a term-MRI scan (acquired at term equivalent age) which was assessed by neonatologists under an MRI scoring system that evaluates the severity of brain injury in white matter, cerebellum, and cortical and deep gray matter [19]. In this work, we analyzed a dataset of 110 preterm infants where 96 patients obtained a normal MRI diagnosis, while 14 preterm infants were defined with an abnormal MRI, according to the clinician's evaluations using the mentioned MRI scoring system. For the prediction of this score in our research, we used a dichotomized version of the scoring proposed by Kidokoro et al. [19], which means that patients who had a score < 8 were considered as having normal to mild term-MRI findings, while patients with a score ≥ 8 were considered as having moderate to severe abnormal term-MRI findings.

2.2   Data Curation and Pre-processing

A total of 60 variables were considered in this study, where 17 Variables had less than 6% of missing values. In this case, variables with missing values were

---

[1] https://github.com/arantxaorle/brain-abnormalities-prediction.

imputed using two simple methods according to the variable type. These imputation approaches are simple and fast methods to perform, however, opinions from the clinicians were considered in order to verify that this imputation will not have an overall effect on results. Variables (features) used in our study were imputed as follows:

- Numerical features were imputed using the mean value. This implies replacing the missing values with the mean value of all known values for each feature. Later, min-max normalization was performed on these features, which means that each numerical feature was scaled to the range 0–1. To accomplish this, the min value of all values for each feature is subtracted from each value and then divided by the difference between max and min (max–min). This scaling was made on the training dataset, while the test dataset was fitted based on the training data.
- Categorical features were imputed using the most frequent value, which implies replacing missing values with the most frequent value of all known values for each feature. The transformation of these features was performed using dummy variable encoding, which means that each category of the variable will be converted into k categories, giving a total of k − 1 [20].

3   Methodology and Experimental Settings

3.1   Methods

In this study, we aimed to test well-known classifiers; a generalized linear model such as Logistic Regression, and a Naive Bayes model such as Gaussian Naive Bayes. Furthermore, non-linear models were tested, such as Decision Tree, and K-nearest Neighbor. Finally, ensemble methods, including ensemble bagging and boosting methods were also tested, such as Random Forest, and AdaBoost.

- Logistic Regression is a model that is able to make a prediction by identifying the relationship between one or multiple features that influence the target feature. The formula can be represented as follows:

$$\pi(\boldsymbol{X}) = \frac{\exp(\beta_0 + \beta_1 \boldsymbol{X}_1 + ... + \beta_k \boldsymbol{X}_k)}{1 + \exp(\beta_0 + \beta_1 \boldsymbol{X}_1 + ... + \beta_k \boldsymbol{X}_k)} \tag{1}$$

where β_0 is the intercept, $\boldsymbol{X}_k$ is the predictor variable and β_k is the corresponding regression coefficient.
- AdaBoost is a meta-estimator that begins by fitting a classifier on the original dataset, by later fitting additional copies of the classifier on the same dataset, but the weights of instances that were incorrectly classified are adjusted so that subsequent classifiers concentrate more on challenging cases. The formula can be represented as follows:

$$H(x) = sign\left(\sum_{t=1}^{T} \alpha_t h_t(x)\right) \tag{2}$$

where H is computed as a weighted majority vote of the weak hypotheses h_t and each hypothesis is designated to weight α_t.

- Decision Tree is a model that develops a prediction of the target variable by learning straightforward decision rules inferred from the properties of the data features. The model resembles an inverted tree, the base is the root node, intermediate nodes, and leaf nodes. In decision trees, the most commons measures are entropy (Eq. 3) and the Gini index (Eq. 4), which are calculated as follows:

$$E(S) = -\sum_{i=1}^{m} p_i \log_2(p_i) \tag{3}$$

where p_i, is the probability of entropy

$$E(S) = 1 - \sum_{i=1}^{m} p_i^2 \tag{4}$$

where p_i is the probability of an element being classified into a class.

- Random Forest is a model that fits a number of decision trees to distinct dataset subsamples and averages them to increase the prediction accuracy. The errors of these decision trees average out and as a result, we have better control over the over-fitting.
- Gaussian Naive Bayes is a model based on the probabilistic approach and Gaussian distribution. The model performs classification by considering that the features have an independent capacity of predicting the target feature and that each feature is likely to be Gaussian. The formula can be represented as follows:

$$P(x_i \mid y) = \frac{1}{\sqrt{2\pi\sigma_y^2}} \exp\left(-\frac{(x_i - \mu_y)^2}{2\sigma_y^2}\right) \tag{5}$$

where σ_y and μ_y are estimated by maximum likelihood.

- K-nearest Neighbor is a model that makes predictions based on the similarity of the data (e.g. Euclidean distance, Manhattan distance, Minkowski distance). In this work, the Euclidean distance was used to fit a KNN model to our data according to Eq. 6.

$$d(p, q) = \sqrt{\sum_{i=1}^{n} (q_i - p_i)^2} \tag{6}$$

3.2 Evaluation Strategy

To obtain reliable results we opted to implement the models at 5 fold cross-validation using train-test splits of 80% and 20% respectively (Fig. 1).

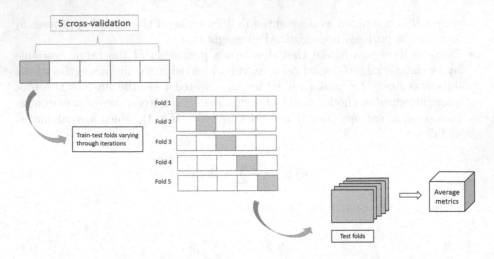

Fig. 1. 5 fold cross-validation and steps followed to obtain evaluation metrics.

Data pre-processing and evaluation of models were performed in Python 3.10.6. Models and default parameters were imported from the library scikit-learn 1.2.1. The following performance statistics metrics were calculated:

- Accuracy refers to all the correct predictions that the model obtained, and is calculated by considering all corrected predictions divided by the total number of observations. When dealing with imbalanced datasets this metric can not be considered as an absolute metric.

$$accuracy = \frac{TP + TN}{TP + FN + FP + TN} \tag{7}$$

- Precision indicates the fraction of correct predictions across all positive predictions.

$$precision = \frac{TP}{TP + FP} \tag{8}$$

- Recall refers to the fraction of correctly classified positive examples across all positive examples.

$$recall = \frac{TP}{TP + FN} \tag{9}$$

- F1 score indicates a harmonic mean of precision and recall, with 1 being the best value which indicates perfect recall and precision. It is a very useful metric when assessing imbalanced classification datasets.

$$f1\,score = 2 \times \frac{precision \times recall}{precision + recall} = \frac{TP}{TP + \frac{1}{2}(FP + FN)} \tag{10}$$

- ROC AUC-score indicates how well the model is able to separate positive and negative samples. When AUC = 1, it indicates that the model perfectly differentiates between the two classes.

$$AUC = \frac{TP\,rate + TN\,rate}{2} \qquad (11)$$

3.3 Dealing with Data Imbalance Issues

Due to the class imbalance that we encounter in our dataset, we performed an oversampling method in our training sets using SMOTE-NC (Synthetic Minority Over-sampling Technique for Nominal and Continuous). This is a variant of SMOTE, for numerical and categorical features [21]. The SMOTE algorithm works by creating new synthetic data points of the minority class, in this case, preterm infants with abnormal MRI, by interpolating between existing examples. It selects a random example from the minority class and finds its k-nearest neighbors. The algorithm selects one of these neighbors and creates a new example by linearly interpolating between the selected example and the neighbor.

4 Results and Discussion

The model's performance for standard and SMOTE-NC prediction are shown in Table 1 and Table 2 respectively, the metrics reflect the average among the 5-fold cross-validation. In this work, the performance results infer that the best models used for our data are Logistic Regression and K-nearest Neighbor using SMOTE-NC. The selection of the best models has been made depending on the main objective of this research. In this case, we aimed to optimize the recall metric (Fig. 2), which measures the proportion of positive examples correctly classified. In this sense, KNN obtained the best results of recall (0.80), followed by Logistic Regression (0.60) (Fig. 3).

Moreover, KNN model obtained considerable ROC AUC score (0.69), a metric that reflects the efficiency of the model to discriminate between positive and negative classes. Hence we can consider by consensus of both Recall and ROC AUC metrics that KNN is the best model. KNN has demonstrated better performance among other models in previous studies, denoting the capacity of this model to work with medical predictions [22–24].

In this work, we want to highlight the use of ML as a complementary tool for the diagnosis of abnormal MRI findings, by using clinical data acquired during a patient's stay at the NICU. Advantages can be taken by using ML for data analysis, due to its ability to manage large datasets and obtain reliable results. The analysis of prenatal, perinatal, and comorbidities factors can provide meaningful insights into potential biomarkers that can be considered in the construction of models that aim to help in the diagnosis of brain injury in preterm infants.

Previous work has found that clinical data correlates to abnormal MRI findings [25,26]. Providing an approximation of potential clinical biomarkers that

Table 1. Performances of classification algorithms.

Models	Accuracy	Precision	Recall	F1 score	ROC AUC
Logistic Regression	0.88	0.30	0.17	0.21	0.58
AdaBoost	0.85	0.30	0.23	0.26	0.59
Decision Tree	0.79	0.15	0.27	0.19	0.57
Gaussian NB	0.67	0.12	0.30	0.16	0.52
KN Neighbors	0.89	0.40	0.13	0.20	0.57
Random Forest	0.87	0.50	0.07	0.10	0.53

Table 2. Performances of classification algorithms using SMOTE-NC.

Models	Accuracy	Precision	Recall	F1 score	ROC AUC
Logistic Regression	0.90	0.63	**0.60**	0.53	0.77
AdaBoost	0.83	0.18	0.27	0.21	0.59
Decision Tree	0.82	0.29	0.23	0.23	0.57
Gaussian NB	0.67	0.12	0.30	0.16	0.52
KN Neighbors	0.61	0.22	**0.80**	0.34	0.69
Random Forest	0.87	0.67	0.17	0.20	0.57

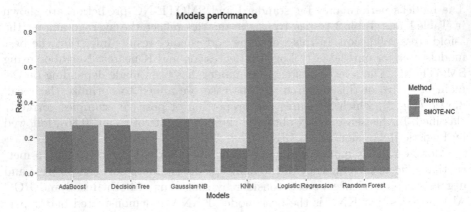

Fig. 2. Model performance based on recall metric, using oversampling and without oversampling methods. Best performance is achieved by KNN.

can help in prioritizing the preterm infants that should go under more rigorous examination, such as the MRI.

MRI is a highly sensitive diagnostic tool for detecting brain abnormalities in preterm infants [27]. However, performing MRI on all preterm infants can be challenging due to potential adverse events on the patients [28], the cost [29], and the possible need for sedation or general anesthesia [30,31], which carries its own risks. These suggestions do not imply that MRI should not be a routine

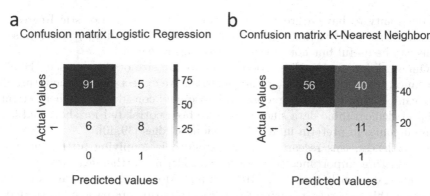

Fig. 3. Confusion matrix of the best models in predicting positive classes. Values in confusion matrix are obtained by the sum total of each metric per iteration.

method for preterm patients, but non-invasive methods such as (CUS) have been widely evaluated to also become a reliable method in detecting brain injuries in preterm infants [32–34].

While in some health facilities, MRI could be a recommended test for preterm infants, it is currently a challenge for the rest of health facilities, especially in low and middle-income countries [9,10,35]. Providing an emerging tool, that with the help of ML methods can help to advise which patients need to go under this type of examination.

Diagnostic performance using clinical data has been performed before for predicting NDI, where they have stated that the accuracy of the prediction has been found by using an optimal set of biomarkers compared to isolated individual biomarkers [26,36]. Identifying those significant prognostic biomarkers can help in constructing models that are able to predict NDI outcomes and help clinicians in decision-making and guide them towards an early intervention [26,36–38]. The intention of this work was to extrapolate these previous studies on earlier steps in the diagnosis of brain injuries in preterm infants.

5   Conclusions and Future Work

In this work, we offer an approach that, to the best of our knowledge, has not been applied before, by using ML to predict abnormal MRI findings using clinical data collected at the NICU. We tested this approach in a cohort of preterm infants from the Hospital Puerta del Mar in Cadiz, Spain. Specifically, we performed a classification task with 6 different models to predict abnormal MRI findings using clinical data, including prenatal, perinatal, and comorbidities records. Due to the class imbalance encountered in this dataset, an over-sampling method (SMOTE-NC) was applied and compared to the initial prediction.

Our results indicate that the best model for predicting abnormal MRI findings in preterm infants was the K-nearest Neighbor algorithm, with the employment of SMOTE-NC. Further models tested in this work have demonstrated the possibility to have a greater prediction accuracy (e.g. Logistic Regression, Random Forest) but in the negative class (normal MRI findings), which in some terms can be useful but not in the scope of this research.

Our work has various limitations, such as the size of the dataset, the class imbalance, and the lack of external validation. We expect that in the future this work can be reproduced using a bigger dataset and considering further perinatal and sociodemographic data since they have been correlated for short and long-term outcomes in preterm infants in previous studies [39, 40].

Nevertheless, this research can be extended by applying further methods, such as feature importance selection or considering further features that have been associated with abnormal MRI findings. Moreover, considering the small dataset in this research, it is expected to incorporate more patients in the model and even external cohorts that can help towards developing an efficient and reliable tool to assist diagnosis.

Acknowledgements. The authors thank Dr. Isabel Benavente-Fernández and Dr. Simón Lubián-López, from Hospital Puerta del Mar, who provided their valuable help in data acquisition and guidance over the study. This work is supported by the PARENT project that has received funding from the European Union's Horizon 2020 research and innovation programme under the Marie Skłodowska-Curie-Innovative Training Network 2020, Grant Agreement N°956394.

References

1. Chawanpaiboon, S., et al.: Global, regional, and national estimates of levels of preterm birth in 2014: a systematic review and modelling analysis. Lancet Glob. Health **7**(1), e37–e46 (2019)
2. Allotey, J., et al.: Cognitive, motor, behavioural and academic performances of children born preterm: a meta-analysis and systematic review involving 64 061 children. BJOG Int. J. Obstet. Gynaecol. **125**, 16–25 (2018). https://doi.org/10.1111/1471-0528.14832
3. Johnson, S., et al.: Neurodevelopmental outcomes following late and moderate prematurity: a population-based cohort study. Arch. Dis. Child. Fetal Neonatal Ed. **100**, F301–F308 (2015). https://doi.org/10.1136/archdischild-2014-307684
4. Rogers, E.E., Hintz, S.R.: Early neurodevelopmental outcomes of extremely preterm infants. Semin. Perinatol. **40**, 497–509 (2016). https://doi.org/10.1053/j.semperi.2016.09.002
5. Bowe, A.K., Lightbody, G., Staines, A., Murray, D.M.: Big data, machine learning, and population health: predicting cognitive outcomes in childhood. Pediatr. Res. **93**, 300–307 (2023). https://doi.org/10.1038/s41390-022-02137-1
6. Hinojosa-Rodríguez, M., et al.: Clinical neuroimaging in the preterm infant: diagnosis and prognosis. NeuroImage Clin. **16**, 355–368 (2017). https://doi.org/10.1016/j.nicl.2017.08.015
7. Kwon, S.H., Vasung, L., Ment, L.R., Huppi, P.S.: The role of neuroimaging in predicting neurodevelopmental outcomes of preterm neonates. Clin. Perinatol. **41**, 257–283 (2014). https://doi.org/10.1016/j.clp.2013.10.003
8. Barkovich, M.J., Williams, C., Barkovich, A.J.: Technical and practical tips for performing brain magnetic resonance imaging in premature neonates. Semin. Perinatol. **45**(7), 151468 (2021). https://doi.org/10.1016/j.semperi.2021.151468

9. Kohli-Lynch, M., Tann, C.J., Ellis, M.E.: Early intervention for children at high risk of developmental disability in low-and middle-income countries: a narrative review. Int. J. Environ. Res. Public Health **16**, 4449 (2019). https://doi.org/10.3390/ijerph16224449

10. Frija, G., et al.: How to improve access to medical imaging in low-and middle-income countries? EClinicalMedicine **38**, 101034 (2021)

11. Krishnan, R., Rajpurkar, P., Topol, E.J.: Self-supervised learning in medicine and healthcare. Nat. Biomed. Eng. **6**, 1346–1352 (2022)

12. He, L., Li, H., Holland, S.K., Yuan, W., Altaye, M., Parikh, N.A.: Early prediction of cognitive deficits in very preterm infants using functional connectome data in an artificial neural network framework. NeuroImage Clin. **18**, 290–297 (2018). https://doi.org/10.1016/j.nicl.2018.01.032

13. Leon, C., Carrault, G., Pladys, P., Beuchee, A.: Early detection of late onset sepsis in premature infants using visibility graph analysis of heart rate variability. IEEE J. Biomed. Health Inform. **25**, 1006–1017 (2021). https://doi.org/10.1109/JBHI.2020.3021662

14. Verder, H., et al.: Bronchopulmonary dysplasia predicted at birth by artificial intelligence. Acta Paediatr. Int. J. Paediatr. **110**, 503–509 (2021). https://doi.org/10.1111/apa.15438

15. Podda, M., Bacciu, D., Micheli, A., Bellù, R., Placidi, G., Gagliardi, L.: A machine learning approach to estimating preterm infants survival: development of the Preterm Infants Survival Assessment (PISA) predictor. Sci. Rep. **8**, 13743 (2018). https://doi.org/10.1038/s41598-018-31920-6

16. Rezaeian, A., Rezaeian, M., Khatami, S.F., Khorashadizadeh, F., Moghaddam, F.P.: Prediction of mortality of premature neonates using neural network and logistic regression. J. Ambient Intell. Human. Comput. **13**, 1269–1277 (2022). https://doi.org/10.1007/s12652-020-02562-2

17. Sheikhtaheri, A., Zarkesh, M.R., Moradi, R., Kermani, F.: Prediction of neonatal deaths in NICUs: development and validation of machine learning models. BMC Med. Inform. Decis. Making **21** (2021). https://doi.org/10.1186/s12911-021-01497-8

18. Tucker, A., Wang, Z., Rotalinti, Y., Myles, P.: Generating high-fidelity synthetic patient data for assessing machine learning healthcare software. NPJ Digit. Med. **3**(1), 1–13 (2020)

19. Kidokoro, H., Neil, J., Inder, T.: New MR imaging assessment tool to define brain abnormalities in very preterm infants at term. Am. J. Neuroradiol. **34**(11), 2208–2214 (2013). https://doi.org/10.3174/ajnr.A3521

20. Butcher, B., Smith, B.J.: Feature engineering and selection: a practical approach for predictive models. Am. Stat. **74**(3), 308–309 (2020). https://doi.org/10.1080/00031305.2020.1790217

21. Chawla, N.V., Bowyer, K.W., Hall, L.O., Kegelmeyer, W.P.: SMOTE: synthetic minority over-sampling technique. J. Artif. Intell. Res. **16**, 321–357 (2002)

22. Zhu, D., et al.: CREDO: efficient and privacy-preserving multi-level medical prediagnosis based on ML-KNN. Inf. Sci. **514**, 244–262 (2020)

23. Li, J.P., Haq, A.U., Din, S.U., Khan, J., Khan, A., Saboor, A.: Heart disease identification method using machine learning classification in e-healthcare. IEEE Access **8**, 107562–107582 (2020). https://doi.org/10.1109/ACCESS.2020.3001149

24. Srividya, M., Mohanavalli, S., Bhalaji, N.: Behavioral modeling for mental health using machine learning algorithms. J. Med. Syst. **42**, 1–12 (2018). https://doi.org/10.1007/s10916-018-0934-5

25. Kidokoro, H., Anderson, P.J., Doyle, L.W., Woodward, L.J., Neil, J.J., Inder, T.E.: Brain injury and altered brain growth in preterm infants: predictors and prognosis. Pediatrics **134**, e444–e453 (2014). https://doi.org/10.1542/peds.2013-2336

26. Rose, J., et al.: Neonatal physiological correlates of near-term brain development on MRI and DTI in very-low-birth-weight preterm infants. NeuroImage Clin. **5**, 169–177 (2014)

27. Guillot, M., Sebastianski, M., Lemyre, B.: Comparative performance of head ultrasound and MRI in detecting preterm brain injury and predicting outcomes: a systematic review. Acta Paediatr. **110**(5), 1425–1432 (2021)

28. Plaisier, A., et al.: Safety of routine early MRI in preterm infants. Pediatr. Radiol. **42**, 1205–1211 (2012). https://doi.org/10.1007/s00247-012-2426-y

29. Edwards, A.D., et al.: Effect of MRI on preterm infants and their families: a randomised trial with nested diagnostic and economic evaluation. Arch. Dis. Child.-Fetal Neonatal Ed. **103**(1), F15–F21 (2018)

30. Heller, B.J., Yudkowitz, F.S., Lipson, S.: Can we reduce anesthesia exposure? Neonatal brain MRI: swaddling vs. sedation, a national survey. J. Clin. Anesth. **38**, 119–122 (2017)

31. Dong, S.Z., Zhu, M., Bulas, D.: Techniques for minimizing sedation in pediatric MRI. J. Magn. Reson. Imaging **50**(4), 1047–1054 (2019). https://doi.org/10.1002/jmri.26703

32. Burkitt, K., Kang, O., Jyoti, R., Mohamed, A.L., Chaudhari, T.: Comparison of cranial ultrasound and MRI for detecting brain injury in extremely preterm infants and correlation with neurological outcomes at 1 and 3 years. Eur. J. Pediatr. **178**, 1053–1061 (2019). https://doi.org/10.1007/s00431-019-03388-7

33. Mohammad, K., et al.: Consensus approach for standardizing the screening and classification of preterm brain injury diagnosed with cranial ultrasound: a Canadian perspective. Front. Pediatr. **9**, 618236 (2021). https://doi.org/10.3389/fped.2021.618236

34. McLean, G., Ditchfield, M., Paul, E., Malhotra, A., Lombardo, P.: Evaluation of a cranial ultrasound screening protocol for very preterm infants. J. Ultrasound Med. **42**(5), 1081–1091 (2022). https://doi.org/10.1002/jum.16121

35. Sutton, P.S., Darmstadt, G.L.: Preterm birth and neurodevelopment: a review of outcomes and recommendations for early identification and cost-effective interventions. J. Trop. Pediatr. **59**(4), 258–265 (2013). https://doi.org/10.1093/tropej/fmt012

36. Ushida, T., et al.: Antenatal prediction models for short- and medium-term outcomes in preterm infants. Acta Obstet. Gynecol. Scand. **100**, 1089–1096 (2021). https://doi.org/10.1111/aogs.14136

37. Ambalavanan, N., et al.: Outcome trajectories in extremely preterm infants. Pediatrics **130**(1), e115–e125 (2012). https://doi.org/10.1542/peds.2011-3693

38. Nakanishi, H., Suenaga, H., Uchiyama, A., Kono, Y., Kusuda, S.: Trends in the neurodevelopmental outcomes among preterm infants from 2003–2012: a retrospective cohort study in Japan. J. Perinatol. **38**, 917–928 (2018). https://doi.org/10.1038/s41372-018-0061-7

39. Benavente-Fernandez, I., et al.: Association of socioeconomic status and brain injury with neurodevelopmental outcomes of very preterm children. JAMA Netw. Open **2**, e192914 (2019). https://doi.org/10.1001/jamanetworkopen.2019.2914

40. Draper, E.S., et al.: EPICE cohort: two-year neurodevelopmental outcomes after very preterm birth. Arch. Dis. Child. Fetal Neonatal Ed. **105**, 350–356 (2020). https://doi.org/10.1136/archdischild-2019-317418

Modelling of Anti-amyloid-Beta Therapy for Alzheimer's Disease

Swadesh Pal[1] and Roderick Melnik[1,2]

[1] M3AI Laboratory, MS2Discovery Interdisciplinary Research Institute,
Wilfrid Laurier University, Waterloo, ON N2L 3C5, Canada
rmelnik@wlu.ca
[2] BCAM - Basque Center for Applied Mathematics, 48009 Bilbao, Spain
http://m3ai.wlu.ca

Abstract. A healthy brain clears different types of debris with the help of specialized glial cells. These cells contiguously tile the entire central nervous system (CNS), exert many essential complex functions in the healthy CNS, and maintain a healthy balance in the brain. However, over age, these cells fail to control the healthy balance of the proteins and cause different neurodegenerative diseases, one of which is Alzheimer's disease (AD). In AD, insoluble amyloid-beta plaques accumulate in the extracellular space along with neurofibrillary tangles (NFTs) inside the brain cells. In this paper, we have developed a model and studied the accumulation of amyloid-beta plaques and NFTs along with an anti-amyloid-beta therapy applied in the treatment of the disease. Based on these studies, we have demonstrated the dynamics of the modelling therapy such that the drug helps clear a subsequent amount of amyloid-beta plaques in each dose. Numerical simulations have been used to show different long-term outcomes of the model. To further analyze the disease progression in the brain and its treatment, we have integrated brain connectome data in the network model as part of our developed modelling framework.

Keywords: Alzheimer's disease · amyloid-beta plaques · phosphorylated tau · drug-controlled treatments · neurodegeneration · cognitive declines · data-driven models · brain connectome

1 Introduction

Alzheimer's disease (AD) is the leading neurodegenerative disease in the current century, and it causes neuronal death in the brain and disables different functional abilities. More than 50 million people worldwide have this dementia, and this number is expected to reach over 150 million in three decades [1]. It is very hard to identify this dementia at the early stage as it slowly develops in the brain. The development of the disease causes damage to the brain, and when it gets noticed, it already affects most parts of the brain. Therefore, it is important to check the disease status on a regular basis. Researchers have been trying to understand the mechanism behind AD progression, but it is still not

© The Author(s), under exclusive license to Springer Nature Switzerland AG 2023
I. Rojas et al. (Eds.): IWBBIO 2023, LNBI 13919, pp. 431–442, 2023.
https://doi.org/10.1007/978-3-031-34953-9_34

fully clear. Not without some controversy, the Food and Drug Administration (FDA) recently granted the first-ever disease-modifying therapy for AD, aducanumab, followed by lecanemab earlier this year. The former one, in particular, is a monoclonal antibody directed against the amyloid-beta protein [2]. Some other approved drugs are available, but none of these drugs prevents neuronal loss [3]. Therefore, disease-modifying therapies play an important role in controlling the brain's AD progression.

While many ingredients participate in AD initiation and progression, the amyloid-beta and tau proteins are widely recognized as the two most active ingredients [4]. Generally, amyloid-beta accumulates in the extracellular space in the form of insoluble plaques, and tau proteins form neurofibrillary tangles (NFTs) inside the brain cells. The normal brain clears these insoluble plaques and NFTs up to a certain level of these productions, but beyond that, these cause disruptions in the normal activities of the brain cell and move towards AD [5]. For instance, specialized glial cells called astrocytes help in clearing amyloid-beta plaques, and they also help in the modulation of memory and learning processes [6–9]. Different disease-modifying therapies support removing some extra amount of these insoluble plaques so that they can maintain a healthy balance in the brain. Applying those in the brain causes side effects, and clinical trials have been going on for some of these therapies. In analyzing a therapy, our better understanding of all the side effects and any necessary modifications of the doses require time. Computational modelling helps researchers in silico trials to go deeper in a limited time and is no exception for AD. Several mathematical models have been developed based on systems biology approaches to AD molecular and cellular pathophysiologic mechanisms [10–13].

Starting with a general reaction-diffusion framework, we have used logistic models for the temporal evolution of amyloid-beta plaques and phosphorylated tau proteins in the brain [2]. In this work, we study the anti-amyloid beta treatment with the help of a newly proposed mathematical model. In addition, we formulate a network mathematical model to integrate the brain connectome data and examine the effect of anti-amyloid beta in the clearance of amyloid-beta plaques in the brain. Each drug has its own reaction time; generally, it increases the body's reaction gradually and then decreases. We have included such types of drug control phenomena in the model. We use different parameter sets estimated by Hao et al. [2] for various typical population groups (e.g., Alzheimer's disease, late mild cognitive impairment, and cognitively normal) captured by the model. We have constructed an anti-amyloid-beta therapy function in the model, which shows a little clearance of amyloid-beta plaques in each dose can help recover from AD in the long term.

The organization of the rest of this paper is as follows. In Sect. 2, we describe the reaction-diffusion and the network models, which use in the brain connectome consideration in the latter case. Simulation results and their discussions are presented in Sect. 3 followed by conclusions in Sect. 4.

2 Model for AD

Amyloid-beta is one of the key factors in Alzheimer's disease. The imbalance between the production and clearance of amyloid-beta causes amyloid-beta plaque accumulation. On the other hand, several studies related to AD show that the accumulation of amyloid-beta plaques enhances the tau protein's phosphorylation. Researchers have been modelling the accumulations of amyloid-beta plaques and the deposition of tau tangles in various mathematical approaches [10–13]; the approach based on partial differential equations is one of them. In this work, we model them with the following system of coupled reaction-diffusion equations [2]:

$$\frac{\partial u}{\partial t} = \nabla \cdot (\mathbf{D}_u \nabla u) + r_u u \left(1 - \frac{u}{K_u}\right), \tag{1a}$$

$$\frac{\partial v}{\partial t} = \nabla \cdot (\mathbf{D}_v \nabla v) + r_v u \left(1 - \frac{v}{K_v}\right), \tag{1b}$$

which we supplement with the non-negative initial conditions $u(\mathbf{x}, T_0) = u_0$ and $v(\mathbf{x}, T_0) = v_0$ at the age T_0. Here, $u(\mathbf{x}, t)$ and $v(\mathbf{x}, t)$ are the densities of amyloid-beta plaques and phosphorylated tau proteins at a spatial point $\mathbf{x} \in \Omega \subset \mathbb{R}^3$ and time t. The first term on the right-hand side in each of the equations incorporates the random movement of the concentrations in the domain Ω. Here, we have considered the diffusion in amyloid-beta plaques because the soluble proteins (oligomers) of amyloid-beta can diffuse through brain tissue [14]. Our coupled model (1) is also supplemented by corresponding boundary conditions. In particular, our results have been obtained under no-flux boundary conditions. The parameters r_u and r_v in (1) are the growth rates of amyloid-beta and tau protein, whereas K_u and K_v are their carrying capacities, respectively. The tau deposition disrupts different cell functions and induces neurodegeneration and cognitive declines. We model this neurodegeneration (n) and cognitive declines (c) by the following equations:

$$\frac{dn}{dt} = r_n v \left(1 - \frac{n}{K_n}\right), \tag{2a}$$

$$\frac{dc}{dt} = (r_{cn} n + r_{cv} v) \left(1 - \frac{c}{K_c}\right), \tag{2b}$$

with non-negative initial conditions $n(\mathbf{x}, T_0) = n_0$ and $c(\mathbf{x}, T_0) = c_0$. Here, r_n is the growth rate of neurodegeneration, and r_{cn} and r_{cv} are the growth rates for cognitive declines due to neurodegeneration and tau protein, respectively. The parameters K_n and K_c in (2) are the carrying capacities for the neurodegeneration and cognitive declines.

2.1 Anti-amyloid-Beta Treatments

Currently, there is a limited number of drugs available for AD treatments. Most recently, the Food and Drug Administration (FDA) approved several new ones

(in particular, aducanumab), but they do not prevent neuronal loss [3]. Therefore, disease-modifying therapies are important in controlling the brain's AD
progression. To study the anti-amyloid-beta therapy, we modify the first equation of (1) to include the drug influence on the evolution of amyloid-beta plaques,
leading to:

$$\frac{\partial u}{\partial t} = \nabla \cdot (\mathbf{D}_u \nabla u) + r_u u \left(1 - \frac{u}{K_u}\right) - \Phi(t)u, \tag{3}$$

with the same initial condition as before. The function $\Phi(t)$ represents the drug-
control function, which clears the amyloid-beta plaques; it is generally non-
constant over time. The drug starts to remove the amyloid-beta plaques after
receiving a dosage. The clearance rate increases initially due to the drug, but
it begins to decay after some time, landing at a certain level before receiving
the next dosage. The AD patient recovers if the latest clearance rate exceeds
the previous rate. However, due to the different side effects of drugs (e.g., brain
edema and hemorrhage), we can not apply the therapy back-to-back to increase
the clearance rate rapidly [15, 16]. Therefore, a slight recovery each time is crucial to achieving the final goal. Taking this into consideration, we can define the
drug-control function $\Phi(t)$ between any two dosages T_p and T_{p+1} ($p = 0, 1, 2, \ldots$),
which satisfies the differential equation:

$$\frac{d\Phi}{dt} = \lambda(t)\Phi(t), \quad t \in (T_p, T_{p+1}), \tag{4}$$

with the positive initial condition $\Phi(T_0) = \phi_0$ as the natural clearance rate for
plaques for a patient before starting the medication. Here, $\lambda(t)$ is the per-capita
clearance rate of the drug. We consider this function as $\lambda(t) = \gamma(t_m - t_l(t))$,
where γ describes the rate of clearance of plaques and $t_l(t)$ is the time since
the last received dosage. The term t_m represents the time when the clearance
rate is maximum, which may vary between drug dosages, but we assume it as
a constant for simplicity. For each $p \geq 1$, the function $\lambda(t)$ is discontinuous
at T_p as $t_l(t) = 0$ for the dosage days. Hence the control function $\Phi(t)$ is not
differentiable at those T_p. In addition, we assume the continuity for the control
function Φ over the whole time period of the treatment. This approach can also
be applied to other treatment plans.

2.2   Network Model for the Brain Connectome

We extend the temporal models defined earlier into a system based on the network model to integrate the brain connectome data [17–21]. This enables us to
characterize the dynamics of the ingredients on a spatio-temporal scale. Suppose
the brain data are represented by a graph $\mathbf{G}$ with V nodes and E edges. For
graph $\mathbf{G}$, we construct the adjacency matrix $\mathbf{A}$, which helps us to assemble the
Laplacian on the graph. We define the (i, j) $(i, j = 1, 2, 3, \ldots, V)$ element of the
matrix $\mathbf{A}$ as

$$A_{ij} = \frac{n_{ij}}{l_{ij}^2},$$

where n_{ij} is the mean fiber number and l_{ij}^2 is the mean length squared between the nodes i and j. We further define the elements of the Laplacian matrix $\mathbf{L}$ as

$$L_{ij} = D_{ij} - A_{ij}, \quad i,j = 1,2,3,\ldots,V,$$

where $D_{ii} = \sum_{j=1}^{V} A_{ij}$ are the elements of the diagonal weighted-degree matrix. Now we can write a network mathematical model on the graph $\mathbf{G}$ with the help of this Laplacian matrix that is given by:

$$\frac{du_j}{dt} = -\rho_u \sum_{k=1}^{V} L_{jk} u_k + r_u u_j \left(1 - \frac{u_j}{K_u}\right) - \Phi(t) u_j, \tag{5a}$$

$$\frac{dv_j}{dt} = -\rho_v \sum_{k=1}^{V} L_{jk} v_k + r_v u_j \left(1 - \frac{v_j}{K_v}\right), \tag{5b}$$

$$\frac{dn_j}{dt} = r_n v_j \left(1 - \frac{n_j}{K_n}\right), \tag{5c}$$

$$\frac{dc_j}{dt} = (r_{cn} n_j + r_{cv} v_j)\left(1 - \frac{c_j}{K_c}\right), \tag{5d}$$

with $\Phi(t)$ satisfying the drug-control equation (4). Here, u_j, v_j, n_j, and c_j denote the concentrations of u, v, n, and c respectively, at the node j. The parameters ρ_u and ρ_v are the diffusion coefficients of amyloid-beta plaques (u) and tau proteins (v), respectively. In the simulations, we use non-negative initial conditions for all the variables.

3   Results and Discussions

In this section, we numerically analyze the model (1) and the anti-amyloid-beta treatment (3). An in-house tool based on Matlab and C-language has been developed and used for the simulations. Authors in [2] estimated the parameter values for the model (1) with integrated therapy (2) for different groups of patients by using cerebrospinal fluid amyloid beta42 biomarker data, and the values are given in Table 1. We use these parameter values in our model along with the drug-control function modified in this work, as described in the previous section.

3.1   Dynamics for the Homogeneous System

First, we note that the homogeneous system corresponding to the network model can be obtained by setting the graph Laplacian to zero. Before studying the drug-control model (3), we start with finding some characteristics of drug function for the FDA-approved drug called 'aducanumab'. During the treatment, aducanumab is administered intravenously (IV) via a 45–60 min infusion every four weeks. After applying it, the drug accelerates the clearance rate of amyloid plaques. The maximum clearance rate must occur at least half-time between

Table 1. The mean initial conditions and parameter values for Alzheimer's disease (AD), late mild cognitive impairment (LMCI) and cognitively normal (CN) groups [2].

Parameter	AD group	LMCI group	CN group
u_0	36.03	41.57	44.92
v_0	12.38	4.21	3.69
n_0	0.26	0.48	0.42
c_0	3.68	6.03	2.58
r_u	18.35×10^{-2}	16.12×10^{-2}	16.82×10^{-2}
K_u	259.44	264.99	276.21
r_v	0.15	0.08	0.12
K_v	123.35	131.66	126.53
r_n	6.90×10^{-3}	7.37×10^{-3}	7.24×10^{-3}
K_n	1.0	1.02	1.03
r_{cn}	1.67	1.26	3.16
r_{cv}	3.83	1.93	2.48
K_c	169.48	129.40	59.89

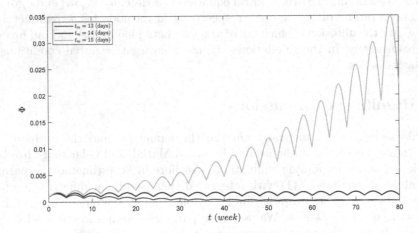

Fig. 1. (Color online) The drug-control function (4) for different t_m with $\gamma = 5.5 \times 10^{-3}$ and $T_p = 4p$ (weeks): magenta - 13 days, blue - 14 days, and cyan - 15 days.

two dosages to cure AD, i.e., t_m should be greater than two weeks. We choose $t_m = 15$ (days) for simulations [see Fig. 1]. On the other hand, based on the aducanumab data released by Biogen, the amyloid PET assessment data at week 78 were decreased by 16.5% compared to the baseline. Using these data and Table 1, we have estimated $\gamma = 5.5 \times 10^{-3}$. We fix $\gamma = 5.5 \times 10^{-3}$ and $t_m = 15$ (days) for all the simulations reported next.

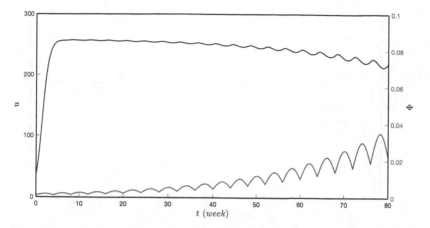

Fig. 2. (Color online) The accumulations of amyloid-beta (blue) for AD group in the presence of drug-control (magenta) with $\gamma = 5.5 \times 10^{-3}, t_m = 15$ (days) and $T_p = 4p$ (weeks). The simulation uses the mean parameter values from Table 1.

Figure 2 depicts the accumulation of amyloid plaques for the AD patient group in the presence of drug control applied every four weeks. It shows that the drug clears the plaques from the highest level. We plot the solutions for phosphorylated tau and cognitive declines in Fig. 3. These solutions show that the anti-amyloid-beta therapy does not affect the phosphorylated tau and cognitive declines; however, they saturate at their carrying capacities. This happens due to the exponential growth in the accumulation of phosphorylated tau inside the brain cells at the initial stage. In addition, it can be shown that a less phosphorylated tau accumulation if we are able to control its growth before it converges to the saturated value. Furthermore, we have observed the same type of behaviour of the solution of neurodegeneration in the model.

We also plot the accumulation and clearance of plaques for the other groups in Fig. 4. It can be seen that the solution behaviours are almost the same for all the groups. Furthermore, it shows that the drug helps to clear the amyloid-beta plaques for all the groups in the long term. However, this clearance cannot control the accumulation of phosphorylated tau, neurodegeneration and cognitive declines. Other interesting things happen if the patient stops the treatment before or after the recovery. To study these scenarios, we stopped the drug after two and three years [see Fig. 5]. The simulations show that the plaques accumulate and are saturated to the carrying capacity. Therefore, the disease will not disappear permanently and could return if we stop the medication.

3.2   Solution for the Network Model

We have used brain connectome data for the network model, which is available at https://braingraph.org/. Our integrated brain data contains 1015 nodes and 16,280 edges. In this brain graph data, each node corresponds to a small area

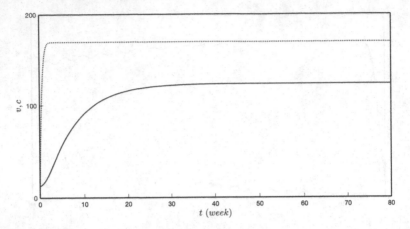

Fig. 3. (Color online) The solutions for phosphorylated tau (solid) and cognitive declines (dotted) for AD group in the presence of drug-control with $\gamma = 5.5 \times 10^{-3}, t_m = 15$ (days) and $T_p = 4p$ (weeks). The simulation uses the mean parameter values from Table 1.

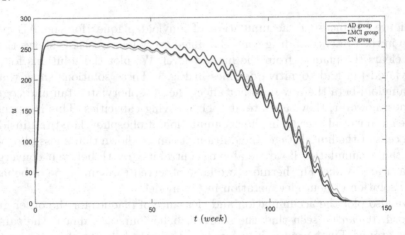

Fig. 4. (Color online) The accumulations of amyloid-beta for each group in the presence of drug-control with $\gamma = 5.5 \times 10^{-3}, t_m = 15$ (days) and $T_p = 4p$ (weeks). For each group, the simulation uses the mean parameter values from Table 1.

(1–1.5 cm^2) of the gray matter, called the region of interest (ROI). An edge may be connected with two nodes if a diffusion-MRI-based workflow finds fibers of axons running between those two nodes in the white matter of the brain [20, 21]. We have calculated the graph Laplacian based on the brain connectome data and used it in the numerical simulation of the network model. We have considered both diffusion coefficients as unity ($\rho_u = \rho_v = 1$).

At the initial stage of AD, the amyloid-beta plaque accumulates in the temporobasal and frontomedial regions in the brain connectome. On the other hand,

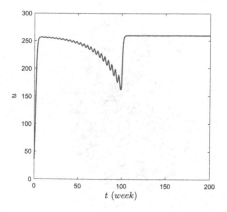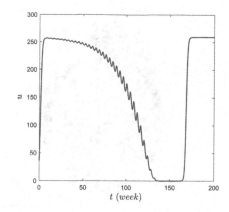

Fig. 5. (Color online) The accumulations of amyloid-beta for the AD group by applying the drug-control for a short-term (right-figure). The simulation uses the mean parameter values from Table 1 with $\gamma = 5.5 \times 10^{-3}, t_m = 15$ (days) and $T_p = 4p$ (weeks).

the phosphorylated tau accumulated in the locus coeruleus and transentorhinal associated regions [see Fig. 6]. We have considered the non-zero initial values mentioned in Table 1 for amyloid-beta plaques and phosphorylated tau in these seeding sites. Furthermore, we have considered the parameter values corresponding to the AD group. Our analysis leads to the conclusion that the solution for the other groups (LMCI and CN) follows the same type of behaviours in the long term.

The developed model can potentially assist in experimental setups for Alzheimer's disease treatments. In such cases, experiments and modelling approaches enrich each other, allowing model validation strategies to be tested. Before applying the anti-amyloid beta therapy to a patient, we need to know the total number of hours/days of the drug's effectiveness. Concurrently, the time when the drug does the maximum effect on a body (e.g., t_m) is also needed. The total incubation period of a drug varies for different patients, and we find their mean (simultaneously with the standard deviation) based on the patient's age or age group. We use this mean value as the hours/days between two dosages for the patient group, i.e., the time between T_p and T_{p+1}. Recall that the mean values for the other parameter values used in our model are given in Table 1. They were initially calculated by Hao et al. [2] using empirical data from the Alzheimer's Disease Neuroimaging Initiative (ADNI) site. Integrating all this information into our mathematical model provides us with the estimate of dosages required to cure the disease and allows evaluation of each dosage's progression.

Finally, we plot the solutions for the network model corresponding to the amyloid-beta plaques and phosphorylated tau for each node presented in Fig. 7. The amyloid-beta plaque accumulation behaviours differ for different nodes at the initial stage. The accumulation grows exponentially from the starting point at some of the nodes; however, at the other nodes, it initially decreases and then

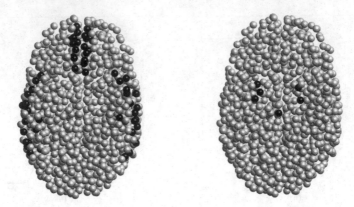

Fig. 6. (Color online) Initial seeding sites for the amyloid-beta plaques (left) and phosphorylated tau (right) in the brain connectome. Black colors represent the non-zero concentration, and gray colors represent the zero concentration.

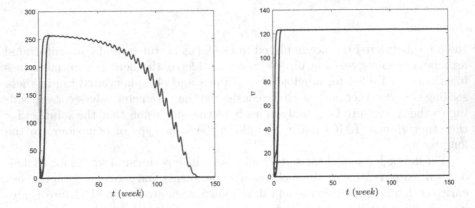

Fig. 7. (Color online) Node-wise accumulations of amyloid-beta plaques (left) and phosphorylated tau (right) for the AD group with respect to time. The simulation uses the mean parameter values from Table 1 with $\gamma = 5.5 \times 10^{-3}, t_m = 15$ (days) and $T_p = 4p$ (weeks).

increases exponentially. These behaviours depend on the degree of the nodes; a node with a higher degree distributes the plaques to all its connected nodes through diffusion, and then all of them grow exponentially. On the other hand, a node with a degree zero keeps away from others and remains unaffected all the time. The anti-amyloid-beta therapy helps to decrease the amyloid-beta plaques from all the nodes in the brain connectome.

4   Conclusions

In this paper, starting with a coupled system of reaction-diffusion equations, we have developed a model incorporating the logistic growth in the accumulation of amyloid-beta plaques and phosphorylated tau proteins. In addition, we have introduced an anti-amyloid-beta therapy in the accumulation of amyloid-beta plaques. A network mathematical model has also been considered to study the accumulations of the ingredients in the brain connectome.

We have observed that the drug function is crucial in the resulting solutions and, therefore, in developing new therapies for AD treatments. Further, a small clearance of plaques in each dosage goes towards the disease-free state for the long term. Our model also shows that the disease could return if we stop the medication after a particular time. Altogether, the developed modelling approach has allowed us to understand better the dynamics of drugs' clearance of the plaques that may drastically affect AD progression. Moreover, the proposed methodology can also be modified to analyze other drug-based therapies for neurodegenerative diseases.

Acknowledgements. The authors thank the NSERC and the CRC Program for their support. RM also acknowledges the support of the BERC 2022-2025 program and the Spanish Ministry of Science, Innovation and Universities through the Agencia Estatal de Investigacion (AEI) BCAM Severo Ochoa excellence accreditation SEV-2017-0718. This research was partly enabled by support provided by SHARCNET (www.sharcnet. ca) and Digital Research Alliance of Canada (www.alliancecan.ca).

References

1. Alzheimer's Association: 2020 Alzheimer's disease facts and figures. Alzheimer's Dementia, pp. 391–460 (2020)
2. Hao, W., Lenhart, S., Petrella, J.R.: Optimal anti-amyloid-beta therapy for Alzheimer's disease via a personalized mathematical model. PLoS Comput. Biol. **18**(9), e1010481 (2022)
3. Vaz, M., Silvestre, S.: Alzheimer's disease: recent treatment strategies. Eur. J. Pharmacol. **887**, 173554 (2020)
4. Hardy, J.A., Higgins, G.A.: Alzheimer's disease: the amyloid cascade hypothesis. Science **256**, 184–186 (1992)
5. Verkhratsky, A., et al.: Astrocytes in Alzheimer's disease. Neurother. J. Am. Soc. Exp. NeuroTherapeutics **7**, 399–412 (2010)
6. Panatier, A., et al.: Glia-derived D-serine controls NMDA receptor activity and synaptic memory. Cell **125**, 775–784 (2006)
7. Ding, S., et al.: Enhanced astrocytic Ca^{2+} signals contribute to neuronal excitotoxicity after status epilepticus. J. Neurosci. **27**, 10674–10684 (2007)
8. González-Reyes, R.E., et al.: Involvement of astrocytes in Alzheimer's disease from a neuroinflammatory and oxidative stress perspective. Front. Mol. Neurosci. **10**, 427 (2017)
9. Trujillo-Estrada, L., et al.: Astrocytes: from the physiology to the disease. Curr. Alzheimer Res. **16**, 675–698 (2019)
10. Ackleh, A.S., et al.: A continuous-time mathematical model and discrete approximations for the aggregation of β-Amyloid. J. Biol. Dyn. **15**, 109–136 (2021)
11. Bertsch, M., et al.: The amyloid cascade hypothesis and Alzheimer's disease: a mathematical model. Eur. J. Appl. Math. **32**(5), 749–768 (2021)

12. Bucci, M., Chiotis, K., Nordberg, A.: Alzheimer's disease profiled by fluid and imaging markers: tau PET best predicts cognitive decline. Mol. Psychiatry **26**, 5888–5898 (2021)
13. Connor, J.P., Quinn, S.D., Schaefer, C.: Sticker-and-spacer model for amyloid beta condensation and fibrillation. Front. Mol. Neurosci. **15**, 962526 (2022)
14. Waters, J.: The concentration of soluble extracellular amyloid-β protein in acute brain slices from CRND8 mice. PLoS One **5**(12), e15709 (2010)
15. Sperling, R.A.: Toward defining the preclinical stages of Alzheimer's disease: recommendations from the National Institute on Aging-Alzheimer's Association workgroups on diagnostic guidelines for Alzheimer's disease. Alzheimers Dement. **7**, 280–292 (2011)
16. Haeberlein, S., et al.: Clinical development of aducanumab, an anti-abeta human monoclonal antibody being investigated for the treatment of early Alzheimer's disease. J. Prev. Alzheimer's Dis. **4**, 255–263 (2017)
17. Thompson, T.B., Chaggar, P., Kuhl, E., Goriely, A.: Protein-protein interactions in neurodegenerative diseases: a conspiracy theory. PLoS Comput. Biol. **16**, e1008267 (2020)
18. Pal, S., Melnik, R.: Pathology dynamics in healthy-toxic protein interaction and the multiscale analysis of neurodegenerative diseases. In: Paszynski, M., Kranzlmüller, D., Krzhizhanovskaya, V.V., Dongarra, J.J., Sloot, P.M.A. (eds.) ICCS 2021. LNCS, vol. 12746, pp. 528–540. Springer, Cham (2021). https://doi.org/10.1007/978-3-030-77977-1_42
19. Pal, S., Melnik, R.: Nonlocal models in the analysis of brain neurodegenerative protein dynamics with application to Alzheimer's disease. Sci. Rep. **12**, 7328 (2021)
20. Kerepesi, C., Szalkai, B., Varga, B., Grolmusz, V.: How to direct the edges of the connectomes: dynamics of the consensus connectomes and the development of the connections in the human brain. PLoS One **11**, e0158680 (2016)
21. Szalkai, B., Kerepesi, C., Varga, B., Grolmusz, V.: High-resolution directed human connectomes and the consensus connectome dynamics. PLoS One **14**, e0215473 (2019)

Using Digital Biomarkers for Objective Assessment of Perfusionists' Workload and Acute Stress During Cardiac Surgery

Roger D. Dias[1,2]([✉]) [iD], Lauren R. Kennedy-Metz[3] [iD], Rithy Srey[4],
Geoffrey Rance[5] [iD], Mahdi Ebnali[1,2], David Arney[1,6] [iD], Matthew Gombolay[7] [iD],
and Marco A. Zenati[1,8] [iD]

[1] Harvard Medical School, Boston, MA, USA
rdias@bwh.harvard.edu
[2] Department of Emergency Medicine, Mass General Brigham, Boston, MA, USA
[3] Department of Psychology, Roanoke College, Salem, VA, USA
[4] Division of Cardiac Surgery, Veterans Affairs Boston Healthcare System, Boston, MA, USA
[5] Department of Cardiac Surgery, Cape Cod Healthcare, Hyannis, MA, USA
[6] Department of Anesthesia, Critical Care and Pain Medicine, Massachusetts General Hospital, Boston, MA, USA
[7] Georgia Institute of Technology, Atlanta, GA, USA
[8] Division of Cardiac Surgery, Veterans Affairs Boston Healthcare System and Medical Robotics and Computer Assisted Surgery Lab, Boston, MA, USA

Abstract. The cardiac operating room (OR) is a high-risk, high-stakes environment inserted into a complex socio-technical healthcare system. During cardiopulmonary bypass (CPB), the most critical phase of cardiac surgery, the perfusionist has a crucial role within the interprofessional OR team, being responsible for optimizing patient perfusion while coordinating other tasks with the surgeon, anesthesiologist, and nurses. The aim of this study was to investigate objective digital biomarkers of perfusionists' workload and stress derived from heart rate variability (HRV) metrics captured via a wearable physiological sensor in a real cardiac OR. We explored the relationships between several HRV parameters and validated self-report measures of surgical task workload (SURG-TLX) and acute stress (STAI-SF), as well as surgical processes and outcome measures. We found that the frequency-domain HRV parameter HF relative power – FFT (%) presented the strongest association with task workload (correlation coefficient: -0.491, p-value: 0.003). We also found that the time-domain HRV parameter RMSSD (ms) presented the strongest correlation with perfusionists' acute stress (correlation coefficient: -0.489, p-value: 0.005). A few workload and stress biomarkers were also associated with bypass time and patient length of stay in the hospital. The findings from this study will inform future research regarding which HRV-based biomarkers are best suited for the development of cognitive support systems capable of monitoring surgical workload and stress in real time.

Keywords: Cognitive Workload · Acute Stress · Perfusionists · Cardiac Surgery · Digital Biomarkers

© The Author(s), under exclusive license to Springer Nature Switzerland AG 2023
I. Rojas et al. (Eds.): IWBBIO 2023, LNBI 13919, pp. 443–454, 2023.
https://doi.org/10.1007/978-3-031-34953-9_35

1 Introduction

The cardiac operating room (OR) is a high-risk, high-stakes environment inserted into a complex socio-technical healthcare system [1, 2]. Despite the tremendous improvements in surgical patients' mortality and morbidity achieved in the past two decades, patient safety and preventable adverse event rates are still suboptimal among cardiac surgery patients [3–5]. During the cardiopulmonary bypass (CPB), the most critical phase of cardiac surgery, the perfusionist has a crucial role within the interprofessional OR team, being responsible for optimizing patient perfusion while coordinating other tasks with the surgeon, anesthesiologist, and nurses [6]. The inherently complex nature of the OR setting, alongside the requirements for effective and sustained situational awareness, decision-making, and teamwork, make perfusionists particularly vulnerable to the deleterious effects that high cognitive workload and stress may have on clinical performance [7, 8].

Extensive previous literature has shown that cognitive overload and high acute stress (distress) impair surgical performance in both simulated and real-life settings [9, 10]. Nonetheless, most of the existing evidence is surgeon-centered, and only a few studies have included other OR team members, particularly the perfusionists. Furthermore, the majority of this previous research has used self-report instruments (e.g., questionnaires) that provide a validated overall measure of the entire surgical procedure but do not support objective monitoring of clinicians' workload and stress in real-time [9, 11].

Among several real-time objective measures, heart rate variability (HRV) is a physiological biomarker that can be unobtrusively captured via digital wearable sensors and has been used as a proxy for cognitive workload and acute stress in various settings, including the OR [9, 12, 13]. Through HRV analysis, many parameters are generated in the time, frequency, and non-linear domains, with each of these parameters reflecting autonomic nervous system activity (sympathetic and/or parasympathetic) at varying levels [14]. HRV has been shown to index high-level cognitive functions (e.g., cognitive workload) and emotional regulation (e.g., stress) [15].

The aim of this study was to investigate objective digital biomarkers of perfusionists' workload and stress derived from HRV metrics captured via a wearable physiological sensor in a real cardiac OR. We explored the relationships between several HRV parameters and validated self-report measures of surgical task workload and acute stress, as well as surgical processes and outcomes measures.

2 Methods

2.1 Study Setting and Design

This was an observational study conducted in the cardiac OR of a tertiary hospital in the United States between January 2021 and May 2022. The research protocol was approved by the Institutional Review Boards (IRB) at the Veterans Affairs (VA) Boston Healthcare System and Harvard Medical School. Informed consent was obtained from all participants, including patients, perfusionists, and OR staff.

2.2 Population

Participants were perfusionists who worked in the cardiac OR and patients who were undergoing open cardiac surgery procedures: coronary artery bypass grafting (CABG) and/or aortic valve replacement (AVR). No exclusion criteria were applied.

2.3 Procedures

At the beginning of each operation, perfusionists were attached to a wireless 3-lead electrocardiogram (ECG) device (*MindWare Mobile Impedance Cardiograph*) that records ECG signals at 500 Hz. After the operation, the ECG files were exported and analyzed in the *Kubios HRV* software [16]. In *Kubios*, an automatic artifact correction was selected.

At the end of the procedure, perfusionists completed two previously validated self-report questionnaires: the Surgery Task Load Index (SURG-TLX) [17], assessing their perceived surgical task workload, and the Spielberger Short-Form State-Trait Anxiety Inventory (STAI-SF) [17, 18], assessing their perceived acute stress.

2.4 Measurements

Heart Rate Variability (HRV). Perfusionist's ECG signals were processed in Kubios [16] to extract a series of 44 different HRV parameters for each 5-min time window, calculated based on established HRV analysis guidelines [16, 19]. The analysis included HRV metrics from time, frequency, and non-linear domains. To gather a measure that reflects the overall surgical procedure, the 5-min widows were averaged across the entire procedure for each HRV parameter.

Perceived Surgical Task Workload. To measure the perceived (self-reported) workload related to the entire surgical procedure, the SURG-TLX was completed by perfusionists [17, 20]. This is a 20-point scale for which the score is calculated by multiplying the number of points selected by 5 (score range: 5–100). The tool covers six different workload domains: mental demands, physical demands, temporal demands, task complexity, distractions, and situational stress. Each domain is rated individually, then averaged to generate the SURG-TLX total score.

Perceived Acute Stress. To measure the perceived (self-reported) acute stress related to the entire surgical procedure, the STAI-SF was completed by perfusionists [18]. This is a 4-point scale for which half of the items have a reverse score. The tool has six different items, asking participants to report how much they feel: calm, tense, upset, relaxed, content, and worried. Each item is rated individually, then the sum of the scores (reversed for items 1, 3, and 5) generates the STAI-SF total score, ranging from 6 to 24 points.

Surgical Processes and Outcome Measures. Patient data was collected from the electronic health record. Surgical processes measures included: total procedure duration, cross-clamp time, and bypass time, all in minutes. Surgical outcome measures included: intrahospital mortality (%), and length of hospital stay after surgery (in days).

3   Results

A total of 37 cardiac surgery patients and 4 perfusionists were included in this study. Twenty-nine (78.4%) patients underwent an isolated CABG procedure, 5 (13.5%) patients received an isolated AVR procedure, and 3 (8.1%) patients had a combined CABG/AVR procedure. Table 1 displays the patient's demographic and surgical characteristics.

Table 1. Patient Demographic and Surgical Characteristics.

Variables	Measures
Sex male – *N (%)*	37 (100.0)
Age – *years*	71.0 (63.0–74.0)
BMI – *kg/m^2*	29.7 (27.3–32.5)
BSA – *m^2*	2.1 (1.9–2.2)
Pre-Operative Risk (VASQIP risk assessment core)	
30-day Mortality Risk – %	0.8 (0.4–1.3)
30-day Morbidity Risk – %	6.9 (6.7–9.6)
30-day SSI Risk – %	1.4 (0.9–2.0)
Surgical Process and Outcome Measures	
Bypass Duration – *min*	112.0 (94.0–134.0)
Cross Clamp Time – *min*	71.0 (56.0–90.0)
Total Procedure Duration – *min*	330.0 (258.0–371.0)
Hospital Length of Stay – *days*	7.0 (6.0–9.0)
Mortality – *N (%)*	0 (0.0)

[*]*BMI: body mass index; BSA: body surface area; VASQIP: Veteran Affair Surgical Quality Improvement Program; SSI: surgical site infection*

A correlation analysis was performed between all objective (HRV parameters) and subjective (self-report instruments) measures of task workload and acute stress. Table 2 (Task Workload) and Table 3 (Acute Stress) show the correlation coefficients and corresponding p-values for all statistically significant associations.

Table 2. Significant Correlations between HRV parameters and SURG-TLX total score.

HRV Parameter	HRV Domain	Correlation Coefficient	p-value
HF relative power – FFT (%)	Frequency-domain	−0.491	0.003
HF absolute power – FFT (n.u.)	Frequency-domain	−0.485	0.004
LF absolute power – FFT (n.u.)	Frequency-domain	0.485	0.004
LF relative power – FFT (%)	Frequency-domain	0.476	0.004
LF/HF ratio – FFT	Frequency-domain	0.472	0.005
VLF peak frequency – FFT (Hz)	Frequency-domain	0.454	0.007
DFA alpha1 – short-term fluctuation slope	Non-linear	0.382	0.026

HF: high-frequency band; LF: low-frequency band; VLF: very-low-frequency band; DFA: detrended fluctuation analysis; FFT: Fast Fourier Transformation

The scatter plots in Figs. 1 and 2 display the relationship between the HRV parameters presenting the strongest correlations with the perfusionist's workload and acute stress. Both parameters – the high-frequency (HF) band and the root mean square of the successive differences between heartbeat intervals (RMSSD) – reflect parasympathetic autonomic nervous activity (vagal modulation). However, HF is influenced by respiratory rate while RMSSD is not [22]. The inverse relationship reflects the fact that high workload and stress situations lead to a decrease in parasympathetic activity while increasing sympathetic tone.

The only HRV parameter associated with patient pre-operative risk scores was the very-low-frequency (VLF) relative power, which presented a correlation coefficient of −0.346, and a p-value of 0.045 with the risk for surgical site infection (SSI). No other relationships with pre-operative risk scores yielded a statistically significant correlation.

Additional correlational analysis between perfusionists' workload and acute stress metrics and surgical processes and outcome measures revealed that the bypass duration was associated with the following metrics: low-frequency (LF) absolute power (log) (correlation coefficient: 0.366, p-value: 0.033); Standard deviation along the line-of-identity (SD2) (correlation coefficient: −0.370, p-value: 0.031); Stress Index (correlation coefficient: −0.346, p-value: 0.045); and VLF absolute power (log) (correlation coefficient: −0.412, p-value: 0.016). Only the HRV parameter Mean RR-intervals (ms), which represent the mean interval between consecutive heartbeats in milliseconds, presented a statistically significant correlation with patient length of stay in the hospital after surgery (correlation coefficient: −0.346, p-value: 0.0045). No other relationships with surgical processes and outcome measures yielded a statistically significant correlation.

Table 3. Significant Correlations between HRV parameters and STAI-SF total score.

HRV Parameter	HRV Domain	Correlation Coefficient	p-value
RMSSD (ms)	Time-Domain	−0.489	0.005
SD1 in Poincaré plot (ms)	Non-linear	−0.488	0.005
HF absolute power – FFT (ms2)	Frequency-Domain	−0.485	0.006
LF peak frequency – FFT (Hz)	Frequency-Domain	−0.469	0.008
HF absolute power – FFT (log)	Frequency-Domain	−0.465	0.008
PNS index	*Kubios* Metric [16]	−0.461	0.009
RPA Lmax – maximum line length (beats)	Non-linear	0.459	0.009
VLF absolute power – FFT (ms2)	Frequency-Domain	−0.459	0.009
pNN50 (%)	Time-Domain	−0.454	0.010
TINN (ms)	Time-Domain	−0.453	0.011
SD2/SD1 ratio	Non-linear	0.436	0.014
SDNN (ms)	Time-Domain	−0.436	0.014
Total power – FFT (ms2)	Frequency-Domain	−0.434	0.015
NN50 (beats)	Time-Domain	−0.431	0.015
DFA alpha1 – short-term fluctuation slope	Non-linear	0.425	0.017
HF relative power – FFT (%)	Frequency-Domain	−0.421	0.018
HF absolute power – FFT (n.u.)	Frequency-Domain	−0.418	0.019
LF absolute power – FFT (n.u.)	Frequency-Domain	0.418	0.019
Stress index	*Kubios* Metric [16]	0.409	0.022
LF relative power – FFT (ms2)	Frequency-Domain	−0.407	0.023
RPA Shannon Entropy	Non-linear	0.401	0.026
D2 – correlation dimension	Non-linear	−0.393	0.029
LF/HF ratio – FFT	Frequency-Domain	0.391	0.030
RPA Determinism (%)	Non-linear	0.379	0.036
RPA Recurrence Rate (%)	Non-linear	0.377	0.037
HRV Triangular Index	Time-Domain	−0.369	0.041
VLF absolute power – FFT (log)	Frequency-Domain	−0.369	0.041
ApEn – Approximate Entropy	Non-linear	−0.358	0.048

(*continued*)

Table 3. (*continued*)

HRV Parameter	HRV Domain	Correlation Coefficient	p-value
SampEn – Sample Entropy	Non-linear	−0.355	0.050

*RMSSD: root mean square of the successive differences between RR intervals; SD1: standard deviation perpendicular to the line-of-identity; SD2: standard deviation along the line-of-identity; HF: high-frequency band; LF: low-frequency band; VLF: very-low-frequency band; PNS: parasympathetic nervous system activity; RPA: recurrent plot analysis; NN50:Number of successive RR interval pairs that differ more than 50 ms; pNN50:NNxx divided by the total number of RR intervals; DFA: detrended fluctuation analysis; FFT: Fast Fourier Transformation; TINN: Baseline width of the RR interval histogram; SDNN: Standard deviation of RR intervals; Stress index: Square root of Baevsky's [21] stress index.

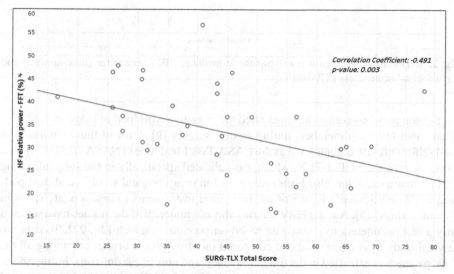

Fig. 1. Relationship between a parasympathetic-mediated HRV parameter (frequency-domain) and perfusionists' workload (SURG-TLX).

4 Discussion

In the present study, we collected data from patients and perfusionists in a real cardiac OR, leveraging an integrative approach based on human factors and cognitive engineering sciences [23]. To investigate objective digital biomarkers of perfusionists' workload and acute stress, we performed HRV analysis from ECG signals and explored the relationships between several HRV parameters and previously validated self-report tools. The findings from this study will inform future research regarding which HRV-based biomarkers are best suited for the development of cognitive support systems capable of monitoring surgical workload and stress in real time.

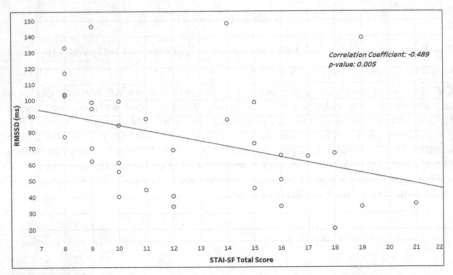

Fig. 2. Relationship between a parasympathetic-mediated HRV parameter (time-domain) and perfusionists' acute stress (STAI-SF).

Several previous studies have assessed the cognitive workload of surgeons and, less often, other OR team members during cardiac surgery [9]. Most of these studies have used self-report instruments such as the NASA Task Load Index (NASA-TLX) [24] and, more recently, the SURG-TLX, which was validated specifically in the surgical setting [9, 11]. Acute stress has also been investigated in many surgical studies, and the Spielberg STAI questionnaire is one of the most used instruments to assess acute stress in various settings [25]. Among HRV metrics, the vast majority of the studies have selected only a few parameters to investigate based on previous research [26, 27]. To date, no surgical study has investigated and compared the strength of correlation among all the HRV parameters extracted in the time, frequency, and non-linear domains. Furthermore, most previous studies in the field have suggested the LF/HF ratio is the best marker of cognitive workload [9]. Nonetheless, our findings show that there are other HRV parameters with equal or even stronger correlation with SURG-TLX total score. Interestingly, our results showed that no time-domain parameter was associated with perfusionists' workload. Regarding acute stress, the strongest correlation was found with RMSSD, which corroborates previous literature that has suggested this time-domain parameter as a good indicator of acute stress [9, 12]. It is also important to notice that many non-linear parameters were significantly associated with perfusionists' workload and stress, and studies exploring these parameters to infer cognitive and emotional functions in the OR are scarce. Future research should include non-linear HRV parameters in their analysis.

Surprisingly, the patient pre-operative risk, which is an indicator of expected task complexity and difficulty [28], was only associated with a few self-report items and HRV parameters. A possible explanation is that the population of patients included in this study did not substantially vary in their pre-operative risk (Table 1), therefore, not allowing the discovery of linear correlations with workload and stress. Another explanation is that high-risk patients may pose an additional workload to the surgeon, particularly in the psychomotor domain, but not necessarily to the perfusionist. The bypass duration, the time in which a heart-lung machine provides the patient perfusion, is a reliable surgical process measure that serves as a surrogate for various post-operative outcomes [29]. Several self-report and HRV parameters were associated with bypass duration, showing that a longer bypass phase is associated with a higher perfusionist's workload and stress. Length of hospital stay, an important patient outcome measure, was also moderately associated with an HRV parameter (mean R-R interval).

The findings from this study are promising, particularly as it relates to validating objective digital biomarkers of perfusionists' workload and stress that can be monitored in near or real-time during cardiac surgery. Although we have averaged the HRV parameters over the entire procedure, these metrics are extracted for each 5-min window or even shorter, such as 1-min, using ultra-short-term HRV analysis [30]. In the same way we currently monitor patient physiology in real-time, informing clinical decision-making in the OR, the continuous monitoring of perfusionists' cognitive and emotional states can be used by intelligent ambient systems capable of detecting cognitive overload situations and re-allocating tasks across the OR team, distributing cognitive demands [31]. Such intelligent systems [32] can also manage unavoidable interruptions to the surgical work-flow, providing the surgical team with a data-driven assessment of clinicians' workload and stress that enable timely interruptions at specific moments when the workload is not highly critical [33].

Although this study reveals important relationships between perfusionists' work-load, stress, and surgical processes and outcome measures, it is important to highlight that these findings cannot infer a cause-effect relationship. In fact, in the same way that task difficulty and complexity can increase cognitive load and stress, these two states can also deteriorate performance, which per se, can increase stress [34]. These relation-ships between task workload, acute stress, and performance are intrinsically complex and should be understood and investigated through a feedback loop perspective. Future research should attempt to model these relationships using non-linear models and net-work analysis in order to capture further the complexity and nuances of such a dynamic system.

5 Conclusion

In conclusion, the findings from this study provide a comprehensive exploratory analysis of the several HRV parameters that have been reported in the literature to index cognitive and emotional states across various industries in which human performance plays a crit-ical role in safety. By investigating the comparative association between these different HRV parameters and established measures of cognitive workload and acute stress in the cardiac OR, our results advance the current scientific knowledge toward the development and validation of objective digital biomarkers of human performance in the surgical set-ting. More importantly, because these metrics are unobtrusively captured in real-time, our findings provide the foundational evidence for future cognitive augmentation capa-bilities [35] that can augment and improve performance in the OR, reducing medical errors and enhancing patient safety.

Acknowledgments. This work was supported by the National Heart, Lung, and Blood Institute (NHLBI) of the National Institutes of Health (NIH) [Grant Numbers: R01HL126896, R01HL157457]. The content is solely the responsibility of the authors and does not necessarily represent the official views of the National Institutes of Health.

References

1. Frank, P., Nurok, M., Sundt, T.: Human factors considerations in cardiac surgery. In: Cohen, T.N., Ley, E.J., Gewertz, B.L. (eds.) Human Factors in Surgery, pp. 131–140. Springer, Cham (2020). https://doi.org/10.1007/978-3-030-53127-0_13
2. Cohen, T.N., Ley, E.J., Gewertz, B.L.: Human Factors in Surgery: Enhancing Safety and Flow in Patient Care. Springer International Publishing, Cham (2020). https://doi.org/10.1007/978-3-030-53127-0
3. Christian, C.K., et al.: A prospective study of patient safety in the operating room. Surgery **139**, 159–173 (2006). https://doi.org/10.1016/j.surg.2005.07.037
4. Martinez, E.A., et al.: Cardiac surgery errors: results from the UK National Reporting and Learning System. Int. J. Qual. Health Care **23**, 151–158 (2011). https://doi.org/10.1093/intqhc/mzq084
5. Gurses, A.P., et al.: Identifying and categorising patient safety hazards in cardiovascular operating rooms using an interdisciplinary approach: a multisite study. BMJ Qual. Saf. **21**, 810–818 (2012). https://doi.org/10.1136/bmjqs-2011-000625
6. Wiegmann, D., Suther, T., Neal, J., Parker, S.H., Sundt, T.M.: A human factors analysis of cardiopulmonary bypass machines. J. Extra Corpor. Technol. **41**, 57–63 (2009)
7. Flin, R., Youngson, G.G., Yule, S.: Enhancing Surgical Performance: A Primer in Nontechnical Skills. CRC Press (2015)
8. Kennedy-Metz, L.R., et al.: Analysis of dynamic changes in cognitive workload during cardiac surgery perfusionists' interactions with the cardiopulmonary bypass pump. Hum. Fact. **63**, 757–771 (2021). https://doi.org/10.1177/0018720820976297
9. Dias, R.D., Ngo-Howard, M.C., Boskovski, M.T., Zenati, M.A., Yule, S.J.: Systematic review of measurement tools to assess surgeons' intraoperative cognitive workload. Br. J. Surg. **105**, 491–501 (2018). https://doi.org/10.1002/bjs.10795
10. Arora, S., Sevdalis, N., Nestel, D., Woloshynowych, M., Darzi, A., Kneebone, R.: The impact of stress on surgical performance: a systematic review of the literature. Surgery **147**(318–330), e1–e6 (2010). https://doi.org/10.1016/j.surg.2009.10.007
11. Kennedy-Metz, L.R., Wolfe, H.L., Dias, R.D., Yule, S.J., Zenati, M.A.: Surgery task load index in cardiac surgery: measuring cognitive load among teams. Surg. Innov. **27**, 602–607 (2020). https://doi.org/10.1177/1553350620934931
12. Peabody, J.E., Ryznar, R., Ziesmann, M.T., Gillman, L.: A systematic review of heart rate variability as a measure of stress in medical professionals. Cureus **15**, e34345 (2023). https://doi.org/10.7759/cureus.34345
13. Kennedy-Metz, L.R., Dias, R.D., Srey, R., Rance, G.C., Furlanello, C., Zenati, M.A.: Sensors for continuous monitoring of surgeon's cognitive workload in the cardiac operating room. Sensors **20**(22), 6616 (2020). https://doi.org/10.3390/s20226616
14. Zammuto, M., Ottaviani, C., Laghi, F., Lonigro, A.: The heart in the mind: a systematic review and meta-analysis of the association between theory of mind and cardiac vagal tone. Front. Physiol. **12**, 611609 (2021). https://doi.org/10.3389/fphys.2021.611609
15. Forte, G., Favieri, F., Casagrande, M.: Heart rate variability and cognitive function: a systematic review. Front. Neurosci. **13**, 710 (2019). https://doi.org/10.3389/fnins.2019.00710

16. Tarvainen, M.P., Niskanen, J.-P., Lipponen, J.A., Ranta-aho, P.O., Karjalainen, P.A., Kubios, H.R.V.: Kubios HRV – Heart rate variability analysis software. Comput. Methods Prog. Biomed. **113**(1), 210–220 (2014). https://doi.org/10.1016/j.cmpb.2013.07.024

17. Wilson, M.R., Poolton, J.M., Malhotra, N., Ngo, K., Bright, E., Masters, R.S.W.: Development and validation of a surgical workload measure: the surgery task load index (SURG-TLX). World J. Surg. **35**, 1961–1969 (2011). https://doi.org/10.1007/s00268-011-1141-4

18. Marteau, T.M., Bekker, H.: The development of a six-item short-form of the state scale of the Spielberger State-Trait Anxiety Inventory (STAI). Br. J. Clin. Psychol. **31**, 301–306 (1992). https://doi.org/10.1111/j.2044-8260.1992.tb00997.x

19. Heart rate variability. Standards of measurement, physiological interpretation, and clinical use. Task Force of the European Society of Cardiology and the North American Society of Pacing and Electrophysiology. Eur. Heart J. **17**, 354–381 (1996). https://www.ncbi.nlm.nih.gov/pubmed/8737210

20. Kennedy-Metz, L., et al.: Human factors analysis of Goal-directed perfusion in cardiac surgery. Healthcare and Medical Devices. AHFE International (2022). https://doi.org/10.54941/ahfe1002120

21. Baevsky, R.M., Chernikova, A.G.: Heart rate variability analysis: physiological foundations and main methods. Cardiometry **10**, 66–76 (2017). https://doi.org/10.12710/cardiometry.2017.10.6676

22. Shaffer, F., Ginsberg, J.P.: An overview of heart rate variability metrics and norms. Front. Public Health **5**, 258 (2017). https://doi.org/10.3389/fpubh.2017.00258

23. Zenati, M.A., Kennedy-Metz, L., Dias, R.D.: Cognitive engineering to improve patient safety and outcomes in cardiothoracic surgery. Semin. Thorac. Cardiovasc. Surg. **32**, 1–7 (2020). https://doi.org/10.1053/j.semtcvs.2019.10.011

24. Wadhera, R.K., et al.: Is the "sterile cockpit" concept applicable to cardiovascular surgery critical intervals or critical events? The impact of protocol-driven communication during cardiopulmonary bypass. J. Thorac. Cardiovasc. Surg. **139**, 312–319 (2010). https://doi.org/10.1016/j.jtcvs.2009.10.048

25. Dias, R.D., Scalabrini, N.A.: Acute stress in residents during emergency care: a study of personal and situational factors. Stress **20**, 241–248 (2017). https://doi.org/10.1080/10253890.2017.1325866

26. Rieger, A., Stoll, R., Kreuzfeld, S., Behrens, K., Weippert, M.: Heart rate and heart rate variability as indirect markers of surgeons' intraoperative stress. Int. Arch. Occup. Environ. Health **87**(2), 165–174 (2013). https://doi.org/10.1007/s00420-013-0847-z

27. Böhm, B., Rötting, N., Schwenk, W., Grebe, S., Mansmann, U.: A prospective randomized trial on heart rate variability of the surgical team during laparoscopic and conventional sigmoid resection. Arch. Surg. **136**, 305–310 (2001). https://doi.org/10.1001/archsurg.136.3.305

28. Lowndes, B.R., et al.: NASA-TLX assessment of surgeon workload variation across specialties. Ann. Surg. **271**, 686–692 (2020). https://doi.org/10.1097/SLA.0000000000003058

29. Nadeem, R., Agarwal, S., Jawed, S., Yasser, A., Altahmody, K.: Impact of cardiopulmonary bypass time on postoperative duration of mechanical ventilation in patients undergoing cardiovascular surgeries: a systemic review and regression of metadata. Cureus **11**, e6088 (2019). https://doi.org/10.7759/cureus.6088

30. Shaffer, F., Meehan, Z.M., Zerr, C.L.: A critical review of ultra-short-term heart rate variability norms research. Front. Neurosci. **14**, 594–880 (2020). https://doi.org/10.3389/fnins.2020.594880

31. Hazlehurst, B., McMullen, C.K., Gorman, P.N.: Distributed cognition in the heart room: how situation awareness arises from coordinated communications during cardiac surgery. J. Biomed. Inform. **40**, 539–551 (2007). https://doi.org/10.1016/j.jbi.2007.02.001

32. Dias, R.D., et al.: Intelligent interruption management system to enhance safety and performance in complex surgical and robotic procedures. In: OR 20 Context Aware Oper. Theatres Comput. Assist. Robot. Endosc. Clin. Image Based Proceed. Skin Image Anal., vol. 11041, pp. 62–68 (2018). https://doi.org/10.1007/978-3-030-01201-4_8

33. Dias, R.D., et al.: Development of an interactive dashboard to analyze cognitive workload of surgical teams during complex procedural care. In: IEEE Int. Interdiscip. Conf. Cogn. Methods Situat. Aware Decis. Support. 2018, pp. 77–82 (2018). https://doi.org/10.1109/COGSIMA. 2018.8423995

34. Teo, G., et al.: Selecting workload and stress measures for performance prediction. Proc. Hum. Fact. Ergon. Soc. Annu. Meet. **61**, 2042–2046 (2017). https://doi.org/10.1177/154193 1213601989

35. Dias, R.D., Yule, S.J., Zenati, M.A.: Augmented cognition in the operating room. In: Atallah, S. (ed.) Digital Surgery, pp. 261–268. Springer, Cham (2021). https://doi.org/10.1007/978-3-030-49100-0_19

Detecting Intra Ventricular Haemorrhage in Preterm Neonates Using LSTM Autoencoders

Idris Oladele Muniru[✉], Jacomine Grobler, and Lizelle Van Wyk

Stellenbosch University, Stellenbosch, South Africa
{25679104,jacominegrobler,lizelle}@sun.ac.za

Abstract. The neonatal period is a critical stage where physiological adaptations for extra-uterine life occur, and newborns are vulnerable to various diseases and disorders. Among these conditions, preterm neonates (PN) born before 37 weeks' gestation are at higher risk of developing intraventricular hemorrhage (IVH), a common complication that can result in severe neurological complications such as cerebral palsy, developmental delays, and cognitive impairments. Early detection and intervention are essential to prevent long-term consequences.

Non-invasive cardiac output monitors (NICOM) have been widely accepted in the neonatal intensive care unit (NICU) for monitoring hemodynamic parameters and have provided vast amounts of data. However, further research is required to explore their predictive tendencies in relation to IVH.

The present study aimed to evaluate the potential of deep learning models to enhance early detection and prevention of IVH in preterm neonates using NICOM parameters. From this study, it was shown that by the LSTM autoencoders are able to predict IVH with moderate precision and accuracy but poor specificity. Nonetheless, this study represents a significant step towards developing a non-invasive, accurate, and timely method for monitoring and preventing IVH in preterm neonates, especially in low-resource settings.

Keywords: LSTM autoencoder · Non-invasive cardiac output monitor · Medical anomaly detection

1 Introduction

During the first 28 days of life, referred to as the neonatal period, newborns go through various physiological changes to adapt to life outside the womb. During this time, they are susceptible to various illnesses and disorders, with preterm neonates (babies born before 37 weeks of gestation) being even more susceptible, particularly during their first week. Premature birth is the leading cause of death among children, accounting for 18% of all deaths among children aged under five years and as much as thirty-five 35% of all deaths among newborns

I. Rojas et al. (Eds.): IWBBIO 2023, LNBI 13919, pp. 455–468, 2023.
https://doi.org/10.1007/978-3-031-34953-9_36

[25]. Intra ventricular haemorrhage (IVH), with an occurrence rate of around 15–20%, is one of the most frequent complications in premature neonates (PN). IVH is associated with several adverse outcomes in preterm neonates, including increased morbidity and mortality and long-term neurodevelopmental impairments such as cerebral palsy, visual impairment, and cognitive delays [8,9]. The severity of these outcomes is often correlated with the severity of the IVH, with higher grades of IVH being associated with worse outcomes.

There are several risk factors for IVH in preterm neonates, including low birth weight, low cardiac output, mechanical ventilation and high blood pressure. Current methods for monitoring and preventing IVH are mainly based on clinical observations and ultrasound scans, which have limitations in terms of accuracy, timeliness, and invasiveness [22].

Non-invasive cardiac output monitors (NICOM) are devices that can continuously measure cardiac parameters. They can provide vital information about the cardiovascular status of preterm neonates and have been proposed as a tool for early identification and prevention of IVH [4,14]. The use of NICOM in the NICU has several potential benefits. Firstly, these devices offer a safe and reliable method for continuously monitoring cardiac function in newborns, particularly those who are preterm or have limited access to invasive monitoring techniques [17]. Secondly, they are non-invasive, meaning that they do not require the insertion of catheters or other invasive devices, which can potentially cause complications such as infection or bleeding. Although NICOM offers a promising tool for the continuous, non-invasive assessment of cardiac function in preterm neonates in the NICU, further research is needed to determine the optimal use of these monitors by leveraging the huge amount of data they generate, for accurate predictions.

Deep learning (DL) models are a type of machine learning that can process large amounts of data and identify complex patterns that may not be visible to the human eye [6]. These models have been used in various medical applications, such as image and signal analysis, and have shown promising results in improving the accuracy and reliability of medical diagnosis and treatment [6,20,24].

In this study, time-aware deep learning autoencoder models were used to analyze time-variant and time-invariant NICOM parameters to detect the IVH occurrence in PM. Ten different models were trained and evaluated. The average evaluation metrics across the ten models as follows: accuracy ($71.0\% \pm 0.121$), precision ($70.8\% \pm 0.13$), negative predictive value ($65.1\% \pm 0.289$), recall ($91.2\% \pm 0.16$), and specificity ($38.3\% \pm 0.343$).

A major contribution of this study is the novel methodology for *data preprocessing and curation* which will contribute to the body of knowledge on medical diagnosis, machine learning, and neonatal care, potentially informing future research in these areas.

The rest of the paper is structured as follows: Sect. 2 provides a brief background on time series, LSTM and autoencoders. The details about the data and deep learning models used are described in Sect. 3. The results of the experiments are reported in Sect. 4. Section 5 discusses the reported results and Sect. 6 provides the conclusion of the paper.

2   Time Series Forecasting, LSTM and Autoencoders

Time series forecasting in healthcare is an important task that can help in predicting future patient outcomes, understanding disease progression, and improving treatment strategies [3,10]. Figure 1 depicts a typical forecasting process. LSTM networks are a type of recurrent neural network (RNN) that is well-suited for handling time series data. LSTMs are designed to capture the temporal dependencies and patterns in data, making them a powerful tool for time series forecasting. LSTM networks consist of memory cells, which store the information and gate units that control the flow of information [26]. These gates determine which information is allowed to pass through and which information is discarded. This allows LSTM to effectively handle vanishing and exploding gradients, which is a common problem in traditional RNNs [26].

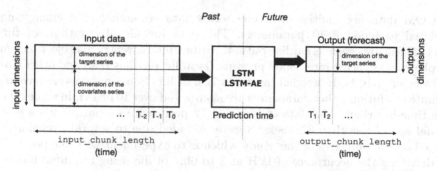

Fig. 1. Training forecasting models on multiple time series (Adapted from [18]) T_{-2}, T_{-1} are the second and first point before prediction time respectively. T_0 is prediction time and T_1, T_2 are first and second points in the future respectively.

Several articles have provided evidence that LSTM has potential applications in healthcare and could be beneficial in terms of reducing mortality and morbidity rates and costs [7,16,19]. Naemi et al. [15] reported a use-case of LSTM to predict patients' illness severity with high accuracy. Whereas [13] used the LSTM model to predict the health status of diabetic patients, the attention module-based LSTM was used to predict cardiovascular disease risk factors by [5]. Similarly, [1] demonstrated that LSTM can be used for the prediction of future trends in diabetes. Shi et al. [21] found that a hybrid approach based on the LSTM model and decision tree can be used for mortality prediction and provider performance evaluation.

Recently, researchers have been exploring the use of LSTM-AEs (Long short-term memory autoencoders) in healthcare. An LSTM-AE is a deep learning model that combines the capabilities of LSTM networks and autoencoder (AE) models. Autoencoders are DL architectures that are used to learn efficient encoding (latent features) of unlabeled data [2]. LSTM-AEs can be used for anomaly detection, imputation of missing values, feature extraction and generating new samples from the time series data. Sagheer and Kotb [20], in their work on predicting heart patient survival, reported that LSTM-based stacked autoencoders outperformed other models and that they are more effective in modelling multivariate time series (MTS) data. With the increasing amount of data available in healthcare, the use of LSTMs and autoencoders will play an important role in improving patient outcomes and understanding disease mechanisms.

3   Methodology

In this section, the data collection, processing and model technique introduced and described in details.

3.1   Data Collection and Preprocessing

NICOM data are multivariate time series data consisting of haemodynamic (related to blood flow) parameters. The data include observations of time-varying features such as cardiac output, cardiac index, ventricular ejection time, stroke volume, heart rate, blood pressure (systolic and diastolic) and other static parameters (e.g body weight) for up to 72 h of life. Some parameters have a 1-minute resolution, other parameters are aggregated over 10 or 15 min while others are time-invariant. The dataset included 72 premature neonates with a gestational age of less than 37 weeks; 3 were excluded due to several irregularities. Based on the objective of this study which is to experiment with the possibility of detecting the occurrence of IVH at 3 to 6hrs of life using the information of the first 1 h after birth, 10 records that are less than 6 h were excluded and 1 duplicate record was also removed. **Therefore, included in this study are**

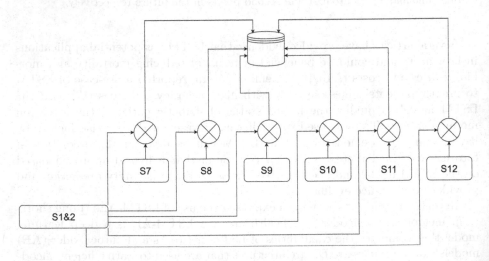

Fig. 2. An Illustration of the Data Curation Pipeline for a Single Patient: A Step-by-Step Process for Creating 6 Samples of Training Data per Patient

those patients with at least 6 h of observations with no occurrence of IVH in the first 3 h of life. The data of 3 patients are labelled as IVH instances confirmed by cranial ultrasound (CUS) and echocardiograph, and 55 neonates have no IVH in the first 6 h of life.

The data preprocessing involves carefully curating a set of samples per patient based on the objective of the study. A unique approach was created to increase a substantial but related sample size per patient without compromising clinical implications. To curate the dataset, 6 h of patient data were partitioned into 30-minute intervals denoted by, S_i, where $i \in \mathbb{R}$, $1 \le i \le 12$. Subsequently, for each patient, S_1 and S_2 were merged with $S_{7,8,9,10,11,12}$ respectively, as depicted in Fig. 2. Based on the expert assessment of the patient at 6 h of life, a binary label (True or False) was assigned to each merged sample, representing the likelihood of intraventricular haemorrhage (IVH) occurrence. This process resulted in 6 new samples per patient.

Next, 10 distinct pairs of training and testing sets were created, each containing samples from 53 patients with no IVH and 5 patients (randomly selected, with 2 having no IVH and 3 with IVH). This was done to evaluate the model's performance on both positive and negative classes.

3.2 Model Training and Testing

An LSTM autoencoder was trained to learn a compressed representation of the multivariate time series parameters from the non-invasive cardiac output monitor. It consists of an encoder and a decoder, with the encoder compressing the input data into a lower-dimensional representation, and the decoder reconstructing the input data from the compressed representation. The goal of an autoencoder is to minimize the reconstruction error between the input and reconstructed data. The autoencoder used in this study was trained only with data from patients with no IVH, hence, it only learns to reconstruct data from normal patients. In principle, it will fail to reconstruct abnormal data (i.e. the reconstruction error will be high). Upon training the autoencoder model, a reconstruction threshold was set at the 99.9th percentile of the reconstruction errors of the training set. The model was then used to reconstruct the test sets. If the reconstruction error of a test sample is greater than the threshold then it is considered as an anomaly (predicted IVH present).

The architecture of the LSTM autoencoder is described in Fig. 3. All 10 models were trained with specific hyperparameters to achieve optimal performance. The hyperparameters used include the optimizer, learning rate, loss function, early stopping criteria, number of epochs, batch size, and validation split. The optimizer used was Adam, with a learning rate of 0.00005. The loss function chosen was *mean absolute error*. Early stopping was implemented with the following parameters; monitor = *val_loss*, patience = *5* and mode = *min* of min. The number of epochs was set to *1000*, and a batch size of 32 was used. Finally, a validation split of 0.1 was employed to ensure generalizability.

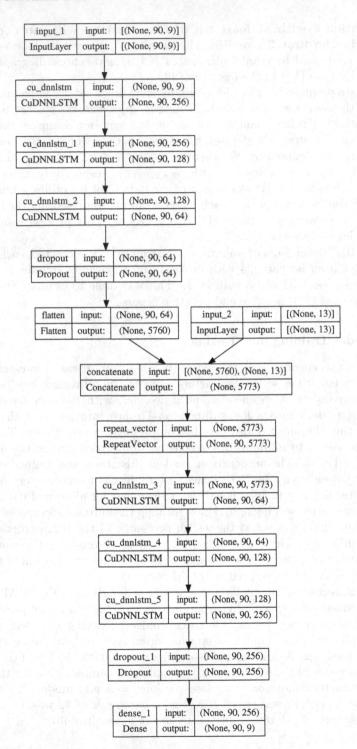

Fig. 3. The LSTM Autoencoder Model Architecture

3.3 Validation Metrics

The performance of the LSTM autoencoders in detecting IHV in the test data was validated using sensitivity, specificity, and accuracy, compared to the expert-labelled cranial ultrasound scans and echocardiography. Sensitivity measures the proportion of true positive predictions (i.e., correctly identified cases of IVH) out of all positive cases. Specificity measures the proportion of true negative predictions (i.e., correctly identified cases without IVH) from negative cases. Accuracy measures the proportion of correct predictions out of all predictions.

4 Results

Figure 4 shows the training performance of the 10 models while the distribution of the reconstruction errors of the training set for each model is shown in Fig. 5. Similarly, Fig. 6 shows the distribution of the reconstruction errors for test set for each model. Finally, Fig. 7 and 8 are plots of the dynamics of six parameters from a sample set predicted as an anomaly (IVH detected) and not anomaly respectively.

Table 1. Evaluation Metrics of 10 Autoencoder Models.

Model	Train MSE	Test MSE	Precision (%)	NPV (%)	Sensitivity (%)	Specificity (%)	Accuracy (%)
1	0.31	0.58	100.0	85.7	88.9	100.0	93.3
2	0.26	0.77	60.0	NaN	100.0	0.0	60.0
3	0.23	0.61	70.8	83.3	94.4	41.7	73.3
4	0.28	0.70	60.0	NaN	100.0	0.0	60.0
5	0.34	0.67	73.7	63.6	77.8	58.3	70.0
6	0.27	0.65	72.0	100.0	100.0	41.7	76.7
7	0.28	0.65	69.2	100.0	100.0	33.3	73.3
8	0.29	0.66	59.3	33.3	88.9	8.3	56.7
9	0.30	0.62	70.8	83.3	94.4	41.7	73.3
10	0.31	0.66	72.0	100.0	100.0	41.7	76.7

Table 1 shows the evaluation metrics of the 10 LSTM autoencoders. The models were evaluated based on their training mean squared error (MSE), test MSE, precision, negative predictive value, sensitivity, specificity, and accuracy. The precision, negative predictive value, sensitivity, specificity, and accuracy are presented as percentages. The summary statistics of these metrics are presented as follows:

- Train MSE = 0.296 ± 0.026
- Test MSE = 0.649 ± 0.051
- Precision = 70.8% + 0.13
- Negative Predictive Value = 65.1% ± 0.289
- Sensitivity (Recall) = 91.2% ± 0.16
- Specificity = 38.3% ± 0.343
- Accuracy = 71.0% ± 0.121

From this study, it was shown that by using, the LSTM autoencoders are able to predict IVH with moderate precision and accuracy but with poor specificity.

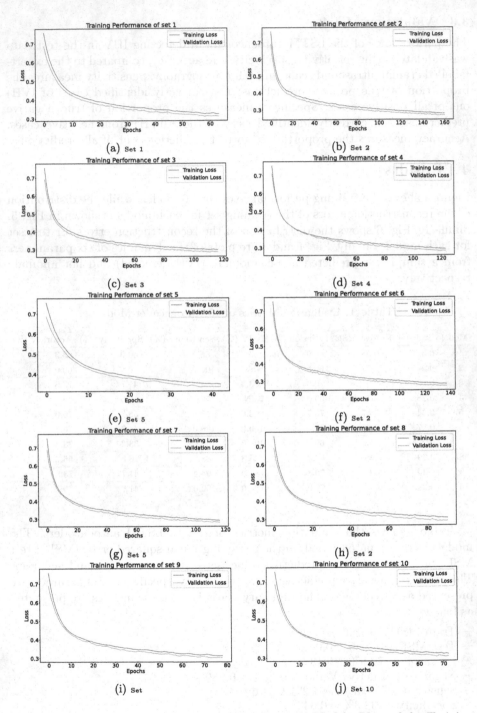

Fig. 4. Comparison of Training Performance Across 10 Models: Plots of the Training and Validation Loss

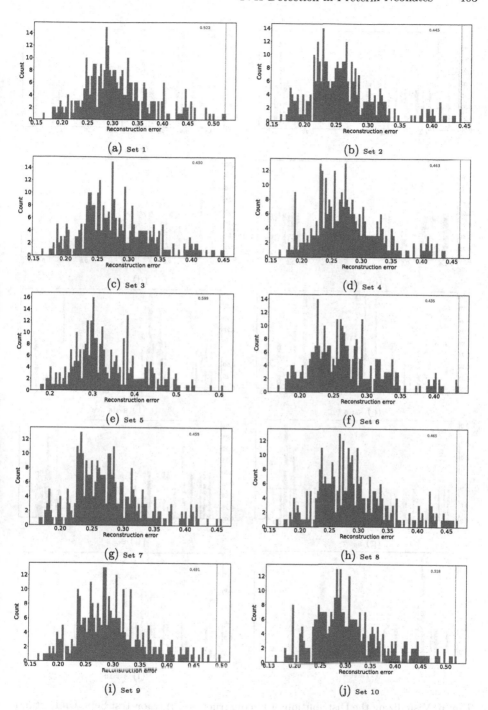

Fig. 5. Visualizing the Distribution of Reconstruction Error for Training Sets: Understanding the Variability Across the 10 Models in Reconstructing Data Points from Training Sets. The vertical line indicates the threshold to be used in classifying if a test sample is an anomaly or not.

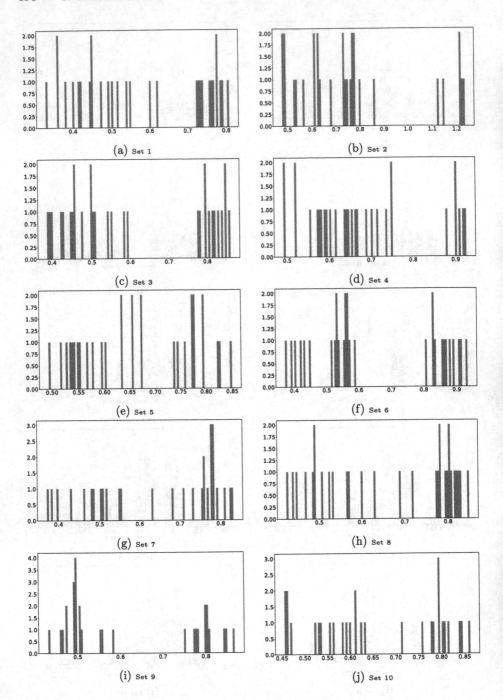

Fig. 6. Visualizing the Distribution of Reconstruction Error for Test Sets: Understanding the Performance of the 10 Models in Reconstructing Data Points from Test Sets.

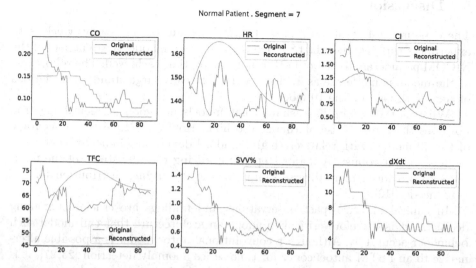

Fig. 7. Original vs Reconstructed Data of patient predicted with IVH This plot shows the observations of selected parameters for the patient predicted to have IVH. HR: Heart Rate, CO: Cardiac Output, CI: Cardiac Index, SVV%: Stroke Volume Variation, $\frac{dX}{dt}$: Instantaneous thoracic bioreactance. In each subplot, the first 60 min represent $S_1 + S_2$ while the last 30 min is S_7

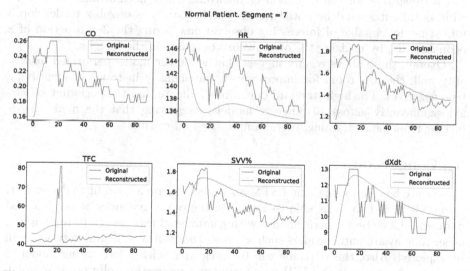

Fig. 8. Original vs Reconstructed Data of patient predicted without IVH This plot shows the observations of selected parameters for the patient predicted to have no IVH. HR: Heart Rate, CO: Cardiac Output, CI: Cardiac Index, SVV%: Stroke Volume Variation, $\frac{dX}{dt}$: Instantaneous thoracic bioreactance. In each subplot, the first 60 min represent $S_1 + S_2$ while the last 30 min is S_7

5   Discussion

The experimental results demonstrate that the aim of this study which is to experiment with whether the LSTM autoencoders can model the dynamics of the NICOM parameters to detect IVH or not was partially achieved. The robustness of the model is moderate as evident in the relatively large standard deviations of the metrics across the 10 models.

Whereas the results have shown that models are good at identifying true positives the rate of false alarms is also high. However, 71.0% mean accuracy of the 10 models with relatively high standard deviations, indicates that more models are not robust. A major step to improving the robustness of the models is to provide a more diverse sample of the underlying distribution for the autoencoder [12].

In clinical practice, past observations and findings are crucial in making decisions. It is therefore critical to adopt an architecture that can model such behaviour accurately. Although a convolutional autoencoder is a possible alternative to an LSTM autoencoder for time series anomaly detection [23,27], it is more suited for capturing spatial information and limited temporal information capture. It will require some extra layer data preprocessing and manipulation such as transforming the time series to spectrogram, RGB images or wavelets to be able to use it, making it more complex [11].

Additionally, a valid concern regarding the present study pertains to its practical applicability, particularly with regard to the source of data concerning the 3–6 h timeframe for the detection of intraventricular haemorrhage (IVH). To address this, it should be noted that a parallel study is ongoing to develop a robust model capable of forecasting observations during the 3–6 h period of a neonate's life, by utilizing the observations obtained in the first 1 h.

Overall, these above results suggest that while the model performed reasonably well, there is room for improvement, particularly in terms of identifying true negatives. The high standard deviations indicate that the performance varied significantly across 10 different models, suggesting that the model needs further refinement, training, more sample size and testing.

6   Conclusion

This study has shown that the LSTM autoencoders are a good fit to detect IVH using NICOM data. It is able to learn the temporal dynamics of the NICOM data, to achieve better performance with compressed features. In future studies, other time-aware autoencoders such as those convolutional neural networks will be explored. Also, this outcome will be integrated with a forecasting model to create a robust end-to-end IVH monitoring and prevention solution. It is worth mentioning that this research is still ongoing and therefore the code is not yet publicly available, however, it can be provided on request.

Acknowledgements. This work is based on the research supported in part by the National Research Foundation of South Africa (Grant Number: 129340).

References

1. Arora, S., Kumar, S., Kumar, P.: Implementation of LSTM for prediction of diabetes using CGM. In: 2021 10th International Conference on System Modeling & Advancement in Research Trends (Smart), pp. 718–722 (2021)
2. Baytas, I.M., Xiao, C., Zhang, X., Wang, F., Jain, A.K., Zhou, J.: Patient subtyping via time-aware LSTM networks. In: Proceedings of the 23rd ACM SIGKDD International Conference on Knowledge Discovery and Data Mining, pp. 65–74 (2017)
3. Che, Z., Purushotham, S., Cho, K., Sontag, D., Liu, Y.: Recurrent neural networks for multivariate time series with missing values. Sci. Rep. 8(1), 6085 (2018)
4. El-Khuffash, A., McNamara, P.J.: Hemodynamic assessment and monitoring of premature infants. Clin. Perinatol. 44(2), 377–393 (2017)
5. Islam, M.D.S., Umran, H.M., Umran, S.M., Karim, M.: Intelligent healthcare platform: cardiovascular disease risk factors prediction using attention module based LSTM. In: 2019 2nd International Conference on Artificial Intelligence and Big Data (ICAIBD), pp. 167–175 (2019)
6. Khedkar, S., Gandhi, P., Shinde, G., Subramanian, V.: Deep learning and explainable AI in healthcare using EHR. In: Dash, S., Acharya, B.R., Mittal, M., Abraham, A., Kelemen, A. (eds.) Deep Learning Techniques for Biomedical and Health Informatics. SBD, vol. 68, pp. 129–148. Springer, Cham (2020). https://doi.org/10.1007/978-3-030-33966-1_7
7. Khorasani, S.T., Cross, J., Maghazei, O.: Lean supply chain management in healthcare: a systematic review and meta-study. Int. J. Lean Six Sigma 11(1), 1–34 (2020)
8. Knüpfer, M., et al.: IVH in VLBW preterm babies-therapy with recombinant activated F VII? Klin. Padiatr. 229(06), 335–341 (2017)
9. Lampe, R., et al.: Assessing key clinical parameters before and after intraventricular hemorrhage in very preterm infants. Eur. J. Pediatr. 179(6), 929–937 (2020). https://doi.org/10.1007/s00431-020-03585-9
10. Lara-Benítez, P., Carranza-García, M., Riquelme, J.: An experimental review on deep learning architectures for time series forecasting. Int. J. Neural Syst. 31(03), 2130001 (2021)
11. Liu, P., Sun, X., Han, Y., He, Z., Zhang, W., Chenxu, W.: Arrhythmia classification of LSTM autoencoder based on time series anomaly detection. Biomed. Signal Process. Control 71, 103228 (2022)
12. Maleki, S., Maleki, S., Jennings, N.R.: Unsupervised anomaly detection with LSTM autoencoders using statistical data-filtering. Appl. Soft Comput. 108, 107443 (2021)
13. Massaro, A., Ricci, G., Selicato, S., Raminelli, S., Galiano, A.: Decisional support system with artificial intelligence oriented on health prediction using a wearable device and big data. In: 2020 IEEE International Workshop on Metrology for Industry 4.0 & IoT, pp. 718–723 (2020)
14. McGovern, M., Miletin, J.: Cardiac output monitoring in preterm infants. Front. Pediatr. 6, 84 (2018)
15. Naomi, A., Schmidt, T., Mansourvar, M., Wiil, U.K.: Personalized predictive models for identifying clinical deterioration using LSTM in emergency departments. Stud. Health Technol. Inf. 275, 152–156 (2020)
16. Nguyen, C.N., Pham, T.T., Le, T.P., Nguyen, K.N.T.: An application of LSTM neural networks to improve the efficiency of monitoring and warning the health status of office workers. J. Mili. Sci. Technol. 81, 3–13 (2022)

17. O'Neill, R., Dempsey, E.M., Garvey, A.A., Schwarz, C.E.: Non-invasive cardiac output monitoring in neonates. Front. Pediatr. **8**, 614585 (2021)
18. Otwarte, S.: Training forecasting models on multiple time series with darts, unit8 (2022). https://unit8.com/resources/training-forecasting-models/
19. Pham, T.D.: Time-frequency time-space LSTM for robust classification of physiological signals. Sci. Rep. **11**(1), 6936 (2021)
20. Sagheer, A., Kotb, M.: Unsupervised pre-training of a deep LSTM-based stacked autoencoder for multivariate time series forecasting problems. Sci. Rep. **9**(1), 1–16 (2019)
21. Shi, P., Gangopadhyay, A., Owens, C., Blunt, B., Grogan, C.: A hybrid model using LSTM and decision tree for mortality prediction and its application in provider performance evaluation. In: 2019 IEEE International Conference on Big Data (Big Data), pp. 2773–2781 (2019)
22. Tataranno, M.L., Vijlbrief, D.C., Dudink, J., Benders, M.J.N.L.: Precision medicine in neonates: a tailored approach to neonatal brain injury. Front. Pediatr. **9**, 634092 (2021)
23. Thill, M., Konen, W., Wang, H., Bäck, T.: Temporal convolutional autoencoder for unsupervised anomaly detection in time series. Appl. Soft Comput. **112**, 107751 (2021)
24. Villarroel, M., et al.: Non-contact physiological monitoring of preterm infants in the neonatal intensive care unit. NPJ Digit. Med. **2**(1), 128 (2019)
25. Walani, S.R.: Global burden of preterm birth. Int. J. Gynecol. Obstet. **150**(1), 31–33 (2020)
26. Yong, Yu., Si, X., Changhua, H., Zhang, J.: A review of recurrent neural networks: LSTM cells and network architectures. Neural Comput. **31**(7), 1235–1270 (2019)
27. Zhang, Y., Chen, Y., Wang, J., Pan, Z.: Unsupervised deep anomaly detection for multi-sensor time-series signals. IEEE Trans. Knowl. Data Eng. (2021)

Measurement of Acute Pain in the Pediatric Emergency Department Through Automatic Detection of Behavioral Parameters: A Pilot Study

Letizia Bergamasco[1,2](✉) , Marco Gavelli[1] , Carla Fadda[3] , Emilia Parodi[4] ,
Claudia Bondone[5] , and Emanuele Castagno[5]

[1] Fondazione LINKS, Via Pier Carlo Boggio 61, 10138 Torino, Italy
{letizia.bergamasco,marco.gavelli}@linksfoundation.com

[2] Dipartimento di Automatica e Informatica, Politecnico di Torino, Corso Duca degli Abruzzi 24, 10129 Torino, Italy

[3] Scuola di Specializzazione in Pediatria, Università degli Studi di Torino, Piazza Polonia 94, 10126 Torino, Italy
carla.fadda@unito.it

[4] S.C. Pediatria e Neonatologia, A.O. Ordine Mauriziano di Torino, Via Magellano 1, 10128 Torino, Italy
eparodi@mauriziano.it

[5] S.C. Pediatria d'Urgenza, Ospedale Infantile Regina Margherita – A.O.U. Città della Salute e della Scienza di Torino, Via Zuretti 23, 10126 Torino, Italy
{cbondone,ecastagno}@cittadellasalute.to.it

Abstract. Acute pain is a frequent symptom in children who access the Emergency Department (ED). Its measurement through validated tools compatible with the time of triage is essential to develop the most appropriate pain-relieving strategy. The algometric scales that can be used in children in whom self-assessment is not possible are based on the evaluation of behavioral and physiological parameters. However, the actual use of algometric scales in the ED is scarce due to environmental factors, heterogeneity of the scales and lack of training, thus making automated pain assessment desirable. In this study, we propose a camera-based system to provide an objective and contactless pain assessment in children aged less than 3 years, through the automatic detection of behavioral parameters from video recordings. To investigate the feasibility of its usage in the ED environment, we collected video recordings of healthy children aged 3–36 months admitted to the ED with acute pain as the main or accompanying symptom, while pain was measured by a healthcare professional according to the Face, Legs, Activity, Cry, and Consolability (FLACC) pain scale. For the recorded videos, we compared the scores for the items Face (F), Legs (L) and Activity (A) given by the operator with the ones given by our system, analyzing the potentiality and limitations of our approach. By showing that automatic pain assessment in young children in the ED could integrate human evaluation to make it easier and faster, without substituting it, we provide the basis for further research in this field.

I. Rojas et al. (Eds.): IWBBIO 2023, LNBI 13919, pp. 469–481, 2023.
https://doi.org/10.1007/978-3-031-34953-9_37

Keywords: Automatic pain assessment · Camera-based approach · Children · Pain · Pediatric emergency department · Algometric scale · FLACC · Google Mediapipe

1 Introduction

Acute pain is a frequent and feared symptom in childhood and is reported in up to 78% of admissions to the Pediatric Emergency Department (ED) [1]. Pain should be adequately considered, measured, and treated whenever it is reported by children or their caregivers, regardless of age, clinical situation, and social role [2]. Acute and repetitive pain experienced at early stages of life can lead to persistent structural and functional changes of the nociceptive system, helping to determine the final architecture of pain system [3]. Many studies have indeed confirmed that untreated pain produces both short and long-term physical and psychological negative effects: healing times are lengthened, complications increase, and long-term sequelae may develop [4, 5].

Accurate assessment and measurement of pain are the cornerstones of pain management and are essential to provide timely adequate analgesic strategy. The measurement of pain in children under the age of 3 years, who cannot provide effective self-assessment, requires standardized, validated tools appropriate to their developmental level, the context, and their prior pain experiences. This can be mostly challenging in the ED setting, in particular on the very first evaluation in triage, where time is limited, anxiety is high, and children and their caregivers are unfamiliar with healthcare professionals and environment [6]. Available validated objective scales are based on the evaluation of both physiological and behavioral parameters. However, literature data suggest that the actual use of algometric scales in the Pediatric ED is limited. Major critical issues are related to environmental factors specific to triage, heterogeneity of scales used, and training deficiencies [7–9]. Among objective pain scales, the Face, Legs, Activity, Cry, and Consolability (FLACC) is based on the detection of behavioral parameters and has been validated for children less than 3 years also in the emergency setting [10].

Due to such reasons, the assessment of pain through objective scales in younger children could be improved by the development of automated machine-based systems aiming to monitor different pain indicators and providing a consistent, minimally biased evaluation of pain. In the past years, there has been an increasing interest in the use of technology for understanding human behavioral responses to pain based on the analysis of facial expressions, and of body or head movements, which are the most important indicators in patients with verbal communication inability [11–13]. Other studies have shown that machine-based systems can be used to detect and analyze physiological changes associated with pain, such as changes in skin color, increase in heart rate [14, 15], and changes in the cerebral hemodynamic of specific brain's regions [16].

Several machine-based approaches have been introduced to analyze infants' body movements for the purpose of diagnosing a specific disease [17, 18]. The development of a machine-based multimodal pain assessment tool that dynamically measures pain in infants has been also proposed, based on the analysis of different behavioral and physiological indicators [19]. Anyway, to our knowledge, no clinical trials have yet been conducted, and most of face detection algorithms are designed and trained for adult faces [19].

Our group has already developed and demonstrated a proof-of-concept computerized tool for the evaluation of pain in newborns, based on the analysis of facial expressions in video recordings [20]. Pain scores obtained from automated analysis have been compared to those assigned by trained healthcare professionals according to three objective neonatal pain scales: the Neonatal Facial Coding System, (NFCS), the Premature Infant Pain Profile (PIPP), and the Douleur Aiguë du Nouveau-né (DAN), showing that manual pain evaluation is challenging and often results in a large variability across scores between different operators, making automated assessment desirable [20].

The main aim of this pilot study was to develop an automated computerized tool for pain evaluation in children aged less than 3 years, using only video recordings acquired from an RGB camera without the aid of sensors on the skin. Second, we aimed to create a dataset and ad hoc registration setting that could be used to demonstrate the feasibility of the usage of such automatic system for pain assessment for children in this specific age in the ED environment. Finally, our goal was to compare the scores of the behavioral parameters assigned by the automatic system with those assigned by a healthcare operator to the items Face (F), Legs (L) and Activity (A) of the FLACC pain scale, analyzing the potentiality and limitations of our approach.

2 Materials and Methods

2.1 Data Collection

In this pilot study, we enrolled healthy children aged 3–36 months admitted to the Pediatric ED of our tertiary teaching Children's Hospital between April and September 2022 with acute pain as main or accompanying symptom. We excluded children with chronic disorders; those for whom face and limbs were not fully visible because of dressing, medications, or any medical device; and those admitted with high triage priority code. Pain was measured in all children by the same healthcare professional using the Italian validated version of the FLACC scale [10, 21] (see Table 1), along with the assessment and recording of heart rate and oxygen saturation for 60 s using the pulse oximeter available in our ED. At the same time, a 60 s video of children's full figure was recorded with an RGB camera with a resolution of 1920 × 1080 pixels and a frame rate of 30 fps. The camera device was placed 100 cm far in front of the subject, in the same light conditions. Written informed consent was obtained from all parents of the involved children. The study protocol was approved by our Local Ethic Committee.

Overall, we have recruited 14 Caucasian children (7 males); the median age was 16 months (range 3–32 months). A total of 22 min of recording were acquired, with a mean length of 1.16 min per child. The acquisition of recordings for a sufficient time was possible for all children, with variable duration of recording fragments suitable for automatic analysis, due to the movements of the child.

2.2 Pain Scale Implementation

Since the FLACC pain scale has been validated for children less than 3 years in the emergency setting, we have selected it as the reference pain scale for the output for our

Table 1. Face, Legs, Activity, Cry and Consolability (FLACC) scale (adapted from [10]).

Categories	Scoring		
	0	1	2
Face	No particular expression	Occasional grimace/frown, withdrawn or disinterested	Frequent/constant quivering chin, clenched jaw
Legs	Normal position or relaxed	Uneasy, restless, tense	Kicking or legs drawn up
Activity	Lying quietly, normal position, moves easily	Squirming, shifting back and forth, tense	Arched, rigid or jerking
Cry	No cry	Moans or whimpers, occasional complaint	Crying steadily, screams or sobs, frequent complaints
Consolability	Content and relaxed	Reassured by occasional touching, hugging or being talked to, distractible	Difficult to console or comfort

automated pain evaluation system. However, in our implementation of the FLACC, cry (C) and consolability (C) categories were excluded, since the audio signal acquired in the emergency department environment turned out to be too noisy to be properly processed and analyzed. Therefore, for this research study a partial version of the FLACC (that we will call pFLACC) has been used, including only face (F), legs (L) and activity (A). The pFLACC score related to an observation period was calculated as defined in Eq. (1):

$$pFLACC_{score} = F_{score} + L_{score} + A_{score} \qquad (1)$$

where F_{score}, L_{score} and A_{score} indicate the scores computed for the single categories (Face, Legs and Activity, respectively) in the considered observation period. Each single-category score ranges between 0 and 2, thus resulting in a total pFLACC$_{score}$ ranging between 0 and 6.

The computation of these single-category scores was based on the analysis of facial expressions and body movements. Some face and body parameters have been identified, deriving from the way the FLACC score is traditionally computed, as described in Table 1. In particular, the identified parameters were 7: mouth opening, brow bulging, eye squeezing, legs outstretching, pedaling, body movement, and arms flailing.

For each parameter, an algorithm has been developed to quantify it using numerical values, as will be better explained in Sect. 2.3. Moreover, a suitable threshold has been defined for each parameter to discriminate the range of values of the parameter representing the "normal" condition, and the range of values representing the "non-normal" condition, related to the pain experience. Threshold values have been defined empirically and have been calibrated based on the available dataset of videos.

All the 7 parameters were continuously measured by the automatic system along all the observation period, at the end of which the $pFLACC_{score}$ was computed. In the case of the videos of our dataset, the observation period was 60 s, corresponding to the duration of the videos. The values of each parameter were compared with the threshold established for that parameter for the duration of the whole observation period; two time thresholds have also been defined empirically (1/3 and 2/3 of the observation time) to check for how long the parameters were in "normal" and "non-normal" range, and therefore assign value 0, 1 or 2 to the pFLACC items. The detailed description of the computation of single-category scores (F, L, A) is reported in the following paragraphs.

Face Score. The computation of F_{score} involved mouth opening, brow bulging and eye squeezing parameters. Such parameters were computed and compared to the relative thresholds along an observation period, as previously defined. A score ranging from 0 to 2 was assigned to F_{score}, according to the methodology illustrated in Table 2.

Table 2. Computation of the face score (F_{score}).

Status of involved parameters	Interpretation	Assigned F_{score}
Mouth opening, brow bulging and eye squeezing exceeding the corresponding thresholds for less than 1/3 of the observation period	Neutral expression	0
Mouth opening, brow bulging or eye squeezing exceeding the corresponding thresholds for more than 1/3 and less than 2/3 of the observation period	Occasional frown	1
Mouth opening, brow bulging or eye squeezing exceeding the corresponding thresholds for more than 2/3 of the observation period	Frequent frown	2

Legs Score. The computation of L_{score} involved legs outstretching and pedaling parameters. Similarly to F_{score}, such parameters were computed and compared to the relative thresholds along an observation period. A score ranging from 0 to 2 was assigned to L_{score}, according to the methodology illustrated in Table 3.

Activity Score. The computation of A_{score} involved body movement and arms flailing parameters. Also in this case, similarly to F_{score} and L_{score}, the involved parameters were computed and compared to the relative thresholds along an observation period, resulting in a score ranging from 0 to 2. The adopted methodology is illustrated in Table 4.

2.3 Face and Body Parameters Computation

As mentioned in Sect. 2.2, there are 7 parameters involved in the computation of the $pFLACC_{score}$: mouth opening, brow bulging, eye squeezing, legs outstretching, pedaling, body movement, and arms flailing. We have calculated these parameters starting from the child's face and body landmarks, that were detected and tracked in the video recording

Table 3. Computation of the legs score (L_{score}).

Status of involved parameters	Interpretation	Assigned L_{score}
Legs outstretching and pedaling exceeding the corresponding thresholds for less than 1/3 of the observation period	Relaxed	0
Legs outstretching or pedaling exceeding the corresponding thresholds for more than 1/3 and less than 2/3 of the observation period	Restless	1
Legs outstretching or pedaling exceeding the corresponding thresholds for more than 2/3 of the observation period	Kicking or outstretching legs	2

Table 4. Computation of the activity score (A_{score}).

Status of involved parameters	Interpretation	Assigned A_{score}
Body movement and arms flailing exceeding the corresponding thresholds for less than 1/3 of the observation period	Normal position	0
Body movement or arms flailing exceeding the corresponding thresholds for more than 1/3 and less than 2/3 of the observation period	Squirming	1
Body movement or arms flailing exceeding the corresponding thresholds for more than 2/3 of the observation period	Jerking	2

along the observation period. To do so, in this work we used Google Mediapipe Holistic (GMH), an open-source framework using machine learning techniques to detect and track face and pose landmarks in real time from the video recording of a person [22, 23].

For each frame of a video, GMH provides 543 landmarks (33 pose landmarks, 21 hand landmarks for each hand, and 468 face landmarks) [23]. It also allows setting a minimum confidence value for the detection and tracking of these points; in our system, we set these values to 0.5, so that we kept as valid landmarks only the ones with confidence higher than 0.5. For a set of landmarks, we will call a frame where those landmarks are valid a "valid frame". For what concerns the format of coordinates of the landmarks, we chose to use the coordinates that GMH provides in the image reference frame. Each landmark is identified by a set of 3D coordinates (x, y, z); in our analysis we used only x and y coordinates, and not the z coordinate, being an estimation of the depth of the point made by GMH. Figure 1 shows an example of a frame of a video in our collected dataset, where landmarks provided by GMH are visualized.

Among all the provided landmarks, we chose the ones that we found most appropriate to quantify the 7 identified parameters, as will be explained in the following paragraphs. In general, all the parameters were computed with a moving window mechanism, with

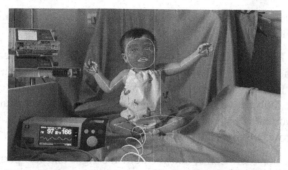

Fig. 1. Example of a frame of a video in our collected dataset, where landmarks provided by GMH are visualized. The image has been postprocessed to anonymize the child's face.

a window length of 1 s. Basically, each video recording was read frame by frame in a simulated real-time fashion. At each new frame that was being analyzed, the current window was shifted by one frame. Some parameters could be calculated for each single valid frame (mouth opening, brow bulging, eye squeezing, legs outstretching), and were therefore averaged over the whole window to get a single value for the window; the other parameters (pedaling, body movement, arms flailing), instead, could be calculated only once for the whole window, by considering the variation of intermediate features along the window. The scheme of the moving window mechanism is illustrated in Fig. 2.

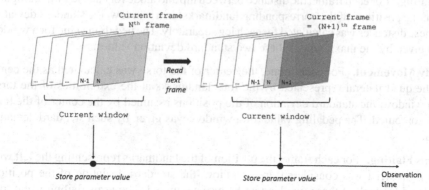

Fig. 2. Scheme of the moving window mechanism for the calculation of face and body parameters.

Mouth Opening. For each frame, the height of the mouth was computed as the height of the minimum bounding rectangle enclosing all the mouth landmarks. Similarly, the height of the whole face was computed as the height of the rectangle enclosing all the face oval landmarks. The mouth opening value for the considered frame was given by the ratio between the mouth height and the face height, converted to a percentage. The mouth opening value for a window was the average of the mouth opening values of the valid frames in the window.

Brow Bulging. For each frame, the distance between eyebrows medial borders was calculated and the width of the whole face was retrieved as the width of the rectangle enclosing all the face oval landmarks. The brow bulging value for the considered frame was given by the ratio between the distance between eyebrows medial borders and the face width, converted to a percentage. The brow bulging value for a window was the average of the brow bulging values of the valid frames in the window.

Eye Squeezing. For each frame, and for both the right and left part of the face, the distance between mid-eyebrow and mid lower eyelid was calculated. Then, the mean between these two distances was retrieved. The face height was computed in the same way as for the mouth opening parameter. The eye squeezing value for the considered frame was given by the ratio between the calculated mean distance and the face height, converted to a percentage. The eye squeezing value for a window was the average of the eye squeezing values of the valid frames in the window.

Legs Outstretching. For each frame, the angle formed by each leg was computed from the positions of the landmarks representing the extremities of the leg and the knee. The legs outstretching value for the considered frame was given by the maximum between the angles formed by each of the two legs. The legs outstretching value for a window was the average of the legs outstretching values of the valid frames in the window.

Pedaling. For each frame, the distance between hip and ankle for each leg was computed from the positions of the corresponding landmarks. In a window, the standard deviation of these distances was calculated for each leg separately. The pedaling value for a window was given by the maximum of these two standard deviation values.

Body Movement. For each frame, the center of the torso was calculated as the center of the quadrilateral represented by the four landmarks at the extremities of the torso. In a window, the standard deviation of the positions assumed by the center of the torso was computed. The pedaling value for a window was given by such standard deviation value.

Arms Flailing. For each frame, the position of the landmarks representing the left wrist and right wrist was considered. In a window, the standard deviation of the positions assumed by each of these two landmarks was computed. The arms flailing value for a window was given by the maximum between the two standard deviation values.

3   Results and Discussion

In Sect. 2, we have illustrated how it is possible to implement a partial version of the FLACC algometric scale to perform automatic pain assessment. In this Section we analyze and discuss the comparison between the pain scores given by the healthcare professional on our collected dataset and the corresponding pain scores given by our automatic system, to investigate the feasibility of its usage with children aged less than 3 years in the ED environment.

As we have shown in our previous work, where we developed a proof-of-concept tool for the automatic evaluation of pain in newborns [20], while comparing scores obtained from an automatic system and scores assigned by healthcare professionals, we should not aim to merely maximize the agreement between them, but we should rather analyze the causes of the differences between the two scores. This is because of the subjectivity and inter-operator variability that are present in pain evaluation made by healthcare professionals, which are motivating us in the investigation of automatic pain assessment.

Therefore, what we can analyze in our comparison are the differences between the values measured by the machine and the healthcare professional for each item, and the reasons behind them. Indeed, for the same observed limb movements and facial expressions, the analysis of what allows the human operator to distinguish pain from restlessness and anxiety is the basis for developing an efficient computer system.

In Table 5, we report the scores for each item (F, L, A) and the total pFLACC scores obtained for all the collected video recordings, as assigned by the healthcare professional and the automatic system. For a quantitative comparison, we also report cosine similarity values in Table 6.

For what concerns the evaluation performed by the healthcare professional, from the comparison of the scores and the feedback given by the healthcare professional, a sort of bias has emerged among analyzed items: facial expression seems to be the most influent to allow healthcare professionals to distinguish restlessness from real pain, followed by activity and lastly by legs movements. In fact, the values assigned to L (legs) and A (activity) quantify movement in general, but do not define the mood in detail. It is therefore difficult to distinguish when this occurs because of pain or due to other reasons.

With respect to the automatic system, instead, the three items (F, L, A) are calculated in a completely independent way, and they have the same weight in the summation that is performed to get the pFLACC score, as described in Eq. (1). If there is a high value for the face item, this does not influence the computation of the legs and activity items. The observations that we reported suggest that future investigations could be carried out on the possibility of assigning different weights to the face, legs and activity items in Eq. (1), so that facial expression has a higher impact on the pFLACC score with respect to legs and body movements.

On the technical side, the major limitation that we observed for the automatic system is that the robustness in the landmarks tracking performed by GMH may not be sufficient when movements occur. In fact, when the child turns to the side or moves the head, the landmarks tracking experiences a degraded performance or is even lost. This may lead to get parameters values that exceed the corresponding thresholds, and, consequently, F_{score}, L_{score} and A_{score} assume values greater than 0 even when the child is quiet.

Using adaptive thresholds instead of fixed thresholds for the face and body parameters could make the system more robust to the variation of facial proportions in children of different ages, and help mitigating the effects of the landmark tracking issues.

If we compare our study with the experience of Parodi et al. [20] with newborns, it is worth noting that pain evaluation in video recordings of children aged less than 3 years is more challenging: newborns not only move less, but also have a different range of facial expressions and movements than infants and older children. Such observation could

Table 5. Scores assigned by the healthcare professional and the automatic system to all the video recordings collected in our dataset.

ID	Healthcare professional				Automatic system			
	F	L	A	pFLACC	F_{score}	L_{score}	A_{score}	$pFLACC_{score}$
1	0	0	0	0	0	0	1	1
2	0	0	1	1	0	0	1	1
3	2	1	1	4	1	0	0	1
4	2	0	1	3	2	1	0	3
5	2	0	1	3	2	0	0	2
6	0	0	0	0	0	0	1	1
7	1	0	0	1	0	0	0	0
8	0	0	0	0	0	0	1	1
9	2	1	1	4	1	0	0	1
10	2	1	1	4	0	0	0	0
11	1	0	1	2	0	0	1	1
12	0	1	0	1	0	1	0	1
13	0	0	0	0	1	0	0	1
14	0	0	0	0	1	0	1	2
15	1	0	0	1	0	1	0	1
16	0	0	0	0	0	0	1	1
17	0	0	0	0	0	0	1	1
18	0	0	0	0	0	0	1	1
19	0	0	0	0	0	0	1	1

Table 6. Cosine similarity results for the comparison of the scores assigned by the healthcare professional and the automatic system to all the video recordings collected in our dataset.

F	L	A	pFLACC
0.72	0.29	0.24	0.58

be considered to enhance the software performance and for the future development of the automatic system, focusing on improving the robustness of landmarks tracking in movement condition and the encoding of face parameters.

Of course, the automatic tool tested in our preliminary research is still not suitable to be used in routine practice, as it represents the first exploratory step within a larger project. Indeed, the analysis that we have performed has been limited by the small population, in particular the low number of infants of our dataset. In the next steps we aim to collect data of other infants, to build a larger dataset that will allow an improved statistical analysis of the results obtained with our system.

Anyway, our results represent the basis to develop a system in which human assessment could be integrated, standardized, and improved thanks to an automatic system. Pain assessment in the pediatric ED is an essential part of triage evaluation and is considered as the fifth vital sign [9]. Rapid and standardized assessment is crucial, but environmental and cultural factors may limit optimal performance on most occasions [7–9]. The standard practice to evaluate pain in children who are not able to make self-assessment is based on the observation made by caregivers and healthcare professionals through validated scales that sometimes fail to meet psychometric standards and requires continuous monitoring. Inter-operator variability is undoubtedly a limit of traditional pain measurement with validated scales. In clinical practice, a more objective approach for pain assessment is desirable to correctly recognize and treat pain.

Different approaches for automated recognition of pain expression have been proposed in the last 20 years; however, most of face detection algorithms have been designed and trained for adult faces and are poorly suitable for infants and young children [19, 24–26]. Our research proposes an automatic system for objective and contactless pain assessment through the automatic detection of behavioral parameters from video recordings, showing the potentiality that machine-based pain evaluation has also in infants and children aged less than 3 years. To the best of our knowledge, there are no other similar studies in the literature that have approached the development of a system of this kind in the context of the Pediatric Emergency Department.

However, on one side the automatic system could reduce the variability in the use of behavioral pain scale in infants and young children, such as FLACC, and make it faster. On the other side, the expertise of trained triage healthcare professionals will continue to be irreplaceable, as it allows to understand and capture aspects that are not yet appreciable by the machine [20]. In fact, a trained observer can pick up nuances beyond the mere detection of specific movements and expressions, and caregivers can usually discern between face expressions of pain, discomfort, or restlessness of their children. In front of a facial grimace (or a frowning expression), the software calculates movements and expressions in an objective way, while a human operator not only analyzes the same parameters, but is also able to better contextualize them. Therefore, it is essential to properly combine both automatic and human pain assessment. Understanding those aspects paves the way for further development and improvement of our pain assessment system in the future.

4 Conclusion and Future Work

Our pilot study explores the possibility to provide automatic, objective and minimally biased assessment of pain in young children, supporting the evaluation made by healthcare professionals, that remains irreplaceable.

The results of our study suggest that the proposed automated pain assessment system is a promising tool that can potentially aid and improve the evaluation of healthcare professionals even in the emergency setting, providing the basis for further research in this field. Anyway, even when it will be validated for routine practice, automatic detection of behavioral parameters should be never intended as a substitute of the observation provided by healthcare professionals. Instead, it could integrate human evaluation and contribute to make it easier and faster.

The main strength of our study is that, for the first time, we have created the assumptions to collect a dataset for the development of an automatic pain detection system in children in the ED. Moreover, we have implemented the recording setting, adapting the one previously elaborated for newborns [20] to children aged less than 3 years. The collaboration between clinicians and engineers allowed us to create a multidisciplinary project for the development of this system.

Further challenges will be the substantial extension of the original recording dataset and the complementation of video processing with the analysis of audio information, thus implementing the full FLACC pain scale. Moreover, future research will include the improvement of the robustness of landmarks tracking in video recordings of children in movement conditions. Another challenge will be to improve the automatic pain assessment approach so to better encode the nuances that the human operator grasps, and integrate them with the detection of physiological parameters, in order to provide a more objective and standardized evaluation of pain.

References

1. Grant, P.S.: Analgesia delivery in the ED. Am. J. Emerg. Med. **24**(7), 806–809 (2006). https://doi.org/10.1016/j.ajem.2006.05.004
2. Raja, S.N., et al.: The revised IASP definition of pain: concepts, challenges, and compromises. Pain **161**(9), 1976 (2020). https://doi.org/10.1097/j.pain.0000000000001939
3. Walker, S.M.: Neonatal pain. Pediatric Anesthesia **24**(1), 39–48 (2014). https://doi.org/10.1111/pan.12293
4. Birnie, K.A., et al.: Hospitalized children continue to report undertreated and preventable pain. Pain Res. Manage. **19**(4), 198–204 (2014). https://doi.org/10.1155/2014/614784
5. Brattberg, G.: Do pain problems in young school children persist into early adulthood? A 13-year follow-up. Eur. J. Pain **8**(3), 187–199 (2004). https://doi.org/10.1016/j.ejpain.2003.08.001
6. Drendel, A.L., Brousseau, D.C., Gorelick, M.H.: Pain assessment for pediatric patients in the emergency department. Pediatrics **117**(5), 1511–1518 (2006). https://doi.org/10.1542/peds.2005-2046
7. Krauss, B.S., Calligaris, L., Green, S.M., Barbi, E.: Current concepts in management of pain in children in the emergency department. Lancet **387**(10013), 83–92 (2016). https://doi.org/10.1016/S0140-6736(14)61686-X
8. Marzona, F., Pedicini, S., Passone, E., Pusiol, A., Cogo, P.: Mandatory pain assessment in a pediatric emergency department: failure or success? Clin. J. Pain **35**(10), 826–830 (2019). https://doi.org/10.1097/AJP.0000000000000743
9. Benini, F., et al.: Consensus on pediatric pain in the emergency room: the COPPER project, issued by 17 Italian scientific societies. Ital. J. Pediatr. **46**, 1–3 (2020). https://doi.org/10.1186/s13052-020-00858-9
10. Merkel, S.I., Voepel-Lewis, T., Shayevitz, J.R., Malviya, S.: The FLACC: a behavioral scale for scoring postoperative pain in young children. Pediatr Nurs **23**(3), 293–297 (1997)
11. Grunau, R.V., Craig, K.D.: Pain expression in neonates: facial action and cry. Pain **28**(3), 395–410 (1987). https://doi.org/10.1016/0304-3959(87)90073-X

12. Brahnam, S., Chuang, C.F., Shih, F.Y., Slack, M.R.: Machine recognition and representation of neonatal facial displays of acute pain. Artif. Intell. Med. **36**(3), 211–222 (2006). https://doi.org/10.1016/j.artmed.2004.12.003

13. Brahnam, S., Nanni, L., Sexton, R.: Introduction to neonatal facial pain detection using common and advanced face classification techniques. In: Yoshida, H., Jain, A., Ichalkaranje, A., Jain, L.C., Ichalkaranje, N. (eds.) Advanced Computational Intelligence Paradigms in Healthcare–1, pp. 225–253. Springer, Heidelberg (2007). https://doi.org/10.1007/978-3-540-47527-9_9

14. Zamzami, G., Ruiz, G., Goldof, D., Kasturi, R., Sun, Y., Ashmeade, T.: Pain assessment in infants: towards spotting pain expression based on infants' facial strain. In: 2015 11th IEEE International Conference and Workshops on Automatic Face and Gesture Recognition (FG), vol. 5, pp. 1–5 (2015). https://doi.org/10.1109/FG.2015.7284857

15. Yoo, S.K., Lee, C.K., Park, Y.J., Kim, N.H., Lee, B.C., Jeong, K.S.: Neural network based emotion estimation using heart rate variability and skin resistance. In: Wang, L., Chen, K., Ong, Y.S. (eds.) ICNC 2005. LNCS, vol. 3610, pp. 818–824. Springer, Heidelberg (2005). https://doi.org/10.1007/11539087_110

16. Ranger, M., Gélinas, C.: Innovating in pain assessment of the critically ill: exploring cerebral near-infrared spectroscopy as a bedside approach. Pain Manag. Nurs. **15**(2), 519–529 (2014). https://doi.org/10.1016/j.pmn.2012.03.005

17. Rahmati, H., Aamo, O.M., Stavdahl, Ø., Dragon, R., Adde, L.: Video-based early cerebral palsy prediction using motion segmentation. In: 2014 36th Annual International Conference of the IEEE Engineering in Medicine and Biology Society, pp. 3779–3783 (2014). https://doi.org/10.1109/EMBC.2014.6944446

18. Stahl, A., Schellewald, C., Stavdahl, Ø., Aamo, O.M., Adde, L., Kirkerod, H.: An optical flow-based method to predict infantile cerebral palsy. IEEE Trans. Neural Syst. Rehabil. Eng. **20**(4), 605–614 (2012). https://doi.org/10.1109/TNSRE.2012.2195030

19. Zamzmi, G., Goldof, D., Kasturi, R., Sun, Y., Ashmeade, T.: Machine-based multimodal pain assessment tool for infants: a review. arXiv preprint arXiv:1607.00331 (2016)

20. Parodi, E., Melis, D., Boulard, L., Gavelli, M., Baccaglini, E.: Automated newborn pain assessment framework using computer vision techniques. In: Proceedings of the International Conference on Bioinformatics Research and Applications, pp. 31–36 (2017). https://doi.org/10.1145/3175587.3175590

21. Di Bari, A., Destrebecq, A., Osnaghi, F., Terzoni, F.: Traduzione e validazione in italiano della scala Revised FLACC per la valutazione del dolore nel bambino con grave ritardo mentale. Pain Nurs. Mag. 2013, 4 (2013)

22. Lugaresi, C., et al.: Mediapipe: a framework for building perception pipelines. arXiv preprint arXiv:1906.08172 (2019)

23. Grishchenko, I., Bazarevsky, V.: Mediapipe holistic (2020). https://ai.googleblog.com/2020/12/mediapipe-holistic-simultaneous-face.html. Accessed 20 Mar 2023

24. Kovac, J., Peer, P., Solina, F.: Human skin color clustering for face detection. In: The IEEE Region 8 EUROCON 2003. Computer as a Tool, vol. 2, pp. 144–148 (2003). https://doi.org/10.1109/EURCON.2003.1248169

25. Asthana, A., Zafeiriou, S., Cheng, S., Pantic, M.: Incremental face alignment in the wild. In: 2014 IEEE Conference on Computer Vision and Pattern Recognition, Columbus, OH, USA, pp. 1859–1866 (2014). https://doi.org/10.1109/CVPR.2014.240

26. Viola, P., Jones, M.: Rapid object detection using a boosted cascade of simple features. In: Proceedings of the 2001 IEEE Computer Society Conference on Computer Vision and Pattern Recognition. CVPR 2001, vol. 1, pp. I-I (2001). https://doi.org/10.1109/CVPR.2001.990517

Clinical Text Classification in Cancer Real-World Data in Spanish

Francisco J. Moreno-Barea[1]([⊠]) [iD], Héctor Mesa[1] [iD], Nuria Ribelles[2] [iD],
Emilio Alba[2] [iD], and José M. Jerez[1] [iD]

[1] Departamento de Lenguajes y Ciencias de la Computación, Escuela Técnica
Superior de Ingeniería Informática, Universidad de Málaga, Málaga, Spain
fjmoreno@lcc.uma.es
[2] Unidad de Gestión Clínica Intercentros de Oncología, Hospitales Universitarios
Regional y Virgen de la Victoria, Málaga, Spain

Abstract. Healthcare systems currently store a large amount of clinical
data, mostly unstructured textual information, such as electronic health
records (EHRs). Manually extracting valuable information from these
documents is costly for healthcare professionals. For example, when a
patient first arrives at an oncology clinical analysis unit, clinical staff
must extract information about the type of neoplasm in order to assign
the appropriate clinical specialist. Automating this task is equivalent to
text classification in natural language processing (NLP). In this study, we
have attempted to extract the neoplasm type by processing Spanish clinical documents. A private corpus of 23,704 real clinical cases has been processed to extract the three most common types of neoplasms in the Spanish territory: breast, lung and colorectal neoplasms. We have developed
methodologies based on state-of-the-art text classification task, strategies
based on machine learning and bag-of-words, based on embedding models in a supervised task, and based on bidirectional recurrent neural networks with convolutional layers (C-BiRNN). The results obtained show
that the application of NLP methods is extremely helpful in performing
the task of neoplasm type extraction. In particular, the 2-BiGRU model
with convolutional layer and pre-trained fastText embedding obtained
the best performance, with a macro-average, more representative than
the micro-average due to the unbalanced data, of 0.981 for precision,
0.984 for recall and 0.982 for F1-score.

Keywords: Text Classification · Natural Language Processing ·
Electronic Health Records · Neoplasm cancer · Spanish

1 Introduction

Public healthcare systems face numerous challenges, including their sustainability, variability in healthcare practice and the need to improve the patient experience, among others. Evidence-based medicine is based on clinical research and

© The Author(s), under exclusive license to Springer Nature Switzerland AG 2023
I. Rojas et al. (Eds.): IWBBIO 2023, LNBI 13919, pp. 482–496, 2023.
https://doi.org/10.1007/978-3-031-34953-9_38

its main tool, randomized clinical trials (RCTs). However, nowadays health outcomes research also includes the collection, compilation and analysis of data generated outside RCTs, in what is known as real-world data (RWD), which in recent years has acquired a growing and renewed interest beyond the classic observational, naturalistic or pragmatic studies, which suffer from significant biases. The widespread, systematic, exhaustive, high quality and transparent collection of data by clinicians in electronic health records, whether these are conventional databases or, more commonly, electronic health records (EHR). Their transformation into useful information of value to the clinician provides a body of knowledge known as Real-World Evidence.

However, the gradual adoption of the EHRs as a key component of healthcare systems raises a number of issues, some of which remain unresolved. EHRs store information of a heterogeneous nature in a variety of formats, including open text documents, such as clinical notes or radiology reports, that contain information related to diagnoses, treatments, or clinical procedures [27]. However, the unstructured nature of these open text fields makes the task of automatically extracting relevant concepts from them particularly difficult, and manual concept extraction is non-reusable, time-consuming and costly [18].

Focusing on a specific medical area, a recurrent problem in oncology clinical analysis units regarding the preparation of the first visit report is the lack of time of the clinical staff to complete the information in the structured fields corresponding to the type of neoplasm, location, histology, etc. This makes subsequent access to the information and exploitation of the results extremely difficult. In other words, the information exists, but it is in text format (not structured) within the EHR information and is not stored in a specific electronic field. The automatic neoplasm type extraction of the text corresponding to the patient's EHR is a key task, allowing the oncology analysis unit to immediately refer the patient to the appropriate specialist.

This process eventually becomes a text classification task. Text classification is a classic problem in natural language processing (NLP). This task is defined as the assignment of text units to one or more categories according to the content and semantics present in the text. These text units can be sentences, questions, paragraphs and, as will be addressed in this study, documents. Text classification is commonly used in marketing, human resources and social analysis tasks such as sentiment analysis (products, companies, online and social media) or news categorisation. Text classification has also proved useful in natural language understanding tasks such as question answering (QA).

Due to the attention this task has received and the increasing amount of textual data, NLP techniques have been applied to the automatic classification of free-text clinical reports in recent years. Approaches to text classification can be divided into three categories: rule-based methods, machine learning (ML) methods and deep learning (DL) methods. The rule-based systems for clinical text classification rely on a large number of manually constructed patterns or rules [15,19]. However, since rule-based methods are not reproducible, studies have focused on ML algorithms for this task. ML methods used include decision tree (DT), naive bayes (NB), support vector machines (SVM) and random forest (RF)

[6, 7, 10, 26]. Finally, with the increased ability to collect large data sets, DL-based methods have become the state-of-the-art (SOTA) for various NPL tasks. Architectures based on convolutional and recurrent neural networks and transformers show impressive results in the text classification task. Models such as long short-term memory (LSMT) [8] and gated-recurrent-units (GRU) [5], including variations such as Bidirectional-LSTM (Bi-LSTM) [12] or Convolutional-LSTM (C-LSTM) [23], and large pre-trained language models with layers of multi-head self-attention architectures [28], have been applied to numerous clinical text classification tasks [2, 11, 14, 29, 33].

However, most of the existing studies in the specific literature refer only to texts in English, due to the scarce availability of linguistic corpora annotated with clinical coding information in other languages. Since Spanish is the second most spoken language in the world in terms of number of native speakers [30], there is a need to apply medical NLP methodologies focused on this language. For this text classification task, we have access to the Galén system [22, 27], a repository of 60, 000 real-world clinical EHRs. The use of this clinical linguistic corpus allows us to obtain reliable information on frequently used words in oncology, as well as grammatical and contextual information in this specific field. Furthermore, the availability of the neoplasm annotations in Gálen for supervised learning allows us to serve as an artificial intelligence laboratory on cancer for the development of NLP models, deploying them in national or international hospitals in Spanish where the neoplasm annotations are not available.

Considering all the above aspects, in this work we propose to advance in the application of NLP models for the automatic extraction of neoplasm type from EHR written in Spanish. The classification algorithms assign to each document the probability of belonging to one of the three most common neoplasms in the Galén information system, as a representation of the Spanish region of Málaga: breast, colorectal and lung; or to another type of neoplasm. ML and DL models studied herein represent SOTA in text classification tasks [13, 32], such as RNNs used in conjunction with CNN and embedding models. However, to the best of our knowledge, this is the first study that examines the application of NLP models to the problem of extracting information about the neoplasm suffered by a patient using real-world medical texts in Spanish.

2   Materials

This section describes the corpus used to perform the text classification task. The automatic classification of clinical texts requires a prior manual analysis of the documents for their collection and correct labelling. In this sense, the research team was able to obtain quality-assured information in a simpler way thanks to the availability of Galén system [22, 27], an integrated software system in oncology centres in the province of Málaga, Spain. The Galén system collects the EHRs of more than 60, 000 oncology patients from the *Hospital Regional Universitario* and the *Hospital Universitario Virgen de la Victoria* in Málaga, Spain, with information completed both in real time and by dedicated staff.

A corpus of EHRs containing an associated neoplasm and containing more than 500 words was selected from the information available in the database, for a total of 23, 704 documents. Each document includes the demographic information, first visit and all information from the remaining episodes (consultation, emergency visit or comments).

After selecting the corpus for the text classification task, the neoplasm labels were processed to group them into breast, lung, colorectal and other neoplasms. The category "other" includes documents on head/neck, liver, prostate, uterus, non-Hodgkin's lymphoma, thyroid, stomach/esophagus and other neoplasms. The selected documents were tokenized, making several decisions to reduce the size of the vocabulary and maximise the inference of contextual relationships. The tokenization was case-insensitive and easily recognisable expressions were replaced by special tokens. In addition, authorised experts obfuscated the documents to maintain anonymity of the real-world EHR for processing. The obfuscation was a bijective transformation of the characters with the additional aim of not losing the properties of n-grams in embedding models.

Table 1 shows the distribution of the neoplasms present in the selected corpus. For the different training, validation and test sets, the columns show the absolute number (abs) of documents for each neoplasms considered and their relative frequency (rel). The majority of neoplasms in the Galén corpus represent the category other (41.9%), although they are not individually sufficiently representative. The most common neoplasm in the corpus is breast cancer (27.4%), while lung and colorectal neoplasms are in the minority but well represented in relation to the rest. In addition, an almost perfect stratification is observed in the training/validation/testing division for the correct evaluation of NLP experimental results.

Table 1. Number and percentage of documents per neoplasm in each corpus subset: training, validation and test.

Neoplasm	Train		Validation		Test	
	abs	rel	abs	rel	abs	rel
Breast	5266	.2743	576	.2699	649	.2738
Lung	2731	.1422	302	.1415	338	.1426
Colorectal	3149	.1640	354	.1659	388	.1637
Other	8054	.4195	902	.4227	995	.4198
Total	19200		2134		2370	

In order to obtain more information about the selected documents and to fine-tune the hyperparameters of the NLP models, an analysis of the text length was performed. Length is measured by the number of tokens present in the text after removing common separators and punctuation marks that do not provide context. Figure 1 shows the number of documents per number of tokens when 100%, 95%, 90% and 75% of the corpus documents are selected. It is observed

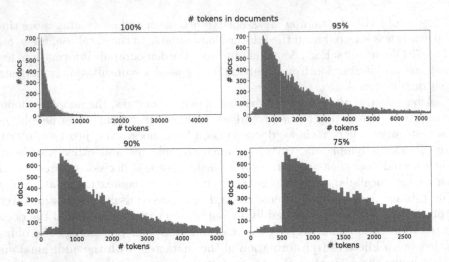

Fig. 1. Distribution of the number of documents by number of tokens when 100%, 95%, 90% and 75% of the corpus documents are considered.

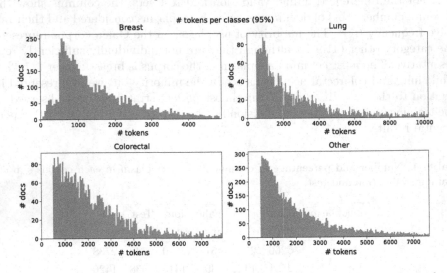

Fig. 2. Distribution of the number of documents by number of tokens for each neoplasm (breast, lung, colorectal, other) when 95% of the corpus documents are considered.

that the maximum number of tokens present in a document is over 40,000, but most documents have less than 10,000 tokens. This is more noticeable when 95% and 90% of the documents are selected. 95% of the documents have less than 7,000 tokens, while 90% have less than 5,000 tokens. Finally, 75% have less than 3,000 tokens, but this clearly implies a lower performance of the NLP models by omitting too large a number of tokens.

When analysing the number of tokens per document belonging to each of the neoplasm classes, 95% of the selected documents is the optimal percentage.

Figure 2 shows these results for breast, lung, colorectal and other neoplasms. There is a clear difference between the number of maximum tokens present in breast and lung neoplasm documents compared to the rest. The 95% of lung documents have less than 10,000 tokens with an average of 3,259.93 tokens, while breast documents are drastically shorter, with less than 5,000 tokens with an average of 1,820.69. Both colorectal neoplasm and other neoplasm documents have a number of tokens less than 7,000 with an average of 2,767.51 and 2,545.71 tokens respectively. Because of these differences, and to ensure that the models are not biased in their choice of neoplasm identified in the EHR solely by the number of tokens present, the hyperparameter fine-tuning with respect to the number of features were set between 5,000 and 7,000 tokens.

3 Methods

This section presents the distinct NLP methodologies developed in this study to tackle the neoplasm type extraction from real-world EHRs in Spanish. The NPL methodologies addressed include ML models, such as NB, SVM and XGBoost; embedding models used in a supervised task; and DL recurrent models, used in conjunction with CNNs and embedding models, such as Word2Vec or fastText.

3.1 Bag-of-Words and Machine Learning Supervised Methods

ML algorithms have been widely used for text processing. However, these methods cannot deal directly with raw text/symbol sequences of variable length, but with numerical feature vectors of fixed size. For this reason, it is necessary to perform a pre-processing of the data for its treatment. Bag-of-Words (BoW) [31] is the most commonly used method for this purpose. BoW transforms documents into a reduced and simplified representation based on criteria such as word frequency, ignoring the order of words and context. BoW creates a dictionary as large as all the different words present in the corpus or limited to the most important or frequent. This dictionary, also known as the vocabulary, is used to vectorise the document, so that the vocabulary is represented as a vector in which each feature is a word stored in it, and its value depends on whether this word occurs in the text and on the criterion chosen.

Count vector and tf-idf are the most common criteria. Count vector is the simplest criterion, where each value associated with a token/word is the number of occurrences of that token, also called term frequency (tf), in the text unit. Term frequency - inverse document frequency (tf-idf) is the combination of tf and inverse document frequency (idf) [25]. The idf assigns a higher weight to words with high or low frequency terms in the document. Thus, the tf-idf value increases in proportion to the number of times a word/token occurs in the document, but is offset by the frequency of the word in the document collection, which reduces the effect of implicitly common words in the corpus.

Once the documents have been vectorised with BoW using the tf-idf criterion, we applied the ML models. The most common ML models considered for this task are NB, SVM, DTs and RF. On one hand, NB classifiers are known for being simple but efficient algorithms. NB classifiers make the naive assumption that all features belonging to the same class are independent and contribute equally to the categorisation result. This assumption is generally not true in real-world situations. NB then calculates the conditional probability of each class, given a set of features, using Bayes' theorem. On the other hand, SVMs aim to obtain a hyperplane that performs a partitioning of the data. For this purpose, SVM maps the input points onto a higher dimensional feature space, so that the decision boundary maximises the margin between the different classes, thereby clustering them. Prediction involves classifying a sample according to the closest cluster.

Finally, RF and DTs were used with eXtreme Gradient Boosting (XGBoost) [4]. XGBoost is a supervised learning method based on DTs and improves on other methods such as RF and Gradient Boosting by using multiple optimisation methods. Like RF, XGBoost uses ensembles of DTs, but differs in using an additive strategy. In this way, each DT is trained by taking into account the residuals, the difference between the predicted value and the observed value, obtained from the previous DT and optimised using regularisation, pruning and parallel learning methods. Each subsequent DT learns from the previous trees and is not given the same weight. In the prediction process, the model output class is calculated by adding the output of each tree multiplied by a learning rate to the initial prediction. The Python package scikit-learn [20] was used to implement the ML and BoW methods.

3.2   Word Embedding and Recurrent Neural Models

The development of more complex models in recent years, has led to the intro-duction of new methods, such as word embedding, which incorporate concepts such as similarity of words and part-of-speech tagging. Word embedding is a learning technique where each word or phrase in the vocabulary is mapped to an N-dimensional vector of real numbers. Word2Vec (W2V) and fastText are two of the most commonly used methods for translating n-grams into understandable input for RNNs models.

W2V model is based on maximum likelihood and conditional probabilities, which can be seen as the probability of a word given some of the surrounding words in the corpus. The distance between two words is very close if they can substitute each other given the context. The general training of a W2V model considers a fixed window to observe a word and the rest around the word within the sentence to obtain a context. Within W2V there are two variations, the continuous Bag-of-Words (CBOW) [16] and the skip-gram [17]. The CBOW model assumes that a word is generated as a function of the words surrounding it in the text sequence. That is, the model considers the conditional probability of generating a core word based on the context words present in the window. Thus, each word in the dictionary has two vectors, one when it is used as a centre word and one when it is used as a context word. The vector associated with the context word is generally used as a representation of the document tokens in the CBOW model. The skip-gram model is similar to the CBOW model, but assumes that a core word can be used to generate the surrounding words in a text sequence. In contrast to CBOW, in skip-gram the core word is usually used as the representation of a word in the transformation of the text unit. It is important to note that both skip-gram and CBOW are self-supervised models, since the supervision comes from untagged data.

The use of n-grams is the main difference between fastText [3] and W2V. FastText operates at a granular level, where words are represented by the sum of the vectors of n-gram frames, whereas Word2Vec only learns vectors for whole words found in the training corpus. This model can produce better vector representations for rare and out-of-vocabulary words because it takes into account the shared parameters of subwords among words with similar structures. The fastText model can also be applied as a reliable text classification algorithm [9]. For this purpose, the structure of the model consists of a hidden layer and an output layer and is quite similar to that of CBOW. The fastText input consists of a sentence with embedded n-gram features averaged as a feature representation of the text. Since the number of n-grams is greater than the number of words, it is impossible to store them all. FastText divides all n-grams into buckets using the hashing track approach, so that they can share an embedding vector. The input layer is summed with the hidden layer, averaged and multiplied by a weight matrix. To produce the output of the model, the hidden layer is then multiplied by another matrix of weights. In order to apply the fastText and W2V methods as embedding models, the Python package gensim [21] was used.

The input stream processed by the embedding is comprehensible to RNNs, which are DL networks specially designed with interconnected units that form an internal memory to deal with problems of temporal structure. RNNs include the GRU [5] and the LSTM [8] networks. The main difference between the two networks is the number and functionality of their internal units. GRU consists of two gates: a reset gate, which determines the amount of past knowledge transferred to the current state; and an update gate, which determines the amount of new information added to the current state. LSTM network consists of three gates: the input gate, which derives the values used to modify the memory; the forget gate, which derives the features to discard; and the output gate, which determines the output based on the input and block memory.

In order to learn the future and past context of the input sequences, both recurrent networks can be structured to form a bidirectional model [12]. This model consists of two layers of recurrent units, GRU or LSTM. To learn the past context, one layer processes the forward sequence based on the current input and the state of the previous hidden unit. To learn the future context, the other layer processes the backward sequence based on the current input and the state of the subsequent hidden unit. The outputs of both recurrent layers are concatenated to feed the other network layers. The model implemented in this study includes a dense layer with ReLU activation function at the output of the recurrent bidirectional layer and a final dense layer with softmax activation function to infer the neoplasm assigned to the document. In addition, to further sensitise the network to the context of the sequence, two bidirectional recurrent layers (2-BiLSTM or 2-BiGRU) can be coupled.

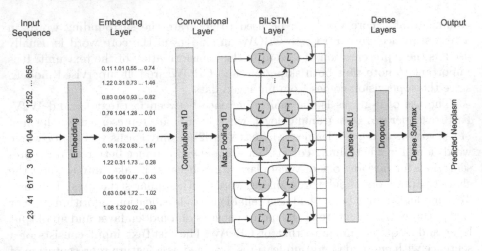

Fig. 3. The CNN + BiLSTM model structure for the text classification system.

In addition, it is possible to add a convolutional layer at the top of these recurrent models [23]. The purpose of this layer is to capture sequence information and reduce input dimensionality in order to feed the recurrent layers. The window of the convolution layer moves across the text representation to extract features, generating sequences that capture the syntax and semantics of the text. The diagram of a BiLSTM model with a convolutional layer and an embedding layer (C-BiRNN) is shown in Fig. 3. The initial sequence is processed by a tokenizer and an embedding model, transforming it into a sequence of tokens with word index values. This input sequence is fed to the embedding, which can be pre-trained or not, to obtain the word vectors that feed the 1D convolutional layer, including a max-pooling. The sequence is fed to the bidirectional recurrent layer and the result is concatenated to feed a dense layer with dropout. Finally, the output layer infers the neoplasm associated with the sequence. Tensorflow [1] package was used to implement these models.

4   Experiments

The experiments performed and the evaluation metrics used in this study are presented in this section. For the experiments, a stratified division of the data into training, validation and test sets is carried out with the data already preprocessed. Table 1 in Sect. 2 describes the result of this division, the number of documents in each set and the corresponding number of neoplasms. In view of the division of data, NLP methods (including BoW, W2V and fastText) are trained using the training set. The validation set is used in a hyperparameter fine-tuning process to achieve maximum classification performance, while the final prediction is performed on the test set. Thus, complete independence is maintained between training, parameter selection and the final prediction performance.

The metrics precision, recall and F1-score were used to evaluate the clinical text classification methods studied. The evaluation metrics are calculated using the true positive (TP), false positive (FP) and false negative (FN) values of the confusion matrix. The precision metric indicates the ratio of correctly predicted documents belonging to a neoplasm to the total number of positively predicted documents. Meanwhile, recall, also known as sensitivity or true positive rate (TPR), is the ratio of correctly predicted documents belonging to a neoplasm to the total number of documents of the actual neoplasm. Equation 1 formally defines the calculation of both metrics. The F_1 score is the harmonic mean of precision and sensitivity (Eq. 2) and provides a reliable measure of the prediction performance achieved in problems where sensitivity is important.

$$\text{Precision} = \frac{TP}{TP + FP} \qquad \text{Recall} = \frac{TP}{TP + FN} \tag{1}$$

$$\text{F1-score} = 2 \cdot \frac{\text{precision} \cdot \text{recall}}{\text{precision} + \text{recall}} = \frac{2TP}{2TP + FP + FN} \tag{2}$$

In addition, considering that we are studying a multi-class problem, the micro- and macro-average were considered [24]. Once the evaluation metrics were defined, the micro-average was calculated by summing the individual TP, FP and FN provided by the prediction for the different classes, and then calculating the precision, recall and F1-measure metrics. In contrast, the macro-average simply performs the average of each of the computed metrics. Since the neoplasm types present in Galén's corpus are slightly unbalanced, the macro-average evaluation is given greater weight.

5   Results

The experimentation process proposed above was followed, and Table 2 shows the neoplasm classification results, precision, recall and F1-score values achieved in macro-average (ovr-average) and micro-average evaluation. The best values achieved are shown in bold, while the second best values are shown in italics. Two main conclusion can be drawn from the results described in Table 2. On the one hand, according to the F1-score metric, the methods based on the application of BoW and ML used in this work obtain, on average, a lower performance. The SVM method is the only one that outperforms the others, surpassing the performance obtained with some RNNs. SVM obtains a micro- and macro-average F1-score of 0.9814 and 0.9788, respectively. On the other hand, RNN-based models with convolutional layers and fastText embedding outperform the other methods in extraction of neoplasm type. It is important to note that the same architectures outperform their versions with W2V embedding, and that RNNs with GRU units outperform LSTM units. Among all the methods, the best performance is achieved by the 2-BiGRU model when a pre-trained CBOW fastText embedding model is applied, obtaining a micro- and macro-average F1-score of 0.9840 and 0.9821, respectively.

Table 2. Micro- and Macro-averaged metrics computed on test set. For each evaluation strategy, precision (P), recall (R) and F1-score (F1) metrics are computed.

Model	micro			macro		
	P	R	F1	P	R	F1
Naive Bayes	.9683	.9654	.9668	.9618	.9652	.9634
SVM	.9814	.9814	.9814	.9768	.9808	.9788
XGBoost (DT)	.9561	.9561	.9561	.9530	.9512	.9520
XGBoost (RF)	.9776	.9776	.9776	.9751	.9749	.9750
fastText	.9814	.9793	.9804	.9790	.9772	.9781
W2V + C-BiLSTM	.9780	.9776	.9778	.9746	.9761	.9753
W2V + C-2-BiLSTM	.9768	.9759	.9764	.9737	.9717	.9727
W2V + C-BiGRU	.9823	.9819	.9821	.9805	.9778	.9791
W2V + C-2-BiGRU	.9789	.9789	.9789	.9767	.9760	.9763
FT + C-BiLSTM	.9827	.9823	.9825	.9791	*.9812*	.9801
FT + C-2-BiLSTM	.9840	*.9831*	*.9835*	**.9837**	.9788	.9812
FT + C-BiGRU	**.9848**	*.9831*	**.9840**	**.9837**	.9797	*.9817*
FT + C-2-BiGRU	*.9844*	**.9835**	**.9840**	*.9806*	**.9838**	**.9821**

Overall, the results in Table 2 do not show a clear effectiveness of obtaining the context of the sequences compared to the other methodologies in text classification. There are two possible reasons for the observed high performance of ML methods with BoW, which do not capture context. One could be the inference of the neoplasm presented in the document by key clinical concepts that are different for each neoplasm, such as the mention of a specific diagnostic test, for example a mammogram in the case of breast neoplasms. This is to be expected and is perfectly acceptable. However, another reason for this behaviour could be the inference of neoplasms by the presence of non-clinical concepts, such as the attending oncology specialist or the medical centre mentioned in the history. The inference based on the presence of these concepts is not desired, as the use of these pre-trained models in medical centres in other regions could lead to errors.

Table 3. Metrics for each neoplasm obtained by the convolutional 2-Bidirectional GRU with fastText embedding (FT + C-2-BiGRU) system on the test set.

Neoplasm	P	R	F1	Support
Breast	.9954	.9969	.9962	649
Lung	.9622	.9793	.9707	338
Colorectal	.9769	.9820	.9794	388
Other	.9878	.9769	.9823	995
micro-avg	.9844	.9835	.9840	2370
macro-avg	.9806	.9838	.9821	2370

With the aim of conducting a thorough analysis of the performance of the convolutional 2-Bidirectional GRU model with pre-trained fastText embedding (FT + C-2-BiGRU), which achieves the best neoplasm extraction results, Table 3 shows the metrics obtained for each of the neoplasms considered separately. The number of texts associated with each neoplasm in the test set is also shown, for a better evaluation of the obtained metrics. The C-2-BiGRU model performs particularly well in the classification of breast neoplasms, with an F1-score of 0.9962 and a recall of 0.9969. As the support value shows, this is the most common independent neoplasm in the Galén corpus (649 documents), which explains the higher performance. For lung and colorectal neoplasms, the results obtained are acceptable, especially for recall, where they outperform precision with values of 0.9793 and 0.9820 respectively. Apart from the F1-score, recall is the most important metric in diagnostic support systems. Taking into account that the category of other neoplasms includes neoplasms that may be related to the previous ones, especially lung and colorectal (e.g. due to the diagnostic tools and the regions of the body in which they are performed), this category performs slightly better, with a value of 0.9823 F1-score.

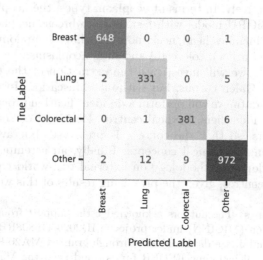

Fig. 4. Confusion matrix obtained with FT + C-2-BiGRU method on the test set.

Finally, Fig. 4 shows the confusion matrix obtained by the C-2-BiGRU model with fastText on the test set. As we can see from the matrix, the number of misclassifications for this particular complex text classification task is low. The total number of misclassified documents is 38 out of 2370 documents, giving an accuracy of 0.984.

6   Conclusions

In this paper we have addressed the problem of the extraction of neoplasm type from real-world clinical documents in Spanish. For this purpose, we have elaborated a corpus of 23, 704 medical cases annotated with the neoplasm presented by the patient, obtained from Galén [22]. The performance of ML and BoW-methods and RNN-based models applied to the text classification task has been analysed. The results obtained show that, on the one hand, BoW-based methods achieve similar results to those that consider the context of the sequences. This is probably caused by two factors: the presence of clinical concepts related to neoplasms, such as mammography and breast cancer, or the presence of non-clinical concepts, such as the names of clinical specialists who treat certain neoplasms. As the corpus was obtained from a small group of oncology centres, further analysis is needed to refute this second idea. On the other hand, bidirectional RNNs with a convolutional layer and pre-trained fastText embedding outperform the others methodologies for neoplasm type extraction. Among these RNN-based systems, the best performance is obtained by the C-2-BiGRU model with fastText, with macro-averaged precision, recall and F1-score of 0.9806, 0.9838 and 0.9821, respectively. In terms of neoplasm types, the neoplasms best classified by the C-2-BiGRU model with fastText are breast neoplasms, followed by other neoplasms (includes head/neck, non-Hodgkin's lymphoma, thyroid, stomach/esophagus and other), colorectal and lung neoplasms.

In future work, we will investigate the extraction of the type of neoplasm from the selected Galén corpus, but without obfuscation. In order to preserve the privacy of the data, we will perform a de-identification process, where private concepts (names, identifiers, medical centres, locations, etc.) will be randomly replaced by others, so that the context is preserved, but avoiding that NLP models learn from non-clinical concepts. Finally, an attempt will be made to validate the developed methodology on external real-world corpora from other Spanish medical centres, given the promising results of this work.

Acknowledgements. The authors acknowledge the support from the Ministerio de Ciencia e Innovación (MICINN) under project PID2020-116898RB-I00, from Universidad de Málaga and Junta de Andalucía through grants UMA20-FEDERJA-045 and PYC20-046-UMA (all including FEDER funds), and from the Malaga-Pfizer consortium for AI research in Cancer - MAPIC.

References

1. Abadi, M., et al.: TensorFlow: large-scale machine learning on heterogeneous systems (2015). https://www.tensorflow.org/
2. Baker, S., Korhonen, A., Pyysalo, S.: Cancer hallmark text classification using convolutional neural networks. In: Proceedings of the 5th Workshop on Building and Evaluating Resources for Biomedical Text Mining (BioTxtM), pp. 1–9 (2016)
3. Bojanowski, P., Grave, E., Joulin, A., Mikolov, T.: Enriching word vectors with subword information. Trans. Assoc. Comput. Linguist. **5**, 135–146 (2017). https://doi.org/10.1162/tacl_a_00051

4. Chen, T., Guestrin, C.: Xgboost: a scalable tree boosting system. In: Proceedings of the 22nd ACM SIGKDD International Conference on Knowledge Discovery and Data Mining, pp. 785–794 (2016). https://doi.org/10.1145/2939672.2939785

5. Chung, J., Gulcehre, C., Cho, K., Bengio, Y.: Empirical evaluation of gated recurrent neural networks on sequence modeling. In: NIPS 2014 Workshop on Deep Learning, December 2014 (2014)

6. Garla, V., Taylor, C., Brandt, C.: Semi-supervised clinical text classification with Laplacian SVMs: an application to cancer case management. J. Biomed. Inform. **46**(5), 869–875 (2013). https://doi.org/10.1016/j.jbi.2013.06.014

7. Hadi, W., Al-Radaideh, Q.A., Alhawari, S.: Integrating associative rule-based classification with naïve bayes for text classification. Appl. Soft Comput. **69**, 344–356 (2018). https://doi.org/10.1016/j.asoc.2018.04.056

8. Hochreiter, S., Schmidhuber, J.: Long short-term memory. Neural Comput. **9**(8), 1735–1780 (1997). https://doi.org/10.1162/neco.1997.9.8.1735

9. Joulin, A., Grave, E., Bojanowski, P., Mikolov, T.: Bag of tricks for efficient text classification. arXiv preprint arXiv:1607.01759 (2016)

10. Kasthurirathne, S.N., et al.: Toward better public health reporting using existing off the shelf approaches: the value of medical dictionaries in automated cancer detection using plaintext medical data. J. Biomed. Inform. **69**, 160–176 (2017). https://doi.org/10.1016/j.jbi.2016.01.008

11. Khadhraoui, M., Bellaaj, H., Ammar, M.B., Hamam, H., Jmaiel, M.: Survey of BERT-base models for scientific text classification: COVID-19 case study. Appl. Sci. **12**(6), 2891 (2022). https://doi.org/10.3390/app12062891

12. Lample, G., Ballesteros, M., Subramanian, S., Kawakami, K., Dyer, C.: Neural architectures for named entity recognition. arXiv preprint arXiv:1603.01360 (2016)

13. Liu, G., Guo, J.: Bidirectional LSTM with attention mechanism and convolutional layer for text classification. Neurocomputing **337**, 325–338 (2019). https://doi.org/10.1016/j.neucom.2019.01.078

14. López-García, G., Jerez, J.M., Ribelles, N., Alba, E., Veredas, F.J.: Detection of tumor morphology mentions in clinical reports in Spanish using transformers. In: Rojas, I., Joya, G., Català, A. (eds.) IWANN 2021. LNCS, vol. 12861, pp. 24–35. Springer, Cham (2021). https://doi.org/10.1007/978-3-030-85030-2_3

15. Mendonça, E.A., Haas, J., Shagina, L., Larson, E., Friedman, C.: Extracting information on pneumonia in infants using natural language processing of radiology reports. J. Biomed. Inform. **38**(4), 314–321 (2005). https://doi.org/10.1016/j.jbi.2005.02.003

16. Mikolov, T., Chen, K., Corrado, G., Dean, J.: Efficient estimation of word representations in vector space. arXiv preprint arXiv:1301.3781 (2013)

17. Mikolov, T., Sutskever, I., Chen, K., Corrado, G.S., Dean, J.: Distributed representations of words and phrases and their compositionality. In: Advances in Neural Information Processing Systems, vol. 26 (2013)

18. Moschitti, A., Basili, R.: Complex linguistic features for text classification: a comprehensive study. In: McDonald, S., Tait, J. (eds.) ECIR 2004. LNCS, vol. 2997, pp. 181–196. Springer, Heidelberg (2004). https://doi.org/10.1007/978-3-540-24752-4_14

19. Nguyen, A.N., et al.: Symbolic rule-based classification of lung cancer stages from free-text pathology reports. J. Am. Med. Inform. Assoc. **17**(4), 440–445 (2010). https://doi.org/10.1136/jamia.2010.003707

20. Pedregosa, F., et al.: Scikit-learn: machine learning in Python. J. Mach. Learn. Res. **12**, 2825–2830 (2011)

21. Řehůřek, R., Sojka, P.: Software framework for topic modelling with large corpora. In: Proceedings of the LREC 2010 Workshop on New Challenges for NLP Frameworks, Valletta, Malta, pp. 45–50. ELRA (2010). http://is.muni.cz/publication/884893/en

22. Ribelles, N., et al.: Galén: Sistema de información para la gestión y coordinación de procesos en un servicio de oncología. RevistaeSalud **6**(21), 1–12 (2010)

23. Shi, X., Chen, Z., Wang, H., Yeung, D.Y., Wong, W.K., Woo, W.C.: Convolutional LSTM network: a machine learning approach for precipitation nowcasting. In: Advances in Neural Information Processing Systems, vol. 28 (2015)

24. Sokolova, M., Lapalme, G.: A systematic analysis of performance measures for classification tasks. Inf. Process. Manag. **45**(4), 427–437 (2009). https://doi.org/10.1016/j.ipm.2009.03.002

25. Sparck Jones, K.: A statistical interpretation of term specificity and its application in retrieval. J. Doc. **28**(1), 11–21 (1972). https://doi.org/10.1108/eb026526

26. St-Maurice, J., Kuo, M.H., Gooch, P.: A proof of concept for assessing emergency room use with primary care data and natural language processing. Methods Inf. Med. **52**(01), 33–42 (2013). https://doi.org/10.3414/ME12-01-0012

27. Urda, D., Ribelles, N., Subirats, J.L., Franco, L., Alba, E., Jerez, J.M.: Addressing critical issues in the development of an oncology information system. Int. J. Med. Inform. **82**(5), 398–407 (2013). https://doi.org/10.1016/j.ijmedinf.2012.08.001

28. Vaswani, A., et al.: Attention is all you need. In: Advances in Neural Information Processing Systems, vol. 30 (2017)

29. Venkataraman, G.R., et al.: Fastag: automatic text classification of unstructured medical narratives. PLoS ONE **15**(6), e0234647 (2020). https://doi.org/10.1371/journal.pone.0234647

30. Vítores, D.F.: El español: una lengua viva. Informe 2019. Instituto Cervantes (2019). https://www.cervantes.es/imagenes/File/espanol_lengua_viva_2019.pdf

31. Wallach, H.M.: Topic modeling: beyond bag-of-words. In: Proceedings of the 23rd International Conference on Machine Learning, pp. 977–984 (2006). https://doi.org/10.1145/1143844.1143967

32. Wang, R., Li, Z., Cao, J., Chen, T., Wang, L.: Convolutional recurrent neural networks for text classification. In: 2019 International Joint Conference on Neural Networks (IJCNN), pp. 1–6. IEEE (2019). https://doi.org/10.1109/ijcnn.2019.8852406

33. Yao, L., Mao, C., Luo, Y.: Clinical text classification with rule-based features and knowledge-guided convolutional neural networks. BMC Med. Inform. Decis. Mak. **19**(3), 31–39 (2019). https://doi.org/10.1186/s12911-019-0781-4

COVID-19 Advances in Bioinformatics and Biomedicine

Stochastic Model of Infection
with the SARS–COV–2 Virus in a Small Group
of Individuals Indoors

Derevich Igor$^{(\boxtimes)}$ (iD) and Panova Anastasiia (iD)

Faculty of Fundamental Sciences, Department of Applied Mathematics, Bauman Moscow State Technical University (National Research University), 2-Ya Baumanskaya Str. 5, Moscow 105005, Russian Federation
DerevichIgor@bmstu.ru

Abstract. A mathematical model of viral infection of a small group of people randomly moving indoors is proposed. The model consists of two different modules. In the first module, based on the standard three-stage cell model, we modified the initial stage of infection. Firstly, the initial degree of immunity is taken into account. Secondly, it is assumed that there is a critical concentration of the pathogen, starting from which there is an active infection of the hosts cells of the human organism. Thirdly, an additional flux of virions from the local atmosphere near the exposed individual is included. The second module presents a model of random movement of people indoors with obstacles. The desired velocity of human consists of a random component and a deterministic velocity that occurs during evacuation from the room. The source of the chaotic desired velocity is modeled by a random Gaussian color process. The actual velocity of individuals is a consequence of the social behavior of the group indoor with obstacles. Physical contacts between individuals and obstacles are described on the basis of effective potential. The numerical study of both the dynamics of the pathogen in an organism in an atmosphere with a random flow of virions and the chaotic movement of individuals indoors is modeled on the basis of a system of stochastic ordinary differential equations.

Keywords: SARS–COV–2 Virus · Stochastic Ordinary Differential Equation · Random Process · Concentration of Virions · Concentration of Pathogen Cells · Effective Potential

1 Introduction

COVID-19 viral infection poses a significant danger to the human organism [1, 2]. The emergence of new strains of viral infections that cause severe consequences in the human organism of an infected person leads to the need to strengthen sanitary and epidemiological standards. However, observance of these rules in the case of real behavior of a group of individuals is not always carried out. This is most clearly manifested when a group of individuals accidentally moves in places of entertainment, in supermarkets

I. Rojas et al. (Eds.): IWBBIO 2023, LNBI 13919, pp. 499–513, 2023.
https://doi.org/10.1007/978-3-031-34953-9_39

and is disrupted during evacuation in a panic. In sanitary and epidemiological norms, a certain critical distance between individuals is determined, with a decrease in which the probability of infection rises. In addition to the critical distance, an important factor is the exposure time of a susceptible person within the radius of an active infection near an infected individual. Of practical interest is the development of mathematical models simulating the initial stage of infection in small groups of individuals that include infected people [3, 4]. The spread of infection occurs by airborne droplets as a result of absorption by the lungs of susceptible members of microdrops exhaled by an infected individual [5–7]. As a result of the random movement of individuals, the concentration of virions in the local atmosphere is random. A modern effective tool for assessing the probability of infection in a group of individuals with infected members is mathematical simulation of both the movement of individuals in various indoor conditions and modeling of the initial stage of infection. In addition, an important practical challenge is to predict the development of the disease after infected individuals leave the critical zone.

The developed model consists of two modules. In the first module, we modified the standard SARS–COV–2 infection model [8–13]. First, we introduced the initial degree of immunity of the organism, assuming that a high level of immunity at the initial stage of infection reduces the probability of an attack by the pathogen of the host cells of the human organism. This mechanism is used to block a new infection when developing promising vaccines [12, 13]. Secondly, it is assumed that there is a critical concentration of the pathogen, starting from which there is an explosive generation of new pathogen cells from infected host cells. We simulate a situation when there are infected and susceptible people in the group. COVID-19 cells enter the human organism by inhaling microdroplets containing virions [14, 15]. Therefore, the third modification is associated with the inclusion of an additional term representing the flux of the pathogen into the human organism in the equation for the concentration of pathogen cells. In the local atmosphere, there is a critical concentration of virions, the excess of which leads to the development of infection in the human organism. We assume that at a relative distance less than the critical one, the concentration of virions in the local atmosphere will be exceeds the critical value from which the explosive growth of the concentration of pathogen cells in the human organism will begin.

The chaotic movement of individuals in the room leads to a random local concentration of virions. We model the concentration of virions in the local atmosphere using a logarithmically normal random process. The system of stochastic ordinary differential equations (SODE) for concentrations of pathogen cells, virus-infected host cells and random virions concentration in the local atmosphere is integrated numerically on the basis of modernized Runge-Kutta algorithms [16].

Hypotheses about the initial immunity and the critical concentration of pathogen cells allow us to describe the duration of the latent phase without involving additional empirical information [14]. There is a satisfactory correspondence between the experimental data and the results of modeling the concentration of the pathogen.

The second module is devoted to modeling the chaotic movement of a small group of people in a room with obstacles. Modern models of social behavior (social force models [17, 18], models of cellular automata [19, 20]) mimic the behavior of large groups of people (several hundred individuals) during the evacuation in a panic. In these models, a compressed mass of people who are in close physical contact moves. While real films and photographic documents show that even during the evacuation, the members of the group strive to maintain a social distance [21]. We propose a new model based on the ideas of molecular dynamics and the modern theory of random processes [22]. The desired velocity of a person is the sum of random and deterministic components that arise during evacuation from a room [23]. The chaotic component of the desired velocity is modeled by a random process structured in time (color noise). The actual velocity of an individual is the result of the interaction of all members of the group with each other and with obstacles. The repulsive forces that arise when people approach each other or the walls, we model on the basis of effective potential.

A system of SODE simulating the dynamics of movement of each member of the group is also integrated using modernized Runge-Kutta algorithms [24]. The calculation results illustrate the dynamics of relative distances between people during chaotic movement indoors without evacuation and during evacuation in a panic. The qualitative difference in evacuation from a room with various internal obstacles is investigated. If there are infected members in the group, the simulation results allow us, in principle, estimate the probability of the spread of viral infection. The developed approach can be useful in assessing the consequences of a bioterrorist attack.

2 Mathematical Modelling of SARS-CoV-2 Infection of Human Host Cells

2.1 Basic Equations of the Standard Model

A three-stage cellular model of infection with the SARS-CoV-2 virus has been proposed in the literature [8–10]. The model is widely used to predict the development of various strains of the COVID-19 virus, modeling the action of inhibitors of the growth of the concentration of pathogenic cells [11–14]. We will refer to this model in the text as a standard one.

At the beginning, we present a slightly modified formulation of the standard cellular model of SARS-Cov-2 infection. Pathogen cells infect the host cells, which become sources of new pathogen cells. At the same time, the number of affected cells of the organism is consumed. The system of equations describing the infection process looks like this

$$
\begin{aligned}
\frac{dX}{dt} &= \frac{1}{T_X}Y - \frac{1}{\tau_X}X, \, X(0) = X_0, \\
\frac{dY}{dt} &= \frac{1}{T_Y}\frac{Z}{Z_0}X - \frac{1}{\tau_Y}Y, \, Y(0) = Y_0, \\
\frac{dZ}{dt} &= -\frac{1}{T_Y}\frac{Z}{Z_0}X - \frac{1}{\tau_Z}Z, \, Z(0) = Z_0.
\end{aligned}
\tag{1}
$$

Here X, Y, Z are the concentrations of pathogen cells, infected cells and host cells attacked by the virus; T_X, T_Y are characteristic generation times of pathogen cells and virus infected cells; τ_X, τ_Y, τ_Z are characteristic lifetimes of pathogen cells, infected cells and host cells; X_0, Y_0, Z_0 are initial values of concentrations.

In the system of Eqs. (1) we proceed to dimensionless concentrations. The dimensionless concentrations of the virus, infected and susceptible cells are equal

$$X^* = \frac{X}{X_{cr}}, \qquad Y^* = \frac{Y}{X_{cr}}, \qquad Z^* = \frac{Z}{Z_0}.$$

Here X_{cr} is the critical value of the virion concentration in the human organism, starting from which the concentration of the pathogen increases. The actual estimate of the critical concentration value will be given below.

As a timescale we choose the characteristic lifetime of pathogen cells in the organism. Dimensionless time is defined as $t^* = t / \tau_X$.

Equations (1) for dimensionless variables take the form

$$\frac{dX^*(t^*)}{dt^*} = \frac{1}{T_X^*} Y^*(t^*) - X^*(t^*), \ X^*(0) = X_0^*,$$

$$\frac{dY^*(t^*)}{dt^*} = \frac{1}{\tau_Y^*} \left\{ \frac{\tau_Y^*}{T_Y^*} Z^*(t^*) X^*(t^*) - Y^*(t^*) \right\}, \ Y^*(0) = Y_0^*, \qquad (2)$$

$$\frac{dZ^*(t^*)}{dt^*} = -\frac{1}{T_Y^* Z_0^*} Z^*(t^*) X^*(t^*) - \frac{1}{\tau_Z^*} Z^*(t^*), \ Z^*(0) = 1.$$

Here $T_X^* = T_X / \tau_X, T_Y^* = T_Y / \tau_Y, \tau_Y^* = \tau_Y / \tau_X$ are dimensionless characteristic times; $Z_0^* = Z_0 / X_{cr}$ is the ratio of the initial concentration of host cells to the critical level of pathogen concentration, $Z_0^* \gg 1$.

The characteristic time of decrease in the concentration of host cells significantly exceeds the lifetime of infected cells. In this case, the quasi-steady approximation is correct

$$Y^*(t^*) \approx \frac{\tau_Y^*}{T_Y^*} Z^*(t^*) X^*(t^*).$$

As a result, the system of Eqs. (2) is simplified

$$\frac{dX^*(t^*)}{dt^*} = \frac{1}{T_X^*} \frac{\tau_Y^*}{T_Y^*} Z^*(t^*) X^*(t^*) - X^*(t^*),$$

$$\frac{dZ^*(t^*)}{dt^*} = -\frac{1}{T_Y^* Z_0^*} Z^*(t^*) X^*(t^*) - \frac{1}{\tau_Z^*} Z^*(t^*).$$

It can be seen that the dynamics of infection depends on three dimensionless parameters. According to experimental studies, these parameters have been determined

$$B_Z = \frac{1}{T_Y^* Z_0^*} = \frac{\beta}{\delta}, \ \Gamma_X = \frac{1}{T_X^*} \frac{\tau_Y^*}{T_Y^*} = \frac{\gamma}{\delta}.$$

Here the numerical values of the constants γ, β, δ are borrowed from [13]. As a result, we obtain the equations of the standard model in dimensionless form

$$\frac{dX^*(t^*)}{dt^*} = \Gamma_X Z^*(t^*) X^*(t^*) - X^*(t^*),$$

$$\frac{dZ^*(t^*)}{dt^*} = -B_Z Z_0^* Z^*(t^*) X^*(t^*) - \frac{1}{\tau_Z^*} Z^*(t^*).$$

2.2 Modified Model

In the modified model, we take into account the initial immunity of the organism, the presence of a critical concentration of pathogen cells, starting from which there is an explosive increase in the concentration of the pathogen in the organism and the flux of virions from the atmosphere (Fig. 1).

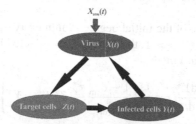

Fig. 1. A sketch of the modernized cellular model of SARS-CoV-2. Pathogen cells entering the organism from the atmosphere and attack the host cells of the human organism. Infected host cells generate new pathogen cells. We assume that active infection of host cells of the organism begins when the concentration of the pathogen in the organism exceeds a certain critical value X_{cr}.

A modified system of equations for calculation the concentration of pathogen cells, the concentration of cells infected with the virus, and the concentration of host cells of the organism has the dimensionless form

$$\frac{dX^*(t^*)}{dt^*} = \frac{1}{T_X^*}Y^*(t^*) + \frac{X_{atm}^*(t^*)}{T_{in}^*} - X^*(t^*),$$

$$\frac{dY^*(t^*)}{dt^*} = \frac{1}{\tau_Y^*}\left\{\frac{\tau_Y^*}{T_Y^*}Z^*(t^*)\left(\frac{X^*(t^*)}{1+\alpha_{im}^*X^*(t^*)} - 1\right)X^*(t^*) - Y^*(t^*)\right\},$$

$$\alpha_{im}^* = \frac{\Gamma_X}{1+\Gamma_X}\alpha_{im},$$

$$\frac{dZ^*(t^*)}{dt^*} = -\frac{1}{T_Y^*Z_0^*}Z^*(t^*)X^*(t^*) - \frac{1}{\tau_Z^*}Z^*(t^*).$$

(3)

Here α_{im} the degree of initial immunity $0 \le \alpha_{im} \le 1$, absolute immunity corresponds to the value $\alpha_{im} = 1$ the absence of immunity $\alpha_{im} = 0$; $X_{atm}^* = X_{atm}/X_{cr}$ is virion concentration in the local atmosphere; $T_{in}^* = T_{in}/\tau_X$ the characteristic time of transport of absorbed virions to the attacked hosts cells of the organism.

In the absence of immunity $\alpha_{im} = 0$, the equation for infected cells of the organism has the form

$$\frac{dY^*}{dt^*} = \frac{1}{\tau_Y^*}\left\{\frac{\tau_Y^*}{T_Y^*}Z^*(X^* - 1)X^* - Y^*\right\}.$$

From this equation it can be seen that the active growth of the concentration of infected host cells of the organism begins when the concentration of pathogenic cells exceeds a critical value $X > X_{cr}$. We define a critical concentration X_{cr} as the concentration of a pathogen with zero immunity $\alpha_{im} = 0$, starting from which the virus actively attacks the target cells of the organism. All concentrations in dimensionless form are normalized to this value. With an increase in the degree of immunity $\alpha_{im} > 0$, the level of the initial concentration of virions in the organism also increases. In the feather text, we will use the term "critical pathogen concentration" for any value of the immunity parameter $0 < \alpha_{im} < 1$, meaning by this the dimensionless value of the virion concentration X^*, above which the active replication of the virus occurs in the affected cells of the organism.

With a non-zero value of the initial degree of immunity $\alpha_{im} \neq 0$ at high concentrations of the pathogen $X^*(t) >> 1$, the equation for infected host cells (3) is similar to the equation of the standard model

$$\frac{dY^*}{dt^*} = \frac{1}{\tau_Y^*}\left\{\frac{1 - \alpha_{im}^*}{\alpha_{im}^*}\frac{\tau_Y^*}{T_Y^*}Z^*X^* - Y^*\right\}.$$

The equation of generation of infected cells (3) includes a term that takes into account the initial level of immunity and the critical concentration. We use the quasi-stationary hypothesis in Eq. (3) and obtain the final system of equations for the concentration of the pathogen in the organism and the concentration of host cells

$$\frac{dX^*(t^*)}{dt^*} = \Gamma_X Z^*(t^*)\left(\frac{X^*(t^*)}{1 + \alpha_{im}^* X^*(t^*)} - 1\right)X^*(t^*) + \frac{X_{atm}^*(t^*)}{T_{in}^*} - X^*(t^*), \quad X^*(0) = X_0^*,$$

$$(4)$$

$$\frac{dZ^*(t^*)}{dt^*} = -B_Z Z_0^* Z^*(t^*)X^*(t^*) - \frac{1}{\tau_Z^*}Z^*(t^*), \quad Z^*(0) = Z_0^*. \tag{5}$$

Analysis of the results of calculations of the system of Eqs. (4) and (5) shows that at the stage of active growth of the pathogen concentration ($Z^*(t) \approx 1$), the system of Eqs. (4), (5) can be reduced to one equation

$$\frac{dX^*(t^*)}{dt^*} = \Gamma_X\left\{\frac{X^*(t^*)}{1 + \alpha_{im}^* X^*(t^*)} - 1\right\}X^*(t^*) + \frac{X_{atm}^*(t^*)}{T_{in}^*} - X^*(t^*). \tag{6}$$

It follows from Eq. (6) that in the absence of pathogens in the atmosphere $X_{atm}^*/T_{in}^* = 0$ and absolute immunity $\alpha_{im} = 1$, any initial concentration of pathogen in the organism will degenerate. It can be said that with absolute immunity, the critical initial concentration of the pathogen tends to infinity.

Next, we will be interested in the initial stage of infection, which is satisfactorily described by Eq. (6). Equation (6) has an analytical solution that is correct at the initial stage of infection. From this solution, critical values of the initial virion concentration in the organism and the critical value of the virion flux from the atmosphere are found. If the flux of virions from the atmosphere exceeds the critical value, then there is a sharp increase in the concentration of pathogen cells in the organism. To save the volume of the article, we do not provide this solution.

2.3 Analysis of the Modified Model

In the first part, we will consider the case of the absence of virions in the atmosphere and a given initial concentration of the pathogen in the organism. If the initial concentration of the pathogen is below the critical value, then the virus will degenerate in the organism.

With an increase in the initial degree of immunity, the critical initial concentration of the pathogen in the organism increases (Fig. 2a). If the initial concentration exceeds the critical value, then there is a significant increase in the concentration of the pathogen in the organism (Fig. 2b). Figure 2 shows that Eq. (6) satisfactorily presents the initial stage of infection. The modified model also describes the latency period, which decreases with an increase in the initial concentration of the pathogen in the organism.

To explain the latent period, various phases of damage to the body's cells are presented in the literature [8, 14]. The inclusion of a new mechanism of infection of host cells of the organism requires the evaluation of new constants. These constants are evaluated empirically.

Innate immunity also cannot be unambiguously assessed [12, 25]. Innate immunity is the result of many individual factors of the human body and can only be considered as a random variable.

We do not take into account the concentration of antibodies, because we are interested in the initial period of infection, when the concentration of antibodies is low.

In the second part, we suggest that the initial concentration of the pathogen in the organism is zero, but there is a constant flux of virions from the local atmosphere.

Figure 3 shows that there is a critical value of the virion flux from the atmosphere, below which a constant concentration of the pathogen in the organism is established. Above critical value of the virion fluxes sharp increase in the concentration of pathogenic cells is observed. The initial stage of infection is satisfactorily described by Eq. (6).

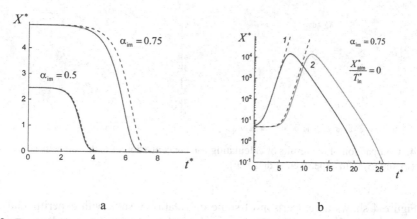

a   b

Fig. 2. Degeneration of the virus at the initial concentration of the pathogen below the critical value (a). The explosion rises of concentration of pathogen at the initial concentration above the critical value (b). Solid curves represent the calculation according to the system of Eqs. (4) and (5), dotted lines indicate the calculation result with Eq. (6).

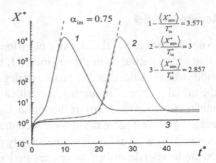

Fig. 3. Dynamics of pathogen concentration in the human organism with constant flux of virions from the atmosphere. Solid curves represent the results of calculations of the system (4) and (5). Dashed curves denote the calculation in accordance with Eq. (6). On curve 3 the calculation results for both methods are almost the same.

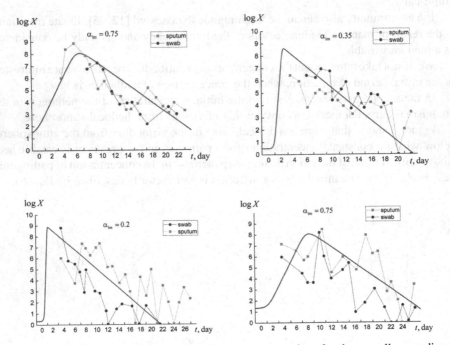

Fig. 4. Comparison of the results of calculating the concentration of pathogen cells according to Eqs. (4) and (5) with experimental data [26].

Figure 4 shows a comparison of some calculation results with experimental data. The constants in Eqs. (4) and (5) were estimated on the base of standard model. In the calculations, we modify the degree of initial immunity and the initial value of the pathogen concentration in the human organism.

The critical concentration of virions, exceeding which leads to intense damage to host cells, currently ranges from 0.1 to 10 copies/ml. During calculations, the critical concentration was set equal $X_{cr} = 1$ copies/ml.

It follows from the results of comparison that the modified model satisfactory reflects the real dynamics of the concentration of pathogen cells during the disease.

2.4   Fluctuations in the Concentration of Virions in the Local Atmosphere

Due to chaotic physical contact between people, among whom there are infected, the local concentration of pathogen cells in the atmosphere changes randomly. We model the flux of virions from the local atmosphere as a logarithmically normal random process

$$\frac{X_{atm}^*(t^*)}{T_{in}^*} = \left\langle \frac{X_{atm}^*}{T_{in}^*} \right\rangle \exp\left[\Xi^*(t^*)\right]. \tag{7}$$

Here $\left\langle X_{atm}^*/T_{in}^* \right\rangle = $ const and angle brackets denote averaging over an ensemble of realizations; $\Xi^*(t^*)$ is color random Gaussian process with autocorrelation function

$$\left\langle \Xi^*\left(t^{*\prime}\right)\Xi^*\left(t^{*\prime\prime}\right) \right\rangle = \left\langle \Xi^{*2} \right\rangle \Psi_\Xi\left(t^{*\prime\prime} - t^{*\prime}\right).$$

Here $\left\langle \Xi^{*2} \right\rangle$ is dispersion a random process; $\Psi_\Xi(t^*)$ is autocorrelation function.

Numerical simulation of a random process is carried out in the form of a solution of a stochastic equation

$$\frac{d\Xi^*(t^*)}{dt^*} = \frac{1}{T_{atm}^*}\left(\eta^*(t^*) - \Xi^*(t^*)\right). \tag{8}$$

Here T_{atm}^* is the integral time macroscale of autocorrelation $\Psi_\Xi(t^*)$; $\eta(t^*)$ is a source of fluctuations (white noise) modeled by a random Gaussian process with an autocorrelation function

$$\left\langle \eta^*\left(t^{*\prime}\right)\eta^*\left(t^{*\prime\prime}\right) \right\rangle = 2\tau_0^*\left\langle \eta^{*2} \right\rangle \delta\left(t^{*\prime\prime} - t^{*\prime}\right),$$

where τ_0^* is time microscale, $\tau_0^* \ll T_{atm}^*$; $\left\langle \eta^{*2} \right\rangle$ is dispersion of the source of fluctuations; $\delta(t^*)$ is Dirac delta function.

Integral time scale T_{atm}^* has the order of the time of existence of a local zone of increased concentration of individuals during their random movement.

Stochastic Eq. (8) gives the dispersion of a random process $\Xi^*(t^*)$ and its autocorrelation function

$$\left\langle \Xi^{*2} \right\rangle = \frac{\tau_0^*}{T_{atm}^*}\left\langle \eta^{*2} \right\rangle, \quad \Psi_\Xi(t^*) = \exp\left(-\frac{t^*}{T_{atm}}\right).$$

The average value of the virion flux from the atmosphere follows from the Eq. (7)

$$\left\langle \frac{X_{atm}^*(t^*)}{T_{in}^*} \right\rangle = \left\langle \frac{X_{atm}^*}{T_{in}^*} \right\rangle \exp\left(\frac{\left\langle \Xi^{*2} \right\rangle}{2}\right).$$

The system of SODE (6) – (8) is integrated on the basis of modernized numerical algorithms of the Runge - Kutta type [16]. Below are presented illustrations of the effect of random flux of virions on the dynamics of pathogen cells in the human organism.

Figure 5 shows two different infection scenarios when virions are in the atmosphere. With a constant flux of virions, less than the critical value, a stationary concentration of pathogen cells is established in the organism. When flux from the atmosphere is exceeded the critical value, an explosive increase of pathogen cells occurs. We observe, that fluctuations in the virion flux in the local atmosphere dramatically change the dynamics of infection.

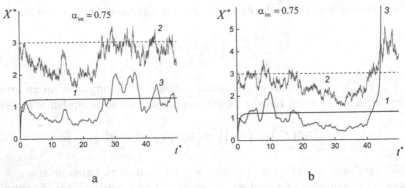

Fig. 5. The dependence of the concentration of pathogen cells in the organism on time in a random atmosphere. Panel (a) shows scenarios without explosion, panel (b) with explosion. Curve 1 indicates the dynamics of the pathogen concentration with a constant flow of virions from the atmosphere. Curve 2 represents random fluctuations of the virion flux; the dotted line represents the average value of the virion flux. Curve 3 shows the dynamics of the pathogen concentration in the organism in a random atmosphere.

Figure 5 shows two examples of significantly different realization of infection dynamics, provided that the average virion flux from the atmosphere is below the critical value. On Fig. 5a during the time under consideration the level of pathogen concentration in the organism does not reach a critical value and no significant increase in concentration is observed. On Fig. 5b the instantaneous flux of virions from the atmosphere has exceeded the critical value and there is a sharp increase in the concentration of the pathogen in the organism.

Figure 6 illustrates the effect of virion flux fluctuations when the average value of the flux exceeds the critical level. It can be seen that fluctuations in the flow have both an inhibitory (Fig. 6a) and an intensifying (Fig. 6b) effect on the dynamics of infection.

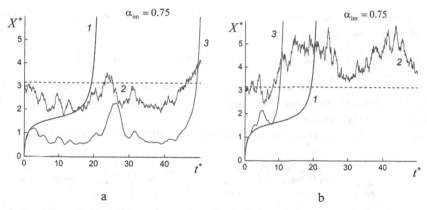

a   b

Fig. 6. The effect of virion flux fluctuations on the dynamics of infection with an averaged flux is higher than the critical value. Panel (a) shows an example of a random trajectory of suppression of the onset of explosive growth of the pathogen concentration. Panel (b) shows an example of a random trajectory with a significant intensification of an explosive increase in the concentration of the pathogen in the organism.

3   A Model of Chaotic Movement of a Group of People Indoors

3.1   Basic Equations

Here we will present a model of chaotic movement of a small group in a room with obstacles. Unlike the popular model of social force, close physical contact between individuals is impossible in our model.

The random desired velocity is a time-structured random process. The integral time scale of fluctuations in the desired velocity reflects the effect of memory of decisions made earlier by the individual about the direction and velocity of movement.

In our dynamic model, the social behavior of individuals and their interaction with walls is modeled on the basis of effective potential, which increases sharply when individuals approach each other or approach borders. The formulated hypotheses allow us to write down the equations of the dynamics of individual α in the form

$$\frac{d\mathbf{V}^{(\alpha)}}{dt} = \frac{1}{\tau^{(\alpha)}}\left(\mathbf{V}_{\text{ini}}^{(\alpha)} + \mathbf{W}^{(\alpha)} - \mathbf{V}^{(\alpha)}\right) - \frac{1}{m^{(\alpha)}}\frac{\partial U_{\text{eff}}^{(\alpha)}(\mathbf{r})}{\partial \mathbf{r}}\bigg|_{\mathbf{r}=\mathbf{R}^{(\alpha)}}, \quad \frac{d\mathbf{R}^{(\alpha)}}{dt} = \mathbf{V}^{(\alpha)}. \quad (9)$$

Here $\mathbf{V}^{(\alpha)}$, $\mathbf{R}^{(\alpha)}$ are random velocity and radius-vector of an individual; $m^{(\alpha)}$, $\tau^{(\alpha)}$ are mass and characteristic relaxation time to the desired velocity; $\mathbf{V}_{\text{ini}}^{(\alpha)}$ is random component of the desired velocity; $\mathbf{W}^{(\alpha)}$ is deterministic drift during evacuation; $U_{\text{eff}}(\mathbf{x})$ is effective potential that takes into account the collective behavior of individuals in a group and their interaction with the boundaries and internal obstacles.

In this section we model the movement of individuals in physical variables without passing to dimensionless quantities. We do not take into account the forces of "friction" that arise when people come into close contact with each other. We investigate relatively small groups of people, when a crowd of people is unlikely even in the case of a panic. The effective potential has the form

$$U_{\text{wall}}(\Delta \mathbf{R}, \delta) = A_{\text{wall}} \exp\{(\delta/|\Delta \mathbf{R}|)^\sigma - (|\Delta \mathbf{R}|/\delta)^\sigma\}, \tag{10}$$

where $\Delta \mathbf{R}$ is vector of the shortest distance between the surface of a person and an obstacle or between the surfaces of two people; δ is characteristic distance between surfaces; $\sigma > 0$ is constant.

In the effective potential, we summarize the potentials of local interaction

$$U_{\text{eff}}^{(\alpha)}(\mathbf{r}) = \sum_{\beta \neq \alpha}^{N_0} U_{\text{wall}}\left(\Delta \mathbf{R}_{\text{wall}}^{(\alpha\beta)}, \delta\right) + \sum_{\text{coln}} U_{\text{wall}}\left(\Delta \mathbf{R}_{\text{coln}}^{(\alpha)}, \delta\right) + \sum_{\text{wall}} U_{\text{wall}}\left(\Delta \mathbf{R}_{\text{wall}}^{(\alpha)}, \delta\right). \tag{11}$$

Here $\Delta \mathbf{R}_{\text{wall}}^{(\alpha\beta)}$ is a distance between the surfaces of individuals α, β; $\Delta \mathbf{R}_{\text{coln}}^{(\alpha)}$ is a distance between the individual and the surface of the column; $\Delta \mathbf{R}_{\text{wall}}^{(\alpha)}$ is a distance between the individual and the border.

During evacuation the deterministic velocity $\mathbf{W}^{(\alpha)}$ is directed to the exit from the room. We model a random component of the desired velocity by color noise as a solution of the stochastic equation

$$\frac{d\mathbf{V}_{\text{ini}}^{(\alpha)}(t)}{dt} = \frac{1}{T^{(\alpha)}}\left[\Xi^{(\alpha)}(t) - \mathbf{V}_{\text{ini}}^{(\alpha)}(t)\right]. \tag{12}$$

Here $T^{(\alpha)}$ is integral time scale of velocity fluctuations; $\Xi^{(\alpha)}(t)$ is source of velocity fluctuation (white noise) generated by a computer.

The correlation of random velocities of the source of fluctuations has the form

$$\left\langle \Xi^{(\alpha)}(t')\Xi^{(\alpha)}(t'')\right\rangle = 2\tau_0 \hat{\mathbf{I}}\left\langle \Xi^{(\alpha)2}\right\rangle \delta(t'' - t'),$$

where τ_0 is time microscale; $\langle \Xi^{(\alpha)2}\rangle$ is dispersion of the source of fluctuations; $\hat{\mathbf{I}}$ is unit matrix.

At $\tau_0 << T^{(\alpha)}$ the random process of desired velocity allows us to describe fluctuations reflecting the effect of memory of decisions made earlier by an individual.

The correlation of fluctuations of the desired velocity has the form

$$\left\langle \mathbf{V}_{\text{ini}}^{(\alpha)}(t')\mathbf{V}_{\text{ini}}^{(\alpha)}(t'')\right\rangle = \hat{\mathbf{I}}\left\langle V_{\text{ini}}^{(\alpha)2}\right\rangle \Psi^{(\alpha)}(t'' - t'),$$

where $\langle V_{\text{ini}}^{(\alpha)2}\rangle$ is the dispersion of the velocity fluctuations; $\Psi^{(\alpha)}(t)$ is the autocorrelation function.

The dispersion and autocorrelation function of the random process $\mathbf{V}_{\text{ini}}^{(\alpha)}(t)$ are equal

$$\left\langle V_{\text{ini}}^{(\alpha)2}\right\rangle = \frac{\tau_0}{T^{(\alpha)}}\left\langle \Xi^{(\alpha)2}\right\rangle, \quad \Psi^{(\alpha)}(t) = \exp(-t/T^{(\alpha)}).$$

The SODE system (9) – (12) is integrated numerically by the Runge-Kutta method, modified to solve stochastic differential equations.

3.2 Numerical Simulation Results

We present the results of numerical simulation of the evacuation of a small group of eight people from a gallery without and with columns. Figure 7a shows that even during evacuation a certain distance is maintained between the members of the group. At the exit from the gallery a zone with an increased concentration of individuals is formed.

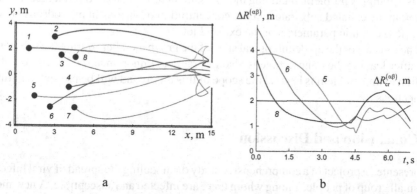

a   b

Fig. 7. Random trajectories of people in the gallery without internal obstacles (a). The relative distance between the infected member of the group and some susceptible individuals, who found themselves in the zone of increased virion concentration (b).

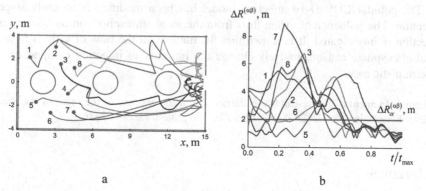

a   b

Fig. 8. Example of random trajectories of movement of people during evacuation from a gallery with columns (a). Change in the relative distance between the infected individual and the rest of the group (b).

We assume that one person designated by number 1 is infected, the other members of the group are susceptible. Figure 7b shows the dependence of the relative distance between infected participant 1 and the rest of the group on time. Figure 7b illustrate the critical relative distance $\Delta R_{cr}^{(\alpha\beta)}$. It is assumed that with a relative distance less than the critical one, the probability of infection increases significantly.

Figure 8 shows random trajectories in a gallery with columns. It can be seen that internal obstacles and collective interaction of individuals lead to a longer-lived zone of increased concentration of input data near the exit (Fig. 8a). During evacuation from the gallery with columns, the probability of infection by individual 1 of the remaining members of the group is significantly higher than during evacuation without columns (Fig. 8b).

To assess the probability of infection, it is necessary to know the flux $X_{\text{atm}}^*(t) \big/ T_{\text{in}}^*$ of virions from the atmosphere into the organism of a susceptible member of the group. This is a complex problem, including the assessment of the respiratory rate, the moisture content in the exhaled substrate and the concentration of virions in microdrops, as well as the aerodynamic parameters of the exhaled jet.

The forecast of the epidemiological state of the group members during the subsequent time after leaving the danger zone is also of practical importance. A description of the solution to this problem is beyond the scope of this report and will be presented in future publications.

4 Conclusion and Discussion

We presented a project of a comprehensive study on modeling the spread of viral infection in a small group of people, among whom there are infected and susceptible. A new model of random movement of individuals is proposed, taking into account the social dynamics in a room with internal obstacles. It is shown that in conditions of panic, the frequency of physical contacts near the exit from the room increases significantly, which can lead to intensive infection of susceptible members of the group.

The popular COVID-19 infection model has been modified at an early stage of infection. The influence of virion flow from the local atmosphere on the dynamics of infection is investigated. It is shown that fluctuations in the flow of virions from the local atmosphere can qualitatively change the intensity of infection compared to the deterministic case.

Acknowledgements. The research was carried out at the expense of the grant of the Russian Science Foundation No. 23–29-00243, https://rscf.ru/project/23-29-00243/

References

1. Janis, A.M., et al.: SARS-CoV-2 infects and replicates in cells of the human endocrine and exocrine pancreas. Nat. Metab. **3**, 149–165 (2021)
2. Wiersinga, W.J., et al: Pathophysiology, transmission, diagnosis, and treatment of coronavirus disease 2019 (COVID-19) a review. JAMA Rev., 12839 (2020)
3. Riley, E.C., Murphy, G., Riley, R.L.: Airborne spread of measles in a suburban elementary school. Am. J. Epidemiol. **107**(5), 421-432 (1978)
4. Mohammadi, A., et al.: Developing levels of pedestrian physical distancing during a pandemic. Saf. Sci. **134**, 105066 (2021).
5. Doremalen, N., et al.: Aerosol and surface stability of SARS-CoV-2 as compared with SARS-CoV-1. N. Engl. J. Med. **382**, 1564–1567 (2020)
6. Santarpia, J.L., et al.: Aerosol and surface contamination of SARS-CoV-2 observed in quarantine and isolation care. Sci. Rep. **10**, 12732 (2020)

7. Noakes, C., Sleigh, P.A.: Mathematical models for assessing the role of airflow on the risk of airborne infection in hospital wards. J. R. Soc. Interface **6**, S791–S800 (2009)
8. Baccam, P., et al.: Kinetics of influenza a virus infection in humans. J. Virol. **80**(15), 7590–7599 (2006)
9. Iwami, S., et al.: Identifying viral parameters from in vitro cell cultures. Front Microbiol. **3**, 312 (2012)
10. Hernandez-Vargas, E.A., Velasco-Hernandez, J.X.: In-host mathematical modelling of COVID-19 in humans. Annu. Rev. Control. **50**, 448–456 (2020)
11. Iwanami, S., et al.: Detection of significant antiviral drug effects on COVID-19 with reasonable sample sizes in randomized controlled trials: A modeling study. PLoS Med **18**(7), e1003660 (2021)
12. Gonçalves, A., et al.: Timing of antiviral treatment initiation is critical to reduce SARS-CoV-2 viral load. CPT Pharmacometrics Syst. Pharmacol. **9**, 509–514 (2020)
13. Kim, K.S., et al.: A quantitative model used to compare within-host SARS-CoV-2, MERS-CoV, and SARS-CoV dynamics provides insights into the pathogenesis and treatment of SARS-CoV-2. PLoS Biol **19**(3), e3001128 (2021)
14. Bernhauerová, V., et al.: Mathematical modelling of SARS-CoV-2 infection of human and animal host cells reveals differences in the infection rates and delays in viral particle production by infected cells. J Theor Biol **21**(531), 110895 (2021)
15. Ahn, J.H., et al.: Nasal ciliated cells are primary targets for SARS-CoV-2 replication in the early stage of COVID-19. J Clin Invest. **131**(13), e1485172021 (2021)
16. Debrabant, K., Rößler, A.: Classification of stochastic Runge-Kutta methods for the weak approximation of stochastic differential equations. Math Comp Simul. **77**(4), 408–420 (2008)
17. Helbing, D., Farkas, I., Vicsek, T.: Simulating dynamical features of escape panic. Nature **407**(28), 487–490 (2000)
18. Helbing, D., Molnar, P.: Social force model for pedestrian dynamics. Phys Rev E **51**(5), 4282–4286 (1995)
19. Wei-Guo, S., et al.: Evacuation behaviors at exit in CA model with force essentials: a comparison with social force model. Physica A **371**, 658–666 (2006)
20. Kirchner, A., Nishinari, K., Schadschneider, A.: Friction effects and clogging in a cellular automaton model for pedestrian dynamics. Phys Rev E **67**, 056122 (2003)
21. Yang, X., Wu, Z., Li, Y.: Difference between real-life escape panic and mimic exercises in simulated situation with implications to the statistical physics models of emergency evacuation: The 2008 Wenchuan earthquake. Physica A **390**, 2375–2380 (2011)
22. Leimkuhler, B., Matthews, C.: Molecular Dynamics with Deterministic and Stochastic Numerical Methods. Springer, Switzerland (2015)
23. Ma, L., et al.: The analysis on the desired speed in social force model using a data driven approach. Physica A **525**, 894–911 (2019)
24. Derevich, I.V., Klochkov, A.K.: Modeling the motion of particles in the potential force field with allowance for the random velocity fluctuations of a medium. J Eng Phys Thermophys. **95**, 1089–1100 (2022)
25. Ramasamy, R.: Innate and adaptive immune responses in the upper respiratory tract and the infectivity of SARS-CoV-2. Viruses **14**, 933 (2022)
26. Wölfel, R., et al.: Virological assessment of hospitalized patients with COVID-2019. Nature **581**, 465–469 (2020)

Training Strategies for Covid-19 Severity Classification

Daniel Pordeus[1]([✉])([iD]), Pedro Ribeiro[2], Laíla Zacarias[4], Adriel de Oliveira[6],
João Alexandre Lobo Marques[3], Pedro Miguel Rodrigues[2], Camila Leite[4],
Manoel Alves Neto[5], Arnaldo Aires Peixoto Jr[5],
and João Paulo do Vale Madeiro[1]

[1] Department of Computing, Federal University of Ceará, Fortaleza, Ceará, Brazil
pordeus@gmail.com, jpaulo.vale@dc.ufc.br
[2] CBQF – Centro de Biotecnologia e Química Fina – Laboratório Associado, Escola
Superior de Biotecnologia, Universidade Católica Portuguesa, Rua de Diogo Botelho
1327, 4169-005 Porto, Portugal
{s-pmsbribeiro,pmrodrigues}@ucp.pt
[3] Laboratory of Applied Neurosciences, University of Saint Joseph,
Macao SAR, China
alexandre.lobo@usj.edu.mo
[4] Graduate Program in Cardiovascular Sciences, Federal University of Ceara,
Fortaleza, Ceará, Brazil
camilafleite@ufc.br
[5] Department of Clinical Medicine, Faculty of Medicine,
Federal University of Ceara, Fortaleza, Ceará, Brazil
arnaldoapj@ufc.br
[6] University for the International Integration of the Afro-Brazilian Lusophony,
Fortaleza, Ceará, Brazil

Abstract. The COVID-19 pandemic has posed a significant public
health challenge on a global scale. It is imperative that we continue
to undertake research in order to identify early markers of disease progression, enhance patient care through prompt diagnosis, identification
of high-risk patients, early prevention, and efficient allocation of medical
resources. In this particular study, we obtained 100 5-min electrocardiograms (ECGs) from 50 COVID-19 volunteers in two different positions,
namely upright and supine, who were categorized as either moderately
or critically ill. We used classification algorithms to analyze heart rate
variability (HRV) metrics derived from the ECGs of the volunteers with
the goal of predicting the severity of illness. Our study choose a configuration pro SVC that achieved 76% of accuracy, and 0.84 on F1 Score in
predicting the severity of Covid-19 based on HRV metrics.

Keywords: Electrocardiogram (ECG) · Heart Rate Variability
(HRV) · signal processing · disease severity classification · COVID-19

Grant FUNCAP (Ceará State Foundation for the Support of Scientific and Technological Development) PS1-0186-00439.01.00/21.

1   Introduction

In Wuhan, China, a new strain of coronavirus was identified in December 2019. As infection rates and deaths increased, the World Health Organization (WHO) recognized it as a Public Health Emergency of International Concern on 30 January 2020. The disease was officially named COVID-19 on 11 February 2020 and declared a pandemic on 11 March 2020 after spreading globally and overwhelming medical resources in many regions.

To diagnose COVID-19, the RT-PCR technique is commonly used on nasopharyngeal and throat swabs, but its sensitivity values range from 30 to 70%. CT scans and X-ray images have been reported to have higher sensitivity values of 98% and 69%, respectively, but RT-PCR remains the standard for COVID-19 diagnosis, despite its slow response time.

Accurate diagnosis of COVID-19 is crucial for controlling the pandemic, and the Nucleic Acid Amplification Test (NAAT) is a laboratory testing technique that can diagnose an active COVID-19 infection in the upper respiratory tract using the Real-Time Polymerase Chain Reaction (RT-PCR) assay [1, 2].

Since the emergence of COVID-19, the scientific community has been searching for faster and more effective diagnostic methods, as time is critical to successful treatment. Medical imaging, such as chest X-ray images, has been proposed in several studies [3, 4]. However, due to the significant variability of symptoms and side effects among COVID-19 patients, other diagnostic and treatment approaches are also being considered.

Specialists are exploring various methods for groups of patients with persistent symptoms, such as fatigue, cardiac arrhythmia, and neurological cognitive impairment. One such approach is the use of heart rate variability (HRV) analysis, which assesses the changes in the duration of intervals between adjacent heartbeats. HRV is an effective technique for evaluating interdependent regulatory systems that operate on different time scales to adapt to challenges and achieve optimal performance [5].

The work presented on [6] reviews the balance between sympathetic and parasympathetic subsystems in the heart and discusses the interpretation of HRV, including criteria for the association between reduced HRV and the risk of disease and mortality, and loss of regulatory capacity. It also highlights the heart-brain connection and the underlying physiological mechanisms of various frequency bands. The most common time and frequency domain measurements and standardized data collection protocols are also reviewed. While fluctuations in HR are apparent on a beat-to-beat basis, they are often overlooked when a mean value over time is calculated. These fluctuations are the result of complex, nonlinear interactions among several physiological systems and are considered a measure of neurocardiovascular function that reflects heart-brain interactions and autonomic nervous system (ANS) dynamics [6].

In this work, we seek to identify that Covid-19 affects cardiac behavior at different levels, verifying whether it is possible to correctly classify an electrocardiogram of patients at different stages of the disease's evolution. A methodology was developed to select the best answer among known algorithms using HRV signal metrics, including frequency-domain, time-domain, and non-linear metrics. These metrics are used as a feature vector for machine learning algorithms to classify patients into two groups: low/moderate and severe/critical. Due to class imbalance, the groups were redefined to improve classification accuracy. Using a specific configuration of SVC, we achieved an accuracy of 76%, precision of 100%, recall of 81%, and F1 score of 0.84, in addition to this, several other configurations achieved maximum scores in recall and precision.

This paper is structured as follows: Sect. 2 describes the dataset and the technique used to extract HRV metrics. Section 4 presents the training strategies and corresponding results. Section 5 discusses the results and draws conclusions.

2   Dataset Acquisition and Data Processing

Hospitalized individuals of both sexes, with clinical symptoms and diagnosis of COVID-19 confirmed by RT-PCR were classified according to the severity shown in Table 1, at Hospital Universitário Walter Cantídio (HUWC) and Hospital Estadual Leonardo da Vinci (HELV), Ceará, Brazil, from May 2021 to January 2022, corresponding to the first, second, and third wave of COVID-19 in Brazil. The database is represented by two different severity levels: 17 patients with low or moderate severity, 33 patients with a severe stage of the disease. The ECG signals were obtained following the recommendations of [8] to reduce circadian influences, and [9] for the correct signal stabilization, both for supine and orthostasis positions. The study followed the ethical precepts of the Declaration of Helsinki [10]. The evaluations were only started after the patient had understood the protocol completely and written informed consent was obtained. The study was approved by the Ethics Committee for Research with Human Beings (CAAE - HELV: 47229221.9.3001.5684; CAAE - HUWC: 47229221.9.0000.5045).

Table 1. COVID-19 severity classification criteria [7].

Low or Moderate	Mild clinical symptoms and no signs of pneumonia on examination image, Presence of fever and respiratory symptoms with evidence radiology of pneumonia
Severe	Respiratory distress (> 30 breaths / min)
	Oxygen saturation < 93% at rest
	Partial pressure of arterial oxygen (PaO2)/ fraction of inspired oxygen (FiO2) <300 mmHg
	Cases with chest images that showed >50% evident lesion progression within 24–48 h

2.1   Preprocessing/Filtering Stage

Due to the presence of multiple sources of noise and interference, discrete wavelet transform (DWT) is applied to suppress high-frequency and low-frequency noise.

DWT can be applied using filter banks and downsampling and upsampling processes. Daubechies10 function ("db10") [11] is used as the mother-wavelet which provides eliminating the coefficients resulting from high-pass filtering after the first decomposition stage, followed by discrete wavelet transform (DWT) and Daubechies4 function('db4) to obtain the coefficients. The ECG signal generated by the filtering process can then be used in the QRS complex detection. The methods include a combination of continuous wavelet transform, Hilbert transform and derived filter techniques. Once the R-R signal generated by the QRS composite detection process was obtained, we must find the first useful sample of the R-R series. R-R series and N-N series results from ectopic elimination processing. Figure 1 illustrates the original RR series (the same as in the previous figure) and the corrected series, which we will now denote NN series (series of intervals between sinus or normal beats).

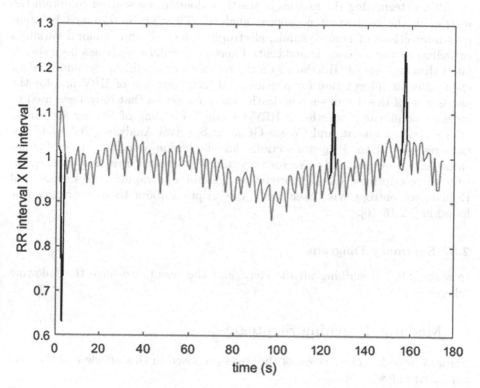

Fig. 1. R-R series and N-N series resulting from ectopic elimination processing.

From this point on, we started obtaining the Heart Rate Variability (HRV) metrics themselves These details can be found in [12].

2.2 Heart Rate Variability Metrics

From the HRV we can extract a large amount of metrics, which can be classified into three categories: Frequency Domain, Time Doman and Non-Linear. Frequency Domain Metrics are peaks of the VLF, LF and HF bands, areas of the individual spectral components, percentages relative to the total area (relative area), LF and HF components in normalized units and LF and HF components. Time Domain Metrics are calculated after computing the frequency domain parameters, proceeding to determine the metrics resulting from the time domain analysis. It is important to emphasize that for the analysis in the time domain, we used a series of NN intervals (after correcting and filling the gap phase caused by discarding out-of-place beats). An interpolated version of the series (where the spacing between samples is uniform) is used for frequency domain analysis only.

After determining the metrics in the time domain, we started to obtain the metrics in the context of non-linear analysis. They are determined by complex interactions of hemodynamic, electrophysiological, and humoral variables as well as by the autonomic and central nervous regulations. It has been speculated that analysis of HRV based on the methods of nonlinear dynamics might elicit valuable information for physiological interpretation of HRV and for the assessment of the risk of sudden death. The parameters that have been used to measure nonlinear properties of HRV include 1/f scaling of Fourier spectra,47 19 H scaling exponent, and Coarse Graining Spectral Analysis (CGSA).48 For data representation, Poincaré sections, low-dimension attractor plots, singular value decomposition, and attractor trajectories have been used. For other quantitative descriptions, the D2 correlation dimension, Lyapunov exponents, and Kolmogorov entropy have been used. A description about these metrics can be found in [12,15,16].

2.3 Summary Diagram

In short, after describing all the steps and the result, we have the following scheme in Fig. 2.

3 Machine Learning Strategies

Figure 3 provides an overview of the steps proposed in this article and they are explained below.

3.1 First Stage

First, the dataset was split into two parts: 39 subjects with 77 samples for training and 11 subjects with 21 samples for test, that we named "training dataset" and "test dataset". The test dataset were ignored during model selection and were used later to test the model's effectiveness. We followed the recommendations by Kunjan et al [13] and separated the training and test sets based on the

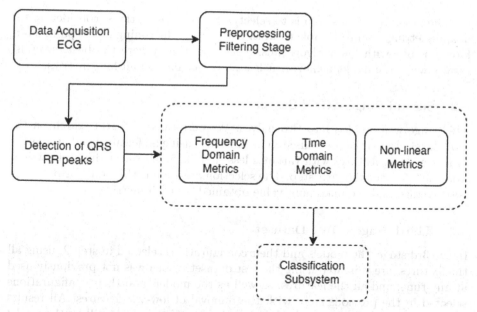

Fig. 2. Overview of the Methodology Applied.

subject rather than the exam. The code was prepared to train respecting this condition and allowing the conventional k-fold training when necessary. Seventeen algorithms was chosen, they are implemented by scikit-learn [14] and they are on the List 1.1.

- BaggingClassifier
- GaussianProcessClassifier(1.0 * RBF(1.0))
- LinearSVC
- ExtraTreesClassifier(n_estimators=300)
- SGDClassifier(max_iter=100, tol=1e-3)
- RandomForestClassifier
- DecisionTreeClassifier(max_depth=5)
- LogisticRegression(solver='lbfgs')
- SVC(gamma='auto')
- MLPClassifier(alpha=1, max_iter=1000)
- GradientBoostingClassifier
- KNeighborsClassifier
- AdaBoostClassifier
- LGBMClassifier
- LinearDiscriminantAnalysis
- OneVsRestClassifier(LinearSVC(random_state=0, dual=False))
- GaussianNB

List 1.1. Models Chosen for this Study.

The objective of this step is to select/filter the best options, considering the default settings and to be able to work better with the tuning of the parameters later on, but with fewer algorithms. Selection criteria were the best "overall" results when the model achieved at least 60% (or 0.6) on all four metrics.

3.2   Second Stage - Grid Search

This stage was divided into two halves. We performed the grid search on the models selected in the previous step in two ways: using all features and at another time, removing features with variance less than 0.15. We select the best settings from each half for the next step. The selection criterion is the same for both: the best average and the maximum value obtained for each metric.

3.3   Third Stage - Test Dataset

In the 3rd stage, the models and their configurations selected in step 2, using all the features, are validated using the test dataset, which was not previously used at any time, and all the features, as well as the models and their configurations selected in the previous step, with the removal of low-var features. All results are compared and the best result considering the four metrics will be the model for this study.

4   Results

4.1   First Stage Results

All 17 algorithms listed in 1.1, implemented by scikit-learn [14], run through 20 rounds, dividing them into training and test sets with Leave One Subject Out Cross Validation for 70% and 30%, and 80% and 20%. They computed the average of accuracy, f1-score, recall, and precision for each algorithm and metric, and the best result for each algorithm and metric was kept. The results are presented in Table 2.

4.2   Second Stage Results

Table 3 presents the grid search configuration that we set up for this step with the 7 previously chosen algorithms. In all cases, the default configuration used in the previous step (see List 1.1) were considered.

For performance reasons, we chose to use GridSearchCV scikit-learn implementation [14]. The issue is the loss of control over the division of sets, giving the chance of not respecting the correct organization between the exams of the same volunteer(subject), but we accept the situation because we believe that the low number of samples per patient, at most two, and the division with small test set would reduce the risk. So we opted for a division of 14 groups of 7 samples each and the results of the best averages and also the maximum of each metric were observed. Table 4 present results for gridsearch applied after remove low var features, and Table 5 brings the results for gridsearch applied with all features.

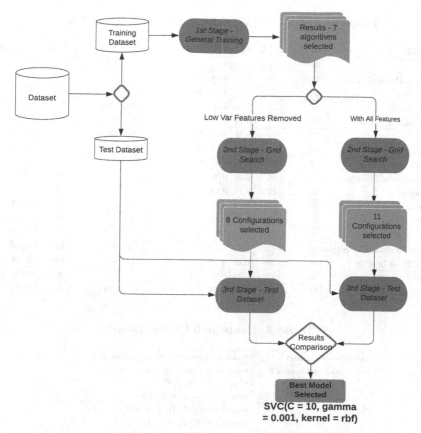

Fig. 3. Process Applied to Select the Best Model and Its Configuration.

4.3 Third Stage Results

This last step ran all the models selected in the previous step, considering when it was chosen, whether using all the features or removing the low variance ones. This part the results are compared. Table 6. Between parentheses is the reference to the configuration of the algorithm that was the best in the previous step. That is, for example, meanAcc indicates that this configuration that obtained the best average accuracy in step 2, and maxAcc was the one that achieved the maximum accuracy, and 'lv' indicates low var, i.e., after removed low variance features, and 'all' is for all features.

Support Vector Machine Classifier(**SVC**) with all features, configured with **kernel RBF, C = 10** and **gamma = 0.0001** obtained the best results for all metrics. This configuration obtained **76% for accuracy, 72% for precision, 100% for recall** and **0.84 on f1-score**.

Table 2. Results Obtained for the First Training Stage.

MODEL	ACC(%)	STD	F1	STD	RECALL(%)	STD	PREC(%)	STD	SELECTED?
SVC	66	12	1.00	0	66	12	79	9	TRUE
GaussianProcessClassifier	17	9	0.29	0.18	39	12	25	13	FALSE
LinearSVC	58	28	0.67	0.27	57	15	60	21	FALSE
SGDClassifier	57	26	0.68	0.25	53	16	58	21	FALSE
KNeighborsClassifier	74	13	0.82	0.11	58	09	70	07	FALSE
LogisticRegression	73	14	0.78	0.13	57	12	69	10	FALSE
BaggingClassifier	72	13	0.76	0.11	62	11	72	10	FALSE
ExtraTreesClassifier	77	13	0.85	0.07	65	11	76	09	TRUE
RandomForestClassifier	75	15	0.90	0.06	65	10	77	08	TRUE
GaussianNB	34	17	0.46	0.19	48	13	45	17	FALSE
DecisionTreeClassifier	69	13	0.77	0.09	59	13	69	12	FALSE
MLPClassifier	63	15	0.78	0.16	57	9	67	09	FALSE
AdaBoostClassifier	73	12	0.78	11	61	9	71	08	FALSE
LinearDiscriminantAnalysis	62	13	0.71	0.13	55	10	64	10	FALSE
OneVsRestClassifier	66	15	0.75	0.11	55	11	65	10	FALSE
LGBMClassifier	79	12	0.82	0.08	65	12	75	09	TRUE
GradientBoostingClassifier	75	12	0.81	0.08	63	11	73	09	FALSE

Table 3. Gridsearch Configuration.

ExtraTreesClassifier	number of estimators: [200, 300, 400]
RandomForestClassifier	number of estimators: [200, 300, 400]
	max depth: [3, 5, 7]
	min samples split: [5, 10]
SVC	kernel linear, C: [1, 2, 5, 10]
	kernel rbf, C: [1, 2, 5, 10]
	gamma: [0.001, 0.0001]
BaggingClassifier	number of estimators: [5, 10, 20]
AdaBoostClassifier	number of estimators: [2, 10, 50]
	learning rate: [0.001, 0.01, 0.1, 1.0]
GradientBoostingClassifier	loss: deviance
	learning rate: [0.01, 0.05, 0.1, 0.15, 0.2]
	min samples split: 0.1 to 0.5, step 5
	min samples leaf: 0.1 to 0.5, step 5
	max depth: [3,5,8]
	max features: [log2, sqrt]
	criterion: [friedman mse, mae]
	subsample: [0.5, 0.85, 1.0]
	number of estimators: [10]
LGBMClassifier	learning rate: [0.005, 0.01]
	n estimators: [8,16,24]
	num leaves: [6,8,12,16]
	boosting type: [gbdt, dart]
	max bin: [255, 510]
	random state: [500]
	colsample bytree: [0.64, 0.65, 0.66]
	subsample: [0.7,0.75]
	reg alpha: [1,1.2]
	reg lambda: [1,1.2,1.4]

Table 4. Results from Second Stage After Removed Low Var Features.

Criteria	Model	Parameters	Score	Std
Best Mean Accuracy Score	GradientBoostingClassifier	max depth=5, min samples split=0.1, learning rate=0.15, criterion='friedman mse', max features='sqrt', min smples leaf=−0.2, subsample=0.85	79%	9
Maximum Accuracy Score	GradientBoostingClassifier	max depth=3, min samples split=0.5, learning rate=0.15, criterion='friedman mse', max features='log2', min smples leaf=−0.2, subsample=0.85	100%	12
Best Mean Precision Score	AdaBoostClassifier	n estimators=10, learning rate=0.01	84%	12
Maximum Precision Score	ExtraTreesClassifier	n estimators=200	100%	12
Best Mean Recall Score	LGBMClassifier	n estimators=16, learning rate=0.01, subsample=0.75, boosting type='gbdt', colsample bytree=0.64, max bin=510, num leaves=6, reg alpha=1, reg lambda=1	100%	0
	SVC	C=1, kernel=rbf, gamma=0.01	100%	0
Maximum Recall Score	LGBMClassifier	n estimators=16, learning rate=0.01, subsample=0.75, boosting type='gbdt', colsample bytree=0.64, max bin=510, num leaves=6, reg alpha=1, reg lambda=1	100%	0
	SVC	C=1, kernel=rbf, gamma=0.01	100%	0
Best Mean F1	GradientBoostingClassifier	max depth=3, min samples split=0.3, learning rate=0.15, criterion='friedman mse', max features='sqrt', min smples leaf=−0.2, subsample=0.5	0.85	0.07
Maximum F1 Score	GradientBoostingClassifier	max depth=3, min samples split=0.1, learning rate=0.2, criterion='friedman mse', max features='sqrt', min smples leaf=−0.3, subsample=1.0	1.0	0.1

Table 5. Results from Second Stage Using All Features

Criteria	Model	Parameters	Score	Std
Best Mean Accuracy Score	GradientBoostingClassifier	max depth=5, min samples split=0.2, learning rate=0.1, criterion='friedman mse', max features='sqrt', min samples leaf=0.2, subsample=1.0	78%	16
Maximum Accuracy Score	SVC	C=10, kernel=rbf, gamma=0.0001	100%	18
Best Mean Precision Score	AdaBoostClassifier	n estimators=2, learning rate=1	83%	20
Maximum Precision Score	AdaBoostClassifier	n estimators=2, learning rate=1	100%	27
Best Mean Recall Score	SVC	C=1, kernel=rbf, gamma=0.001	100%	0
Best Mean Recall Score	SVC	C=1, kernel=rbf, gamma=0.01	100%	0
Best Mean Recall Score	SVC	C=10, kernel=rbf, gamma=0.01	100%	0
Best Mean Recall Score	SVC	C=5, kernel=rbf, gamma=0.01	100%	0
Best Mean Recall Score	SVC	C=2, kernel=rbf, gamma=0.01	100%	0
Maximum Recall Score	LGBMClassifier	n estimators=8, learning rate=0.01, subsample=0.7, boosting type='gbdt', colsample bytree=0.66, max bin=255, num leaves=12, reg alpha=1.2, reg lambda=1	100%	0
Best Mean F1	GradientBoostingClassifier	max depth=8, min samples split=0.4, learning rate=0.15, criterion='friedman mse', max features='log2', min samples leaf=0.1, subsample=0.85	85%	12
Maximum F1 Score	SVC	C=10, kernel=rbf, gamma=0.0001	100%	0.14

Table 6. Final Classification

Model	Accuracy	Precision	Recall	F1	Avg
SVC (maxAcc all)	76.19%	100.00%	81.25%	0.84	0.85
SVC (meanRec 1) all	61.90%	100.00%	100.00%	0.76	0.85
SVC (meanRec 2) all	61.90%	100.00%	100.00%	0.76	0.85
SVC (meanRec 3) all	61.90%	100.00%	100.00%	0.76	0.85
SVC (meanRec 4) all	61.90%	100.00%	100.00%	0.76	0.85
SVC (meanRec 5) all	61.90%	100.00%	100.00%	0.76	0.85
LGBMc (lv)	61.90%	100.00%	100.00%	0.76	0.85
SVC (lv)	61.90%	100.00%	100.00%	0.76	0.85
GBc (meanF1 lv)	57.14%	92.31%	100.00%	0.73	0.81
GBc (maxAcc lv)	57.14%	92.31%	100.00%	0.73	0.81
GBc (meanAcc lv)	57.14%	92.31%	100.00%	0.73	0.81
GBc (maxF1 lv)	52.38%	84.62%	100.00%	0.69	0.76
ET (lv)	57.14%	84.62%	91.67%	0.71	0.76
GBc (meanAcc all)	47.62%	76.92%	100.00%	0.65	0.72
Ada (prec all)	47.62%	76.92%	100.00%	0.65	0.72
GBc (meanF1 all)	47.62%	76.92%	100.00%	0.65	0.72
Ada (lv)	47.62%	76.92%	100.00%	0.65	0.72

5   Discussions and Conclusions

This study showed that there are differences between the degrees of severity of COVID-19 on cardiac behavior and it illustrates the potential benefits of using machine learning to predict the severity of COVID-19 by analyzing up to 5 min of electrocardiogram (ECG) data. The extracted metrics from the ECG can provide valuable information to assist in patient care, improving the accuracy in predicting Covid-19 severity and its treatment.

In Sect. 4, we demonstrate the effectiveness of machine learning algorithms in classifying COVID-19 severity based on linear and non-linear ECG metrics. The training accuracy rates ranged around 70% for various training strategies, with F1-Score values above 0.80, precision above 90%, and recall close to 90%. Despite the small size and class imbalance of the dataset, which consisted of 50 volunteers, 98 samples, and 38 features, our proposed system achieved promising results, with the Support Vector Machine Classifier showing the best performance.

Although we performed some training rounds without separating by subject due to the limited number of samples, we believe that the contamination effect was minimal. However, as the dataset grows, differences in the results may become more apparent, highlighting the need for careful learning focused on the disease and boosting confidence in the results.

A plausible tool for binary classification of the degree of severity of hospitalized COVID-19 patients could be developed, with one group consisting of patients classified as low or moderate and the other group consisting of patients classified as severe. Due to the limited sample size and imbalance, this separation became necessary. Our proposed solution achieved an accuracy of 76% and an F1-score of 0.84, which we consider a good result, showing that COVID-19 affects cardiac behavior in different ways, as we hypothesized.

We approached this research with caution, given the sensitivity of the subject matter, which has increased our confidence in the results obtained, but keeping chance for further improvements.

References

1. Hosseini, E., et al.: The novel coronavirus Disease-2019(COVID-19): mechanism of action, detection and recent therapeutic strategies. Virology **551**, 1–9 (2020)
2. Lai, C., Lam, W.: Laboratory testing for the diagnosis of COVID-19. Biochem. Biophys. Res. Commun. **538**, 226–230 (2021)
3. Wang, S., et al.: A fully automatic deep learning system for COVID-19 diagnostic and prognostic analysis. Eur. Respir. J. **56**(2), 2000775 (2020). https://doi.org/10.1183/13993003.00775-2020
4. Khuzani, Z., et al.: COVID-classifier: an automated machine learning model to assist in the diagnosis of COVID-19 infection in chest X-ray images. Sci. Rep. J. **6**(11) (2021). https://doi.org/10.1038/s41598-021-88807-2
5. Shaffer, F., McCraty, R., Zerr, C. L.: A healthy heart is not a metronome: an integrative review of the heart's anatomy and heart rate variability. Front. Psychol. **5**, 1040. https://doi.org/10.3389/fpsyg.2014.01040
6. Mccraty, R., Shaffer, F.: Heart rate variability: new perspectives on physiological mechanisms, assessment of self-regulatory capacity, and health risk. Glob. Adv. Health Med. **4**(1), 46–61 (2015). https://doi.org/10.7453/gahmj.2014.073
7. Yan, X., et al.: Clinical characteristics and prognosis of 218 patients with COVID-19: a retrospective study based on clinical classification. Front. Med. **7**, 485 (2020)
8. Felber Dietrich, D., Schindler, C., Schwartz, J., et al.: Heart rate variability in an ageing population and its association with lifestyle and cardiovascular risk factors: results of the SAPALDIA study. Europace **8**(7), 521–529 (2006). https://doi.org/10.1093/europace/eul063
9. Brown, S.J., Brown, J.A.: Resting and postexercise cardiac autonomic control in trained master athletes. J. Physiol. Sci. **57**(1), 23–29 (2007). https://doi.org/10.2170/physiolsci.RP012306
10. World Medical Association: World Medical Association Declaration of Helsinki: ethical principles for medical research involving human subjects. JAMA **310**(20), 2191–2194 (2013). https://doi.org/10.1001/jama.2013.281053
11. Vonesch, C., Blu, T., Unser, M.: Generalized daubechies wavelet families. IEEE Trans. Signal Process. **55**(9), 4415–4429 (2007). https://doi.org/10.1109/TSP.2007.896255
12. Madeiro, J.P.V.: An innovative approach of QRS segmentation based on first-derivative, Hilbert and Wavelet Transforms. Med. Eng. Phys. **34**(9), 1236–1246 (2012)

13. Kunjan, S., et al.: The necessity of leave one subject out (LOSO) cross validation for EEG disease diagnosis. In: Mahmud, M., Kaiser, M.S., Vassanelli, S., Dai, Q., Zhong, N. (eds.) BI 2021. LNCS (LNAI), vol. 12960, pp. 558–567. Springer, Cham (2021). https://doi.org/10.1007/978-3-030-86993-9_50
14. Pedregosa, F., et al.: Scikit-learn: machine learning in python. J. Mach. Learn. Res. **12**, 2825–2830 (2011)
15. Shaffer, F., Meehan, Z.M., Zerr, C.L.: A critical review of ultra-short-term heart rate variability norms research. Front. Neurosci. **14**, 594880 (2020)
16. Silva, L., et al.: Heart rate variability as a biomarker in patients with Chronic Chagas Cardiomyopathy with or without concomitant digestive involvement and its relationship with the Rassi score. Biomed. Eng. Online **21**(1), 44 (2022). https://doi.org/10.1186/s12938-022-01014-6

The Effectiveness of Quarantine in Viral and Bacterial Epidemics: New Evidence Provided by the Covid-19 Pandemic

Andreu Martínez-Hernández[1] and Vicente Martínez[2(✉)]

[1] Hospital General de Castelló, 12004 Castelló de la Plana, Spain
martinez_andher@gva.es
[2] Institut de Matemàtiques i Aplicacions de Castelló, Departament de Matemàtiques, Universitat Jaume I, 12071 Castelló de la Plana, Spain
martinez@uji.es

Abstract. The effectiveness of confining the population has been observed for centuries. However, this effectiveness has now been demonstrated with data during the COVID-19 pandemic, for which an enormous amount of data and studies are available. In this sense, this paper identifies the determination of the number of people susceptible to contracting the disease, which is present in many epidemic transmission models, as the fundamental variable for understanding the dynamics of infections. We primarily consider the SIRD model, but the data are also contrasted with models and techniques used in pandemic analysis. In addition, the facts and conditions of the COVID-19 pandemic are compared with others that occurred historically.

Keywords: Pandemic · Viral and bacterial epidemics · Covid-19 · Mathematical modeling

1 Introduction

The harsh COVID-19 pandemic that humanity is suffering has put the national health systems of all countries around the world to the test, and there are many lessons that scientists are learning. Much of the evidence that has been amassed over the centuries has also been verified. We will focus on the evidence regarding preventive quarantines to avoid infection, among other reasons, because the scientific methodology used in recent decades has made it possible to demonstrate the effectiveness of quarantine with data and simulations.

In this article, we will provide data that have led us to identify the determination of the number of people likely to be infected as a fundamental variable for the spread or control of the infection. These data have been obtained from a wide variety of countries, considering various circumstances and using various methods. Furthermore, the conclusions have been validated through numerical simulations and real data, which will allow us to ensure that if the number of

I. Rojas et al. (Eds.): IWBBIO 2023, LNBI 13919, pp. 528–541, 2023.
https://doi.org/10.1007/978-3-031-34953-9_41

exposed people is not controlled, the infection will spread. Therefore, having control over this variable will help mitigate the harmful effects of the pandemic.

In Sect. 2, we introduce a review of the state of the art on the effectiveness of quarantine in preventing disease transmission, as well as a table comparison of similar papers published during 2021–2022.

We will pay more attention to the SIRD model [1], which, as a mathematical model, quantifies variables more clearly and also allows us to perform simulations of future scenarios (Sect. 3).

In Sect. 4, we will analyze all the known data about the COVID-19 pandemic in order to test infectious disease dynamic models with the enormous amount of data that this pandemic has provided. Thus, we can obtain conclusions about the behavior of similar epidemics that may occur in the future.

In addition, we will make a comparison with other pandemics that have occurred throughout history, such as the pandemic that hit Europe in the last century, known as *The Spanish Flu*, or the small amount of data that have reached us about *The Black Death* in the 15th century, which was also a devastating pandemic (Sect. 5).

Finally, we will analyze some conclusions that will allow us to support the thesis presented in previous paragraphs (Sect. 6).

2   Literature Review

During the last two years there have been a large number of studies that have evaluated the effectiveness of quarantine and other isolation measures with the aim of stopping the transmission of the SARS-CoV-2 virus. The conclusions obtained are based on a wide variety of techniques and methods, among them we can find: Stochastic methods; Dynamic systems; The principle to characterize the optimal controls; Approximating the infectiousness density function by a Gamma function; Estimated by Kalman filter; SEIR and SIR models; etc.

In Table 1 we can see a summary of a set of works that support the hypothesis about the effectiveness of the quarantine. Obviously they are not all, but those mentioned here provide us with a view of the recent panorama of the state of the art in this topic. The main achievement that can be named from these works listed focuses on the effectiveness of mandatory quarantine for a large amount of data obtained in various countries and treated with a wide range of methods and procedures. Obviously the expansion of an epidemic depends to a large extent on the people susceptible to contracting the virus who are also exposed to it. Mandatory quarantine has proven its effectiveness for centuries, but never before in history has there been as much contrasted and reliable data as is available today, which will allow us to assure our affirmation without any doubt.

3   The SIRD Model of Infectious Diseases

The SIRD model was proposed in 1927 [1] by Kermack and McKendrick, and since then, it has been widely used [19–24]. The reason for considering this method here is because it allows us to carry out simulations in an agile and efficient way. The model considers four types of people:

(S) Susceptible—those who could become infected.
(I) Infected—those who are infected at that moment.
(R) Recovered—those who have had the disease and are now healthy.
(D) Deceased—those who have died of the disease.

Considering that these variables depend on time, the SIRD model is governed by the following nonlinear system of differential equations:

Table 1. Latest references about quarantine in the COVID-19 pandemic.

Author(s)	Year	Country(ies)	Methods	Confirmed effectiveness
Aggarwal, D. *et al.* [2]	2022	5 EU countries	Stochastic method	Yes
Al Zabadi, H. *et al.* [3]	2021	Palestine	Logistic regression	Yes
Bennett, M. *et al.* [4]	2021	Chile	Control Methods	Yes
Bo, Y. *et al.* [5]	2021	190 countries	Meta-analysis	Yes
Bou-Karroum, L. *et al.* [6]	2021	31 countries	Meta-analysis	Yes
Chiyaka, E.T. *et al.* [7]	2022	Zimbabwe	Electronic repositories Empirical evidence	Yes
Foncea, P. *et al.* [8]	2022	Chile	Sample-to-answer reporting time	Yes
Gu, Y. *et al.* [9]	2022	Pakistan	Dynamic systems Parameter estimation	Yes
James, A. *et al.* [10]	2021	New Zealand	Time-dependent kernel	Yes
Johansson, M.A. *et al.* [11]	2021	USA	Approximates density function	Yes
Kouidere, A. *et al.* [12]	2021	Morocco	Pontryagin's principle Optimal control	Yes
Kristjanpoller, W. *et al.* [13]	2021	Chile	Verhulst-Richards models Probabilistic approach	Yes
Li, F. *et al.* [14]	2021	China	SEIR model & Monte Carlo	Yes
Quilty, B.J. *et al.* [15]	2021	UK	Empirical evidence	Yes
Tu, H. *et al.* [16]	2021	China	Standard SIR model	Yes
Wang, R. *et al.* [17]	2021	Japan	Stochastic models	Yes
Zhang, A.Z. *et al.* [18]	2022	China	Transmission dynamics	Yes

$$
\begin{aligned}
S'(t) &= -\beta\, S(t) I(t), \\
I'(t) &= \beta\, S(t) I(t) - \alpha\, I(t) - \gamma\, I(t), \\
R'(t) &= \alpha\, I(t), \\
D'(t) &= \gamma\, I(t),
\end{aligned}
\tag{1}
$$

where α (recovery rate per unit of time), β (infected rate per unit of time), and γ (death rate per unit of time).

On the other hand, there is another important parameter (the reproduction number), referred to as the ρ_0 number, which is given by

$$\rho_0(t) = \frac{\beta}{\alpha + \gamma} S(t). \tag{2}$$

If this parameter is greater than 1, the epidemic is out of control. However, when it is less than 1, the epidemic tends to subside. Readers can find more details about this parameter in [25, 26].

In [27] Eq. 1 has been solved using numerical methods for solving systems of differential equations. The following section shows the numerical simulations of the possible scenarios of the dynamics of the COVID-19 pandemic in Spain.

4   Some Remarks About the COVID-19 Pandemic

Since the new coronavirus SARS-CoV-2 was detected in December 2019 in China, causing a global epidemic coined by the World Health Organization (WHO) as COVID-19, as of October 21, 2021, this disease has presented 241,411,380 confirmed cases and 4,912,112 deaths globally. Some notable data (see [28]) from the worst hit countries are shown in Table 2.

Since December 2019, a multitude of scientific studies have been carried out thanks to our current research capacity, and several vaccines have been obtained in just one year, a sign of the great technological development in this century.

The paper [27] describes how this model was adjusted to the first wave of the COVID-19 pandemic in Spain. This work, among other conclusions, provided us with an important piece of information: the spread of the pandemic depends on the number of people exposed to the virus (variable S, susceptible people in the SIRD model). Interested readers can find in [27] all the details about the SIRD method, including its stability and convergence.

In Spain, the pandemic has had a greater or lesser impact depending on the strictness of the isolation measures. Figure 1 shows the evolution of the reproduction number ρ_0 during the latency period of the first wave of the COVID-19 pandemic. In the period between March 22 and April 13, a strong decrease is observed in this parameter. This occurred when the quarantine measures were imposed by the Spanish Government [29–31]. On April 13, there was a slowdown in the decrease due to relaxation of isolation measures.

On the other hand, if we observe Eq. 2, we can deduce that when we are immersed in an epidemic with given parameters α, β, and γ, the only way to decrease ρ_0 is to decrease the number of people at risk of infection (variable S). For this reason, social isolation is effective. The SIRD model and the data obtained confirm this conclusion. In [27] it is estimated how catastrophic it would have been if the Spanish Government had not taken social isolation measures on March 14, 2020 [29]. This model estimates that the number of people infected with active disease in Spain would have reached 800,000 by April 12, 2020. But,

Table 2. A sample of the countries most affected by the COVID-19 pandemic. Official data provided by the World Health Organization and the Spanish Government as of October 21, 2021.

Country	Confirmed cases	Deceased
United States	45,070,875	728,125
India	34,108,996	452,651
Brazil	21,651,910	603,465
United Kingdom	8,589,737	139,031
Argentina	5,273,463	115,704
Colombia	4,982,575	126,886
Mexico	3,758,469	284,477
Peru	2,190,396	199,882
France	7,102,079	117,376
Italy	4,725,887	131,688
Germany	4,417,708	94,875
Spain	318,183	87,102

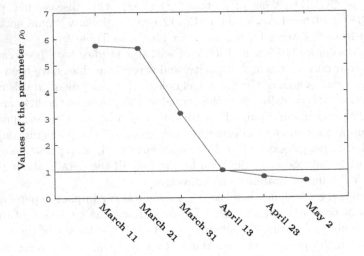

Fig. 1. Values of the parameter ρ_0 over time in Spain during the first wave of the COVID-19 Pandemic.

once the isolation measures were taken, the number of infected people was 87,280. Only 11% of these people were infected without isolation measures.

Figure 2 shows the evolution of the pandemic in Spain: the new cases reported daily, the polynomial fit and the day on which the quarantine began.

Recently, [32] has shown the transmission dynamics of the COVID-19 pandemic in Bangladesh during the third wave caused by the Indian delta variant. This article uses the SIRD model and reaches similar conclusions. Comparable

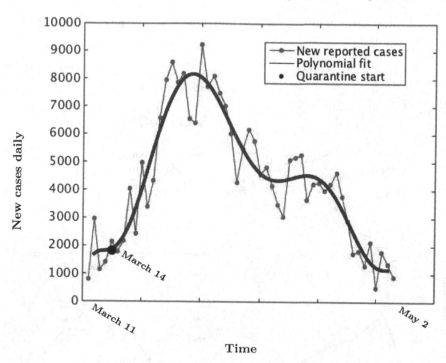

Fig. 2. Official data provided by Spanish Government of new cases daily from 11 March to 2 May 2020.

conclusions were also reached in [33] based on what happened in Italy during the first wave of this pandemic.

March and April 2020 was when the first wave of the pandemic mainly occurred. Italy decreed a state quarantine on March 4 [34] (later, on March 9, it took further measures [35]). France, on 14 March [36], Germany on March 10 [37] (later on March 22 it took further action [37]) and the United Kingdom on March 23 [38,39]. It can be seen that the few days delay in the UK taking its strict isolation measures led to a more dramatic situation: infections took longer to stabilize and the decline in infections was slower, resulting in a higher number of infected people.

Figure 3 shows a comparison of the evolution of the first wave of the COVID-19 pandemic in the four most populated countries of Europe, with the exception of Russia. A similar behavior is observed in the stabilization of new cases reported daily by these countries to WHO [28]. The observed pattern always holds true: the new cases reported daily stabilize between 10 and 14 days after a strict quarantine has been decreed. This delay is due to the incubation period of the virus in people already infected [40].

Finally, we will cite two reports, one from Imperial College London and the other from the British Parliament. The first of them, [41], quantifies the deaths that would have occurred in different countries if no quarantine had

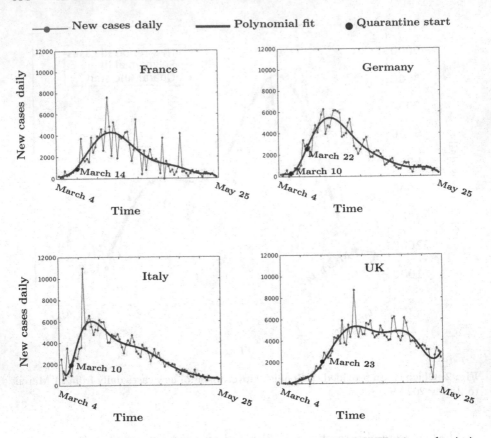

Fig. 3. Comparison of the evolution of the first wave of the COVID-19 pandemic in some countries of Europe (March-April 2020).

been imposed. The second, [39], by two important parliamentary committees (Health and Social Assistance and Science and Technology) after a year of interrogation of the Government's decision-makers, concludes that *a full lockdown was inevitable and should have come sooner*. It criticizes the fact that a lockdown was not ordered until March 23, 2020. Considering that the first case of COVID-19 in the United Kingdom was dated January 31, this decision was made too late and only when the estimated number of deaths was socially unacceptable. The decision to pursue herd immunity without isolation measures and without vaccines was a serious mistake.

In addition, interested readers can observe similar behaviors in these estimates in a wide variety of models for the study of the COVID-19 pandemic: using Bayesian and stochastic techniques [42–44], including mobility [45], isolation and quarantine [46,47], fractional models [48], and logistic models [49].

5   Other Pandemics in History

If we look at the pandemics that have occurred throughout the history of human-ity, perhaps the most similar to the current one is the Spanish Flu Pandemic, which took place between 1918 and 1920 [50–52]. This pandemic caused more than 40 million victims worldwide, and Spain was one of the most badly-affected countries with 8 million people infected and 300,000 deaths. Quarantine, isola-tion measures, and the use of masks were also established to avoid the number of people exposed to the virus.

The Guardian (https://www.theguardian.com/artanddesign/2020/may/03/) published some photographs showing a very popular slogan among many con-scientious people of that time: *No mask go to jail*.

Throughout history, there have been many outbreaks of influenza with a wide variety of viral mutations. In 1918, censorship and a lack of resources meant that the deadly focus of the virus was not investigated. It was not until 1933, during a flare-up, that Wilson Smith, Christopher Howard Andrewes, and Patrick Playfair Laidlaw at the National Institute for Medical Research in London were successful in isolating and identifying the virus. The original article was published in The Lancet [53] (later reproduced in [54] with permission from ©The Lancet Ltd.). We now know that the Spanish flu was caused by an outbreak of influenza virus A (H1N1 subtype). This nomenclature was proposed by the WHO in order to clarify the different types of influenza [55]. Interested readers can find more details in [51,56–59].

The therapy of applying isolation against a disease is not new. In the past, it has been applied against plagues and all kinds of infectious diseases. The Bible reveals the panic that people showed against leprosy, sufferers of which were forced into compulsory isolation. The truth is that this type of therapy has worked for centuries and is based on the collective consciousness of humanity.

If we go back five centuries, we find the work of an illustrious Valencian named Lluís Alcanyís, a professor at the recently created University of Valencia, Spain (1499). He enjoyed great professional prestige and therefore had a decisive influence on the medical and surgical practice of the time. Unfortunately, he ended his days burned at the stake by the Inquisition accused of encouraging Judaism.

The work of Alcanyís [60] published in Catalan in 1490, "Preservative and curative regiment of pestilence" (original name: *Regiment preservatiu e curatiu de la pestilència*), was motivated by a serious plague that the city of Valencia suffered during that year and the following year. It was the first medical work published in the Catalan language and not in Latin, which was the usual scientific language at that time.

Alcanyís advocated a preventive regimen against the cause of divine punish-ment. Among other nutritional and hygiene considerations, he advised staying away from infected people and also from those who came from infected places, albeit timidly due to the deep religious convictions of that time. He could not help mentioning the evil vices of living in sin.

The plague that afflicted the city of Valencia in 1499 was an outbreak of the terrible plague that devastated West Asia, the Middle East, North Africa, and Europe between 1346 and 1353 (*The Black Death*). This terrible disease led to the death of 80% of the people who contracted it, which caused panic among the population and led to very harsh isolation measures. Those who were suspected of being infected were subjected to a strict 40-day quarantine. Sanitary cordons were established, and armed guards guarded the entrances to cities. Punishments of 100 lashes or a 10-year sentence to the galleys were common for people who broke the rules.

In later centuries, there were more outbreaks, and the black rat was identified as the carrier agent. However, it was not until 1894, during one of these outbreaks in the city of Canton (China), that the plague spread to the British colony of Hong Kong. It was then that scientists started to investigate the infectious agent. This task was entrusted to the Japanese Kitasato Shibasaburo and the Frenchman Alexandre Yersin, who were sent to Hong Kong by their respective governments. They detected the bacterium *Yersinia Pestis* as the cause of the infection (it was subsequently named after Alexander Yersin, probably because his findings were more widely publicized among the scientific community than the work of Kitasato Shibasaburo). Later, in 1903, the English entomologist W. Glen Liston identified the rat flea as the transmitting vector. This delay is what explains why, for centuries, there was no treatment or vaccine available. Readers who wish to study *The Black Death* in more depth can consult [61–63].

To get an idea of how devastating epidemics are, we can analyze some figures about the population of the American continent before and after European people arrived there and transmitted the diseases they carried. The indigenous population in Mexico was approximately 25 million people prior to the arrival of Hernán Cortés, and it was reduced to just over a million by 1635. Some researchers have estimated that the population of America would have been between 90 and 112 million inhabitants before Europeans arrived there, and it was reduced to just under 5 million by 1650 (See [64]).

6   Remarks

For centuries, the effectiveness of social isolation has proved to prevent and mitigate the effects of epidemics caused by viruses and bacteria. However, it has been in recent years with the COVID-19 pandemic that scientific evidence has consistently demonstrated such efficacy. In this sense, the validation of mathematical models with real data has been fundamental. In recent decades, this evidence that has been present for centuries has been supported by a multitude of scientific works, a small sample of which are cited in the bibliography.

The variable S, which denotes the people susceptible to infection, is the fundamental variable that must be controlled, as pointed out by common sense and the simulations carried out in [27]. The disappointing data from the COVID-19 pandemic have not been worse than *The Spanish Flu* of 1918 thanks, in particular, to current scientific progress and international collaboration in the fight against the COVID-19 pandemic. It is thanks to the development of several vaccines to protect the population in just one year that the effects have not been worse. Vaccination is a fundamental instrument to reduce the S variable, especially if the vaccine is sterilizing, that is to say, in addition to preventing severe cases, it prevents transmission of the disease.

Currently, without the scientific and technical means at our disposal, the S variable considered worldwide in the current globalized social system from a commercial, tourist, or migratory point of view would have been enormous and difficult to control. If we analyze the measures that have been effective in previous centuries, especially quarantine, it may appear that we have not made much progress in understanding the preventive measures that must be used to protect ourselves from pandemics, but the truth is that we must recognize that the arguments discussed in the preceding paragraphs about scientific and technical advances have prevented many deaths. However, we have an important challenge ahead of us, the challenge of solidarity; we must bring vaccines to the least developed countries. The pandemic needs to be controlled at a global level; otherwise, it will not be controlled.

The impact of the lockdown can be compared to alternative mitigation strategies. In Sweden (https://en.wikipedia.org/wiki/COVID-19_pandemic_in_Sweden) a mild co-confinement was attempted and good results were achieved, also in Japan (https://en.wikipedia.org/wiki/COVID-19_pandemic_in_Japan), they were paying special attention to schools and universities. But in both countries, the evolution of the pandemic was monitored at all times, in case it had become necessary to apply more rigorous measures. It makes sense to recall here the case of the United Kingdom, already commented previously, where at first they acted with the intention of achieving group immunity with a high contagion rate, but when they detected that the health system was not going to be able to withstand the pressure, then they had to change strategy in the middle of the pandemic. It is necessary to consider that this type of strategy was implemented in developed countries and it makes sense to consider that it would have occurred in countries without a strong health system.

References

1. Kermack, A.O., McKendrick, A.G.: A contribution to the mathematical theory of epidemics. Proc. R. Soc. Lond. **115**, 700–721 (1927); reprinted in Bull. Math. Biol. **53**, 33–55 (1991)
2. Aggarwal, D., Page, A.J., Schaefer, U., et al.: Genomic assessment of quarantine measures to prevent SARS-CoV-2 importation and transmission. Nat. Commun. **13**, 1012 (2022). https://doi.org/10.1038/s41467-022-28371-z
3. Al Zabadi, H., Yaseen, N., Alhroub, T., Haj-Yahya, M.: Assessment of quarantine understanding and adherence to lockdown measures during the COVID-19 pandemic in Palestine: community experience and evidence for action. Front. Public Health **9**, 570242 (2021). https://doi.org/10.3389/fpubh.2021.570242
4. Bennett, M.: All things equal? Heterogeneity in policy effectiveness against COVID-19 spread in Chile. World Dev. **137**, 105208 (2021). https://doi.org/10.1016/j.worlddev.2020.105208

5. Bo, Y., Guo, C., Lin, Ch., Zeng, Y., et al.: Effectiveness of non-pharmaceutical interventions on COVID-19 transmission in 190 countries from 23 January to 13 April 2020. Int. J. Infect. Dis. **102**, 247–253 (2021). https://doi.org/10.1016/j.ijid.2020.10.066

6. Bou-Karroum, L., Khabsa, J., Jabbour, M., Hilal, N., et al.: Public health effects of travel-related policies on the COVID-19 pandemic: a mixed-methods systematic review. J. Infect. **83**, 413–423 (2021). https://doi.org/10.1016/j.jinf.2021.07.017

7. Chiyaka, E.T., Chingarande, G., Dzinamarira, T., Murewanhema, G., et al.: Prevention and control of infectious diseases: lessons from COVID-19 pandemic response in Zimbabwe. COVID **2**, 642–648 (2022). https://doi.org/10.3390/covid2050048

8. Foncea, P., Mondschein, S., Olivares, M.: Replacing quarantine of COVID-19 contacts with periodic testing is also effective in mitigating the risk of transmission. Sci. Rep. **12**, 3620 (2022). https://doi.org/10.1038/s41598-022-07447-2

9. Gu, Y., Ullah, S., Khan, M.A., Alshahrani, M.Y.: Mathematical modeling and stability analysis of the COVID-19 with quarantine and isolation. Results Phys. **34**, 105284 (2022). https://doi.org/10.1016/j.rinp.2022.105284

10. James, A., Plank, M.J., Hendy, S., Binny, R., et al.: Successful contact tracing systems for COVID-19 rely on effective quarantine and isolation. PLoS ONE **16**(6), e0252499 (2021). https://doi.org/10.1371/journal.pone.0252499

11. Johansson, M.A., Wolford, H., Paul, P., Diaz, P.S., et al.: Reducing travel-related SARS-CoV-2 transmission with layered mitigation measures: symptom monitoring, quarantine, and testing. BMC Med. **19**, 94 (2021). https://doi.org/10.1186/s12916-021-01975-w

12. Kouidere, A., El Youssoufia, L., Ferjouchia, H., Balatif, O., et al.: Optimal control of mathematical modeling of the spread of the COVID-19 pandemic with highlighting the negative impact of quarantine on diabetics people with cost-effectiveness. Chaos Solitons Fractals **145**, 110777 (2021). https://doi.org/10.1016/j.chaos.2021.110777

13. Kristjanpoller, W., Michell, K., Minutolo, M.C.: A causal framework to determine the effectiveness of dynamic quarantine policy to mitigate COVID-19. Appl. Soft Comput. J. **104**, 107241 (2021). https://doi.org/10.1016/j.asoc.2021.107241

14. Li, F., Jin, Z., Zhang, J.: Assessing the effectiveness of mass testing and quarantine in the spread of COVID-19 in Beijing and Xinjiang, 2020. Hindawi Complex. (2021). Article ID 5510428. https://doi.org/10.1155/2021/5510428

15. Quilty, B.J., Clifford, S., Hellewell, J., Russell, T.W.: Quarantine and testing strategies in contact tracing for SARS-CoV-2: a modelling study. Lancet Public Health **6**, e175–e183 (2021). https://doi.org/10.1016/S2468-2667(20)30308-X

16. Tu, H., Hu, K., Zhang, M., Zhuang, Y.: Effectiveness of 14 day quarantine strategy: Chinese experience of prevention and control. BMJ **375**, e066121 (2021). https://doi.org/10.1136/bmj-2021-066121

17. Wang, R.: Measuring the effect of government response on COVID-19 pandemic: empirical evidence from Japan. COVID **1**, 276–287 (2021). https://doi.org/10.3390/covid1010022

18. Zhang, A.Z., Enns, E.A.: Optimal timing and effectiveness of COVID-19 outbreak responses in China: a modelling study. BMC Public Health **22**, 679 (2022). https://doi.org/10.1186/s12889-022-12659-2

19. Brauer, F., Castillo-Chavez, C.: Mathematical Models in Population Biology and Epidemiology, 2nd edn. Springer, New York (2001)

20. Brauer, F., Feng, Z., Castillo-Chávez, C.: Discrete epidemic models. Math. Biosci. Eng. **7**, 1–15 (2010). https://doi.org/10.3934/mbe.2010.7.1

21. Haefner, J.W.: Modeling Biological Systems. Springer, New York (2005). https:// doi.org/10.1007/b106568
22. Martcheva, M.: An Introduction to Mathematical Epidemiology. Springer, New York (2015). https://doi.org/10.1007/978-1-4899-7612-3
23. Sameni, R.: Mathematical Modeling of Epidemic Diseases; A Case Study of the COVID-19 Coronavirus. arXiv (2020). arXiv:2003.11371
24. Fernández-Villaverde, J., Jones, C.I.: Estimating and Simulating a SIRD Model of COVID-19 for Many Countries, States, and Cities; Working Paper 27128; National Bureau of Economic Research: Cambridge (2020). https://doi.org/10.3386/w27128
25. Diekmann, O., Heesterbeek, J.A.P.: Mathematical Epidemiology of Infectious Diseases: Model Building, Analysis and Interpretation. Wiley, New York (2000)
26. Heesterbeek, J.A.P.: A brief history of R_0 and a recipe for its calculation. Acta. Biotheor. **50**, 189–204 (2002). https://doi.org/10.1023/A:1016599411804
27. Martínez, V.: Modified SIRD model to study the evolution of the COVID-19 pandemic in Spain. Symmetry **13**, 723 (2021). https://doi.org/10.3390/sym13040723
28. Spanish-Government. Web of Instituto de Salud Carlos III. https://covid19.isciii. es. Accessed March-May 2020
29. Spanish-Government. Royal Decree 463/2020, of March 14, Declaring the State of Alarm for the Management of the Health Crisis Situation Caused by COVID-19. https://www.boe.es/eli/es/rd/2020/03/14/463. Accessed 20 Mar 2020. (in Spanish)
30. Spanish-Government. Royal Decree 476/2020, of March 27, Extending the State of Alarm Declared by Royal Decree 463/2020, of March 14, Declaring the State of Alarm for the Management of the Situation of Health Crisis Caused by COVID-19. https://www.boe.es/eli/es/rd/2020/03/27/476/con. Accessed 20 Apr 2020. (in Spanish)
31. Spanish-Government. Royal Decree 487/2020, of April 10, Which Extends the State of Alarm Declared by Royal Decree 463/2020, of March 14, Which Declares the State of Alarm for the Management of the Situation of Health Crisis Caused by COVID-19. https://www.boe.es/eli/es/rd/2020/04/10/487. Accessed 20 Apr 2020. (in Spanish)
32. Faruk, O., Kar, S.: A data driven analysis and forecast of COVID-19 dynamics during the third wave using SIRD model in Bangladesh. COVID **1**, 503–517 (2021). https://doi.org/10.3390/covid1020043
33. Remuzzi, A., Remuzzi, G.: COVID-19 and Italy: what next? Lancet **395**, 1225–1228 (2020). https://doi.org/10.1016/S0140-6736(20)30627-9
34. Italian-Government. Decree of The President of The Council of Ministers of The Italian Republic, of March 4, 2020. https://www.gazzettaufficiale.it/ gazzetta/serie_generale/caricaDettaglio?dataPubblicazioneGazzetta=2020-03-04&numeroGazzetta=55. Accessed 20 Dec 2021. (in Italian)
35. Italian-Government. Decree of The President of The Council of Ministers of The Italian Republic, of March 9, 2020. https://www.gazzettaufficiale.it/ gazzetta/serie_generale/caricaDettaglio?dataPubblicazioneGazzetta=2020-03-09&numeroGazzetta=62. Accessed 20 Dec 2021. (in Italian)
36. French-Government. Decree of The President of The French Republic, of March 13, 2020. https://www.legifrance.gouv.fr/loda/id/JORFTEXT000041721820/. Accessed 20 Dec 2021. (in French)
37. https://www.dw.com/en/covid-how-germany-battles-the-pandemic-a-chronology/a-58026877 . Accessed 20 Dec 2021
38. British-Government. Prime Minister's Office. https://www.gov.uk/government/ organisations/prime-ministers-office-10-downing-street. Accessed 20 Dec 2021

39. House of Commons, United Kingdom. Coronavirus: lessons learned to date. Report of The Health and Social Care and Science and Technology Committees. https://committees.parliament.uk/publications/7496/documents/78687/default/. Accessed 20 Oct 2021

40. McAloon, C., et al.: Incubation period of COVID-19: a rapid systematic review and meta-analysis of observational research. BMJ Open **10**, e039652 (2020). https://doi.org/10.1136/bmjopen-2020-039652

41. Flaxman, S., et al.: Report 13: estimating the number of infections and the impact of non-pharmaceutical interventions on COVID-19 in 11 European countries. arXiv (2020). arXiv:2004.11342

42. Berihuete, A., Sánchez-Sánchez, M., Suárez-Llorens, A.: A Bayesian model of COVID-19 cases based on the gompertz curve. Mathematics **9**, 228 (2021). https://doi.org/10.3390/math9030228

43. Taghizadeh, L., Karimi, A., Heitzinger, C.: Uncertainty quantification in epidemiological models for the COVID-19 pandemic. Comput. Biol. Med. **125**, 104011 (2020). https://doi.org/10.1016/j.compbiomed

44. Umar, M., Sabir, Z., Zahoor Raja, M.A., Shoaib, M., Gupta, M., Sánchez, Y.G.: A stochastic intelligent computing with neuro-evolution heuristics for nonlinear SITR system of novel COVID-19 dynamics. Symmetry **12**, 1628 (2020). https://doi.org/10.3390/sym12101628

45. Aràndiga, F., et al.: A spatial-temporal model for the evolution of the COVID-19 pandemic in Spain including mobility. Mathematics **8**, 1677 (2020). https://doi.org/10.3390/math8101677

46. Tang, B., et al.: The effectiveness of quarantine and isolation determine the trend of the COVID-19 epidemics in the final phase of the current outbreak in China Int. J. Infect. Dis. (2020). https://doi.org/10.1016/j.ijid.2020.03.018

47. De la Sen, M., Ibeas, A., Agarwal, R.P.: On confinement and quarantine concerns on an SEIAR epidemic model with simulated parameterizations for the COVID-19 pandemic. Symmetry **12**, 1646 (2020). https://doi.org/10.3390/sym12101646

48. Ndaïrou, F., Area, I., Nieto, J.J., Silva, C.J., Torres, D.F.M.: Fractional model of COVID-19 applied to Galicia, Spain and Portugal. Chaos Solitons Fractals **144**, 110652 (2021). https://doi.org/10.1016/j.chaos.2021.110652

49. Cherniha, R., Davydovych, V.: A mathematical model for the COVID-19 outbreak and its applications. Symmetry **12**, 990 (2020). https://doi.org/10.3390/sym12060990

50. Belser, J.A., Tumpey, T.M.: The 1918 flu, 100 years later. Science **359**, 255 (2018). https://science.sciencemag.org/content/359/6373/255. Accessed 23 Apr 2020

51. Morens, D.M., Fauci, A.S.: The 1918 influenza pandemic: insights for the 21st century. J. Infect. Dis. **195**, 1018–1028 (2007). https://doi.org/10.1086/511989

52. Pulido, S.: The Spanish Flu: The 1918 Pandemic that Did not Start in Spain. https://gacetamedica.com/investigacion/la-gripe-espanola-la-pandemia-de-1918-que-no-comenzo-en-espana-fy1357456/. Accessed 20 Apr 2020. (in Spanish)

53. Smith, W., Andrewes, C.H., Laidlaw, P.P.: A Virus obtained from influenza patients. Lancet 66–68 (1933)

54. Smith, W., Andrewes, C.H., Laidlaw, P.P.: A virus obtained from influenza patients. Med. Virol. **5**, 187–191 (1995). https://doi.org/10.1002/rmv.1980050402

55. World Health Organization. A revision of the system of nomenclature for influenza viruses: a WHO memorandum. Bulletin of the World Health Organization **58**(4), 585–591 (1980). https://apps.who.int/iris/handle/10665/262025. Accessed Dec 2021

56. Ghendon, Y.: Introduction to pandemic influenza through history. Eur. J. Epidemiol. **10**(4), 451–453 (1994). https://doi.org/10.1007/BF01719673
57. Nguyen-Van-Tam, J.S., Hampson, A.W.: The epidemiology and clinical impact of pandemic influenza. Vaccine **21**, 1762–1768 (2003). https://doi.org/10.1016/S0264-410X(03)00069-0
58. Rafart, J.V.: Epidemiology of influenza A (H1N1) in the world and in Spain. Arch. Bronconeumol. **46**(2), 3–12 (2010). (In Spanish)
59. Centers for Disease Control and Prevention. Types of Influenza Viruses. Department of Health and Human Services of USA. https://www.cdc.gov/flu/about/viruses/types.htm. Accessed Dec 2021
60. Alcanyís, L.L.: Regiment Preservatiu e Curatiu de la Pertilència. Reprint of the University of Valencia (Spain). Introduction by López-Piñero, J. M. Study and edition by Ferrando, A. Valencia University Press (1999). (in Catalan)
61. Ziegler, Ph.: The Black Death. Sutton Publishing Ltd. (2003)
62. Byrne, J.P.: Encyclopedia of the Black Death. ABC-CLIO, Llc. (2012)
63. Benedictow, O.J.: The Complete History of the Black Death. Boydell Press; Revised edition (2021)
64. Sherburne, F., Borah, W.: The historical demography of aboriginal and colonial america: an attempt at perspective. In: Denevan, W.M. (ed.) The Native Population of the Americas in 1492, Second Revised Edition. University of Wisconsin Press, Madison (1992)

Physiological Polyphosphate: A New Molecular Paradigm in Biomedical and Biocomputational Applications for Human Therapy

Werner E. G. Müller[✉], Shunfeng Wang, Meik Neufurth, Heinz C. Schröder, and Xiaohong Wang[✉]

ERC Advanced Investigator Grant Research Group at the Institute for Physiological Chemistry, University Medical Center of the Johannes Gutenberg University, Duesbergweg 6, 55128 Mainz, Germany

{wmueller,wang013}@uni-mainz.de

Abstract. Inorganic polyphosphates (polyP) consist of linear chains of orthophosphate units linked together by high-energy phosphoanhydride bonds. The family of polyP molecules are evolutionarily old biopolymers and found from bacteria to man. PolyP is exceptional, no other molecule concentrates as much (bio)chemically usable energy as polyP in animals, including humans. Before this discovery, we found that the long-neglected polymer provides orthophosphate units required for bone (hydroxyapatite) synthesis. Hence, polyP is a cornerstone for bone synthesis and repair, especially in higher animals. Besides its importance for regenerative medicine, especially for the reconstitution of osteoarticular impairments/defects, a further imperative property could be attributed the polyP. This polymer is the only extracellular generator of metabolic energy in the form of ATP. While the mitochondria synthesize ATP in large amounts intracellularly, it is polyP, which functions as the storage for extracellular ATP. After enzymatic hydrolysis of polyP by alkaline phosphatase (ALP) the released free energy is partially stored in ADP (formed from AMP), which in the second step is up-phosphorylated to ATP by adenylate kinase (ADK). In turn, the two enzymes ALP and ADK are the biocatalytic proteins that conserve the released free energy and store it in ATP, especially in the extracellular space. In a proof-of-concept, we could demonstrate that polyP is an essential component for human regeneration processes, especially in those regions, which are poorly vascularised, like in bone, cartilage and wounds (including chronic wounds).

Keywords: Regenerative medicine · polyphosphate · biomaterial · metabolic energy · morphogenetic activity · tissue regeneration · bone and cartilage · chronic wounds

1 Introduction

During the last years, a group of polymers has attracted increasing attention due to their unique ability to provide metabolic energy, which is essentially needed for tissue regeneration and repair. It is the physiological polymer inorganic polyphosphate (polyP).

I. Rojas et al. (Eds.): IWBBIO 2023, LNBI 13919, pp. 542–559, 2023.
https://doi.org/10.1007/978-3-031-34953-9_42

The polyP molecules are polyelectrolytes composed of multiple phosphate (P_i) residues linked together by high-energy phosphoanhydride bonds. Structurally, the polyP chains are composed of tetrahedrally coordinated P_i units that are interconnected to one another via shared oxygen atoms [1, 2] (Fig. 1).

PolyP has been identified in yeast already during 1880 and its importance was disclosed later in bacteria by Lohmann, Langen, Holzer/Lynen, Kulaev/Belozerskij, and Kornberg (for a description of the history of the polymer, see [3]). Our group focused on polyP already in 1998, when we found that *in vitro* `polyP` modulates the biomineralization process [4]. Subsequently, we described that the biominerals present in skeletal elements such as bone and teeth or in the silica skeletons of the siliceous sponges are synthesized by enzymes, a finding, which also opened a new window for therapeutic approaches for human repair processes. If the enzymes for a given mineral tissue are known, then these proteins become suitable targets for compounds that act either inhibitory or activating on the biological catalysts. These compounds have the potential to become new drugs that are also suitable for use in humans.

The first enzyme in this field was silicatein [5], which synthesizes the skeleton of the most basal animals, the sponges. After that, focusing on human bone, we pinpointed that for bone formation both alkaline phosphatase (ALP) and carbonic anhydrase are needed [6, 7]. After these observations it became possible to sketch the bone anabolic processes with the following sequence shown in Fig. 2. During endochondral ossification, the hyaline cartilage acts as a template mold for the initial mineralization, most likely of calcium carbonate. In parallel, an ingrowth of blood vessels occurs, followed by the formation of the primary ossification centers in the diaphysis [2]. Later, spongy bone is formed in the epiphyses at the secondary ossification centers. In these two regions of the hyaline cartilage that remain on the surface of the epiphysis (articular cartilage) and the epiphyseal plate (growth region) between the epiphysis and the diaphysis final hydroxyapatite (HA) bone formation takes place. Appositional growth of the bone proceeds in the absence of a cartilage template [1, 2].

The distinguished feature of all the repair processes involved in tissue regeneration, from bone via cartilage to wounds, is the involvement of polyP a physiological polymer synthesized in all cells and in high concentrations in the blood platelets [1, 2, 8]. The unique property of this ancient "simple" polymer polyP is that it entirely consists inorganic phosphate (P_i) units that are linked by high-energy phosphoanhydride bonds [9]. PolyP is synthesized intracellularly in the mitochondria from ATP [1, 2]. It is enriched in the acidocalcisomes, small cell organelles found in all cells, including bone mineral forming osteoblasts [10–12] (Fig. 3A). Especially in the platelets, high amounts of polyP are found in the acidocalcisomes ("dense granules"), which is secreted after activation [11]. Extracellularly, polyP is present either as a soluble polymer together with bound counterions or as nanoparticles formed with divalent cations bound to the platelet surface. Nanoparticles of different polyP salts can also be obtained synthetically, e.g. Ca-polyP nanoparticles (Fig. 3B), which can be prepared by a biomimetic procedure from soluble polyP (Na-polyP) and a Ca^{2+} salt solution ($CaCl_2$) [13]. The EDX spectrum of Ca-polyP nanoparticles reveals the presence of the Ca, P and O atomic constituents (Fig. 3C). In addition to Ca-polyP nanoparticles, a series of nanoparticles of polyP with further (mostly divalent) counterions has been prepared, e.g. Sr-polyP nanoparticles [14].

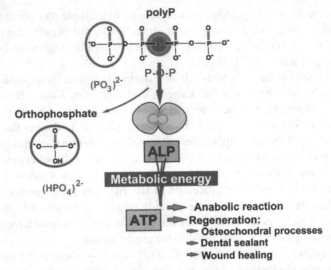

Fig. 1. Structure of polyphosphate (polyP) and release of orthophosphate and transfer of the chemical energy stored in the energy-rich P-O-P phosphoanhydride bonds of the polymer under formation of ATP needed for metabolic energy-consuming anabolic reactions and regeneration/repair processes.

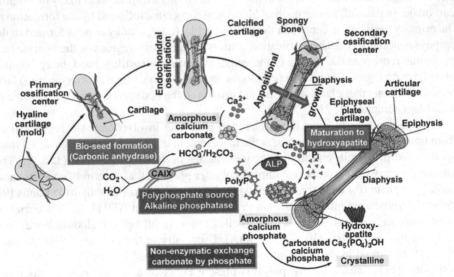

Fig. 2. Steps during bone formation (scheme). The mineralization starts with endochondral ossification, a process that is initially driven by carbonic anhydrase IX (CAIX). Mineral deposition involves two enzymes; first, carbonic anhydrase and second, alkaline phosphatase (ALP). The latter enzyme hydrolyzes physiological polyP to orthophosphate (P_i), which replaces carbonate in ACC under formation of amorphous calcium phosphate (ACP). ACP then maturates to hydroxyapatite (HA). Details are given in the text (Taken from ref. [2] with permission).

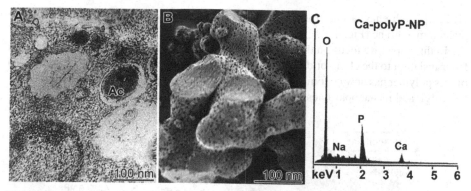

Fig. 3. Natural and synthetic polyP particles/depositions. (**A**) Acidocalcisomes (Ac) in osteoblast-like SaOS-2 cell, containing accumulated polyP deposits; TEM. (**B**) Ca-polyP nanoparticles; SEM. (**C**) EDX spectrum of Ca-polyP nanoparticles (Ca-polyP-NP) showing the pronounced peaks of Ca, P and O. The Na peak is due to residual amounts of sodium from Na-polyP used for preparing the particles.

After the disclosure of the bio-mineralization processes in bone it was demonstrated that amorphous Ca-phosphate (ACP) particles, stabilized by inorganic polyP, are a suitable matrix for cell growth and attachment and show pronounced osteoblastic and vasculogenic activity in *in vitro* and also *in vitro* angiogenesis events in the tube forming assay. A possible involvement of an ATP gradient generated by polyP during tube formation of human umbilical vein endothelial cells was confirmed by ATP-depletion experiments [15]. In order to assess the morphogenetic activity of the hybrid particles *in vivo*, experiments in rabbits using the calvarial bone defect model were performed. The particles were encapsulated in poly(D,L-lactide-*co*-glycolide) microspheres. In contrast to crystalline Ca-phosphate amorphous ACP even caused pronounced osteoinductive activity already after a six-week healing period [14, 16].

A further important property of polyP was determined in *in vitro* experiments – the generation of extracellular metabolic activity [17]; Fig. 4. Using the appropriate cells, human umbilical vein endothelial cells (HUVEC), it was disclosed that these cells require a considerable amount of metabolic energy, of ATP. During incubation of these cells with polyP in the extracellular space, this polymer was hydrolyzed *via* ALP and, together with adenylate kinase (ADK), ATP is finally formed, which provides the energy for cell migration during initial vascularization. This process was abolished by apyrase, which eliminates extracellular ATP. ATP was also identified as a signal for the chemotactic migration of cells during vascularization. The data demonstrated that polyP is the storage for extracellular ATP and also a signaling molecule for an autocrine chemotactic pathway of ATP with polyP during endothelial cell migration [16].

The joint action of the two enzymes ALP and ADK prevents this chemical energy stored in polyP from being dissipated as heat, as in the case of hydrolysis of the polymer, instead being retained in the form of ATP (via phosphotransfer of energy-rich phosphate groups). A mechanism has been proposed by which ALP forms ADP from AMP, which is then used as a substrate for the ADK-catalyzed interconversion reaction to form ATP

[1, 2, 8]. Both enzymes are membrane-associated and located on the outer side of the plasma membrane (Fig. 4).

In this paper, we focus on those potential applications of polyP that are most relevant for translation to the clinic or daily use by consumers in the near future. The application of this polymer has now become realistic since a GMP-compliant manufacturing process for polyP and its nanoparticles has been developed and is available by us.

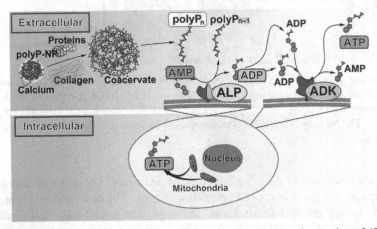

Fig. 4. Extracellular ATP generation from polyP. Extracellularly, a major portion of ATP needed for tissue regeneration and repair is delivered by the blood platelets that release polyP in the form of particulate or soluble polyP after activation, particularly at the sites of tissue damage. If not available in sufficient amounts, under certain pathogenic conditions (viral infections) or high demand for the polymer, synthetically obtained calcium-polyP nanoparticles (Ca-polyP-NP) can be administered. In the presence of proteins such as in wound fluids, the particles are transformed into a coacervate, which, after disintegration, releases polyP. This polymer is then used by the two cell membrane-bound enzymes ALP (phosphotransfer from polyP to AMP) and adenylate kinase (ADK; interconversion of the resulting ADP to AMP and ATP) to form ATP. Intracellularly, ATP is mainly produced by the mitochondria.

2   Methods

2.1   Preparation of Ca-polyP Nanoparticles and Coacervate

Amorphous nanoparticles of Ca-polyP were prepared from soluble polyP (sodium salt; Na-polyP) with an average chain length of 40 P_i units and calcium chloride as described [13, 16].

2.2   Fabrication of Compressed Collagen-Based Mats

The collagen-based mats were fabricated by plastic compression either without or after supplementation with Ca-polyP nanoparticles as described [18]. Bovine tendon collagen type I was used. The mats were stored in 70% (v/v) ethanol.

2.3 Formulation of polyP-Containing Hydrogel

The polyP-containing hydrogel was prepared by adding a sterile-filtrated suspension of soluble polyP and particulate polyP (Na-polyP to Ca-polyP-nanoparticle weight ratio of 10:1) to an autoclaved suspension/solution of hydroxyethyl cellulose as outlined [19].

2.4 Cultivation of Bacteria

For cultivation onto the hydrogel, the bacteria (*Staphylococcus aureus*) were cultivated in liquid cultures with tryptic soy broth medium as described [19]. Aliquots of liquid overnight cultures were spread on dishes layered with the hydrogel and incubated over various time periods.

2.5 Treatment of Chronic Wounds

The studies on patients suffering from chronic wounds (off-label studies) were conducted after written and signed informed consent and respecting the rules of Good Clinical Practice and the Helsinki declaration (following §37 of this Declaration).

2.6 Preparation of polyP Dentifrice

The composition of the polyP-supplemented dentifrice has been given earlier [20]. Na-polyP and Ca-polyP microparticles were present at a final concentration of 3% [w/w] and 1% [w/w]respectively.

2.7 Treatment of the Teeth

For brushing of teeth an electric toothbrush was used as described [20].

2.8 Analytical Procedures

Fourier transformed infrared spectroscopic (FTIR) analysis was performed with ground powder in an Agilent ATR (attenuated total reflectance)-FTIR spectroscope/Varian 660-IR spectrometer, an EDAX Genesis System attached to a scanning electron microscope was used. Observations by transmission electron microscopy (TEM), scanning electron microscopy (SEM) and environmental scanning electron microscopy (ESEM) were performed as described [16, 19]. Light microscopical images were taken with a Keyence VHX600 Digital Microscope.

2.9 In Silico Simulation

For the molecular dynamics simulation studies, the BIOVIA Discovery Studio 3.1 program [21] was used, assuming a polyP chain length of 20 P_i units [16].

3   Results

3.1   polyP – Basic Studies: Nanoparticle and Coacervate Formation and Function

In the presence of Ca^{2+} ions in aqueous solution, polyP can form either a coacervate or a nano/microparticle, depending on the pH conditions. At neutral pH (pH 7) a coacervate is obtained, while in alkaline solution (pH 10) Ca-polyP nano- or microparticles are formed. The size of the particles (nano or micro) can be adjusted by selecting the appropriate preparation protocol (reaction time, mixing velocity, size separation, etc.). During the reaction, the polyanionic polyP molecules (Na-polyP) interact with the oppositely charged cationic Ca^{2+} ions. However, this interaction does not only occur via electrostatic attractions, but also the ion-dipole attractive forces occurring in the aqueous solution must be considered, as discussed recently [16]. Therefore, molecular dynamics simulations were performed that allow to predict and visualize the changes in the spatial arrangement of the three partners polyP, Ca^{2+} and water during coacervate and nanoparticle formation [16]. Initially, both at pH 7 and pH 10, a clustering of stochastically arranged polyP molecules and Ca^{2+} ions is seen (Fig. 5 Left, A and B). However, at pH 10, after 1000 cleaning cycles, a separation of the polyP molecules and Ca^{2+} ions takes place, leading to the formation of a polyP-rich core surrounded by a Ca^{2+}-rich periphery (Fig. 5 Left, C and D). In contrast, the simulation at pH 7 did not show a charge separation during Ca-polyP coacervate formation (not shown in Fig. 5). In Fig. 5 Left, B and D, the distribution of the water molecules (omitted in Fig. 5 Left, A and C) is shown; it is obvious, that the water molecules are not displaced from the center of the formed aggregates, but are still present or even enriched.

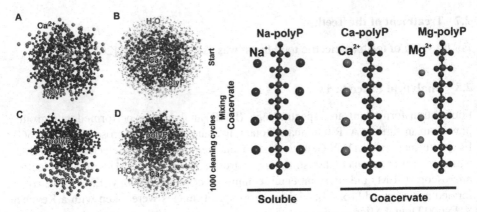

Fig. 5.   Left: Simulation of polyP nanoparticle formation by interaction of polyP molecules (chain length of 20 P_i units) and Ca^{2+} and water. Initially, at pH 10, a clustering of stochastically arranged polyP molecules and Ca^{2+} ions is observed (**A** and **B**), which, after 1000 cleaning cycles, separate into a polyP-rich core and a Ca^{2+}-rich periphery (**C** and **D**). At neutral pH (pH 7), no charge separation occurs, leading to formation of a Ca-polyP coacervate (not shown). **Right:** Structure of the polyP salts with monovalent and divalent cations. The simulation studies showed that the monovalent ions (Na^+) are arranged alternately, while the divalent cations (Ca^{2+} and Mg^{2+}) are located only on one side of the polyP strand.

Further results, again based on the simulation studies, revealed that in Na-polyP the monovalent Na^+ ions are arranged alternately, while in the polyP salts with Ca^{2+} and Mg^{2+} the divalent cations are located only on one side of the polymer (Fig. 5 Right). Based on these results a model for polyP coacervate formation and polyP nanoparticle formation has been proposed [16]. It is assumed the Ca^{2+} ions sitting at one side of the polymer form a ribbon around a polyP core, leading – at pH 7 – to the formation of small liquid droplets that fuse together trapping water within the associates formed.

In contrast to nanoparticle formation, coacervate formation is a very fast process [19]. This can be demonstrated by dropping saline onto a polyP-containing viscous hydrogel, as shown in Fig. 6A-C. Immediately a formation of schlieren is observed. This process continues during addition of further salt solution (Fig. 6B and C). The coacervate within the overlay solution is formed from the Na-polyP present in the hydrogel, which has a pH of 6.5 [19, 22]. The amount of the coacervate product can be further increased if a protein-containing (10% fetal calf serum) medium is used for overlaying the hydrogel. Then the Ca-polyP nanoparticles, which are present in the hydrogel, also undergo coacervation in the soluble phase besides the soluble Na-polyP ingredient [19].

Coacervate formation can lead to the entrapment (and inactivation) of virus particles as well as of bacteria. This we could demonstrate using the respiratory virus SARS-CoV-2 or the pathogenic bacteria *S. aureus* and *Streptococcus mutans*. The incorporation of polyP and Ca-polyP particles allowed the development of antiviral masks preventing corona virus infection not only by killing of SARS-CoV-2 but also by engulfment of virus particles by coacervate formation of polyP with proteins present in the mucus-mucin- and viral-protein-containing aerosol particles [23, 24]. The trapping of *S. aureus* bacteria, which play a dominant role in wound infections, is illustrated in Fig. 6D-F. In the experiment shown the bacteria were placed onto Petri dishes covered with a hydrogel without or with polyP ingredient. If the bacteria were incubated in medium onto the polyP-lacking hydrogel, they remain suspended during the whole 2 h incubation period (Fig. 6D). If, however, the bacteria were kept onto the surface of the polyP-containing hydrogel, they begun to aggregate already during a 1 h incubation period (Fig. 6E) and became entrapped into the formed polyP coacervate [19] (Fig. 6F).

3.2   polyP – Application in Dentistry and Dental Care

There is a huge number of products on the market for dental care, especially denti-frices, as well as specialty products, but there are still many deficiencies and unresolved issues that await novel solutions. In fact, we were able to show that amorphous Ca-polyP microparticles are able to efficiently repair cracks/fissures in the tooth enamel and dentin and to reseal open dentinal tubules exposed on the tooth surface [20, 25]. PolyP microparticles with a size of 100–400 nm were found to be most effective. In addition, we found that amorphous Ca-polyP microparticles exhibit morphogenetic activity and induce the expression of the ALP gene in precursor odontoblasts [25]. Furthermore, it turned out that these particles, especially when enriched with retinyl acetate, cause a strong increase in collagen type I expression in the periodontal tissue surrounding the teeth [20]. Moreover, it was found that polyP elicits a strong antibacterial effect against the cariogenic *S. mutans* [20]. This effect is mainly caused by the soluble polyP,

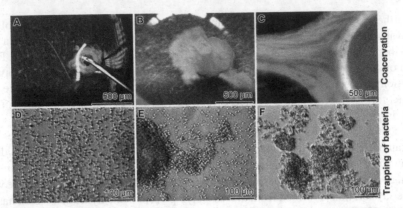

Fig. 6. Coacervate formation of polyP leading to entrapment of bacteria. (**A-C**) Formation of a coacervate on the surface of Na-polyP- and Ca-polyP nanoparticles-supplemented hydrogel. Overlaying the hydrogel with saline added *via* a needle results in the formation of a coacervate of the readily soluble Na-polyP; microscopic view (time course from 0 to 30 min). (**D–F**) Trapping of bacteria (*S. aureus*) into the polyP coacervate formed on the hydrogel surface. (D) Bacteria incubated in medium on the polyP-lacking hydrogel are still suspended after 2 h. (E) Bacteria on the surface of the polyP-containing hydrogel tend to aggregate and become entrapped into coacervate (Coa) clusters; light microscopy. (Taken from ref. [16] with permission).

which, unless administered directly as the sodium salt (Na-polyP), is formed by ALP-mediated disintegration of the microparticles in the oropharyngeal cavity. The inhibition of *S. mutans*, the main causative agent of caries (tooth decay), shows a comparatively high selectivity compared to the broad-spectrum antibiotic triclosan, which only slightly affects the cariogenic bacterium but inhibits beneficial bacteria [20].

Based on these results, a biomimetic dentifrice supplemented with 1% [w/w] Ca-polyP microparticles was developed, which exploits the biological effects of the physiological polymer polyP [20, 25]. In order to demonstrate the sealing activity of this dentifrice, a series of tests was performed in which the polyP-supplemented dentifrice was compared with a dentifrice of the same composition but lacking polyP. Human teeth were brushed with both dentifrices over a period of one week (twice daily) and the surfaces of the dentin and enamel zones were analyzed electron microscopically. It was found that after treating the teeth with the polyP-free control paste for one week, the deep carious lesions on the tooth surface (Fig. 7A) were not closed by a homogenous covering (Fig. 7B). In contrast, the teeth treated with the Ca-polyP microparticle-containing dentifrice for the same period of time exhibited smooth surfaces with no chips or residual cracks as in the controls (Fig. 7C). More detailed studies combining SEM and EDX revealed that the cement and dentin surfaces of teeth incubated with amorphous Ca-polyP microparticles, but not controls, were covered by a 50-μm thick and nearly homogenous, coacervate-like polyP layer [25], most likely caused by fusion of the microparticles by ALP present in the tooth mineral [26]. The mechanical properties of this polyP coating were close to natural enamel with a Martens hardness of 4.33 ± 0.69 GPa and a reduced elastic modulus of 101.61 ± 8.52 GPa; already after a 3 h exposure of teeth to the amorphous Ca-polyP microparticles, these values amounted to 3.85 ± 0.64 GPa and $94.72 \pm$

8.54 GPa, respectively [25]. The sustainability of the dental covering obtained with the polyP-based dentifrice was examined by ultrasound treatment. The sealing was found to be stable and resisted high power sonication [20]. In addition, the regular use of the developed dentifrice with the incorporated amorphous polyP microparticles led to an efficient reduction in dental biofilm formation.

Dentifrice: minus polyP   **plus polyP**

Fig. 7. Sealing of the tooth surface with polyP-supplemented dentifrice. (**A**) Tooth with the enamel (ena) and dentin (den) regions. (**B**) Tooth surface treated with a control dentifrice lacking polyP. (**C**) Tooth surface after treatment with the Ca-polyP microparticle-containing dentifrice. Only with the dentifrice supplemented with 1% [w/w] Ca-polyP microparticles all cracks are sealed and a smooth coacervate film on the tooth surface is observed.

The amorphous Ca-polyP microparticles can even be incorporated into the bristles of a toothbrush to develop their regenerative, resealing and anti-caries activities during brushing of teeth, even in the absence of polyP in the dentifrice used. Such innovative microsized Ca-polyP containing toothbrush bristles have recently been fabricated by extrusion of polyester filaments containing embedded amorphous Ca-polyP microparticles.

A further interesting, application-relevant property of the polyP component of the dentifrice and the toothbrush bristles is their anti-coronavirus SARS-CoV-2 activity. We discovered that polyP specifically binds to the spike proteins of the virus particles, thereby preventing the attachment of the virus to the host cell receptor [24, 27]. A dentifrice (or a toothbrush) that is protective against infection by SARS-CoV-2 is new and has not been realized before. PolyP is even effective against virus mutants [28]. Based on this antiviral activity we also developed nasal spray (inhalant) containing both soluble (Na-polyP) and particulate polyP (Ca-polyP nanoparticles) [23].

3.3 polyP – Application in Wound Healing

As a further strong proof-of-concept, we succeeded to show that polyP is a powerful curative molecule for wound healing [18, 19]. Wound healing is a multi-stage process consisting of several partially overlapping phases that are run through in a fixed order: hemostasis, inflammation, granulation, and epithelialization. Disturbances in this process can lead to the emergence of chronic wounds that are characterized by long duration, impaired vascularization and high risk of infection. Such difficult-to-treat chronic or non-healing wounds place a significant burden on patients and also on healthcare systems. They can be caused and arise from various pathophysiological conditions such

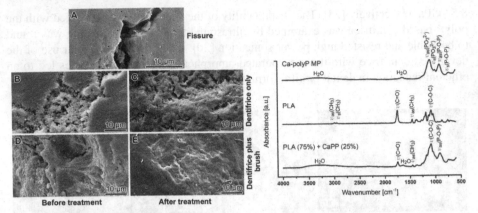

Fig. 8. Efficiency of daily treatment of teeth with a conventional dentifrice applied with a toothbrush with Ca-polyP-microparticle-supplemented bristles compared to the dentifrice only. **Left:** (**A**) Fissure/crack on tooth. (**B** and **C**) Treatment with dentifrice only. (**D** and **E**) Application of dentifrice together with toothbrush with Ca-polyP-microparticle-containing bristles. Only in the latter case, a smooth surface sealing of tooth fissures/cracks are seen. **Right:** FTIR analysis of the toothbrush bristles consisting of Ca-polyP-microparticles ("CaPP"; 25%) incorporated into PLA (75%), compared to the individual components.

as metabolic disorders (e.g. diabetic foot ulcers), cardiovascular diseases (ischemia), persistent pressure on skin (bedsores), tumor diseases, infectious diseases or increasing age. In the United States, around 15% of all Medicare beneficiaries suffer from at least one type of wound. Among them diabetic wounds have the second highest prevalence; the annual cost expenditures for wound care range between $30 and $100 billion.

A hallmark of impaired wound healing is a chronic hypoxic condition due to impaired oxygen and nutrient supply to the wound tissue. Wound regeneration is a highly metabolic energy-demanding (ATP-dependent) process [29]. ATP is required not only intracellularly but also extracellularly for the formation of new blood vessels, the structural organization of the extracellular matrix (ECM) and the function of extracellular chaperon(-like) proteins (e.g., clusterin) and enzymes (e.g., kinases and cis-trans isomerases) [8, 30]. The energy requirement for these process increases dramatically during wound healing. Therefore, adequate oxygenation is essential to maintain the ATP production required for tissue regeneration in the wound area. The lack of metabolic energy in chronic wounds also leads to disturbances in angiogenesis. The initial stage of angiogenesis, microvascularization, requires an energy-dependent migration of cells along a chemotactic gradient. Using HUVEC cells as a model, we were able to show that ATP acts as a chemotactic signal for endothelial cells [15].

The ECM lacks the cell's main energy factories, the mitochondria, as well as other ATP-generating cellular pathways such as glycolysis, and typically the amounts of ATP released by the cells are rather small. However, as we have shown, polyP can serve as an efficient source of ATP in the extracellular space, which is formed via the enzyme pair ALP and ADK bound to the outer cell membrane [12, 31]. Physiologically, polyP is distributed to the sites of tissue injury throughout the body via the blood platelets and released upon activation. The crucial role of platelets in wound healing has long been known [32, 33]. It was therefore reasonable to assume that the administration of exogenous polyP would be a rational solution, particularly for the treatment of chronic wounds. In fact, we were able to show that polyP markedly promotes wound repair. Animal experiments with healthy and diabetic mice with impaired wound healing showed that the polyP nanoparticles significantly accelerate wound healing [28]. Exogenous polyP can replace the missing or insufficient polyP in the area of the chronic wound. It can be administered either in a readily soluble form, as Na-polyP, or in particulate form, as amorphous Ca-polyP nanoparticles, a more slowly degradable depot form of the polymer. The enzymes ALP and ADK necessary for ATP formation are present at the site of the damaged tissue and are also found in the wound exudate.

The wound healing studies in mice treated with the nanoparticles revealed, besides an increased formation of granulation tissue, an induction of microvessel formation and a rise of the ATP level in the wound area [34]. The essential role of ATP was confirmed by exogenous application of the nucleotide to wounds [35]. In addition, an upregulation of collagen types I and III expression and the expression of the genes encoding for α-smooth muscle actin and plasminogen activator inhibitor-1 was found [34].

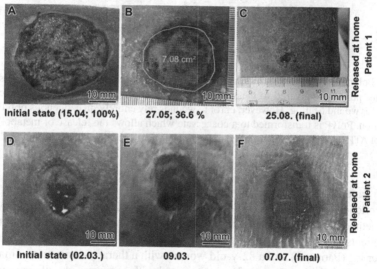

Fig. 9. Successful healing of chronic wounds in patients using polyP-containing compressed collagen mats. (**A** to **C**) Healing progression of the chronic wound of Patient 1. (A) Wound before debridement and (B) 6 weeks and (C) 18 weeks after start of treatment. The size of the wound (circled in B) is expressed as percentage of the initial wound area (100% in A). (**D** to **F**) Healing of the chronic wound of Patient 2. (D) Initial state and state after one (E) and 18 weeks (F). After this period, the treatment of both patients with the polyP-based mats could be terminated. (Taken from ref. [18] with permission).

Based on these results, we succeeded to provide a protocol for a complete healing of chronic wounds in patients. In the "bench to bedside" process, our protocol was introduced into the clinical application. The polyP was engineered into collagen-based mats and applied onto human wounds [18]. Alternatively, polyP produced in the form of a gel can be used [19]. In addition to amorphous Ca-polyP nanoparticles, which serve as a depot form, the readily water-soluble and immediately available Na-polyP is included in the formulations or added separately. The results of first clinical studies (off-label) on patients with chronic wounds are available and have been consistently successful.

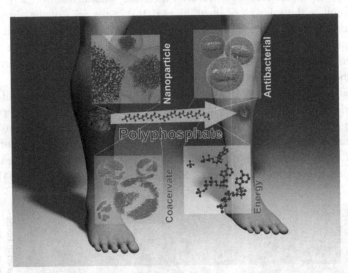

Fig. 10. Polyphosphate, the first biomaterial to heal chronic wounds. Under normal physiological conditions, polyP is delivered to the wound via the bloodstream and supplies sufficient polyP required for wound healing. However, chronic wounds rely entirely on exogenous polyP to allow regeneration. PolyP is transformed to a coacervate which allows the release of metabolic energy in form of ATP. This polymer is supplied to the injury via mats of gels. In addition polyP is killing bacteria.

Two patients are described in more detail. The treatment of these patients with the polyP-enriched collagen mats started in March/April 2021. The application of conventional methods of wound treatment had previously been unsuccessful in both cases. Both patients, a 79-y-old man suffering from a chronic wound on the ventral tibia of the lower leg (Patient 1) and a 82-y-old woman with a therapy-resistant traumatic ulcer on the lateral malleolus (Patient 2), were treated with compressed collagen containing incorporated amorphous Ca-polyP nanoparticles [18]. In addition to the mats, a wetting solution containing immediately available soluble polyP (Na-polyP), in addition to Ca-polyP nanoparticles, was regularly applied to the wound area. Application of the polyP-supplemented wound mats to the wound of Patient 1, an ulcer approximately 5 cm in diameter (Fig. 9A) developed after surgical tumor resection, impressively accelerated the re-epithelialization rate, with a reduction of the wound area to 36.6% after 6 weeks (Fig. 9B); complete healing was achieved by 18 weeks and no further treatment was required (Fig. 9C). The condition of the wound of Patient 2 prior to treatment is shown in Fig. 9D. After just one week, a marked reduction of the peripheral redness (inflammation) and the formation of a marginal epithelial rim surrounding the wound was observed

(Fig. 10E) and treatment with the polyP mats was successfully terminated after 18 weeks (Fig. 9F). A further sketch is given in Fig. 10.

3.4 polyP – Manufacturing and Safety

PolyP can be produced at low-cost – and also with GMP conformity. In view of the planned registration and large-scale production, a GMP-qualified fabrication method for amorphous Na-polyP and amorphous Ca-polyP nanoparticles has been developed (to be published). An advantage of this method is that it is a one-pot procedure, allowing to synthesize Na-polyP and Ca-polyP nanoparticles from simple chemical precursors almost simultaneously, and it is very fast (only a few hours). PolyP, the product, is a safe compound. No adverse effects has been reported. PolyP shows no carcinogenicity, reproductive or developmental toxicity in rats, and no genotoxicity [36, 37]. Also no allergic reactions have been observed [38]. In addition, this polymer is a widely used food additive and has been assessed by the US Food and Drug Administration (FDA), the European Union (EU) and the joint FAO/WHO Expert Committee on Food Additives (JECFA) (World Health Organization (WHO) 2002; European Food Safety Authority 2013) [39, 40]. The EU allows the addition of polyP to food in accordance with EU Directive 95/2/EC (amended by 98/72/EC); it is listed under the E numbers E452i (Na-polyP), E452iii (Na/Ca-polyP) and E452iv (Ca-polyP). The maximum tolerable daily intake (MTDI) of polyP (expressed as P_i) is 70 mg kg^{-1} body weight [41]. The LD$_{50}$ for Na-polyP is >10,000 mg kg^{-1} for rats (Weiner et al. 2001) and 3700 mg kg^{-1} for mice [42].

4 Discussion

There is an increasing need for regeneratively active materials for tissue engineering/repair and for prevention of degenerative processes, particularly in relation to osteochondral (cartilage and bone including teeth) tissues. A major reason is the demographic development with the increasing proportion of older people worldwide, but also changes in lifestyle (sedentary activities or changes in diet) and an increasing health awareness. For example, the estimated number of osteoporosis patients in Germany alone is over 6 million, and late complications of osteoporosis, especially fractures due to reduced bone mineral density, are a major cost driver for the health system. In Germany, around 80,000 osteoporotic vertebral body fractures and around 130,000 femoral neck fractures are counted every year. There is also a growing (but unmet) need in dentistry for materials that promote bone regeneration. Up to 20% of patients requiring a dental implant will require a sinus lift (sinus augmentation) after tooth extraction. The materials currently used are inert and do not induce natural bone formation. In view of the high number of sinus lifts performed annually (approx. 2 million, with an estimated world market of approx. 1.5 billion euros) and the long healing time until stable osseointegration of the dental implant is possible (6–12 months), the availability of regeneratively active materials such as amorphous Ca-polyP, which stimulate ingrowth, differentiation and mineralization of bone-forming cells, would significantly reduce patient suffering and discomfort. In the oral care market, too, there is an increasing interest in regeneratively active materials, particularly specialty products with these properties, due to changing consumer behavior. Similarly, there is a huge and rapidly growing market for wound dressings worldwide. The prevalence of various types of wounds requiring therapy such as surgical wounds, ulcers or burns is globally more than 500 million people every year. In particular, the therapy of chronic, non-healing or delayed healing wounds is still a major challenge. In the United States alone, 3–6 million people suffer from non-healing wounds. Products with properties similar to the polyP-based materials described in this paper are not available.

In our studies, we focus on inorganic materials showing regenerative activity. Three inorganic material/polymers turned out to exhibit such property: amorphous silica, amorphous calcium carbonate (ACC) and amorphous polyphosphate (Na-polyP or Ca-polyP nanoparticles) [8, 43, 44]. It is crucial that these materials are amorphous. Only in the amorphous state, they are able to stimulate cell differentiation and proliferation, as well as mineralization of bone-forming cells and the expression or activity of the proteins or enzymes involved in tissue regeneration/repair. And - only - the amorphous materials are biodegradable and able to adapt to the specific situation, e.g. during maturation of hard tissues.

5   Conclusion

We applied new concepts for biology-based processes in human biomedicine from nature, preferably from metazoans. Both bioinspired and biomimetic approaches are possible. It is demonstrated that polyP, a physiological inorganic polymer, is a new powerful material in biotechnology that solves important needs in human biomedicine, such as for bone repair and wound healing, as worked out by us. A key point is that this material is physiological and therefore biocompatible and free from side effects. Furthermore, this polymer is able to induce the physiological repair processes. The damaged tissue or hard substances are thus replaced by newly formed endogenous tissue or biominerals. This is important because it is the only way to ensure full functionality, an efficient immune defense and the ability to adapt to changing physiological conditions and stress, for example in the course of the aging process.

Acknowledgements. W.E.G. Müller is the holder of an ERC Advanced Investigator Grant (Grant No.: 268476). In addition, W.E.G. M. has obtained three ERC-PoC Grants (Si-Bone-PoC, Grant No.: 324564; MorphoVES-PoC, Grant No.: 662486; and ArthroDUR, Grant No.: 767234). This work was also supported by the International Human Frontier Science Program and the BiomaTiCS research initiative of the University Medical Center, Mainz. Further support came from a Grant from BMBF (SKIN-ENERGY, Grant No.: 13GW0403A/B).

References

1. Müller, W.E.G., Schröder, H.C., Wang, X.H.: Inorganic polyphosphates as storage for and generator of metabolic energy in the extracellular matrix. Chem. Rev. **119**, 12337–12374 (2019)

2. Wang, X.H., Schröder, H.C., Müller, W.E.G.: Amorphous polyphosphate, a smart bioinspired nano-/bio-material for bone and cartilage regeneration: towards a new paradigm in tissue engineering. J. Mat. Chem. B **6**, 2385–2412 (2018)
3. Langen, P.: Research on inorganic polyphosphates: the beginning. In: Schröder, H.C., Müller, W.E.G. (eds) Inorganic Polyphosphates - Biochemistry, Biology, Biotechnology. Prog. Mol. Subcell. Biol. 23, pp. 19–26. Springer, Berlin (1999)
4. Leyhausen, G., et al.: Inorganic polyphosphate in human osteoblast-like cells. J. Bone Miner. Res. **13**, 803–812 (1998)
5. Krasko, A., Lorenz, B., Batel, R., Schröder, H.C., Müller, I.M., Müller, W.E.G.: Expression of silicatein and collagen genes in the marine sponge Suberites domuncula is controlled by silicate and myotrophin. Europ. J. Biochem. **267**, 4878–4887 (2000)
6. Müller, W.E.G., Schröder, H.C., Schlossmacher, U., Grebenjuk, V.A., Ushijima, H., Wang, X.H.: Induction of carbonic anhydrase in SaOS-2 cells, exposed to bicarbonate and consequences for calcium phosphate crystal formation. Biomaterials **34**, 8671–8680 (2013)
7. Du, T., et al.: Orthophosphate and alkaline phosphatase induced the formation of apatite with different multilayered structures and mineralization balance. Nanoscale **14**, 1814–1825 (2022)
8. Suess, P.M., Smith, S.A., Morrissey, J.H.: Platelet polyphosphate induces fibroblast chemotaxis and myofibroblast differentiation. J Thromb Haemost. **18**, 3043–3052 (2020)
9. Kulaev, I.S., Vagabov, V., Kulakovskaya, T.: The Biochemistry of Inorganic Polyphosphates, 2nd edn. John Wiley, Chichester (2004)
10. Docampo, R., Ulrich, P., Moreno, S.N.: Evolution of acidocalcisomes and their role in polyphosphate storage and osmoregulation in eukaryotic microbes. Philos. Trans. R. Soc. Lond. Ser. B Biol. Sci. 365, 775–784 (2010)
11. Morrissey, J.H., Choi, S.H., Smith, S.A.: Polyphosphate: an ancient molecule that links platelets, coagulation, and inflammation. Blood **119**, 5972–5979 (2012)
12. Müller, W.E.G., et al.: Amorphous Ca^{2+} polyphosphate nanoparticles regulate the ATP level in bone-like SaOS-2 cells. J. Cell Sci. **128**, 2202–2207 (2015)
13. Müller, W.E.G., et al.: A new polyphosphate calcium material with morphogenetic activity. Mater. Lett. **148**, 163–166 (2015)
14. Müller, W.E.G., Schröder, H.C., Suess, P., Wang, X.H. (eds) Inorganic polyphosphates - from basic research to medical application. In: Progress Molecular Subcellular Biology, Springer, Cham, vol. 61; p. 189 (2022)
15. Müller, W.E.G., et al.: Role of ATP during the initiation of microvascularization. Acceleration of an autocrine sensing mechanism facilitating chemotaxis by inorganic polyphosphate. Biochemist. J. **475**, 3255–3273 (2018)
16. Müller, W.E.G., Neufurth, M., Lieberwirth, I., Wang, S.F., Schröder, H.C., Wang, X.H.: Functional importance of coacervation to convert calcium polyphosphate nanoparticles into the physiologically active state. Mater. Today Bio. **16**, 100404 (2022)
17. Müller, W.E.G., et al.: Rebalancing β-amyloid-induced decrease of ATP level by amorphous nano/micro polyphosphate: suppression of the neurotoxic effect of amyloid β-protein fragment 25–35. Int. J. Mol. Sci. **18**, 2154 (2017)
18. Schepler, H., et al.: Acceleration of chronic wound healing by bio-inorganic polyphosphate: *In vitro* studies and first clinical applications. Theranostics **12**, 18–34 (2022)
19. Müller, W.E.G., et al.: The physiological polyphosphate as a healing biomaterial for chronic wounds: Crucial roles of its antibacterial and unique metabolic energy supplying properties. J. Mater. Sci. Technol. **135**, 170–185 (2023)
20. Müller, W.E.G., et al.: Bifunctional dentifrice: Amorphous polyphosphate a regeneratively active sealant with potent anti-Streptococcus mutans activity. Dent. Mater. **33**, 753–764 (2017)
21. Biovia, D.S.: Discovery Studio Modeling Environment. Dassault Syst. Release, San Diego (2015)

22. Müller, W.E.G., et al.: Transformation of amorphous polyphosphate nanoparticles into coacervate complexes: an approach for the encapsulation of mesenchymal stem cells. Small **14**, e1801170 (2018)
23. Müller, W.E.G., et al.: Triple-target stimuli-responsive anti-COVID-19 face mask with physiological virus-inactivating agents. Biomater. Sci. **9**, 6052–6063 (2021)
24. Müller, W.E.G., Schröder, H.C., Neufurth, M., Wang, X.H.: An unexpected biomaterial against SARS-CoV-2: bio-polyphosphate blocks binding of the viral spike to the cell receptor. Mater. Today **51**, 504–524 (2021)
25. Müller, W.E.G., et al.: A biomimetic approach to ameliorate dental hypersensitivity by amorphous polyphosphate microparticles. Dent. Mater. **32**, 775–783 (2016)
26. Müller, W.E.G., et al.: Molecular and biochemical approach for understanding the transition of amorphous to crystalline calcium phosphate deposits in human teeth. Dental Mater. **38**, 2014–2029 (2023)
27. Neufurth, M., et al.: The inorganic polymer, polyphosphate, blocks binding of SARS-CoV-2 spike protein to ACE2 receptor at physiological concentrations. Biochem. Pharmacol. **182**, 114215 (2020)
28. Schepler, H., Wang, X.H., Neufurth, M., Wang, S.F., Schröder, H.C., Müller, W.E.G.: The therapeutic potential of inorganic polyphosphate: a versatile physiological polymer to control coronavirus disease (COVID-19). Theranostics **11**, 6193–6213 (2021)
29. Im, M.J., Hoopes, J.E.: Energy metabolism in healing skin wounds. J. Surg. Res. **10**, 459–464 (1970)
30. Müller, W.E.G., et al.: Fabrication of amorphous strontium polyphosphate microparticles that induce mineralization of bone cells *in vitro* and *in vivo*. Acta Biomater. **50**, 89–101 (2017)
31. Wang, X.H., Schröder, H.C., Müller, W.E.G.: Polyphosphate as a metabolic fuel in Metazoa: a foundational breakthrough invention for biomedical applications. Biotechnol. J. **11**, 11–30 (2016)
32. Nurden, A.T.: The biology of the platelet with special reference to inflammation, wound healing and immunity. Front. Biosci. (Landmark Ed.) **23**, 726–751 (2018)
33. Müller, W.E.G., et al.: Fabrication of a new physiological macroporous hybrid biomaterial/bioscaffold material based on polyphosphate and collagen by freeze-extraction. J. Mat. Chem. B **5**, 3823–3835 (2017)
34. Sarojini, H., et al.: Rapid tissue regeneration induced by intracellular ATP delivery-a preliminary mechanistic study. PLoS One **12**, e0174899 (2017)
35. Lanigan, R.S.: Final report on the safety assessment of sodium metaphosphate, sodium trimetaphosphate, and sodium hexametaphosphate. Int. J. Toxicol. **20**(Suppl. 3), 75–89 (2001)
36. Food and Drug Administration (FDA): Phosphates; proposed affirmation of and deletion from GRAS status as direct and human food ingredients. Fed. Regist. **44**(244), 74845–74857 (1979)
37. Wang, S.F., et al.: Acceleration of wound healing through amorphous calcium carbonate, stabilized with high-energy polyphosphate. Pharmaceutics **15**, 494 (2023)
38. World Health Organization (WHO): Evaluation of certain food additives and contaminants. In: Fifty-Seventh Report of the Joint FAO/WHO Expert Committee on Food (2002)
39. European Food Safety: Authority assessment of one published review on health risks associated with phosphate additives in food. EFSA J. **11**, 3444 (2013)
40. Joint FAO/WHO Expert Committee on Food Additives (JECFA): Evaluation of certain food additives and contaminants (Twenty-sixth report of the Joint FAO/WHO Expert Committee on Food Additives). WHO Tech. Rep. Ser. 683 (1982)
41. International Program on Chemical Safety (IPCS): Toxicological evaluation of certain food additives. World Health Organization (WHO) Food Addit. Ser. 17, 1–22 (1982)

42. Wang, X.H., et al.: Amorphous polyphosphate/amorphous calcium carbonate implant material with enhanced bone healing efficacy in a critical-size defect in rats. Biomed. Mater. **11**, 035005 (2016)

43. Schröder, H.C., et al.: Inorganic polymeric materials for injured tissue repair: biocatalytic formation and exploitation. Biomedicines **10**, 658 (2022)

Quantitative EEG Findings in Outpatients with Psychosomatic Manifestations After COVID-19

Sergey Lytaev$^{(\boxtimes)}$ ⓘ, Nikita Kipaytkov ⓘ, and Tatyana Navoenko

St. Petersburg State Pediatric Medical University, St. Petersburg 194100, Russia
physiology@gpmu.org

Abstract. EEG is considered an important tool in the diagnostic and treatment process of patients with neurological manifestations of COVID-19, especially with encephalopathy, seizures, and status epilepticus. The present research was aimed at quantitative and visual analysis of the EEG of 85 neuropsychiatric outpatients with COVID-19 history and with psychosomatic complaints at the time of the examination. The control group consisted of 35 healthy subjects. Three types of EEG patterns have been established: polymorphic low-frequency activity; low-frequency polymorphic activity with a predominance of delta, theta rhythms; high frequency EEG with a visible dominant of the beta1 range. The correlation index in the alpha range is stable for the EEG in the control group, where in 90% of the subjects the correlation coefficients in the alpha range were more than 0.6. On the contrary, patients have a polymorphic picture, stable indicators with a coefficient of more than 0.6 for all the studied connections, both between the hemispheres and within the hemispheres were registered only in 25% of the subjects. Analysis of the coherence coefficients in patients, on the contrary, shows a higher stability of interhemispheric connections and various options for reducing connections within the hemispheres, which often have a "mirror character".

Keywords: COVID-19 · Quantitative EEG · Biosignal Patterns · Polymorphic EEG Activity

1 Introduction

In 2019, the coronavirus infection (COVID-19) divided medicine into "pre-COVID era" and "COVID era". The explosion of research in the field of pharmacology, infectious diseases, virology and epidemiology has gradually shifted to narrow specialized areas of medicine. More and more specialists are forced to discover not only the pathogenesis and clinic of coronavirus infection, but also a wide variety of specific and non-specific manifestations of the consequences, both close to the end of the disease, and very distant.

Today, the scientific debate continues as to whether non-specific but similar findings in a published case series with COVID-19 encephalopathy (disorganized background activity with diffuse slowing of frontal predominance, many of them non-reactive) can

I. Rojas et al. (Eds.): IWBBIO 2023, LNBI 13919, pp. 560–572, 2023.
https://doi.org/10.1007/978-3-031-34953-9_43

be considered hallmarks of COVID-19 or, on the contrary, it is due to a multifactorial origin, not caused directly by the virus [9].

Interest in the results of EEG studies in COVID-19 is due to similar findings with a number of neurological diseases - viral encephalitis/encephalopathy associated with COVID-19, epilepsy, as well as demyelinating and neurodegenerative processes [9, 11, 25]. Neurological complications occur in one 1/3 (34.6%) of patients with severe COVID-19 and include stroke, headache, and seizures [14]. EEG reports in COVID-19 patients range from isolated cases [11, 28] to studies involving hundreds of EEGs [32, 39]. The results of these studies are varied and include both normal EEG [3] and diffuse slowing in all cases [13, 22, 31]. Some studies note the predominance of frontal findings [6, 9, 11, 33].

The most common EEG sign, reported in 2/3 (68.6%) of patients, is diffuse background slowing. This may indicate that diffuse nonspecific encephalopathy is the most common brain anomaly in this condition. Other EEG features indicative of diffuse encephalopathy include generalized rhythmic delta activity and generalized periodic discharges with triphasic morphology. Lateral periodic and rhythmic disturbances are also recorded, indicating concomitant focal dysfunction in some patients. Epileptiform discharges are not uncommon (13.0%), indicating an underlying cortical excitability predisposing to seizures. In fact, 5.5% had seizures or status epilepticus [2, 34].

Occasionally, studies note characteristic frontal abnormalities, suggesting that this correlates with the presumed entry of COVID-19 into the brain [6, 11, 13, 22, 33]. It is believed that the early clinical manifestations of COVID-19, such as anosmia and ageusia, are associated with the penetration of the virus into the nasal and oral mucosa, which is facilitated by ACE-2 receptors [7, 8, 10]. Subsequent spread to the orbitofrontal region [7] via afferent nerves leads to predominant involvement of the olfactory bulb and orbitofrontal/frontal regions, which may explain the formation of frontal EEG findings [34].

A number of researchers suggest using frontal epileptiform discharges as biomarkers for COVID-19 [4, 11, 13, 28, 36]. Approximately half of all reported epileptic statuses (10 out of 22) are formed in the frontal lobes compared with 18% (4 out of 22) in the occipital lobe. In addition to the frontal lobe, COVID-19 can also affect the piriform cortex, limbic structures, thalamus, hypothalamus, brainstem, and autonomic structures [12, 21, 38].

Researchers are attempting to relate the varied EEG findings in patients to viral transmission patterns or other factors such as lung injury followed by hypoxia, cardiac arrest, metabolic changes, sedation, systemic inflammatory response syndrome, hypercoagulability, vasculitis and strokes, etc..d. In support of systemic causes, evidence suggests that the severity of EEG disturbances correlates with oxygen saturation at admission, but not with neuroimaging data [36]. EEG changes can also be associated with pre-existing neurological diseases such as epilepsy [6, 19, 20].

Thus, today there is no common EEG pattern that is formed in a short period since the onset of COVID-19. There is no doubt that the spectrum of neurophysiological changes will be very wide and polymorphic, however, a statistically reliable analysis of a large amount of data will open up directions for future analysis. In addition, curious specific findings are not excluded.

The present study was aimed at a quantitative and visual analysis of the EEG of outpatient neuropsychiatric patients who had a history of COVID-19 and had psychosomatic complaints at the time of the examination.

2 Methods

We studied outpatients undergoing routine diagnostic and treatment studies in a neuropsychiatric clinic. 2 groups were formed – A and B. Group A consisted of 85 patients (age 18–65 years, 45 women and 40 men) with a history of COVID-19 in 2020–2022 and those who first applied for psychotherapeutic help for psychosomatic complaints that arose after suffering COVID-19 from 3 to 9 months ago. Group B (control) consisted of 35 healthy subjects (age and gender ratios similar to group A) undergoing a study for admission to professional activities without a history of coronavirus infection.

The study according to the guidelines of the Declaration of Helsinki was performed and by the local ethical committee of the St. Petersburg State Pediatric Medical University (protocol no. 14/1, 10 Dec 2020) was approved. All participants of the research project signed a voluntary informed consent.

EEG was recorded according to the international system 10/20 from 16 monopolar sites (Fp_1, Fp_2, F_3, F_4, F_7, F_8, C_3, C_4, T_3, T_4, T_5, T_6, P_3, P_4, O_1, O_2) with reference electrodes on the earlobes (AvL, AvR) using the Mizar-EEG-202 software and hardware complex. Standard tests were used with closing/opening of the eyes, variable-frequency photo stimulation (1–15 Hz), and hyperventilation. Computer analysis was carried out using WinEEG, EEG Studio, and Loreta-Key Viewer programs. Initially, the EEG was assessed visually by frequency and amplitude parameters, as well as by the distribution of the rhythm on the convexital surface and the degree of desynchronization of the recording upon opening the eyes and/or presenting single flashes of light.

By apply quantitative EEG methods, the spectral power was determined in the main frequency ranges, and correlation coefficients were also considered when comparing intrahemispheric and interhemispheric interactions. The parameters for determining the spectral power were as follows: a fragment of 10 s, an analysis epoch duration of 5 s, an analysis epoch overlap of 50%, a Hann time window, and a low-frequency signal range of 0.25–1.25. In addition, for all recording channels for the selected EEG segments free from artifacts, the average spectrum was calculated. The results obtained are converted into percentages so that the total sum over all frequency ranges for each lead under study is 100%.

To visualize local changes in the spectral power, a low-resolution electromagnetic tomography application (LORETA, soft Loreta-Key Viewer 04) was used with the construction of 2D distribution graphs on anatomical structures. The obtained data were processed using the Microsoft Office Excel program, the SPSS Statistics 17.0 statistical software package. The reliability of the obtained data was calculated by Student's criterion. Statistical differences were considered significant at $p < 0.05$ [27].

3 Outcomes

Visual EEG assessment shows a decrease in the alpha rhythm index in outpatients who have had COVID-19. The manifestation of the alpha rhythm was slow both at the beginning of registration and during recovery after the test with the opening/closing of the eyes or after single flashes of light. It is known that the registration of a stable alpha rhythm of a high index at the beginning of an EEG recording under standard conditions (blackout, muscle relaxation, silence, etc.) has a significant time variation in healthy subjects (from 1 s to 1 min in 90% of the subjects). Therefore, for comparison, we chose the test with the "return of the alpha rhythm" after the test with the opening and subsequent closing of the eyes, where the reference point is the time stamp in the EEG file "eyes closed".

In healthy subjects (control group), a stable alpha rhythm is recorded from 1–2 s in 85% of subjects. In the study group A, the "return of the alpha rhythm" in 65% of the subjects was registered for 5–10 s. On Fig. 1 shows a fragment of the original EEG illustrating this phenomenon. The patient, man (48 year), complained of poor sleep, inability to concentrate, decreased attention and memory. He suffered a coronavirus infection 4 months ago, was treated on an outpatient basis, the acute period was at least 20 days in total, and pneumonia typical of COVID-19 was diagnosed on CT.

It should also be noted that in patients (group A), the alpha rhythm is generally reduced in amplitude, and in terms of time characteristics it approaches the upper frequency limit (11–12 Hz in patients and 9–10 Hz in the control group).

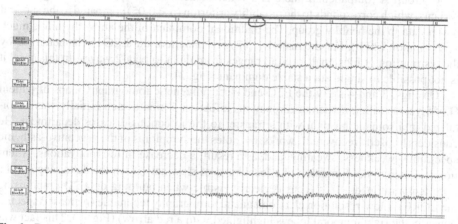

Fig. 1. Deceleration of alpha rhythm recovery after eye closure in a post-COVID-19 patient. Markers note a steady increase in the alpha range index from 5 s.

In the present study, when comparing the EEG spectral power in two groups, the following regularity was noted. The EEG in healthy subjects demonstrates a clear dominant alpha range both in visual analysis and in determining the spectral power of the signal. On Fig. 2, a more "lighter" spectrum of red-yellow color is noted, corresponding to high levels of the presence of alpha-range waves, and its symmetrical distribution is noted with a maximum index in the occipital region. Such a "pattern" of the spectral power distribution is typical for 80% of healthy subjects. Visually, such EEGs are evaluated as patterns with a dominant alpha rhythm.

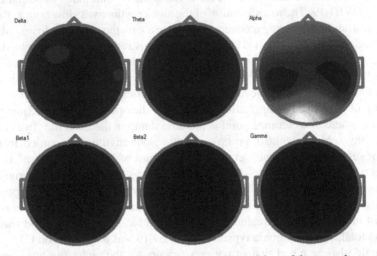

Fig. 2. EEG spectral power distribution in a healthy subject of the control group B.

In group A (outpatients), there is a wider distribution of spectral subrange power variants, which corresponds to the reliability criterion ($p < 0.05$) compared to the spectral power data in the healthy group. The traditional dominant of the alpha rhythm is much less common (20%). For the majority (62% of patients), the EEG is characterized by the distribution of biosignals over all ranges from delta to beta-2 waves. In this case, it is not possible to single out a stable dominant (Fig. 3).

On Figs. 3 and 4 show the EEG and electromagnetic tomography maps of patient G., 55 years, with a moderate coronavirus infection 5 months ago initially was hospitalized in a somatic hospital with a transfer to the infectious diseases department. He considers himself "recovered", works, but notes the symptoms of the vegetative-asthenic spectrum – weakness, sweating, and, as a result, a decrease in life motivations. Attention is drawn to the formation of a local burst of slow activity in the delta range in the right temporal leads. This focus is clearly visible when building on a 2D graph using the LORETA low-resolution electromagnetic tomography technique (increased spectral power in the right temporal lobe is highlighted with a red arrow) (Fig. 4.). This fact was noted in 12% of patients from group A. Visual assessment refers such EEG to the type of low-amplitude polymorphic activity.

EEG variant III (about 18%) of patients is characterized by a shift in the spectral power to the beta1 and beta2 ranges with the formation of two "centers of activity" – in the alpha and beta regions (Fig. 5). Visually, such EEG variants are assessed as high frequency, with a visible dominant of the beta1 range and an extremely low alpha rhythm index. In the conclusion of a specialist, the phrase "flat EEG" is often used. It should be noted that this type III is more characteristic of psychosomatic patients, which is difficult to say about type 2 EEG spectral powers detected among patients.

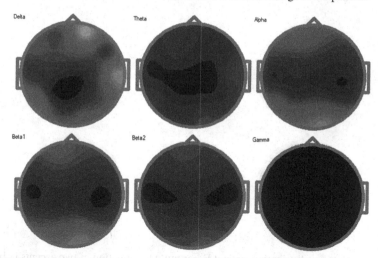

Fig. 3. EEG spectral power in patient G. Signal distribution over all ranges from delta to beta2.

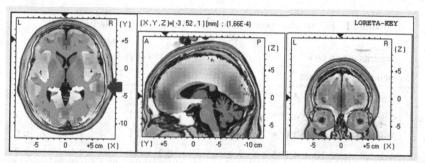

Fig. 4. Low-resolution electromagnetic tomography LORETA – visualization of local changes in the right temporal region (patient G.).

In Table 1, summarized data on the spectral power in (uV²) in the studied ranges are presented. Red shading marks the cells demonstrating locality in the right temporal leads.

Among the systemic findings (Table 1), noteworthy is the significant discrepancy between the spectral power of the EEG between patients and healthy subjects in terms of slow waves – delta and theta rhythms. With a decrease in the amplitude of these rhythms in patients, their spectral severity is significantly higher. This fact has a generalized character with some predominance in the right temporal points of registration. The spectral power of the EEG alpha rhythm in healthy people is higher than in patients. However, significant differences are recorded only in the frontal and occipital regions.

After determining the spectral power of the EEG, an analysis of coherence was carried out, aimed at comparing interhemispheric or intra-hemispheric interactions. In healthy subjects (group B), high correlation coefficients are consistently observed both

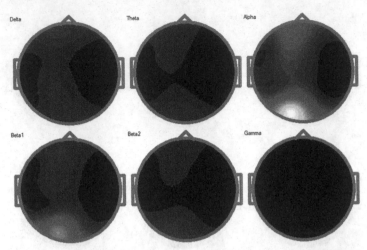

Fig. 5. Spectral power distribution, patient E. (group A). Formation of two accents in beta1 and alpha ranges.

Table 1. EG Spectral power in (uV^2) in Delta, Theta, Alfa EEG bands

	Patients group (A)			Control group (B)		
	Delta	Theta	Alfa	Delta	Theta	Alfa
Fp1	5 ± 1,2**	10 ± 4,2**	12 ± 4,5**	0,2 ± 0,1	5 ± 2,2	21 ± 7,8
Fp2	6 ± 2,1**	11 ± 4,6**	17 ± 6,4	0,3 ± 0,1	5 ± 2,3	19 ± 7,5
F3	5 ± 2,2**	9 ± 3,9**	16 ± 5,4	0,1 ± 0,1	4 ± 1,9	17 ± 6,8
F4	4 ± 1,9**	10 ± 4,1**	17 ± 6,5	0,2 ± 0,1	3 ± 1,6	19 ± 7,6
F7	5 ± 2,3**	11 ± 4,5**	15 ± 5,0	0,3 ± 0,1	3 ± 1,7	16 ± 6,5
F8	4 ± 1,7**	8 ± 3,8*	19 ± 7,9	0,3 ± 0,1	4 ± 1,8	22 ± 8,1
C3	2 ± 1,4*	10 ± 4,1**	11 ± 3,9	0,3 ± 0,1	3 ± 1,6	12 ± 4,7
C4	1 ± 0,6*	9 ± 3,7**	10 ± 3,6	0,2 ± 0,1	3 ± 1,6	10 ± 4,4
T3	3 ± 1,6**	11 ± 4,3**	16 ± 5,6	0,3 ± 0,1	2 ± 1,4	17 ± 6,7
T4	6 ± 4,5**	15 ± 5,9**	14 ± 4,8	0,4 ± 0,1	3 ± 1,5	15 ± 4,3
T5	3 ± 1,7**	10 ± 4,5**	18 ± 7,6	0,3 ± 0,1	3 ± 1,6	20 ± 7,5
T6	7 ± 5,6**	14 ± 5,7**	16 ± 5,4	0,2 ± 0,1	2 ± 1,4	18 ± 6,9
P3	4 ± 1,9**	9 ± 3,8**	19 ± 7,8*	0,4 ± 0,2	3 ± 1,6	25 ± 10
P4	4 ± 1,9**	10 ± 4,1**	20 ± 8,1	0,3 ± 0,1	3 ± 1,5	22 ± 9,5
O1	6 ± 4,3**	11 ± 4,4**	22 ± 8,8**	0,2 ± 0,1	4 ± 1,8	35 ± 11,3
O2	6 ± 4,5**	10 ± 4,2*	24 ± 10**	0,3 ± 0,1	5 ± 2,0	32 ± 11,1

Note. * – $p < 0,05$; ** – $p < 0,01$

between the hemispheres (in pairs Fp_1-Fp_2, C_3-C_4, F_7-F_8, T_3-T_4, T_5-T_6, P_3-P_4, O_1-O_2), and within hemispheres (in pairs Fp_1-T_3, Fp_1-C_3, Fp_2-C_4, Fp_2-T_4, T_3-O_1, C_3-O_1, T_4-O_2, C_4-O_2) (Fig. 6).

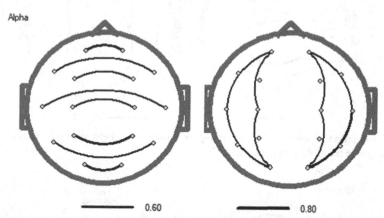

Fig. 6. Correlation of inter- and intra-hemispheric interaction between sites according to the 10/20 system in the alpha range in the examined control group.

The correlation index in the alpha range is stable for 10 s EEG segments in the control group, where in 90% of the subjects the correlation coefficients in the alpha range were more than 0.6. On the contrary, patients have a polymorphic EEG pattern – stable indicators with a coefficient of more than 0.6 for all the studied connections both between hemispheres and within the hemispheres were noted only in 25% of the subjects. Which corresponds to the criterion of significance ($p < 0.05$) in comparison with the indicators in the group of healthy people. In other cases, there is a decrease in conjugation in a number of compared pairs of leads, mainly in the temporal points T_3 and/or T_4.

On Fig. 7 shows a variant of coherence analysis, patient C., 45 years, with a moderate form of COVID-19 with hospitalization, treatment in intensive care. Six months after recovery, the patient complains of headache, decreased performance, and increased anxiety. EEG digital processing shows a reduction in the correlation coefficient between the hemispheres at recording points T_3-T_4, as well as a reduction in intra-hemispheric interactions in leads Fp_1-T_3, T_4-O_2.

In general, when analyzing the coherence coefficients in patients of group A, there is a higher stability of interhemispheric connections and various options for reducing connections within the hemispheres, which often have a "mirror character". On Fig. 8 shows a variant of coherence analysis, patient K., 55 years, with a coronavirus infection 5 months ago in a moderate form, with hospitalization at the beginning in a somatic hospital with subsequent transfer to the infectious diseases department. He considers himself "recovered", works, however, notes the symptoms of the vegetative-asthenic spectrum – weakness, sweating, and, as a result, a decrease in life motivations. A decrease in the correlation coefficient between the hemispheres is registered only in leads T_3-T_4 and a decrease in intra-hemispheric interactions in four pairs Fp_1-C_3, T_3-O_1, Fp_2-C_4, T_4-O_2.

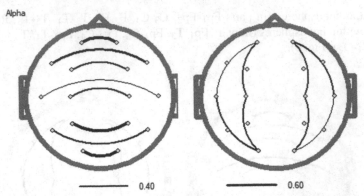

Fig. 7. Correlation of inter- and intra-hemispheric interaction between leads according to the 10/20 system in the alpha range, patient C.

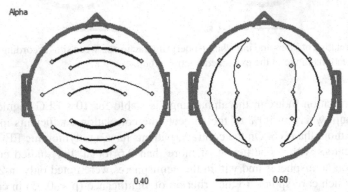

Fig. 8. Correlation of inter- and intra-hemispheric interaction between leads according to the 10/20 system in the alpha range, patient K.

4 Discussion

Thus, the analysis of the present results of the EEG study of patients with COVID-19 and literature data for 2020–2022 indicates that the main EEG findings are focused on a decrease in the voltage (amplitude) of biosignals and various variants of polymorphic activity with a predominance of certain EEG rhythms in different parts of the brain.

Postponed coronavirus infection in patients with a neuropsychiatric profile is characterized on the EEG by a number of features. This is a decrease in the reactivity of the cortex and a change in the parameters of the recorded alpha rhythm. A significant change in the pattern of the EEG spectrum is noted due to the expansion of the dominant areas of activity to the frequencies closest to the alpha range – beta, theta. When conducting a coherent analysis, high coefficients in interhemispheric interaction are preserved and, on the contrary, a decrease in intra-hemispheric connections, often affecting the temporal leads. As an interesting, possibly specific detail of the analysis, we can note the characteristic shift of the spectral power in the slow-wave bands with an emphasis in the right temporal region.

Low-voltage EEG patterns are recorded in about 10% of normal healthy subjects [2, 27, 29]. They do not generate an alpha dominant rhythm, show reduced reactivity, and beta and alpha frequencies are mixed. In a series of studies, 100% of patients recorded an EEG with voltage at or below 20 uV with rare patterns of reactivity. A standard sleep EEG showed that when the patient fell asleep spontaneously, the spindles were absent or abnormal. Follow-up registers performed in patients with a favorable outcome revealed the appearance of a posterior dominant theta rhythm, which excluded the physiological absence of a posterior alpha rhythm in patients [2, 3, 5, 7, 23, 24].

Patterns of low EEG voltage or generalized suppression are often associated with extensive cortical and subcortical lesions such as anoxic encephalopathy or severe traumatic brain injury. This pattern preceded an unfavorable prognosis (4 points according to the qualitative classification of EEG changes) or irreversible processes [1, 16]. Similar processes have been reported in ischemic encephalopathy [15, 17], with the exception of hypothyroid encephalopathy, which is considered reversible, except for cretinism [2]. Also low EEG voltage has been described in diffuse neuronal dysfunction due to hypothermia, intoxication with drugs that depress the central nervous system, or anesthesia [17].

An analysis of the literature notes a large amount of disparate data on the sensitivity of the spectral power to both acquired and congenital conditions that pathologically change the activity of the brain. An increase in the power of slow delta and theta ranges in patients with acute cerebrovascular accidents was shown with a clear positive dynamics by day 14 in the absence of local foci of slow activity [24, 26]. A number of authors note an increase in the spectral power in the delta range in patients with congenital organic brain damage [24–26]. At the same time, significant changes in the spectral characteristics of the EEG in schizophrenia are detected only when analyzing the response to the presentation of sound or visual stimuli [5, 23]. We must not forget that the brain cannot function "in isolation" from the rest of the body. Changes in the mode of operation of many internal organs through psychosomatic connections form changes in the work of the brain, which means changes in the spectral power of the EEG. For example, an increase in spectral power in the beta1 and alpha ranges is described in various endocrine pathologies [2].

A number of articles on EEG and COVID-19 mention background activity stress [18, 30, 35]. Some authors find low voltage but attribute it to metabolic or ischemic causes. It is believed that the problem may lie in the fact that the studies are carried out according to a methodology different from the standard one – using less than 10 electrodes instead of 21 (to reduce the risk of COVID infection). The currently accepted normal voltage values, as well as the American Society of Neurophysiological Criteria for Standardized Critical Care EEG Terminology, are derived using 21-electrode EEG caps, considered the gold standard. It is known that the distance between the electrodes affects the voltage, so records with a small number of electrodes double the inter electrode distance and get a higher voltage, so they should have their standard values [37]. However, these considerations are more typical for bipolar registration, but not monopolar.

A nonspecific EEG pattern has also been described in patients with COVID-19, which differs from those found in other infectious or metabolic encephalopathies. As a distinctive finding, a low voltage was recorded, resembling severe encephalopathy conditions associated with a poor prognosis. It has been demonstrated that a non-reactive low voltage pattern in patients with COVID-19 is not necessarily associated with a poor prognosis. A similar pattern was observed both in intensive care and in standard wards, as well as in patients with favorable development or progressive deterioration. At the same time, in patients with a favorable outcome, anomalies showed a tendency to reduce to a normal EEG. Based on these EEG results, a metabolic/hypoxic mechanism seems unlikely. Low-voltage EEG, possibly due to desynchronization, may explain the low prevalence of epileptic activity described in patients with COVID-19 [34].

Thus, the EEG is an important tool in the treatment and diagnostic process of patients with neurological manifestations of COVID-19, especially with encephalopathy, seizures, and status epilepticus. Abnormalities, if present, include deceleration, periodic discharges, epileptiform discharges, seizures, and status epilepticus, indicating the presence of localized dysfunction, nonspecific encephalopathy, and cortical excitability in this condition. The degree of EEG abnormalities correlates with COVID-19 diagnosis, length of follow-up, pre-existing neurological conditions such as epilepsy, and disease severity. EEG abnormalities affecting the frontal lobes appear to be common in COVID-19 encephalopathy and are suggested as a potential biomarker if recorded consistently.

5 Conclusion

We have performed a visual analysis of the EEG, an analysis of the spectral power of rhythms, and a coherent analysis, which allowed us to speak about the following. There is a wide distribution of variants of spectral power subranges. Imaging subdivides the EEG with COVID-19 sequelae into three groups. Firstly, the reduced dominant of the alpha rhythm, which is noted in 20% of cases. For the majority (62% of patients), the EEG is characterized by a wide distribution of biosignals over all ranges from delta to beta-2 waves. In this case, it is not possible to single out a stable dominant. This is a variant of polymorphic low-frequency activity. Secondly, EEG by the type of low-frequency polymorphic activity with a predominance of delta and theta rhythms. Thirdly, a high frequency EEG with a visible dominant of the beta1 range, but an extremely low alpha rhythm index is a "flat EEG".

The correlation index in the alpha range is stable for the EEG in the control group, where in 90% of the subjects the correlation coefficients in the alpha range were more than 0.6. On the contrary, patients have a polymorphic picture, stable indicators with a coefficient of more than 0.6 for all the studied connections both between the hemispheres and within the hemispheres were registered only in 25% of the subjects. Analysis of the coherence coefficients in patients, on the contrary, shows a higher stability of interhemispheric connections and various options for reducing connections within the hemispheres, which often have a "mirror character".

References

1. Amodio, P., Marchetti, P., Del Piccolo, F., et al.: Spectral versus visual EEG analysis in mild hepatic encephalopathy. Clin Neurophysiol. **110**, 1334–1344 (1999)

2. Antony, A.R., Haneef, Z.: Systematic review of EEG findings in 617 patients diagnosed with COVID-19. Seizure: Eur. J. Epilepsy. **83**, 234–241 (2020)

3. Asadi-Pooya, A.A., Simani, L.: Central nervous system manifestations of COVID-19: a systematic review. J Neurol Sci. **413**, 116832 (2020)

4. Balloy, G., et al.: Non-lesional status epilepticus in a patient with coronavirus disease 2019. Clin. Neurophysiol. **131**(8), 2059–2061 (2020). https://doi.org/10.1016/j.clinph.2020.05.005

5. Belskaya, K.A., Lytaev, S.A.: Neuropsychological analysis of cognitive deficits in Schizophrenia. Hum Physiol. **48**, 37–45 (2022)

6. Campanella, S., Arikan, K., Babiloni, C., et al.: Special report on the impact of the COVID-19 pandemic on clinical EEG and research and consensus recommendations for the safe use of EEG. Clin. EEG Neurosci. **52**, 3–28 (2021)

7. Canham, L.J.W., Staniaszek, L.E., Mortimer, A.M., et al.: Electroencephalographic (EEG) features of encephalopathy in the setting of Covid-19: a case series. Clin Neurophysiol. Pract. **5**, 199–205 (2020)

8. Chougar, L., et al.: Retrospective observational study of brain MRI findings in patients with acute SARS-CoV-2 infection and neurologic manifestations. Radiology **297**(3), E313–E323 (2020). https://doi.org/10.1148/radiol.2020202422

9. De Stefano, P., Nencha, U., De Stefano, L., et al.: Focal EEG changes indicating critical illness associated cerebral microbleeds in a Covid-19 patient. Clin. Neurophysiol. Pract. **5**, 125–129 (2020)

10. DosSantos, M.F., Devalle, S., Aran, V., et al.: Neuromechanisms of SARS-CoV-2: a review. Front Neuroanat. **14**, 37 (2020)

11. Flamand, M., Perron, A., Buron, Y., Szurhaj, W.: Pay more attention to EEG in COVID-19 pandemic. Clin Neurophysiol. **131**, 2062–2064 (2020)

12. Franceschi, A.M., Ahmed, O., Giliberto, L., Castillo, M.: Hemorrhagic posterior reversible encephalopathy syndrome as a manifestation of COVID-19 infection. AJNR Am. J. Neuroradiol. (2020). https://doi.org/10.3174/ajnr.A6595

13. Galanopoulou, A.S., Ferastraoaru, V., Correa, D.J., et al.: EEG findings in acutely ill patients investigated for SARS-CoV-2/COVID-19: a small case series preliminary report. Epilepsia Open **5**, 314–324 (2020)

14. Huang, C., Wang, Y., Li, X., et al.: Clinical features of patients infected with 2019 novel coronavirus in Wuhan China. Lancet **395**, 497–506 (2020)

15. Juan, E., Kaplan, P.W., Oddo, M., Rossetti, A.O.: EEG As an indicator of cerebral functioning in postanoxic coma. J. Clin. Neurophysiol. **32**, 465–471 (2015)

16. Kaplan, P.W.: The EEG in metabolic encephalopathy and coma. J. Clin. Neurophysiol. **21**, 307–318 (2004)

17. Kaplan, P.W., Rossetti, A.O.: EEG Patterns and imaging correlations in encephalopathy: encephalopathy part II. J. Clin. Neurophysiol. **28**, 233–251 (2011)

18. Koutroumanidis, M., Gratwicke, J., Sharma. S., et al.: Alpha coma EEG pattern in patients with severe COVID-19 related encephalopathy. Clin Neurophysiol. **132**, 218–225 (2021)

19. Kremer, S., Lersy, F., de Sèze, J., et al.: Brain MRI findings in severe COVID-19: a retrospective observational study. Radiology **297**, e242–e251 (2020)

20. Lambrecq, V., Hanin, A., Munoz-Musat, E., et al.: Cohort COVID-19 Neurosciences (CoCo Neurosciences) study group. Association of Clinical, Biological, and Brain Magnetic Resonance Imaging Findings with Electroencephalographic Findings for Patients with COVID-19. JAMA Netw Open **4**, e211489 (2021)

21. Leitinger, M., Beniczky, S., Rohracher, A., et al.: Salzburg consensus criteria for non-convulsive status epilepticus—approach to clinical application. Epilepsy Behav. **49**, 58–163 (2015)

22. Louis, S., Dhawan, A., Newey, C., et al.: Continuous electroencephalography characteristics and acute symptomatic seizures in COVID19 patients. Clin Neurophysiol. **131**, 2651–2656 (2020)

23. Lytaev, S., Belskaya, K.: Integration and Disintegration of Auditory Images Perception. In: Schmorrow, D.D., Fidopiastis, C.M. (eds.) Foundations of Augmented Cognition. LNCS (LNAI), vol. 9183, pp. 470–480. Springer, Cham (2015). https://doi.org/10.1007/978-3-319-20816-9_45

24. Lytaev, S.: Modern neurophysiological research of the human brain in clinic and psychophysiology. In: Rojas, I., CastilloSecilla, D., Herrera, L.J., Pomares, H. (eds.) Bioengineering and Biomedical Signal and Image Processing. LNCS, vol. 12940, pp. 231–241. Springer, Cham (2021). https://doi.org/10.1007/978-3-030-88163-4_21

25. Lytaev, S.: PET-neuroimaging and neuropsychological study for early cognitive impairment in parkinson's disease. In: Rojas, I., Valenzuela, O., Rojas, F., Herrera, L.J., Ortuño, F. (eds) Bioinformatics and Biomedical Engineering. IWBBIO 2022. Lecture Notes in Computer Science 13346. Springer, Cham. 143–153 (2022). https://doi.org/10.1007/978-3-031-07704-3_12

26. Lytaev, S.: Modern human brain neuroimaging research: analytical assessment and neurophysiological mechanisms. In: Stephanidis, C., Antona, M., Ntoa, S. (eds.) HCI International 2022 Posters, pp. 179–185. Springer International Publishing, Cham (2022). https://doi.org/10.1007/978-3-031-06388-6_24

27. Lytaev, S.: Psychological and neurophysiological screening investigation of the collective and personal stress resilience. Behav. Sci. **13**, 258 (2023)

28. Mao, L., Jin, H., Wang, M., et al.: Neurologic manifestations of hospitalized patients with Coronavirus disease 2019 in Wuhan. JAMA Neurol. **77**, 683–690 (2020)

29. Pasini, E., Bisulli, F., Volpi, L., et al.: EEG findings in COVID-19 related encephalopathy. Clin Neurophysiol. **131**, 2265–2267 (2020)

30. Pastor, J., Vega-Zelaya, L., Abad, E.M.: Specific EEG encephalopathy pattern in SARS-CoV-2 patients. J Clin Med. **9**, e1545 (2020)

31. Paterson, R.W., Brown, R.L., Benjamin, L., et al.: The emerging spectrum of COVID-19 neurology: clinical, radiological and laboratory findings. Brain **143**, 3104–3120 (2020)

32. Pellinen, J., Carroll, E., Friedman, D., et al.: Continuous EEG findings in patients with COVID-19 infection admitted to a New York academic hospital system. Epilepsia **61**, 2097–2105 (2020)

33. Petrescu, A.-M., Taussig, D., Bouilleret, V.: Electroencephalogram (EEG) in COVID-19: a systematic retrospective study. Neurophysiol. Clin. **50**, 155–165 (2020)

34. Sáez-Landete, I., Gómez-Domínguez, A., Estrella-León, B., et al.: Retrospective analysis of EEG in patients with COVID-19: EEG recording in acute and follow-up phases. Clin EEG Neurosci. **53**, 215–228 (2022)

35. Skorin, I., Carrillo, R., Perez, C.P., et al.: EEG findings and clinical prognostic factors associated with mortality in a prospective cohort of inpatients with COVID-19. Seizure **83**, 1–4 (2020)

36. Thompson, R.: Pandemic potential of 2019-nCoV. Lancet Infect. Dis. **20**, 280 (2020)

37. TjepkemaCloostermans, M.C., Hofmeijer, J., Hom, H.W., et al.: Predicting outcome in postanoxic coma: are ten EEG electrodes enough? J. Clin. Neurophysiol. **34**, 207–212 (2017)

38. Vespignani, H., Colas, D., Lavin, B.S., et al.: Report on electroencephalographic findings in critically ill patients with COVID-19. Ann. Neurol. **88**, 626–630 (2020)

39. Wu, Y., Xu, X., Chen, Z., et al.: Nervous system involvement after infection with COVID-19 and other coronaviruses. Brain Behav. Immun. **87**, 18–22 (2020)

Author Index

I. Rojas et al. (Eds.): IWBBIO 2023, LNBI 13919, pp. 573–576, 2023.
https://doi.org/10.1007/978-3-031-34953-9

Printed in the United States
by Baker & Taylor Publisher Services